VOLUME ONE HUNDRED AND EIGHTEEN

PROGRESS IN MOLECULAR BIOLOGY AND TRANSLATIONAL SCIENCE

The Molecular Biology of Arrestins

VOLUME ONE HUNDRED AND EIGHTEEN

PROGRESS IN MOLECULAR BIOLOGY AND TRANSLATIONAL SCIENCE

The Molecular Biology of Arrestins

Edited by

LOUIS M. LUTTRELL

Departments of Medicine and Biochemistry & Molecular Biology, Medical University of South Carolina, and the Research Service of the Ralph H. Johnson Veterans Affairs Medical Center, Charleston, South Carolina, USA

ELSEVIER

AMSTERDAM • BOSTON • HEIDELBERG • LONDON
NEW YORK • OXFORD • PARIS • SAN DIEGO
SAN FRANCISCO • SINGAPORE • SYDNEY • TOKYO
Academic Press is an imprint of Elsevier

Academic Press is an imprint of Elsevier
The Boulevard, Langford Lane, Kidlington, Oxford, OX5 1GB, UK
32 Jamestown Road, London NW1 7BY, UK
Radarweg 29, PO Box 211, 1000 AE Amsterdam, The Netherlands
225 Wyman Street, Waltham, MA 02451, USA
525 B Street, Suite 1800, San Diego, CA 92101-4495, USA

First edition 2013

Library of Congress Cataloging-in-Publication Data
A catalog record for this book is available from the Library of Congress

British Library Cataloguing in Publication Data
A catalogue record for this book is available from the British Library

ISBN: 978-0-12-394440-5
ISSN: 1877-1173

Printed and bound by CPI Group (UK) Ltd, Croydon, CR0 4YY

Transferred to digital print 2013

13 14 15 12 11 10 9 8 7 6 5 4 3 2 1

CONTENTS

Contributors xi
Preface xv

Part I
Perspective—The Duality of Arrestin Function

1. Arrestins Come of Age: A Personal Historical Perspective **3**

Robert J. Lefkowitz

1. Introduction 4
2. "Prehistory" of Arrestins 4
3. A Family of 7 Transmembrane Receptors 5
4. The GRK and Arrestin Families 5
5. Arrestins and Desensitization 8
6. Arrestins and Endocytosis 8
7. Arrestins and Signaling 9
8. Biased Signaling 10
9. Bar Code Hypothesis 12
10. Other Receptors, Other Functions 13
11. Future Perspectives 14

References 14

Part II
The Molecular Biology of Arrestins

2. True Arrestins and Arrestin-Fold Proteins: A Structure-Based Appraisal **21**

Laurence Aubry and Gérard Klein

1. Introduction 22
2. True Arrestins: A Structure Adapted to Multiple Scaffolding 23
3. Novel Arrestin-Related Proteins 30
4. Other Arrestin-Fold Proteins 35
5. Are Arrestins and Arrestin-Fold Proteins Related by a Shared Mechanism for Their Function? 42
6. Conclusion and Perspectives 46

References 49

3. Structural Determinants of Arrestin Functions **57**

Vsevolod V. Gurevich and Eugenia V. Gurevich

1. Introduction 58
2. What the Crystal Structure Reveals, and What it Does Not 59
3. How do Arrestins Fit Receptors? 60
4. Interactions with Other Signaling Proteins 72
5. Designing Signaling-Biased Arrestin Mutants 77
6. Conclusions: Where do we go from Here? 83

Acknowledgments 83
References 83

4. Arrestins: Role in the Desensitization, Sequestration, and Vesicular Trafficking of G Protein-Coupled Receptors **93**

Cornelia Walther and Stephen S.G. Ferguson

1. Introduction 93
2. Arrestins in GPCR Desensitization 95
3. Arrestins in GPCR Trafficking 99
4. Conclusions 107

Acknowledgments 107
References 107

5. Arrestins as Regulators of Kinases and Phosphatases **115**

Louis M. Luttrell and William E. Miller

1. Introduction 116
2. Arrestins as GPCR Effectors 118
3. Positive and Negative Regulation of Kinase Pathways 124
4. Arrestin-Regulated Kinase and Phosphatase Pathways 126
5. Conclusions 140

Acknowledgments 141
References 141

6. β-Arrestins: Modulators of Small GTPase Activation and Function **149**

Audrey Claing

1. Introduction 150
2. Ras Family GTPases 154
3. Rho Family GTPases 157
4. Rab Family GTPases 162
5. ARF Family GTPases 164

6. Ran Family GTPases 166
7. Perspectives and Future Directions 167
8. Conclusions 169
References 169

7. Arrestins and Protein Ubiquitination 175

Reddy Peera Kommaddi and Sudha K. Shenoy

1. Introduction 176
2. Ubiquitination of Arrestins 182
3. Deubiquitination of Arrestins 187
4. Arrestins Act as Adaptors for Ubiquitination 189
5. Arrestins and Seven-Transmembrane Receptor Deubiquitination 193
6. Arrestin-Like Proteins 194
7. Conclusions 198
Acknowledgment 199
References 199

8. Arrestins in Actin Reorganization and Cell Migration 205

Kathryn A. DeFea

1. Introduction 206
2. β-Arrestins as Regulators of Gradient Sensing for Chemokine Receptors 206
3. β-Arrestins as Regulators of Actin Assembly 209
4. Regulation of Kinase Activities by β-Arrestins 214
5. Additional Roles for β-Arrestins and Chemotaxis *In Vivo* 217
6. Role of β-Arrestin-Dependent Chemotaxis in Health and Disease 218
7. Concluding Remarks 219
References 220

Part III
The Physiological Roles of Arrestins

9. The Role of Arrestins in Development 225

Melanie Philipp, Tama Evron, and Marc G. Caron

1. Introduction 226
2. β-Arrestins in Model Organisms 228
3. The Function of Arrestins in Invertebrates 229
4. β-Arrestins in Vertebrate Development 230
5. Conclusions 238
Acknowledgments 238
References 238

10. The Role of Arrestins in Visual and Disease Processes of the Eye **243**

W. Clay Smith

1. Visual Arrestin Nomenclature 244
2. The Function of Visual Arrestins in Quenching Phototransduction 244
3. Splice Variants of Arrestin1 245
4. Arrestins in Cone Photoreceptors 246
5. Translocation of Visual Arrestins 247
6. New Roles for Arrestin1 in the Retina 252
7. Arrestins in Disease Processes in the Eye 257
8. The Future for Visual Arrestins 259

References 259

11. β-Arrestins in the Central Nervous System **267**

Camille Latapy and Jean Martin Beaulieu

1. Introduction: Arrestins and GRK in the Central Nervous System 268
2. Arrestins and GRK in Dopaminergic Neurotransmission 270
3. Arrestins in Serotonergic Neurotransmission 275
4. Arrestins in Noradrenergic Neurotransmission 278
5. Arrestins in Opioid Receptor Signaling 281
6. Arrestins and Corticotropin Receptors 283
7. Other β-arrestin-Dependent Behavior 285
8. Conclusions 286

Acknowledgment 287

References 287

12. Arrestins in the Cardiovascular System **297**

Anastasios Lymperopoulos and Ashley Bathgate

1. Introduction 298
2. Cardiovascular Adrenergic Receptors and βarrs 299
3. Cardiovascular Angiotensin II Receptors and βarrs 309
4. Other Cardiovascular GPCRs and βarrs 315
5. Cardiovascular Non-GPCRs and βarrs 320
6. Biased Agonism/Antagonism of Cardiovascular GPCR Ligands 322
7. Perspectives and Future Directions 326
8. Conclusions 327

References 328

13. Arrestins in Bone **335**

Brittany N. Bohinc and Diane Gesty-Palmer

1. Introduction 336
2. Arrestin Signaling: A New Dimension to GCPR Signaling in Bone 337
3. Functional Selectivity in Bone 341
4. Arrestin Signaling Effects in Bone 345
5. Perspectives and Future Directions 352
6. Conclusions 353

Acknowledgments 353
References 353

14. β-Arrestins in the Immune System **359**

Dianhua Jiang, Ting Xie, Jiurong Liang, and Paul W. Noble

1. Introduction 360
2. β-Arrestins in Innate Immunity 361
3. β-Arrestins in Adaptive Immunity 364
4. β-Arrestins and Structural Cells 367
5. β-Arrestins Regulate Immune Signaling Pathways 367
6. Role of β-Arrestins in Human Diseases 374
7. β-Arrestins in Therapeutic Development for Inflammatory Diseases 383

References 384

15. The Role of β-Arrestins in Cancer **395**

Philip Michael Sobolesky and Omar Moussa

1. Introduction 396
2. ARRBs and Cancer-Associated Cell Phenotypes 397
3. The Role of ARRBs in Cancer 400
4. Conclusions 407

Acknowledgments 407
References 407

16. Arrestins in Metabolic Regulation **413**

Jian Zhao and Gang Pei

1. A Short Introduction to Metabolic Regulation 413
2. Functional Roles of β-Arrestins in Regulation of Whole-Body Energy Balance and Body Weight Control 415
3. Functional Role of β-Arrestins in Regulation of Carbohydrate and Lipid Homeostasis 417

4. Conclusions 423
References 424

Part IV
The Future—The Potential for Arrestin-based Therapeutics

17. Systems Analysis of Arrestin Pathway Functions **431**

Stuart Maudsley, Sana Siddiqui, and Bronwen Martin

1. G Protein-Coupled Receptor Signaling Activity 432
2. Signaling Diversity Among GPCRs 433
3. Systems Analysis of Receptor Signaling Systems 438
4. Functional Analyses of Arrestin Signaling Paradigms 457
5. Conclusions 459

Acknowledgments 459
References 459

18. Arrestin Pathways as Drug Targets **469**

Louis M. Luttrell

1. Introduction 470
2. Biased GPCR Agonism 471
3. The Many Faces of Arrestin-Dependent Biased Agonism 477
4. The Conserved Arrestin-Dependent Signaling Repertoire 485
5. Conclusions 491

Acknowledgments 492
References 492

Index *499*

CONTRIBUTORS

Laurence Aubry
CEA, IRTSV, Laboratoire Biologie à Grande Echelle; INSERM, U1038, F-38054, and Université Joseph Fourier, F-38000, Grenoble, France

Ashley Bathgate
Department of Pharmaceutical Sciences, Nova Southeastern University College of Pharmacy, Fort Lauderdale, Florida, USA

Jean Martin Beaulieu
Département de Psychiatrie et de Neurosciences, Faculté de Médecine, Université Laval, Pavillon Ferdinand-Vandry, and Institut Universitaire en Santé Mentale de Québec (IUSMQ), 2601, chemin de la Canardière, Québec City, Quebec, Canada

Brittany N. Bohinc
Department of Medicine, Division of Endocrinology and Metabolism, Duke University Medical Center, Durham, North Carolina, USA

Marc G. Caron
Department of Cell Biology; Department of Medicine, and Department of Neurobiology, Duke University Medical Center, Durham, North Carolina, USA

Audrey Claing
Department of Pharmacology, Université de Montréal, Montréal, Quebec, Canada

Kathryn A. DeFea
Biomedical Sciences Division, School of Medicine, University of California Riverside, Riverside, California, USA

Tama Evron
Department of Cell Biology, Duke University Medical Center, Durham North Carolina, USA

Stephen S.G. Ferguson
J. Allyn Taylor Centre for Cell Biology, Robarts Research Institute, and Department of Physiology and Pharmacology, Western University Canada, London, Ontario, Canada

Diane Gesty-Palmer
Department of Medicine, Division of Endocrinology and Metabolism, Duke University Medical Center, and Parexel International, Research Triangle Park, Durham, North Carolina, USA

Eugenia V. Gurevich
Department of Pharmacology, Vanderbilt University, Nashville, Tennessee, USA

Vsevolod V. Gurevich
Department of Pharmacology, Vanderbilt University, Nashville, Tennessee, USA

Dianhua Jiang
Division of Pulmonary, Allergy, and Critical Care Medicine, Department of Medicine, Duke University School of Medicine, Durham, North Carolina, USA

Gérard Klein
CEA, IRTSV, Laboratoire Biologie à Grande Echelle; INSERM, U1038, F-38054, and Université Joseph Fourier, F-38000, Grenoble, France

Reddy Peera Kommaddi
Department of Medicine, Duke University Medical Center, Durham, North Carolina, USA

Camille Latapy
Département de Psychiatrie et de Neurosciences, Faculté de Médecine, Université Laval, Pavillon Ferdinand-Vandry, and Institut Universitaire en Santé Mentale de Québec (IUSMQ), 2601, chemin de la Canardière, Québec City, Quebec, Canada

Robert J. Lefkowitz
Department of Medicine; Department of Biochemistry, and the Howard Hughes Medical Institute, Duke University School of Medicine, Durham, NC, USA

Jiurong Liang
Division of Pulmonary, Allergy, and Critical Care Medicine, Department of Medicine, Duke University School of Medicine, Durham, North Carolina, USA

Louis M. Luttrell
Department of Medicine; Department of Biochemistry & Molecular Biology, Medical University of South Carolina, and The Research Service of the Ralph H. Johnson Veterans Affairs Medical Center, Charleston, South Carolina, USA

Anastasios Lymperopoulos
Department of Pharmaceutical Sciences, Nova Southeastern University College of Pharmacy, Fort Lauderdale, Florida, USA

Bronwen Martin
Metabolism Unit, National Institute on Aging, Baltimore, Maryland, USA

Stuart Maudsley
Receptor Pharmacology Unit, National Institute on Aging, Baltimore, Maryland, USA

William E. Miller
Department of Molecular Genetics, Biochemistry & Microbiology, University of Cincinnati College of Medicine, Cincinnati, Ohio, USA

Omar Moussa
Department of Pathology and Laboratory Medicine, Medical University of South Carolina, Charleston, South Carolina, USA

Paul W. Noble
Division of Pulmonary, Allergy, and Critical Care Medicine, Department of Medicine, Duke University School of Medicine, Durham, North Carolina, USA

Gang Pei
State Key Laboratory of Cell Biology, Institute of Biochemistry and Cell Biology, Shanghai Institutes for Biological Sciences, Chinese Academy of Sciences, and School of Life Science and Technology, Tongji University, Shanghai, PR China

Melanie Philipp
Institute of Biochemistry and Molecular Biology, Ulm University, Ulm, Germany

Sudha K. Shenoy
Department of Medicine, and Department of Cell Biology, Duke University Medical Center, Durham, North Carolina, USA

Sana Siddiqui
Receptor Pharmacology Unit, National Institute on Aging, Baltimore, Maryland, USA

W. Clay Smith
Department of Ophthalmology, University of Florida, Gainesville, Florida, USA

Philip Michael Sobolesky
Department of Pathology and Laboratory Medicine, Medical University of South Carolina, Charleston, South Carolina, USA

Cornelia Walther
J. Allyn Taylor Centre for Cell Biology, Robarts Research Institute, and Department of Physiology and Pharmacology, Western University Canada, London, Ontario, Canada

Ting Xie
Division of Pulmonary, Allergy, and Critical Care Medicine, Department of Medicine, Duke University School of Medicine, Durham, North Carolina, USA

Jian Zhao
State Key Laboratory of Cell Biology, Institute of Biochemistry and Cell Biology, Shanghai Institutes for Biological Sciences, Chinese Academy of Sciences, Shanghai, PR China

PREFACE

The arrestins, a small family of cytosolic proteins, have emerged as central players in the regulation of many facets of G protein-coupled receptor signaling. Originally discovered for their role in desensitization of the visual photoreceptor, rhodopsin, it soon became apparent that arrestin binding regulates the desensitization, internalization, and intracellular trafficking of nearly every G protein-coupled receptor. The subsequent discovery that arrestins serve as ligand-regulated signaling scaffolds added a new dimension to G protein-coupled receptor signal transduction. By binding and recruiting a host of catalytically active proteins to the receptor, arrestins function as alternative signal transducers, conferring novel G protein-independent signaling functions. It is increasingly apparent that the complementary desensitizing and signaling roles of arrestins are critical to many physiologic processes, from embryologic development to brain, cardiovascular, and immune system function, and to pathophysiologic processes such as cancer. Moreover, the finding that arrestin-mediated functions can be regulated independent of G protein signaling has made it possible to envision arrestin pathway-targeted therapeutics.

This volume represents an effort to present the full spectrum of arrestin biology, from structure–function relationships to translational applications for arrestin pathway-selective "biased" agonists. Beginning with a historical perspective and overview of the field written by 2012 Nobel Laureate in Chemistry Dr. Robert J. Lefkowitz (Chapter 1), the volume first covers the molecular biology of arrestins from high-resolution crystallographic structure, through molecular determinants of arrestin function, to mechanisms of arrestin dependent G protein-coupled receptor desensitization and signaling (Chapters 2–8). The next section considers the physiological functions of arrestins in different organs and tissues, from embryologic development and metabolic regulation to roles in the eye, brain, bone, cardiovascular and immune systems, and cancer (Chapters 9–16). The volume closes with a glimpse of the future, discussing the analysis of arrestin function *in vivo* using large microarray and proteomic data sets and the opportunities and pitfalls of arrestin selectivity in drug design (Chapters 17 and 18). Hopefully, this compendium will provide a useful and enduring introduction to a rapidly expanding field of investigation that offers substantial translational potential.

Louis M. Luttrell

PART I

Perspective—The Duality of Arrestin Function

CHAPTER ONE

Arrestins Come of Age: A Personal Historical Perspective

Robert J. Lefkowitz[*,†]
[*]Departments of Medicine and Biochemistry, Duke University Medical Center, Durham, NC, USA
[†]Howard Hughes Medical Institute, Duke University Medical Center, Durham, NC, USA

Contents

1. Introduction 4
2. "Prehistory" of Arrestins 4
3. A Family of 7 Transmembrane Receptors 5
4. The GRK and Arrestin Families 5
5. Arrestins and Desensitization 8
6. Arrestins and Endocytosis 8
7. Arrestins and Signaling 9
8. Biased Signaling 10
9. Bar Code Hypothesis 12
10. Other Receptors, Other Functions 13
11. Future Perspectives 14
References 14

Abstract

Visual arrestin and the two β-arrestins (1 and 2) were originally discovered 25–30 years ago in the context of their ability to desensitize phosphorylated G protein-coupled receptors (rhodopsin and the β2-adrenergic receptor, respectively). A fourth retinal-specific member of the family (X-arrestin) was discovered later. Over the past 10–15 years, however, it has become clear that these versatile molecules subserve a host of other roles in modulating and mediating the function of most GPCRs as well as other types of receptors. Functioning as multifunctional adaptor proteins, the β-arrestins also play prominent roles in receptor endocytosis, signaling, trafficking, and ubiquitination among others. Here, I provide a brief personal perspective on how the field has evolved since its inception and speculate on future directions.

Progress in Molecular Biology and Translational Science, Volume 118
ISSN 1877-1173
http://dx.doi.org/10.1016/B978-0-12-394440-5.00001-2

1. INTRODUCTION

For more than 40 years, I have had the privilege of watching and participating in one of the most critical areas of biomedical research, that devoted to understanding the nature, function, and regulation of the huge family of G protein-coupled receptors (GPCRs). From my own point of view, much of what we have learned about the activity and regulation of these receptors relates to their ability to interact, in a stimulus-dependent fashion, with three families of proteins: the heterotrimeric guanine nucleotide-binding proteins (G proteins), for which they are named; the arrestins; and the G protein-coupled receptor kinases (GRKs). This is not to deny the importance of other families of proteins that, in turn, regulate the G proteins themselves (e.g., RGS) or numerous downstream signaling elements. However, the G proteins, arrestins, and GRKs are the only proteins currently known which interact almost universally with the stimulus- or agonist-modified receptors and which are thus poised to most directly translate extracellular signals impinging on the receptors into desired cellular perturbations.

The functions and activities of the arrestins and GRKs are inextricably interrelated, and it is impossible to tell the story of one without the other. Most remarkably, over the past 10–15 years, a major paradigm shift has occurred in our understanding of how these families of proteins regulate and mediate the activities of GPCRs. Both protein families were discovered in the context of attempts to understand how the two prototypic GPCRs, rhodopsin and the β2-adrenergic receptor (β2AR), were desensitized, or turned off, after stimulation. However, as this volume attests, this was just the beginning. Today, the arrestins are known to carry out multiple functions including mediation of myriad signaling pathways as well as endocytic and trafficking functions, among others. They have emerged as GPCR adaptor proteins par excellence, parlaying their agonist-dependent association with the receptors into all manner of downstream protein–protein interactions with important cellular consequences. The reader will find all of these topics reviewed in this volume. Here, I provide a very brief historical introduction to the field hanging the thread of the story largely on work from my own laboratory over the past several decades.

2. "PREHISTORY" OF ARRESTINS

The protein we know today as arrestin (the visual arrestin, a.k.a. arrestin1) was originally discovered in the 1970s as a highly immunogenic soluble retinal antigen that was involved in experimental allergic uveitis, an autoimmune disease.[1] It was referred to as S-antigen. Meanwhile, Hermann Kuhn described a protein of 48 kDa which, together with rhodopsin kinase and several others, bound to photoreceptor membranes in a light-dependent fashion.[2,3] Its function at the time was completely unknown. It was subsequently shown by several groups that this protein quenches the light-stimulated retinal cGMP-phosphodiesterase activity by binding to phosphorylated photoexcited rhodopsin.[4–6] With the discovery of this "arresting" function in retinal signaling, the protein was renamed arrestin. In 1985, Kuhn *et al.* demonstrated, using several biochemical and immunological criteria, that S-antigen and 48K protein were identical.[7]

3. A FAMILY OF 7 TRANSMEMBRANE RECEPTORS

Prior to 1986, it was not appreciated that rhodopsin and other GPCRs were members of the same large gene family.[8] In retrospect, this seems hard to fathom. It was known that the rhodopsin signaling pathway, consisting of the photosensor rhodopsin, a heterotrimeric G protein, transducin, and an effector, cGMP phosphodiesterase, was clearly analogous to hormonal systems such as the β2AR, G_s, and adenylate cyclase.[9] Nonetheless, nobody envisaged any structural relationship between rhodopsin and the β2AR. Rather rhodopsin, the sequence of which had been determined by conventional protein sequencing in 1982, was analogized with bacteriorhodopsin, a light-sensitive proton pump found in *Archaebacteria*.[10,11] This was because the sequence of both proteins predicted seven transmembrane domains and both were light sensitive. Thus, at the outset, seven membrane spans were thought to represent a signature feature of light-sensitive proteins. Only with the cloning of the β2AR in 1986,[8] the α2AR a year later,[12] and then many others did it become clear that GPCRs constituted a huge family of 7 transmembrane receptors (7TMRs).[13]

4. THE GRK AND ARRESTIN FAMILIES

Through the 1970s and mid 1980s, two contemporaneous streams of research indicated that multiple phosphorylation of rhodopsin and the β2AR on their cytoplasmic C-terminal tails was occurring during the process of desensitization or dampening of G protein-mediated signaling. The enzyme responsible for rhodopsin phosphorylation was identified as novel and named rhodopsin kinase in the late 1970s.[14,15] Meanwhile, a body of evidence had been developed which indicated that the multiple phosphorylation of the β2AR which accompanied so-called homologous- or "agonist-specific" desensitization of the β2AR was carried out by a novel cAMP-independent kinase, initially named the β-adrenergic receptor kinase or βARK.[16] The discovery of this enzyme in 1986 was part of a rapidly moving story which would, within the space of just a few years, firmly establish a very general paradigm for a mechanism of desensitization of G protein-coupled receptor signaling.

Jeff Benovic, then a very talented graduate student in my lab, was busy purifying βARK from bovine brain during this period.[17] To assay its catalytic activity through various stages of purification, he measured its ability to phosphorylate β2AR (purified from endogenous sources, no cloned or overexpressed material was available during this era) reconstituted in phosphatidylcholine vesicles. He could then assess the "uncoupling" or desensitization of the receptor by following its ability to mediate stimulation of the GTPase activity of co-reconstituted G_s. Paradoxically, the more he purified the enzyme and enhanced its specific catalytic kinase activity, the lower became its "desensitizing" ability.[18] We feared we were losing some necessary cofactor or accessory component. However, a variety of biochemical approaches failed to yield the putative missing element.

While we pondered this situation in 1986, Kuhn published a report that 48K protein binds to photoexcited rhodopsin and quenches phosphodiesterase activation.[6] This report really caught our attention, since the 48K protein (arrestin) seemed to match the profile of our missing desensitizing element. However, we knew what we were looking for could not be arrestin itself since the expression of arrestin was known to be limited to the retina and pineal gland. Undeterred, I called Kuhn and persuaded him to send me some of his arrestin protein. Benovic was then able to demonstrate that when added to his reconstituted system (albeit at high arrestin/G_s molar ratios), it restored the desensitizing ability of the purified βARK.[18] This led us to the hypothesis that what we were purifying away from βARK might be a protein analogous to and possibly homologous with visual arrestin.

When, in 1987, Shinohara cloned "48K" protein,[19] we obtained his clone and Martin Lohse in our lab was able to use it to clone a 70% sequence identical protein we named β-arrestin.[20] Not long thereafter, another fellow in the lab, Håvard Attramadal cloned another member of this family and we named it β-arrestin2, now referring to β-arrestin as β-arrestin1.[21] In contrast to visual arrestin, βarr1 and 2 were found to have much higher affinity for phosphorylated β2AR than for phosphorylated rhodopsin.[21] Moreover, unlike arrestin, they are ubiquitously expressed. Today, we know there is a fourth member of the family called X-arrestin or arrestin4, expression of which also seems largely limited to the retina.[22] As described in subsequent chapters, all members of this small family share the same fold consisting of two domains comprised almost entirely of antiparallel β-sheets connected by a hinge region and have many shared, though also individual, properties.[23]

A brief word about the somewhat confusing nomenclature in this field seems in order. The genesis of the name arrestin for the visual "48K" protein, as a molecule which arrests signaling, is described above. Hence, when we cloned the second member of the family as a protein which worked in concert with βARK to desensitize the β2AR, it seemed natural to name it β-arrestin.[20] At the time, we had no evidence that this mechanism of receptor regulation was a very general one. Two years later, in September 1992, we reported the cloning of a third family member and named it β-arrestin2.[21] The following summer, in July of 1993, Benovic's group reported the cloning of the same molecule (albeit from cow rather than rat) and the fact that there were two alternatively spliced forms with or without an 11-amino acid insert.[24] His group proposed that βarr1 now be renamed arrestin2 and that β-arrestin2 be renamed arrestin3. Over the subsequent 20 years, both sets of terminology have persisted and are used interchangeably.

Further evidence of the conserved nature of the receptor "turnoff" mechanism involving GRKs and arrestins was our demonstration also in 1986 that rhodopsin kinase could phosphorylate the agonist-occupied β2AR and that βARK could phosphorylate light-bleached rhodopsin, albeit each with much slower kinetics than their natural substrates.[25]

In 1988, we cloned the cDNA for βARK and realized that it was the founding member of a novel subfamily of kinases, most closely related to the AGC kinase family of second messenger-dependent kinases.[26] After persuading Kris Palczewski to purify a preparation of rhodopsin kinase over his Christmas holiday in 1989 before moving to Seattle from his *post doc* in Paul Hargrave's lab, we were able to obtain some amino acid sequence on this protein, which allowed us to design probes to clone its cDNA. Rhodopsin kinase turned out to be a member of the same new gene family as βARK, sharing a similar tridomain structure.[27]

Subsequently, we and others cloned additional members of this kinase family such that today we know that there are seven members, referred to as GRK1–7. Rhodopsin kinase is GRK1 and βARK is GRK2. GRKs 2, 3, 5, and 6 are quite ubiquitously expressed, whereas GRK1 and 7 are limited to the retina, and GRK4 also has a very restricted distribution (reviewed in Refs. 28,29).

John Tesmer has solved crystal structures of several members of the family, all of which share a conserved central catalytic domain flanked by two more divergent regulatory domains.[30,31] In GRK2 and 3, the carboxyl terminal domain is a PH or pleckstrin homology domain that binds free Gβγ as a means of stimulus-dependent translocation to the plasma membrane from the cytosol. GRKs 5 and 6, however, are constitutively membrane bound through other mechanisms.[28]

5. ARRESTINS AND DESENSITIZATION

Both visual arrestin and β-arrestin1 were discovered in the specific context of their ability to functionally turn off or desensitize rhodopsin or β2AR signaling, respectively. When we cloned βarr1 and 2, we took as the ultimate validation of the cloned molecules' functionality the ability of the authentic recombinant materials to block β2AR activation of G_s, in reconstituted systems. Discovery of GRK2 and the β-arrestins capped my 20-year search to understand how stimulation of the β2AR (and presumably the other GPCRs) led to a rapid waning of the response to further stimulation. I had been particularly interested in the so-called homologous desensitization wherein stimulation of the receptor desensitizes only that receptor and not others (the latter phenomenon being referred to as heterologous desensitization).[32] The GRK–βarr mechanism is a major mediator of homologous desensitization, while phosphorylation of the receptor by second messenger kinases such as PKA and PKC is a major mechanism of heterologous desensitization of the receptors.[32,33] Numerous other mechanisms can lead to desensitization of GPCRs, especially the heterologous form.

6. ARRESTINS AND ENDOCYTOSIS

That arrestins might be involved in the very general phenomenon of agonist-induced GPCR endocytosis was initially suggested by evidence that receptor phosphorylation, in particular, by GRKs, played a role in this process.[34] Then it was demonstrated that β-arrestins promoted receptor internalization when overexpressed in cells and that "dominant negative" β-arrestin mutants inhibited this process.[35] Shortly thereafter, it was demonstrated that β-arrestins can bind to clathrin, immediately offering a potential mechanism for this new function of arrestins.[36] Subsequently, it has been found that β-arrestins interact with other components of the clathrin-coated pit machinery including AP2[37] and ARF6.[38] More recently, this endocytic adaptor function has been extended to non-GPCRs such as the tyrosine kinase IGF1 receptor.[39]

The involvement of β-arrestins in receptor trafficking is now known to be complex and multifaceted. β-Arrestins appear to be involved in receptor recycling as well as trafficking to other cellular compartments.[40] The ability of β-arrestins to serve as adaptors linking the receptor to several cellular motors is also undoubtedly involved in these processes.[41]

7. ARRESTINS AND SIGNALING

Having been discovered and named in the context of their roles in desensitization, the idea that the arrestins might actually serve as signal transducers in their own right was not initially considered. However, it was observed that in some systems GPCR activation of ERK appeared to require receptor endocytosis.[42] Since the idea that β-arrestins were involved in clathrin-mediated 7TMR endocytosis was gaining credence at this time (late 1990s), it was hypothesized that perhaps β-arrestins were involved at some stage of the assembly of multiprotein signaling complexes such as the MAP kinase cascade.[43] In fact, the first signaling pathway to be demonstrated to be activated in a β-arrestin-dependent fashion was c-Src downstream of the β2AR.[44] β-Arrestin1 was shown to function as an adaptor protein recruiting c-Src to the activated receptor and facilitating activation of downstream ERK 1 and 2. Subsequently, over the next few years, β-arrestin was shown to scaffold MAP kinase complexes of the ERK cassette,[45,46] p38,[47] and Jnk 2 and 3.[48] These signaling pathways appear to function in parallel with those mediated by G proteins but lead to activation of distinct pools of the enzymes. For example, activation of ERK1/2 downstream of G proteins generally leads to its translocation into the nucleus and phosphorylation and activation of various transcription factors. In contrast, activation of ERK1 and 2 by scaffolding by β-arrestin2 leads to retention of the activated ERK in the cytosol due to the fact that β-arrestin2 contains a nuclear exclusion signal, and hence, a completely different set of substrates, in this case cytosolic, are phosphorylated, with quite different cellular consequences.[49] Moreover, the kinetics of G protein versus β-arrestin-mediated ERK activation are often distinct with G protein signaling being rapid in onset and quite transient (due to β-arrestin-mediated desensitization).[49] In contrast, β-arrestin-mediated ERK activation is slower in onset and more persistent.

These mechanisms have turned out to be quite general though the specifics with respect to the individual roles of β-arrestin1 and 2 differ substantially from one receptor to another. Detailed biochemical studies of β-arrestin interaction with components of the ERK cascade enzymes have been published and strongly support a scaffolding mechanism.[45,50]

Initial studies of "β-arrestin-mediated" signaling by 7TMRs were not interpreted as necessarily suggesting wider roles for β-arrestins. However, subsequent proteomics studies analyzed the β-arrestin "interactome"[51] and also provided a quantitative global phosphorylation analysis downstream of β-arrestin activation by the angiotensin II type 1A receptor.[52] These studies indicated a surprisingly rich network of β-arrestin-binding partners and linked downstream signaling elements. Subnetworks engaged downstream of β-arrestins include, in addition to a multiplicity of MAP kinase pathways, elements of systems involved in PI3kinase/AKT signaling, DNA damage repair, cell cycle and development, and cytoskeletal reorganization among others.[52] Hundreds of papers published over the past few years attest to the diversity and number of signaling mechanisms in which the β-arrestins are involved.[53]

An interesting and as yet unexplained feature of this signaling is the involvement, or lack of involvement, of G proteins. For most cases, where the 7TMR appears to be G_s or G_q coupled, β-arrestin-mediated signaling appears to be independent of G proteins. However, in the case of many G_i-coupled receptors, signaling appears to simultaneously require involvement of G_i (can be blocked by pertussis toxin) as well as β-arrestin (blocked by β-arrestin siRNA). The explanation for this is not known.

8. BIASED SIGNALING

Biased agonism, also known as ligand-directed signaling or functional selectively, refers to the ability of ligands for a single receptor to demonstrate differential efficiency for different responses. The phenomenon was first noted in the 1990s in the context of several GPCRs that could stimulate more than one G protein, for example, G_s and G_i, or G_s and G_q.[54] Some ligands show preferential activity for one signaling pathway over the other. However, after the discovery of β-arrestin-mediated signaling, this phenomenon received increasing attention as ligands were discovered which were either G protein or β-arrestin biased.[55] Such biased signaling strongly implies the existence of distinct conformations of the receptor, preferentially stabilized by distinct ligands, which lead to the different signaling pathways.

This of course is in contrast to the classical simple 2-state model of receptor signaling $A + R \leftrightarrows AR^*$ where R and R^* are unitary inactive and active states of the receptor, respectively, and A is an agonist. Despite a number of recent biophysical studies suggesting such a multiplicity of receptor active states,[56–58] their existence remains to be firmly established.

The existence of G protein or β-arrestin-biased ligands has now been demonstrated for dozens of receptors.[59] In most cases, the "biased" ligands are synthetic molecules. However, in at least one case, the CCR7 receptor, one endogenous chemokine ligand, CCL21, is G-biased, whereas its other endogenous ligand CCL19 is unbiased.[60] Other examples will likely be forthcoming. At least one case of an endogenously expressed, entirely β-arrestin-biased chemokine receptor, CXCR7, has been described.[61]

The existence of such biased ligands has potential therapeutic implications. This is true because biased signaling allows for the possibility of designing drugs with greater specificity and fewer side effects. The details of course depend upon which effects of a receptor are desired ones, which are considered side effects, and which are mediated by β-arrestins versus G proteins.

Two quick examples will illustrate the point. Angiotensin receptor blockers (ARBs) are widely used in clinical medicine because they block the potentially deleterious G protein (G_q)-mediated effects of the angiotensin II type 1A receptor to increase blood pressure via vasoconstriction. However, they also block potentially advantageous β-arrestin-mediated effects of the receptor. These include increased cardiac performance and anti-apoptosis.[62,63] A β-arrestin-biased angiotensin analog has been shown to slow the progression of heart failure in animal models and might be useful for the treatment of congestive heart failure.[64] Such a biased agonist of course is itself a competitive antagonist of G protein-mediated signaling (hence, it lowers blood pressure like a standard ARB) while simultaneously activating β-mediated signaling (unlike standard ARBs).[65]

A situation where a G-biased ligand might be therapeutically desirable is the μ-opioid receptor that mediates the pain relieving or antinociceptive effects of opiate analgesics. These effects have been shown to be G_i mediated. However, prominent narcotic side effects such as tolerance, constipation, and respiratory depression seem to be mediated largely through β-arrestin-mediated signaling.[66,67] Thus, one might speculate that a G-biased μ-opioid ligand would have preserved antinociceptive efficacy with markedly reduced β-arrestin-dependent side effects. The potential for such biased ligands to provide added therapeutic benefit is very attractive but remains to be clinically demonstrated.

9. BAR CODE HYPOTHESIS

The remarkable diversity of β-arrestin functions raises the issue of how, once bound to a GRK phosphorylated receptor, the β-arrestin "knows" what it is supposed to do. This situation suggests that there might be some code in the receptor that leads to β-arrestin assuming a variety of conformations with distinct functional capacities. In fact, several approaches including limited proteolysis of β-arrestins 1 and 2[68,69] and studies with an intramolecular β-arrestin2 BRET probe among others[70] have confirmed the existence of such differing β-arrestin conformations. Moreover, the BRET studies clearly indicate that different conformations of β-arrestin are stabilized after receptor stimulation by unbiased versus β-arrestin-biased agonists.[70]

These results in turn beg the question of how different receptor conformations, presumably formed in response to agonists of differing bias, can impart this information to the arrestins. This question has given rise to what has been called the "bar code hypothesis." The idea is that GRK isoforms differentially recognize various receptor conformations stabilized by different types of agonists and then phosphorylate them on different sites. These distinctive phosphorylation patterns would then form a "bar code" of sorts, read by the β-arrestin, thereby determining its conformation and functional activity.[71,72] Data consistent with this hypothesis have already been published for the β2AR,[73] angiotensin II type 1a receptor,[71] V2 vasopressin receptor,[72] M3 muscarinic acetylcholine receptor,[74] and CCR7 receptor.[60] For example, in the case of the β2AR expressed in HEK cells, stimulation by the unbiased ligand isoproterenol leads to phosphorylation on two GRK6 sites and six GRK2 sites, whereas stimulation by the weak β-arrestin-biased ligand carvedilol leads to phosphorylation only on the two GRK6 sites.[73] Phosphorylation of the GRK2 sites appears to inhibit phosphorylation on the GRK6 sites. Moreover, phosphorylation by the different GRKs promotes distinct functional profiles of the subsequently bound β-arrestin. Phosphorylation by GRK2 or 6 supports desensitization of G protein signaling though with greater effectiveness noted for GRK2. Similarly, both kinases can support β-arrestin-mediated endocytosis though again with an advantage for GRK2. However, whereas GRK6 is essential for β-arrestin-mediated ERK activation, GRK2 phosphorylation actually inhibits this process.[73] Moreover, this antithetical relationship of the two GRKs for β-arrestin-mediated ERK activation has been noted for a number of different receptors.[75] Whether this is simply a reflection of GRK2 phosphorylation inhibiting GRK6 phosphorylation or whether additional factors play a role is not currently known. Recently, it has been proposed that a delicate interplay between different GRKs (e.g., 2 and 6) may regulate the balance of G protein and β-arrestin-mediated signaling pathways downstream of 7TMRs.[75]

10. OTHER RECEPTORS, OTHER FUNCTIONS

In addition to their paradigmatic roles in the regulation of classical 7TMRs, β-arrestins have also been found to regulate some receptors from other diverse families. These include so-called atypical 7TMRs such as frizzled[76] and smoothened,[77] the receptor tyrosine kinases for IGF1[39] and TrkA,[78] the nicotinic cholinergic receptor,[79] and even some cytokine receptors such as the TGF-βRIII.[80] Depending on the particular receptor, the β-arrestins may mediate endocytosis, signaling, or other functions.

Another role that has emerged for the β-arrestins in recent years is that of an E3-ubiquitin ligase adaptor which facilitates interaction of ligases of widely varying structures with their substrates. This role was first appreciated for the β2AR[81] though at the time the relevant ligase was unknown (it was subsequently shown to be NEDD4).[82] Later, it was shown that other 7TMRs are regulated in this way by a variety of ligases (reviewed in Ref. 82). Moreover, it is now known that diverse families of proteins are also ubiquitinated by ligases using β-arrestins as adaptor proteins.[83] β-Arrestin-mediated ubiquitination has been shown to play roles in the internalization of membrane proteins such as 7TMRs,[81] routing them to lysosomes for destruction, in proteosomal degradation and signaling.[83] β-Arrestins themselves are ubiquitinated by MDM2, and this ubiquitination appears to be necessary for their endocytic and at least some of their signaling functions.[81,84] Whether this ubiquitination plays a role in β-arrestin—partner recognition and/or other mechanisms is not clearly understood at present.

Recently, a much wider family of proteins containing predicted "arrestin domains" has been discovered in organisms from yeast to mammals.[85] These "arrestin-like" proteins have little or no sequence similarity to the mammalian arrestins but are predicted to share the same fold. These proteins have been referred to as "α-arrestins."[86] Members include ARRDC1-5 and TXNIP.[85] In yeast, there is a family of nine arrestin-related trafficking adaptors.[87] These proteins have some sequence similarities to ARRDCs and play roles in the ubiquitination, endocytosis, and lysosomal trafficking of membrane cargoes.[87] In mammalian cells, ARRDCs play various roles in trafficking of membrane proteins which are still being defined.

11. FUTURE PERSPECTIVES

As the chapters in this volume attest, research on arrestins has been remarkably robust and multidisciplinary over the past decade. And the horizons continue to expand at an exponential rate. What might we anticipate in the next decade or so?

1. Cellular and *in vivo* experiments will continue to expand our understanding of new functions of arrestins in signaling and trafficking not only of 7TMRs but also of other types of receptors and other categories of macromolecules.
2. The molecular mechanisms of β-arrestin-mediated signaling, endocytosis, and other functions will be elucidated by detailed biochemical and biophysical studies. The key to much of this will be application of continuously improving structural biology approaches, such as X-ray crystallography, NMR, EPR, HDX, and many others to the study of arrestins and their cellular partners. Structures of arrestin complexes with their binding partners, for example, clathrin or MAP kinase components, or of receptors in complex with arrestins, should elucidate how arrestins mediate their diverse activities with atomic level detail.
3. The therapeutic potential of targeting the arrestins, either through the receptors with biased ligands or directly with small molecules or other approaches, will be explored in numerous systems. Recent studies indicating roles for arrestins in diverse diseases, such as chronic myelogenous leukemia,[88] idiopathic pulmonary fibrosis,[89] and various cancers[90] have recently presented a rationale for this latter approach.

As exciting as the first 25 years of research on arrestins has been, the next 25 should be even more so and proceed at an ever-increasing rate.

REFERENCES

1. Wacker WB, Donoso LA, Kalsow CM, Yankeelov Jr JA, Organisciak DT. Experimental allergic uveitis. Isolation, characterization, and localization of a soluble uveopathogenic antigen from bovine retina. *J Immunol.* 1977;119:1949–1958.
2. Kuhn H. Light-regulated binding of rhodopsin kinase and other proteins to cattle photoreceptor membranes. *Biochemistry.* 1978;17:4389–4395.
3. Wilden U, Wust E, Weyand I, Kuhn H. Rapid affinity purification of retinal arrestin (48 kDa protein) via its light-dependent binding to phosphorylated rhodopsin. *FEBS Lett.* 1986;207:292–295.
4. Miller JL, Fox DA, Litman BJ. Amplification of phosphodiesterase activation is greatly reduced by rhodopsin phosphorylation. *Biochemistry.* 1986;25:4983–4988.
5. Zuckerman R, Cheasty JE. A 48 kDa protein arrests cGMP phosphodiesterase activation in retinal rod disk membranes. *FEBS Lett.* 1986;297:35–41.

6. Wilden U, Hall SW, Kuhn H. Phosphodiesterase activation by photoexcited rhodopsin is quenched when rhodopsin is phosphorylated and binds the intrinsic 48 kDa protein of rod outer segments. *Proc Natl Acad Sci USA*. 1986;83:1174–1178.
7. Pfister C, Chabre M, Plouet J, et al. Retinal S antigen identified as the 48K protein regulating light-dependent phosphodiesterase in rods. *Science*. 2012;228:891–893.
8. Dixon RAF, Kobilka BK, Strader DJ, et al. Cloning of the gene and cDNA for mammalian β-adrenergic receptor and homology with rhodopsin. *Nature*. 1986;321:75–79.
9. Stryer L. Cyclic GMP cascade of vision. *Annu Rev Neurosci*. 1986;9:87–119.
10. Ovchinnikov YA. Rhodopsin and bacteriorhodopsin: structure-function relationships. *FEBS Lett*. 1982;148:179–191.
11. Hargrave PA, McDowell JH, Curtis DR, et al. The structure of bovine rhodopsin. *Biophys Struct Mech*. 1983;9:235–244.
12. Kobilka BK, Matsui H, Kobilka TS, et al. Cloning, sequencing, and expression of the gene coding for the human platelet α_2-adrenergic receptor. *Science*. 1987;238:650–656.
13. Dohlman HG, Caron MG, Lefkowitz RJ. A family of receptors coupled to guanine nucleotide regulatory proteins. *Biochemistry*. 1987;26:2657–2664.
14. Weller M, Virmaux N, Mandel P. Light-stimulated phosphorylation of rhodopsin in the retina: the presence of a protein kinase that is specific for photobleached rhodopsin. *Proc Natl Acad Sci USA*. 1975;72:381–385.
15. Wilden U, Kuhn H. Light-dependent phosphorylation of rhodopsin: number of phosphorylation sites. *Biochemistry*. 1982;21:3014–3022.
16. Benovic JL, Strasser RH, Caron MG, Lefkowitz RJ. β-Adrenergic receptor kinase: identification of a novel protein kinase which phosphorylates the agonist-occupied form of the receptor. *Proc Natl Acad Sci USA*. 1986;83:2797–2801.
17. Benovic JL, Mayor Jr F, Staniszewski C, Lefkowitz RJ, Caron MG. Purification and characterization of the β-adrenergic receptor kinase. *J Biol Chem*. 1987;262:9026–9032.
18. Benovic JL, Kuhn H, Weyand I, Codina J, Caron MG, Lefkowitz RJ. Functional desensitization of the isolated β-adrenergic receptor by the β-adrenergic receptor kinase: potential role of an analog of the retinal protein arrestin (48K protein). *Proc Natl Acad Sci USA*. 1987;84:8879–8882.
19. Shinohara T, Dietzschold B, Craft CM, et al. Primary and secondary structure of bovine retinal S antigen (48-kDa protein). *Proc Natl Acad Sci USA*. 1987;84:6975–6979.
20. Lohse MJ, Benovic JL, Codina J, Caron MG, Lefkowitz RJ. β-Arrestin: a protein that regulates β-adrenergic receptor function. *Science*. 1990;248:1547–1550.
21. Attramadal H, Arriza JL, Aoki C, et al. β-Arrestin2—a novel member of the arrestin/β-arrestin gene family. *J Biol Chem*. 1992;267:17882–17890.
22. Craft CM, Whitmore DH, Wiechmann AF. Cone arrestin identified by targeting expression of a functional family. *J Biol Chem*. 1994;269:4613–4619.
23. Gurevich VV, Gurevich EV. The structural basis of arrestin-mediated regulation of G-protein-coupled receptors. *Pharmacol Ther*. 2006;110:465–502.
24. Sterne-Marr R, Gurevich VV, Goldsmith P, et al. Polypeptide variants of β-arrestin and arrestin3. *J Biol Chem*. 1993;268:15640–15648.
25. Benovic JL, Mayor Jr F, Somers RL, Caron MG, Lefkowitz RJ. Light dependent phosphorylation of rhodopsin by the β-adrenergic receptor kinase. *Nature*. 1986;321:869–872.
26. Benovic JL, DeBlasi A, Stone WC, Caron MG, Lefkowitz RJ. β-Adrenergic receptor kinase: primary structure delineates a multigene family. *Science*. 1989;246:235–240.
27. Lorenz W, Inglese J, Palczewski K, Onorato JJ, Caron MG, Lefkowitz RJ. The receptor kinase family: primary structure of rhodopsin kinase reveals similarities to the β-adrenergic receptor kinase. *Proc Natl Acad Sci USA*. 1991;88:8715–8719.
28. Pitcher JA, Freedman NJ, Lefkowitz RJ. G protein-coupled receptor kinases. *Annu Rev Biochem*. 1998;67:653–692.

29. Krupnick JG, Benovic JL. The role of receptor kinases and arrestins in a G protein-coupled receptor regulation. *Annu Rev Pharmacol Toxicol.* 1998;38:289–319.
30. Lodowski DT, Pitcher JA, Capel WD, Lefkowitz RJ, Tesmer J. Keeping G proteins at bay: a complex between G protein-coupled receptor kinase 2 and Gβγ. *Science.* 2003;300:1256–1262.
31. Gurevich EV, Tesmer JJ, Mushegian A, Gurevich VV. G protein-coupled receptor kinases: more than just kinases and not only for GPCRs. *Pharmacol Ther.* 2012;133:40–69.
32. Hausdorff WP, Caron MG, Lefkowitz RJ. Turning off the signal: desensitization of β-adrenergic receptor function. *FASEB J.* 1990;11:2881–2889.
33. Benovic JL, Pike LJ, Cerione RA, et al. Phosphorylation of the mammalian β-adrenergic receptor by cyclic AMP-dependent protein kinase: regulation of the rate of receptor phosphorylation and dephosphorylation by agonist occupancy and effects on coupling of the receptor to the stimulatory guanine nucleotide regulatory protein. *J Biol Chem.* 1985;260:7094–7101.
34. Menard L, Ferguson SG, Barak LS, et al. Members of the G protein-coupled receptor kinase family that phosphorylate the β2-adrenergic receptor facilitate sequestration. *Biochemistry.* 1996;35:4155–4160.
35. Ferguson SSG, Downey III WE, Colapietro AM, Barak LL, Menard L, Caron MG. Role of β-arrestin in mediating agonist-promoted G protein-coupled receptor internalization. *Science.* 1996;271:363–365.
36. Goodman Jr OB, Krupnick JG, Santini F, et al. β-Arrestin acts as a clathrin adaptor in endocytosis of the β2-adrenergic receptor. *Nature.* 1996;383:447–450.
37. Laporte SA, Oakley RH, Zhang J, et al. The β2-adrenergic receptor/β-arrestin complex recruits the clathrin adaptor AP-2 during endocytosis. *Proc Natl Acad Sci USA.* 1999;96:3712–3717.
38. Claing A, Chen W, Miller WE, et al. β-Arrestin-mediated ARF6 activation and β_2-adrenergic receptor endocytosis. *J Biol Chem.* 2001;276:42509–42513.
39. Lin F-T, Daaka Y, Lefkowitz RJ. β-Arrestins regulate mitogenic signaling and clathrin-mediated endocytosis of the insulin-like growth factor I receptor. *J Biol Chem.* 1998;273:31640–31643.
40. Shenoy SK, Lefkowitz RJ. β-Arrestin-mediated receptor trafficking and signal transduction. *Trends Pharmacol.* 2011;32:521–533.
41. Kovacs JJ, Whalen EJ, Liu R, et al. β-Arrestin mediated localization of smoothened to the primary cilium. *Science.* 2008;320:1777–1781.
42. Daaka Y, Luttrell LM, Ahn S, et al. Essential role for G protein-coupled receptor endocytosis in the activation of mitogen-activated protein kinase. *J Biol Chem.* 1998;273:685–688.
43. Lefkowitz RJ. G protein-coupled receptors. New roles for receptor kinases and β-arrestins in receptor signaling and desensitization. *J Biol Chem.* 1998;273:18677–18680.
44. Luttrell LM, Ferguson SSG, Daaka Y, et al. β-Arrestin-dependent formation of β2-adrenergic receptor/Src kinase complexes and mitogenic signaling. *Science.* 1999;283:655–661.
45. Luttrell LM, Roudabush FL, Choy EW, et al. Activation and targeting of extracellular signal-regulated kinases by β-arrestin scaffolds. *Proc Natl Acad Sci USA.* 2001;98:2449–2454.
46. DeFea KA, Zalevsky J, Thoma MS, Dery O, Mullins RD, Bunnett NW. β-Arrestin-dependent endocytosis of proteinase-activated receptor 2 is required for intracellular targeting of activated ERK1/2. *J Cell Biol.* 2000;148:1267–1281.
47. Sun Y, Cheng Z, Ma L, Pei G. β-Arrestin2 is critically involved in CXCR4-mediated chemotaxis, and this is mediated by its enhancement of p38 MAPK activation. *J Biol Chem.* 2002;277:49212–49219.
48. McDonald PH, Chow C-W, Miller WE, et al. β-Arrestin2: a receptor-regulated MAPK scaffold for the activation of JNK3. *Science.* 2000;290:1574–1577.
49. Ahn S, Shenoy SK, Wei H, Lefkowitz RJ. Differential kinetic and spatial patterns of β-arrestin2 and G protein-mediated ERK activated by the angiotensin II receptor. *J Biol Chem.* 2004;279:35518–35525.

50. Coffa S, Breitman M, Spiller BW, Gurevich VV. A single mutation in arrestin-2 prevents ERK1/2 activation by reducing c-Raf1 binding. *Biochemistry*. 2011;50:6951–6958.
51. Xiao K, McClatchy DB, Shukla AK, et al. Functional specialization of β-arrestin interactions revealed by proteomic analysis. *Proc Natl Acad Sci USA*. 2007;104:12011–12016.
52. Xiao K, Sun J, Kim J, et al. Global phosphorylation analysis of β-arrestin-mediated signaling downstream of a seven transmembrane receptor (7TMR). *Proc Natl Acad Sci USA*. 2010;107:15299–15304.
53. DeWire SM, Ahn S, Lefkowitz RJ, Shenoy SK. β-Arrestins and cell signaling. *Annu Rev Physiol*. 2007;69:483–510.
54. Kenakin T. Functional selectivity through protean and biased agonism: who steers the ship? *Mol Pharmacol*. 2007;72:1393–1401.
55. Rajagopal S, Rajagopal K, Lefkowitz RJ. Teaching old receptors new tricks: biasing seven-transmembrane receptors. *Nat Rev Drug Discov*. 2010;9:1–12.
56. Kahsai AW, Xiao K, Rajagopal S, et al. Multiple ligand-specific conformations of the β2-adrenergic receptor. *Nat Chem Biol*. 2011;7:692–700.
57. Seifert R, Gether U, Wenzel-Seifert K, Kobilka BK. Effects of guanine, inosine, and xanthine nucleotides on β2-adrenergic receptor/G(s) interactions: evidence for multiple receptor conformations. *Mol Pharmacol*. 1999;56:348–358.
58. West GM, Chien EY, Katritch V, et al. Ligand-dependent perturbation of the conformational ensemble for the GPCR β2-adrenergic receptor revealed by HDX. *Structure*. 2011;19:1424–1432.
59. Whalen E, Rajagopal S, Lefkowitz RJ. The therapeutic potential of β-arrestin and G protein-biased agonists. *Trends Mol Med*. 2011;17:126–139.
60. Zidar DA, Violin JD, Whalen EJ, Lefkowitz RJ. Selective engagement of G protein-coupled receptor kinase (GRKs) encodes distinct functions of biased ligands. *Proc Natl Acad Sci USA*. 2009;106:9649–9654.
61. Rajagopal S, Kim J, Ahn S, et al. β-Arrestin-but not G-protein-mediated signaling by the "decoy" receptor CXCR7. *Proc Natl Acad Sci USA*. 2010;107:628–632.
62. Rajagopal K, Whalen EJ, Violin JD, et al. β-Arrestin2-mediated inotropic effects of the angiotensin II type 1A receptor in isolated cardiac myocytes. *Proc Natl Acad Sci USA*. 2006;103:16284–16289.
63. Ahn S, Kim J, Hara MR, Ren X-R, Lefkowitz RJ. β-Arrestin-2 mediates anti-apoptotic signaling through regulation of BAD phosphorylation. *J Biol Chem*. 2009;284:8855–8865.
64. Boerrigter G, Lark MW, Whalen EJ, Soergel DG, Violin JD, Burnett Jr JC. Cardiorenal actions of TRV120027, a novel β-arrestin-biased ligand at the angiotensin II type I receptor, in healthy and heart failure canines: a novel therapeutic strategy for acute heart failure. *Circ Heart Fail*. 2011;6:770 778.
65. Violin JD, DeWire SM, Yamashita D, et al. Selectively engaging β-arrestins at the angiotensin II type 1 receptor reduces blood pressure and increases cardiac performance. *J Pharmacol Exp Ther*. 2010;335:572–579.
66. Bohn LM, Lefkowitz RJ, Gainetdinov RR, Peppel K, Caron MD, Lin F-T. Enhanced morphine analgesia in mice lacking βarrestin-2. *Science*. 1999;286:2495–2498.
67. Raehal KM, Walker JK, Bohn LM. Morphine side effects in β-arrestin 2 knockout mice. *J Pharmacol Exp Ther*. 2005;314:1195–1201.
68. Xiao K, Shenoy SK, Nobles K, Lefkowitz RJ. Activation dependent conformational changes in β-arrestin2. *J Biol Chem*. 2004;279:55744–55753.
69. Nobles KN, Ziqiang G, Kunhong X, Oas TG, Lefkowitz RJ. The active conformation of β-arrestin1: direct evidence for the phosphate sensor in the N-domain and conformational differences in the active states of β-arrestins1 and 2. *J Biol Chem*. 2007;282: 21370–21381.
70. Shukla AK, Violin JD, Whalen EJ, Gesty-Palmer D, Shenoy SK, Lefkowitz RJ. Distinct conformational changes in β-arrestin report biased agonism at seven-transmembrane receptors. *Proc Natl Acad Sci USA*. 2008;105:9988–9995.

71. Kim J, Ahn S, Ren X-R, et al. Functional antagonism of different G protein-coupled receptor kinases for β-arrestin-mediated angiotensin II receptor signaling. *Proc Natl Acad Sci USA*. 2005;102:1442–1447.
72. Ren X-R, Reiter E, Ahn S, Kim J, Chen W, Lefkowitz RJ. Different G protein-coupled receptor kinases govern G protein and β-arrestin-mediated signaling of V2 vasopressin receptor. *Proc Natl Acad Sci USA*. 2005;102:1448–1453.
73. Nobles K, Xiao K, Ahn S, et al. Distinct G protein-coupled receptor kinase phosphorylation sites on the β-adrenergic receptor: a bar code for differential encoding of β-arrestin functions. *Sci Signal*. 2011;4:1–10.
74. Butcher AJ, Prihandoko R, Kong KC, et al. Differential G-protein-coupled receptor phosphorylation provides evidence for a signaling bar code. *J Biol Chem*. 2011;286: 11506–11518.
75. Heitzler D, Durand G, Gallay N, et al. Competing G protein-coupled receptor kinases balance G protein and β-arrestin signaling. *Mol Syst Biol*. 2012;8:590.
76. Chen W, ten Berge D, Brown J, et al. Dishevelled2 recruits βarrestin2 to mediate Wnt5A-stimulated endocytosis of frizzled. *Science*. 2003;301:1391–1394.
77. Chen W, Ren X, Nelson CD, et al. Activity-dependent internalization of smoothened mediated by β-arrestin 2 and GRK2. *Science*. 2004;306:2257–2260.
78. Rakhit S, Pyne S, Pyne NJ. Nerve growth factor stimulation of p42/p44 mitogen-activated protein kinase in PC12 cells: role of G(i/o), G protein-coupled receptor kinase 2, β-arrestin 1, and endocytic processing. *Mol Pharmacol*. 2001;60:63–70.
79. Dasgupta P, Rastogi S, Pillai S, et al. Nicotine induces cell proliferation by β-arrestin-mediated activation of Src and Rb-Raf-1 pathways. *J Clin Invest*. 2006;116:2208–2217.
80. Chen W, Kirkbride KC, How T, et al. β-Arrestin2 mediates endocytosis of type III TGF-β receptor and down regulation of its signaling. *Science*. 2003;301:1394–1397.
81. Shenoy SK, McDonald PH, Kohout TA, Lefkowitz RJ. Regulation of receptor fate by ubiquitination of activated β_2-adrenergic receptor and β-arrestin. *Science*. 2001;294: 1307–1313.
82. Shenoy SK, Xiao K, Venkataramanan V, Snyder PJ, Freedman NJ, Weissman AM. Nedd4 mediates agonist-dependent ubiquitination, lysosomal targeting, and degradation of the β2-adrenergic receptor. *J Biol Chem*. 2008;283:22166–22176.
83. Shukla AK, Xiao K, Lefkowitz RJ. Emerging paradigms of β-arrestin-dependent seven transmembrane receptor signaling. *Trends Biochem Sci*. 2011;36:457–469.
84. Shenoy SK, Barak LS, Xiao K, Ahn S, Luttrell LM, Lefkowitz RJ. Ubiquitination of β-arrestin links 7-transmembrane receptor endocytosis and ERK activation. *J Biol Chem*. 2007;282:29549–29562.
85. Aubry L, Guetta D, Klein G. The arrestin fold: variations on a theme. *Curr Genomics*. 2009;10:133–142.
86. Alvarez CE. On the origins of arrestin and rhodopsin. *BMC Evol Biol*. 2008;8:222.
87. Lin CH, MacGurn JA, Chu T, Stefan CJ, Emr SD. Arrestin-related ubiquitin-ligase adaptors regulate endocytosis and protein turnover at the cell surface. *Cell*. 2008;135:714–725.
88. Fereshteh M, Ito T, Kovacs J, et al. β-Arrestin2 mediates the initiation and progression of myeloid leukemia. *Proc Natl Acad Sci USA*. 2012;109:12532–12537.
89. Lovgren AK, Kovacs JJ, Xie T, et al. β-Arrestin deficiency protects against pulmonary fibrosis in mice and prevents fibroblast invasion of extracellular matrix. *Sci Trans Med*. 2011;3:1–10.
90. Bonnans C, Flaceliere M, Grillet F, et al. Essential requirement for β-arrestin2 in mouse intestinal tumors with elevated Wnt signaling. *Proc Natl Acad Sci USA*. 2012;109: 3047–3052.

PART II

The Molecular Biology of Arrestins

CHAPTER TWO

True Arrestins and Arrestin-Fold Proteins: A Structure-Based Appraisal

Laurence Aubry[*,†,‡], **Gérard Klein**[*,†,‡]
[*]CEA, IRTSV, Laboratoire Biologie à Grande Echelle, F-38054, Grenoble, France
[†]INSERM, U1038, F-38054, Grenoble, France
[‡]Université Joseph Fourier, F-38000, Grenoble, France

Contents

1. Introduction 22
2. True Arrestins: A Structure Adapted to Multiple Scaffolding 23
   2.1 Function of arrestins 23
   2.2 Crystal structure of arrestins 24
   2.3 Binding sites for the endocytic machinery on arrestin's C-terminal tail 26
   2.4 Sites of posttranslational modifications 27
   2.5 Binding sites for inositol phosphates 28
   2.6 Conservation analysis of essential residues among visual and β-arrestins 30
3. Novel Arrestin-Related Proteins 30
   3.1 Mammalian ARRDCs 30
   3.2 ARTs in yeast and fungi 33
   3.3 Amoebal ADCs 35
4. Other Arrestin-Fold Proteins 35
   4.1 VPS26, a component of the retromer 35
   4.2 DSCR3, a VPS26-related protein 39
   4.3 RGP1, yet another arrestin-fold protein 40
5. Are Arrestins and Arrestin-Fold Proteins Related by a Shared Mechanism for Their Function? 42
6. Conclusion and Perspectives 46
References 49

Abstract

Arrestin-clan proteins are folded alike, a feature responsible for their recent grouping in a single clan. In human, it includes the well-characterized visual and β-arrestins, the arrestin domain-containing proteins (ARRDCs), isoforms of the retromer subunit VPS26, and DSCR3, a protein involved in Down syndrome. A new arrestin-fold-predicted protein, RGP1, described here may join the clan. Unicellular organisms like the yeast *Saccharomyces cerevisiae* or the amoeba *Dictyostelium discoideum* harbor VPS26, DSCR3,

Progress in Molecular Biology and Translational Science, Volume 118
ISSN 1877-1173
http://dx.doi.org/10.1016/B978-0-12-394440-5.00002-4

and RGP1 isoforms as well as arrestin-related trafficking adaptors or ADCs, but true arrestins are missing. Functionally, members of the arrestin clan have generally a scaffolding role in various membrane protein trafficking events. Despite their similar structure, the mechanism of cargo recognition and internalization and the nature of recruited partners differ for the different members. Based on the recent literature, true arrestins (visual and β-arrestins), ARRDCs, and yeast ARTS are the closest from a functional point of view.

ABBREVIATIONS

7TM seven-transmembrane spanning
ADC amoebal arrestin domain-containing protein
ARRDC arrestin domain-containing protein
ART arrestin-related trafficking adaptor
BAR Bin/amphyphisin/rvs
CI-MPR cation-independent mannose-6-phosphate receptor
DSCR3 Down syndrome critical region gene 3
ESCRT endosomal sorting complex required for transport
GARP Golgi-associated retrograde protein
GEF guanylate nucleotide exchange factor
GPCR G-protein-coupled receptor
GRK G-protein-coupled receptor kinase
$InsP_6$ nositol hexakisphosphate
PM plasma membrane
RGP1 retrograde Golgi transport protein
SNX sorting nexin
TGN trans-Golgi network
TXNIP thioredoxin-interacting protein
VDUP vitamin D3-upregulated protein
VPS vacuolar protein sorting

1. INTRODUCTION

Visual/β-arrestins are scaffolding proteins regulating G-protein-coupled receptor (GPCR)-dependent signaling. They are folded almost exclusively as 20 β-strands plus a single α-helix organized as pseudosymmetrical N- and C-domains.[1–8]

Recently, other proteins involved in trafficking and distantly related to arrestins on the basis of their sequence (about 10% identity) have been modeled (arrestin domain-containing proteins (ARRDCs), arrestin-related trafficking adaptors (ARTs), DSCR3)[9,10] or crystallized (vacuolar protein sort (VPS)26).[11,12] Surprisingly, their tertiary structure is strongly reminiscent of that of arrestins. Altogether, they now constitute the arrestin clan (CL0135).

We review here the structure and the functional and mechanistic determinants of the diverse members of the arrestin clan, from the data published in the literature and obtained by modeling. We conclude that they share indeed a stable convergent conformation and generally act as scaffolds. However, the family is heterogeneous in terms of functional properties. ARTs and ARRDCs share some important functional characteristics of arrestins, whereas VPS26 functions in a different type of trafficking, with a different mode of recognition and different partners. The nature of the arrestin-fold protein (arrestins vs. ARRDCs) recruited by some GPCRs also determines their fate toward recycling or lysosomal degradation pathways.

2. TRUE ARRESTINS: A STRUCTURE ADAPTED TO MULTIPLE SCAFFOLDING

2.1. Function of arrestins

The search for proteins involved in the desensitization of the GPCRs rhodopsin or β-adrenergic receptor led to the discovery of visual- and β-arrestins, here referred to as true arrestins.[13–17]

In response to agonist binding, GPCRs experience a conformational change and interact with heterotrimeric G proteins, on which they act as a guanylate nucleotide exchange factor (GEF). GTP-loaded Gα dissociates from Gβ/γ and both entities transduce the receptor-dependent signal to downstream second-messenger producing effectors. Activation of GPCRs immediately activates a negative feedback loop to dampen persistent activation of the triggered signaling pathways. This loop involves the phosphorylation of the cytoplasmic tail of the GPCRs by G-protein-coupled receptor kinases (GRKs), the recruitment of cytoplasmic arrestins to the activated, phosphorylated receptors, and the steric uncoupling of the GPCRs from the G proteins.[18–23]

Arrestins further interfere with receptor-dependent signaling by promoting internalization of activated receptors. The binding of arrestins to the phosphorylated receptors destabilizes their polar core and releases cryptic binding sites for proteins of the endocytic machinery, adaptins, and clathrin. The fate of internalized receptors, either recycling or routing to degradation in lysosomes, depends on their ubiquitination promoted by the scaffolding of HECT-type ubiquitin ligases by arrestins.[24,25] The chemokine receptor

CXCR4 promotes the interaction of β-arrestin 1 with the ubiquitin ligase AIP4.[26] Interestingly, AIP4 can also bind CXCR4 directly. Two serine residues within the C-terminal tail of CXCR4 are essential for this interaction and the subsequent ubiquitination of CXCR4.[27] CXCR4-bound β-arrestin 1 recruits endosomal sorting complex required for transport (ESCRT)-0 via an initial interaction between ubiquitin and hepatocyte growth factor-regulated tyrosine kinase substrate (Hrs), and possibly between β-arrestin 1 and STAM-1. Scaffolding of AIP4 modulates the ubiquitination of Hrs and its subsequent dissociation following an internal interaction between ubiquitin and the ubiquitin-binding domain of Hrs, thereby controlling the amount of CXCR4 that will be degraded.[28]

Another aspect of arrestin function pertains to their capacity to recruit, on the endocytic vesicles, G protein-independent signaling machinery and serve thus as signal transducers on their own. The first kinase scaffold that was identified comprises mitogen-activated protein kinase (MAPK) modules: Raf-1, MEK1/2, ERK1/2; and ASK1, MKK4, and Jnk3. Arrestins signal also to other kinases such as Src kinase, protein kinase B/Akt, and PI(3) kinase.[24,29–35]

The role of β-arrestins in the regulation of GPCR signaling, endocytosis, and degradation has also been demonstrated for other plasma membrane (PM) proteins; for example, atypical 7TM receptors, tyrosine kinases, and ion exchangers.[36–39] This extended repertoire of targets, together with the variety of interfaces that β-arrestins are able to scaffold, explains their important roles in multiple cellular processes in the retina, nervous system, immune system, bone remodeling, and cancer.[32] A consequence of arrestin recruiting signaling molecules, some of them involved in actin reorganization, is their contribution in cell polarity and migration.[30,40] β-Arrestins have been implicated in networks belonging to the Wingless, Hedgehog, Notch, and TGFβ pathways, and their ablation has major developmental consequences.[41] β-Arrestins partially localize on the centrosome and the primary cilium.[42–44] Within cilia, β-arrestins participate in the GPCR-dependent signaling pathways related to the sensory function of this antenna.

2.2. Crystal structure of arrestins

The crystal structure of visual arrestins and β-arrestins was solved not long after their discovery.[1–8] Arrestin proteins are structured as two head-to-tail oriented modules, the arrestin-N (PFAM signature PF00339) and arrestin-C (PFAM signature PF02752) domains connected by a flexible hinge

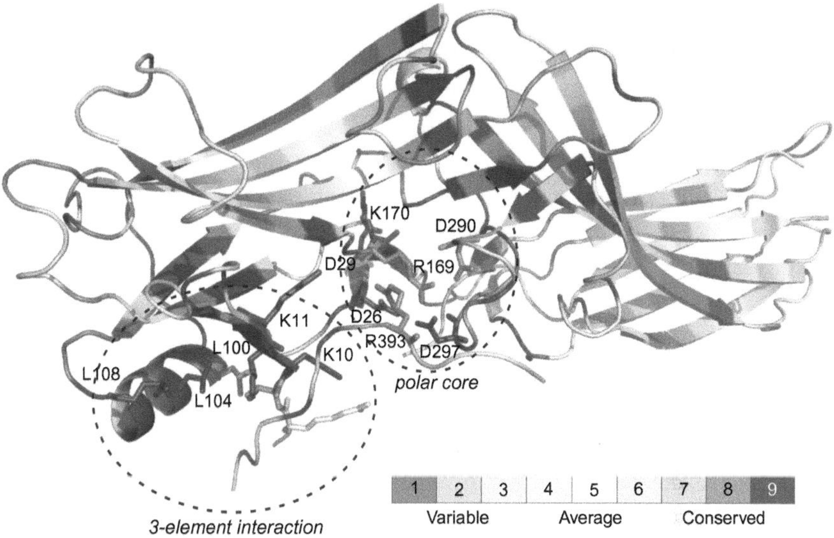

Figure 2.1 Functional conservation within the arrestin family. The sequences of true arrestins were first aligned with Clustal Omega.[45] The multiprotein alignment was then processed with the Consurf software.[46,47] Residue conservation among the aligned members of the arrestin family was represented with PyMOL using the PDB 1ZSH crystal structure of bovine β-arrestin 1 as the template and colored using the Consurf plugin for PyMOL. The following sequences were used: human rod and cone arrestins and β-arrestin 1 and 2; *Drosophila melonogaster* arrestins 1 and 2 and kurtz; *Ciona intestinalis* BAB60819.1; *Caenorhabditis elegans* NP_508133. Sequences are colored from Bordeaux red (highly conserved) to cyan (variable stretches). Full conservation of the amino acids of the polar core and of the 3-element interaction domain (indicated as sticks) highlights their conserved functional role. (See Color Insert.)

(Fig. 2.1). Each module is about 150 amino acids long. This structure, which includes 20 β-strands and an α-helix, reveals two deeply curved sandwiches related by a pseudo-twofold rotation axis, each made of two antiparallel β-sheets structured as fibronectin type III module.[48] The N-domain is made of strands 1–10, strand 20, and helix 1, whereas the C-domain contains strands 11–19.[3,5]

The functionality of arrestins resides in their mode of recognition of activated receptors and the subsequent responses it triggers. The binding of visual arrestin to phosphorylated rhodopsin follows a two-step mechanism. First, a 3-element interaction cluster consisting of β-strand 1, α-helix 1, and the C-tail senses the phosphorylation status of the receptor's cytoplasmic tail; second, the breaking of a salt bridge within this 3-element interaction cluster upon binding the phosphorylated receptor guides the phosphate groups in the direction of the central polar core and induces a global conformational change of arrestin, resulting in release of the C-terminal tail[20] (Fig. 2.1). This recognition mechanism is conserved for all four human true arrestins.[49] In the case of β-arrestins, exposure of the C-tail allows the recruitment of the endocytic machinery (see below). Disruption of the polar core in β-arrestin 1 by a $R^{169}E$ mutation or a deletion of the C-terminal tail generates a constitutively active β-arrestin, able to bind β_2-adrenergic receptor in a phosphorylation-independent manner.[50]

Additional receptor-binding multielements exist on the concave sides of N- and C-arrestin domains, which are organized as a saddle to accommodate the receptor. The primary binding site of phosphorylated GPCRs is located on β-strands 5 and 6 (N-domain) for rod arrestin. For β-arrestin 1, it is distributed on β-strands 5 and 6 and their connecting loop, β-strands 9 and 10 and their connecting loop (N-domain), and β-strands 15 and 16 (C-domain).[5,51] Segment swapping between visual and β-arrestins strongly modifies the binding specificity toward the recognized receptors.[2,8] Cone arrestin binds green cone opsin and nonvisual receptors and conversely β-arrestin 1 is able to bind both visual receptors in addition to nonvisual receptors.[6]

2.3. Binding sites for the endocytic machinery on arrestin's C-terminal tail

The polar core typically consists of three positive residues (R^{169}, K^{170}, and R^{393} for bovine β-arrestin 1) balanced by four negative residues (D^{26}, D^{29}, D^{290}, and D^{297}) in the resting arrestin. These charged residues are contributed by arrestin's N-domain, C-domain, and its C-terminal tail (Fig. 2.1). Binding of arrestins to activated, phosphorylated receptors leads to the transition to their active state. Two strategic lysines in the 3-element interaction cluster (K^{10} and K^{11} in bovine β-arrestin 1) are able to sense the phosphate groups of the receptor that are channeled toward the polar core where they disrupt the charge balance. This induces a conformational change in the arrestin that liberates its C-terminal tail normally engaged in the polar core. The binding sites for clathrin (LIELD) and the β-subunit of adaptor protein AP-2 (comprising R^{393}–R^{395}) localized on the C-tail are then free to recruit the endocytic machinery. This mechanism is wonderfully adapted to bind arrestin's partners following the activation of the receptor in order to promote its internalization in clathrin-coated vesicles.[2,3,18,31,49,52–54]

Analysis of an alternatively spliced long isoform of β-arrestin 1 reveals a second clathrin-binding site ($[L/I]_2GxL$) in the splice loop.[55] Human visual arrestins, which lack the classical clathrin box (LΦxΦ[D/E]), harbor this alternate binding site. This site in the looser form ΦΦGxΦ is also present in other species and may contribute to the recruitment of clathrin and the endocytosis of rhodopsin that have been described in flies.[56–58]

2.4. Sites of posttranslational modifications

Nonvisual arrestins are regulated by multiple posttranslational modifications such as phosphorylation, nitrosylation, ubiquitination, and SUMOylation.[34] The exact temporal sequence and functional cross talk of these posttranslational modification events are not yet clearly established.

In the cytosol, both β-arrestins 1 and 2 are phosphorylated on sites located on their C-terminal tail in proximity of the binding sites for clathrin and adaptin; β-arrestin 1 is phosphorylated on S^{412} by ERK; and β-arrestin 2 is phosphorylated on S^{361} and T^{383} by casein kinase II. The recruitment of β-arrestins 1 and 2 by activated receptors at the PM leads to their rapid dephosphorylation by an as yet unknown phosphatase. This dephosphorylation does not affect the desensitization of the receptor but is essential for the scaffolding of clathrin and internalization of the receptor. Once on an endocytic vesicle, β-arrestins 1 and 2 are rapidly rephosphorylated.[59–62] Analogously, visual arrestin is phosphorylated at a similar position (S^{366}) by a calcium/calmodulin-dependent protein kinase upon light activation,[63] and dephosphorylation is required for the arrestin–clathrin interaction to proceed.[64] Following activation of some G protein-independent receptors, β-arrestin 1 can also be phosphorylated on S^{412} by GRK5.[65]

Besides the Ser/Thr phosphorylation, the interaction between β-arrestin 1 and the μ-subunit of AP-2 is negatively regulated by phosphorylation of Y^{54} by Src. Introducing a $Y^{54}F$ mutation into β-arrestin 1 results in improved arrestin–adaptin interaction and enhanced β_2-adrenergic receptor internalization.[66]

Bovine β-arrestin 2 is nitrosylated at the level of its C-terminal cysteine. Following β-adrenergic receptor activation, S-nitrosylation of β-arrestin 2 promotes its binding to the clathrin/adaptin endocytic machinery and accelerates receptor internalization.[67] This residue is conserved in β-arrestin 2 paralogues from *Xenopus laevis* and *Danio rerio* to mammals, within a [D,E][D,E]xΨC consensus sequence, where Ψ is a hydrophobic residue and C is the C-terminal nitrosylated cysteine residue. Whether this type of modification is general to all arrestins is not clear, as visual arrestins and β-arrestin 1 have no cysteine at their C-terminus and S-nitrosylation at other predicted sites has not yet been documented.

At the membrane, the stability of the association of β-arrestin 2 with GPCRs is dependent on its ubiquitination of K^{11} and K^{12} by the ubiquitin E3 ligase Mdm2.[68] A stable receptor–arrestin complex on the endosome is essential for the arrestin-dependent signaling, for example, ERK phosphorylation and sequestering in the cytoplasm.[69] As discussed earlier, these two lysines serve as phosphate-sensors when β-arrestins bind activated and phosphorylated GPCRs. Use of the same residues to recognize the receptor and to stabilize the interaction by ubiquitination may provide a sequential switch to temporally order interaction events. The deubiquitinating protease USP33 binds to N- and C-domains of β-arrestin 2 and promotes its deubiquitination. The reciprocal activities of Mdm2 and USP33 exerted on β-arrestins are essential for the regulation of GPCR trafficking and downstream signaling.[68]

Bovine β-arrestin 1 and 2 are also modified by SUMO-1 upon activation of the β_2-adrenergic receptor. The main SUMOylation site in β-arrestin 2 is K^{400} close to the C-terminus and this modification regulates GPCR internalization.[70] This lysine is not conserved in a multiprotein alignment of β-arrestins from different species. Interestingly, a second lysine (K^{295}) can be modified by SUMO-1.[70] This lysine resides in a consensus motif [L,I] KxE found as such in all but one arrestin of the metazoan clade (Fig. 2.2), suggesting that both visual and β-arrestins could potentially be SUMOylated. This point awaits experimental verification.

2.5. Binding sites for inositol phosphates

Soluble phosphoinositols and membrane-inserted phosphoinositides are major regulators of the inactive/active states of visual and nonvisual arrestins. Arrestins bind $InsP_6$ (also known as inositol hexakisphosphate or phytic acid) on patches of positive residues in their N- and C-domains. A first $InsP_6$ binding site (high-affinity site) is located on the concave surface of the C-domain (K^{232}, R^{236}, K^{250}, K^{324}, K^{326} in bovine β-arrestin 1). A low-affinity site, also formed from positively charged residues (K^{157}, K^{160}, R^{161}), is present in the N-domain.[4,5] $InsP_6$ and another polyanionic compound, heparin, act as mimics of the phosphorylated tail of GPCRs. Upon binding to the N-domain of rod arrestin, they are able to release its C-terminal tail.[74] $InsP_6$ promotes the homo-oligomerization of β-arrestin 1 and its hetero-oligomerization with β-arrestin 2.[4] Given the high cytosolic concentration of $InsP_6$,[75] cytosolic basal-state arrestins are mostly stored in an oligomeric $InsP_6$-bound form. The binding of arrestins

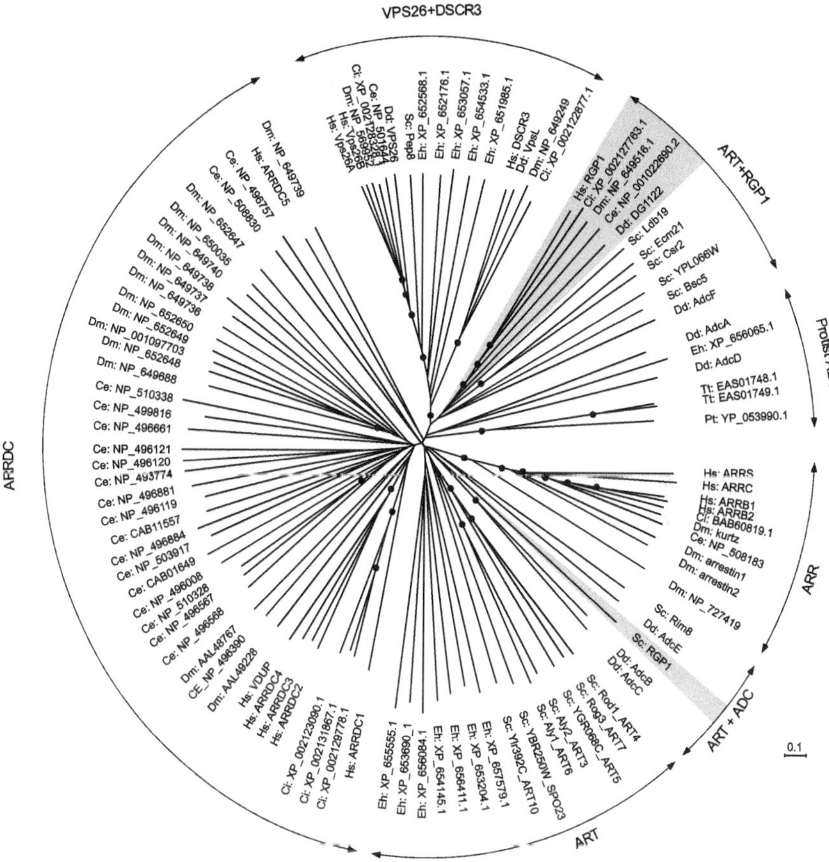

Figure 2.2 The arrestin-fold protein family tree. The sequences of mammalian arrestins (ARR) and ARRDCs and yeast ARTs were blasted against the genome of *C. elegans*, *D. melanogaster*, *D. discoideum*, *C. intestinalis*, *Tetrahymena thermophila*, and *Paramecium tetraurelia*. Alternately, the modeled structures of human β arrestin 1 and yeast Aly1 were searched against the genome of human and yeast, respectively, using the BackPhyre option of the Phyre 2 (Protein Homology/analogY Recognition Engine V 2.0) web engine[71] in order to search for new proteins sharing the arrestin fold. The entire list of hits (Blast + BackPhyre) was aligned using Clustal Omega[45] and a phylogenetic tree constructed with Splitstree 4 on the basis of this alignment.[72] The shown tree corresponds to a Neighbor Joining tree, for which 1000 bootstrap steps were performed. All splits supported by a bootstrap score above 95% were considered "correct"[73] and are shown with a black dot. RGP1 homologs clustered with ART proteins (gray sector). The scale bar shows the number of substitutions per site.

to phosphorylated GPCRs requires the displacement of $InsP_6$. This step may involve the binding of arrestins to phosphoinositides, the highest affinities being observed for $PtdIns(4,5)P_2$ and $PtdIns(3,4,5)P_3$. Phosphoinositides may act as coreceptors for coincidence detection, increasing the binding affinity/selectivity of activated arrestins for a phosphorylated GPCR. Mutations in arrestin's phosphoinositide-binding site impair β_2-adrenergic receptor endocytosis.[76]

2.6. Conservation analysis of essential residues among visual and β-arrestins

Analysis of amino acid conservation among proteins from the same family highlights the importance of each position for the protein's structure or function. If the amino acid conservation of visual arrestins/β-arrestins paralogues in several species (human, fly, worm, *Ciona*) is scored on the online Consurf server,[46,47] a very strong consensus is obtained for all amino acids involved in the polar core and the electrostatic interaction domain (Fig. 2.1), evidencing the mechanism of receptor recognition and binding already present in the oldest arrestin forms (*Ciona*) and conserved through evolution. And, as expected from the very different cognate receptors, the remaining of the molecule is weakly conserved in terms of amino acid identity.

3. NOVEL ARRESTIN-RELATED PROTEINS

Recently, novel arrestin-related proteins have been found in the genomes from protists to mammals. All these proteins harbor an arrestin core, with both arrestin N- and C-domains.

However, this arrestin core is extended, in some cases, on both extremities by sequence appendages likely to provide novel specificities for partner interactions or function.

3.1. Mammalian ARRDCs

Besides the true arrestins, the genome from mammals encodes six proteins containing the arrestin N- and C-domains: ARRDC1–5 and TXNIP (thioredoxin-interacting protein) also known as VDUP (vitamin D3 upregulated protein). Interestingly, paralogues for ARRDCs and TXNIP can be found in the crown of higher eukaryotes outside mammals, namely, fish, frog, and bird. Fly and worm harbor an ARRDC5 paralogue only.

No crystal structure of ARRDCs has been solved so far, although a preliminary X-ray analysis has been conducted on TXNIP crystals.[77, Note added in Proof] At the date of submission of this review, the structure of TXNIP–TRX complex has been deposited at the PDB by Hwang and Kim but has not yet been released. We have modeled the putative structure of the different human ARRDCs and VDUP and extended our previous work.[9] The relevance of modeled structures is highly dependent on the coverage of the modeled structure and the confidence level. The core of all human ARRDCs (including TXNIP) could be modeled with a high confidence level (>99%). The coverage of the modeled structures ranged from 65% to over 90% of the full-length proteins due to extensions matching with no known structural domains (Fig. 2.3A). In all cases, a structure composed of a two-module β-strand sandwich was recognizable (Fig. 2.3B). No reliable information could be obtained at the level of the interface between the N- and C-domains, especially in terms of neighboring charged amino acids (polar core).

Because of their recent discovery, little is known about the molecular function of ARRDCs and available data are restricted to the human isoforms, which play a physiological role in the regulation of metabolism.[78]

Among all arrestin domain-containing proteins, only ARRDC2 and its paralogues in other species harbor a C-terminal cysteine predicted to be S-nitrosylated (http://sno.biocuckoo.org/). Human ARRDCs do not share consensus SUMOylation motifs.

TXNIP and all ARRDCs except ARRDC5 harbor two [L,P]PxY sites that are able to bind WW domains generally found in HECT-type ubiquitin ligases. ARRDC1–4 and TXNIP do interact with several E3 ubiquitin ligases: WWP1, WWP2, Nedd4, and Itch (except ARRDC2 and ARRDC4 that lack interaction with WWP2 and Itch).[79] ARRDC1 is ubiquitinated by WWP1[79] and WWP2.[80] Besides binding ubiquitin ligases and being a target for their ubiquitination, ARRDCs also serve as a platform for the ubiquitination of their cargoes. ARRDC3 binds to the activated, phosphorylated form of β-4 integrin and controls its endocytosis, ubiquitination, and final degradation.[81] ARRDC3 also recruits the ubiquitin E3 ligase Nedd4 to mediate ubiquitination of the β_2-adrenergic receptor, an essential step in its degradation.[82] An analysis of glucose uptake showed that it was inhibited by TXNIP and ARRDC4, suggesting their possible role in the endocytosis of glucose transporter, but their WW-binding domains were dispensable for that inhibition.[83]

ARRDC1–4 and TXNIP proteins show interactions with subunits of the ESCRT machinery, namely, Alix and Tsg101 (known as VPS23 in yeast). The current hypothesis is that ARRDC1 is involved at the PM in PPxY-dependent viral budding by bridging HECT ubiquitin ligases and ESCRT-III components.[79] Interestingly, cells are able to shed in a virus-independent process microvesicles (ARMMs, ARRDC1-mediated

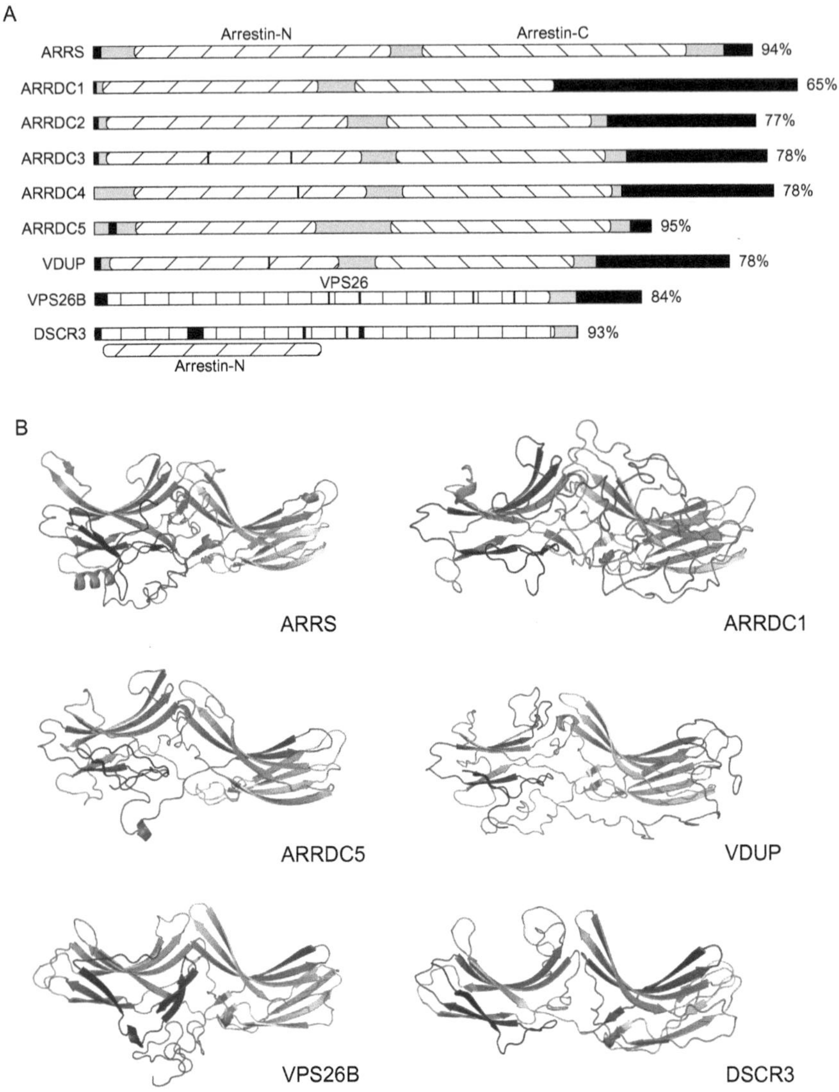

Figure 2.3 Human ARRDCs share an arrestin fold. The structure of human α-arrestins, β-arrestins, VPS26B, and other arrestin-fold proteins was individually modeled on the Phyre2 web server in the intensive mode. Best templates used for modeling were 1SUJ, 1G4M, 1AYR, 1CF1, 1JSY, 2FAU for the different ARRDCs and 2FAU for VPS26B and DSCR3. (A) Regions of the modeled proteins are schematized as follows: gray regions correspond to the regions modeled with a confidence level above 99% on crystallized templates and black regions correspond to structures modeled *ab initio* that are not reliable. The percentage of reliable coverage of the modeled structure is indicated on the side. PFAM-predicted domains are superimposed upon modeled regions with the following code: Right slanted lines correspond to a predicted arrestin-N domain;

microvesicles) that differ from exosomes. This budding is driven by the specific interaction between a PSAP sequence of TSG101/VPS23 and ARRDC1.[80]

In addition to their role at the PM, ARRDC3 and ARRDC4 are associated with endosomes and lysosomes, whereas TXNIP and ARRDC2 are essentially nuclear.[84]

3.2. ARTs in yeast and fungi

Despite the presence of a few GPCRs in yeast (three for *Saccharomyces cerevisiae,* nine for *Schizosaccharomyces pombe*), no gene coding β-arrestins can be found in these organisms. Instead, the *Saccharomyces* genome encodes 10 arrestin-related proteins also called ARTs.[10]

ARTs share a central arrestin core with sometimes-long extensions on either side, increasing their size up to about 1100 amino acids for the longest members. The percentage of identity with a visual arrestin is around 10% and this percentage rises maximally to about 20% when identities and homologies are considered. Threading their structures with the PSIPRED software (http://bioinf.cs.ucl.ac.uk/psipred) indicates secondary structures essentially organized as β-strands. They aligned with arrestin-clan proteins with a p-value of 3×10^{-8} (Rod1 vs. β-arrestin 2) or 5×10^{-8} (Rim8 vs. VPS26). The arrestin core of most ART proteins (9 of 10 ARTs) was successfully modeled with the Phyre2 online software using the structures of mammalian arrestins as templates (Fig. 2.4). The lower β-sheet of the arrestin-N domain is, in general, poorly modeled in ARTs. In fact, ARTs differ from genuine arrestins in the C-terminal portion of their arrestin core that includes β-strand 20. As β-strand 20 contributes to the N domain, the structure of ARTs could not be reliably modeled at this level.

As with ARRDCs, most ART proteins have been shown to serve as scaffolds for E3 ubiquitin ligases through their WW domain-binding sites and to participate in the endocytosis of PM transporters.[10,85,86]

The paralogue of ART9, called PalF in *Aspergillus nidulans* (known as *Emericella nidulans* in its sexual reproductive stage), is involved in a pH signaling pathway. The function of PalF is related to the endocytosis of the 7TM pH sensor PalH and is regulated by phosphorylation and

left-slanted lines correspond to a predicted arrestin-C domain; vertical hatches correspond to a predicted VPS26 domain. (B) Ribbon representation (generated with the PyMOL software) of representative proteins of the human arrestin clan. Coloring goes from blue N-terminus to red C-terminus. (See Color Insert.)

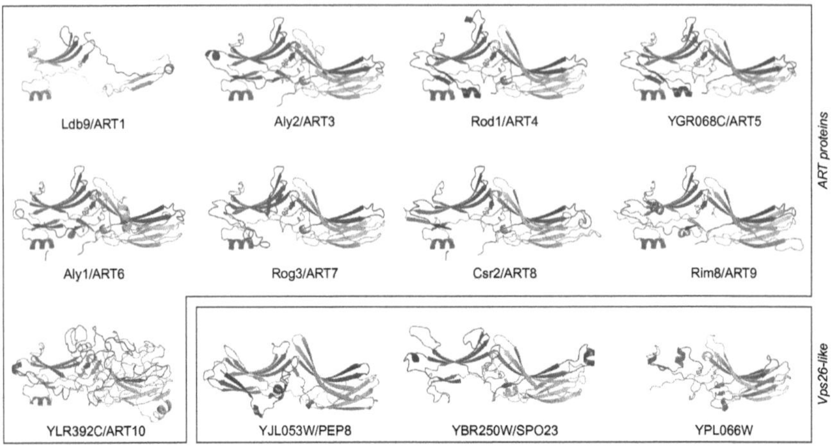

Figure 2.4 Modeling of the yeast ART and VPS26 families. The structure of *S. cerevisiae* ARTs and VPS26 proteins was individually modeled on the Phyre2 web server in the normal mode. Best templates used for modeling were 1G4M for Ldb9/ART1 (22% coverage, 98% confidence) and YPL066W (47% coverage, 99.4% confidence), 1SUJ for Aly2/ART3 (31% coverage, 99.9% confidence), Rod1/ART4 (38% coverage, 100% confidence), YGR068C/ART5 (51% coverage, 100% confidence), Aly1/ART6 (36% coverage, 99.9% confidence), Rog3/ART7 (43% coverage, 100% confidence), Csr2/ART8 (27% coverage, 99.4% confidence), Rim8/ART9 (54% coverage, 100% confidence), 2R51 for Pep8 (75% coverage, 100% confidence), 1AYR for YBR250W/SPO23 (58% coverage, 100% confidence). YLR392C/ART10 (68% coverage, 99% confidence) was modeled on multi-templates (1SUJ, 1G4M, and 2FAU) in the intensive mode. The coverage is rather low because of important extensions on both sites of the arrestin core for most ARTs. Ecm21/ART2 could not be modeled. Ribbon structures were generated with PyMOL. Coloring goes from blue N-terminus to red C-terminus. (See Color Insert.)

ubiquitination.[87] In *S. cerevisiae*, ART9 (previously known as Rim8) binds both Rim21, the paralogue of PalH and VPS23, a subunit of the ESCRT-I subcomplex. Binding of VPS23 proceeds through a SXP^{536} motif on Rim8 and the ubiquitin E2 variant domain of VPS23, and recruits *in fine* the whole ESCRT-I subcomplex on Rim8. This interaction is essential for Rim signaling. Rim8 itself is monoubiquitinated on K^{521} near the SXP motif by ubiquitin ligase Rsp5 bound to a PKY^{508} motif. Overexpressed Rim8 and VPS23 colocalize on cortical punctate structures.[88]

The role of ARTs has been demonstrated in the ubiquitination and the endocytosis of PM transporters targeted to the vacuole for degradation in response to stress situations or excess of substrate.[10,85,86,89] There are no data for a possible role of ARTs in cargo recycling.

3.3. Amoebal ADCs

The amoeba *Dictyostelium discoideum* possesses 55 GPCRs,[90] but no true arrestin. Instead, it harbors six genes encoding arrestin domain-containing proteins (AdcA-F) as well as VPS26 and DSCR3 (VPS26L) genes.[9,91] The amoebal ADC proteins are characterized by extensions on both sides of the central arrestin core, with various protein–protein or protein–lipid interaction domains depending on the isoforms, such as C2, LIM, MIT, or FYVE domains.[9] The expression of AdcA-F is developmentally regulated (http://dictybase.prg). AdcA, -C, and -D are found in the proteome of the vegetative endocytic pathway.[92]

The multiple domains of AdcA structure have successfully been modeled on several templates with the Phyre2 software in the intensive mode (Fig. 2.5). The assembled structure shows the three independent domains: the His domain, the arrestin core, and the FYVE domain. Their respective positioning based essentially on *ab initio*-modeled residues is not yet supported by experimental evidence. In arrestins, β-strand 20 is folded as to form the N-domain together with β-strands 1–10, resulting in the relative proximity of N- and C-domains in the inactive state. The model of AdcA built on the arrestin structure copies this neighborhood. The N-terminal side triple-polyhistidine cluster is involved in the oligomerization of AdcA (Guetta D., Klein G. & Aubry L., unpublished results) and its C-terminal FYVE domain targets the protein to early endosomes.[91] It is tempting to assume that the association of the C-terminal FYVE domain with endosomal PI(3)P or the oligomerization of the histidine rich N-terminus influences the overall structure of the protein, and possibly the accessibility of binding partners to the arrestin core. Whether AdcA regulates membrane cargoes is not known, even though its massive presence at the endosomal membrane fits with such a hypothesis.

4. OTHER ARRESTIN-FOLD PROTEINS

4.1. VPS26, a component of the retromer

The retromer was first described in yeast and characterized as a coat complex required for the retrieval of the lysosomal hydrolase receptor VPS10 from endosomes and its retrograde transport to the trans-Golgi network (TGN).[93] It was later described in mammals, where it shuttles the cation-independent mannose-6-phosphate receptor, CI-MPR, from endosomes to the TGN.[94] The composition of retromers is conserved from yeast to

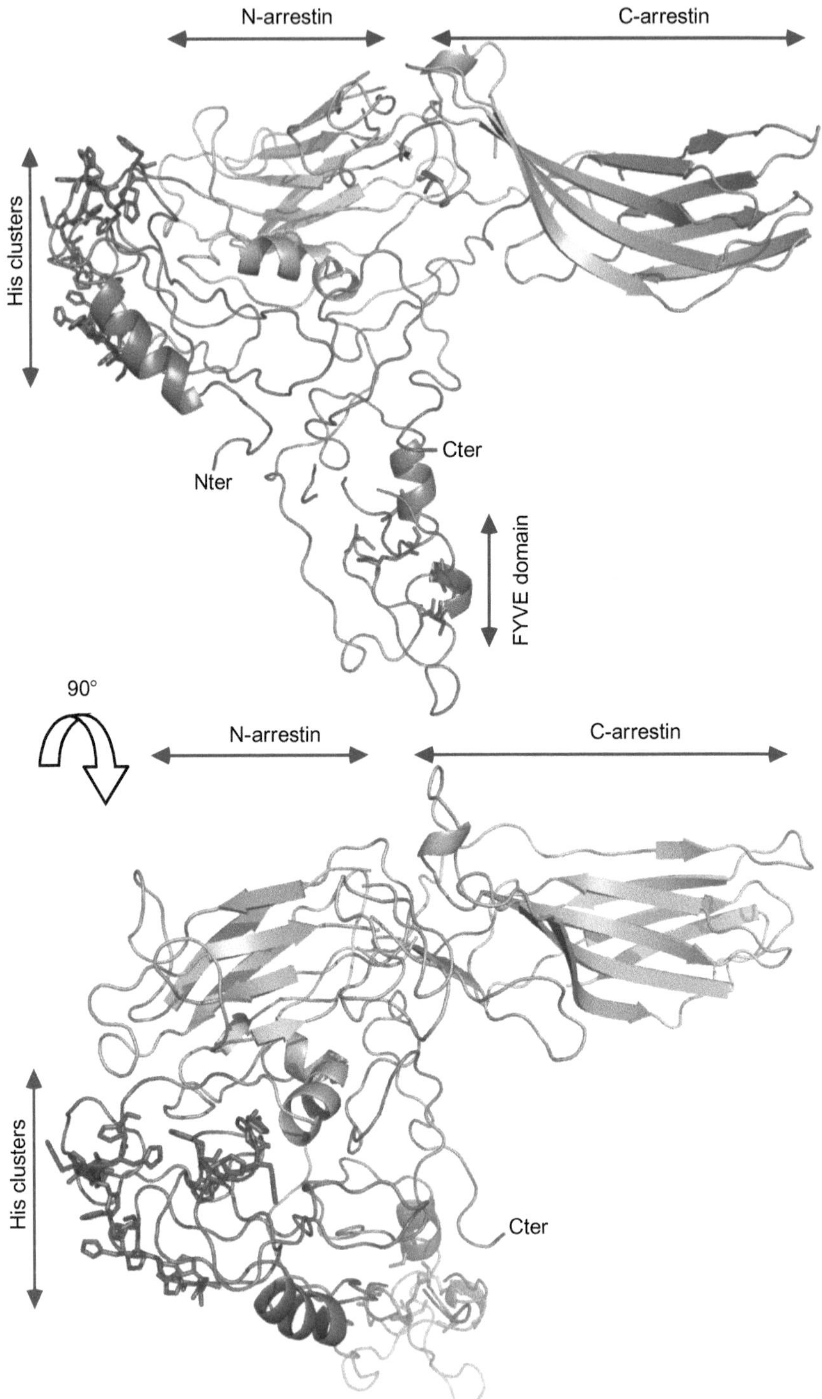

Figure 2.5 See legend on opposite page.

mammals and comprises five subunits divided into a trimeric cargo-selective subcomplex (VPS35, VPS29, and VPS26) and a membrane-deforming subcomplex (sorting nexins—SNXs harboring PX- and BAR domains).

4.1.1 Crystal structure of retromer subunits

The crystal structure of the retromer subunits has given clues to its functional architecture.

VPS26 has a strong structural relationship to arrestins[11,12] with the characteristic paired β-sandwich subdomains formed of antiparallel β-strands[48] (Fig. 2.3). The arrestin fold of crystallized VPS26A first appeared as completely unexpected, as human VPS26 and arrestins share only about 10% identity and 20% overall homology.

VPS26 not only shares the arrestin fold but also the polar core and an extended C-terminal tail.[11,12] There is, however, no evidence that the mechanism of protein partner recognition functions in the same way, as the salt bridge (Arg294–Glu117) at the subdomain interface of VPS26 is buried in the structure. A conformational change in this polar core would require an extensive change in the interaction network at the interface, and no signal or interactor to induce such a change has yet been found. The current view of VPS26 structure proposes relative rigidity of the paired subdomains with limited flexibility at their interface.[95]

VPS26 binds VPS35, the cargo-recognition subunit of the retromer that binds the cytosolic tail of the cargoes through their conserved sorting motif [F,W]L[M,V].[96] The VPS26 region interacting with VPS35 is an exposed peripheral loop between strands β15 and β16 in the C-terminal subdomain, far from the polar core.[11,95,97] VPS35 is predicted to be composed of 34

Figure 2.5 Modeled structure of *Dictyostelium* arrestin domain-containing protein AdcA. The multistructure and *ab initio* model of *Dictyostelium* AdcA was calculated on Phyre 2 web server in the intensive mode. It was modeled on 14 templates harboring either an arrestin domain (bovine rod arrestin 1AYR and 1CF1, bovine β-arrestin-1 1G4M, bovine β-arrestin-2 1JSY, *Ambystona* cone arrestin 1SUJ), or a FYVE domain (early endosomal antigen 1 1JOC, hepatocyte growth factor-regulated tyrosine kinase substrate 1DVP and 3ZYQ, endofin 1DVPA2 and 3T7L, *Leishmania* lm5-1 FYVE domain 1Z2Q), with a confidence value of 99–100%. The N-terminal region was modeled on structures 1RW2, 3FW0, 1ZLG with a confidence of 58–61%. One hundred and thirty-six residues were modeled *ab initio* and their structure is not reliable. The arrestin core is colored in orange and extensions in olive green. Histidine residues of the N-terminal His clusters (blue) and Zn-coordinating cysteines of the FYVE domain (purple) are shown as sticks. A side view and a 90 ° tilted view are shown for a better appreciation of the arrestin core and the histidine and cysteine clusters. (See Color Insert.)

α-helices forming 17 HEAT-like repeats.[95,98] Such an α-solenoid arrangement is found in other coat proteins (components of the AP adaptor complex, clathrin) involved in trafficking. VPS35 binds also VPS29 via its C-terminal ends and masks its metal-binding site. VPS29 harbors a fold strikingly like that of Ser/Thr phosphatases.[99,100] Although the hypothesis of VPS29 acting in control of the phosphorylated state of CI-MPR is very appealing, it seems now established that VPS29 is not an active metallophosphatase but serves as a scaffold for the binding of accessory proteins.[101] VPS35 binds directly to the membrane-deforming SNXs.[93,102–104] The PX domains of SNX-BAR proteins are structured as a wedge with a three-stranded β-sheet followed by a lipid-binding α-helical pocket accommodating phosphoinositides. The specificity of retromer-associated SNXs for PI(3)P generally ascribed to the central positively charged amino acid in the R[R,K]Y sequence[105] seems to hold for VPS5 and SNX1-2 but is not obvious for VPS17, SNX5, and SNX6 where this position corresponds to either S or Q. Their lipid specificity has been compiled elsewhere.[106]

4.1.2 Regulation of the retromer's function

The list of retromer cargoes and functions has steadily grown.[97,107,108] Retromer functions now extend to polarized transport and apical–basal polarity, cell surface recycling, transport from mitochondria to peroxisomes, amyloid precursor protein sorting and processing, generation of morphogen gradients, trafficking of pathogens, and many more.[107]

Recruitment of the cargo-selective retromer complex to the endosomal membrane is regulated by the small GTPase Rab7.[109] VPS29[110] and VPS26[111] interact with a Rab GTPase-activating protein (GAP), TBC1D5. The two retromer accessory proteins Rab7 and TBC1D5 represent thus a molecular switch that catalyzes/downregulates the recruitment of the VPS35–VPS29–VPS26 complex to the membrane.[112] The diversity of the retromer cargoes and direction of trafficking are mediated by the use of various SNXs and the recruitment of accessory proteins. A functional analysis of VPS26A and VPS26B showed that they localize on early and late endosomes, respectively, and that their differential cargo recognition is linked to the variable C-terminal part of VPS26B.[111]

Entamoeba histolytica is peculiar with five paralogues of VPS26 (Fig. 2.2) and VPS35, but only one copy of VPS29. This situation is not unique, and some Excavata (*Leishmania major*, *Naegleria gruberi*, *Trichomonas vaginalis*, *Giardia intestinalis*, *Trypanosoma brucei*) also harbor up to 6 VPS26 paralogues with multiple VPS29 and VPS35 isoforms.[108] The reason for such complexity in these unicellular organisms is unclear.

SNXs used in retromer-dependent retrograde transport are members of the SNX-BAR subfamily and contain two membrane-binding domains: a phosphatidyl inositide-binding PX (Phox homolog) and a membrane-curvature sensing domain BAR (Bin/Amphyphisin/Rvs). BAR domains are found in endophilins and retromer-associated SNXs. They usually form a dimer that senses and generates membrane curvature. The function of the BAR domain is to drive tubulation of the endosomal membrane, an early step in the formation of the endosomal vesicles due to retrograde transport toward the TGN.[106] Recently, new sorting nexins (SNX3, SNX27) that contain no BAR domain have been shown to associate with the cargo-selective subunits of the retromer and to confer cargo specificity.[107,113–115]

SNX-coated tubule formation and fission of cargo-enriched vesicles due to retrograde transport are dependent on the actin cytoskeleton to generate the mechanical force for tubulation and scission.[110,116,117] SNXs directly interact with $p150^{glued}$, a component of the dynactin–dynein microtubule motor.[118,119] Microtubules provide the long-range tracks for retrograde transport.[114]

Recent evidence suggests a role for clathrin in retrograde trafficking of the *Shiga* toxin and CI-MPR via a partnership with the DNA-J domain-containing protein RME-8.[120–122] A RME-8-dependent association of SNX1 of the retromer with the ESCRT-0 subunit Hrs may provide the possibility for a cargo to follow the retrograde transport pathway or to be sorted by the ESCRT machinery for degradation, depending on the adequate stimulus.[120,123]

At the TGN, the mammalian GARP (Golgi-associated retrograde protein) complex plays a role in the delivery of Golgi proteins sorted by the retromer, at least for the CI-MPR and the Shiga toxin.[124]

4.2. DSCR3, a VPS26-related protein

Down syndrome is a consequence of a partial or full trisomy of chromosome 21, characterized by facial morphological features, muscular hypotonia, and mental retardation. DSCR3 (for Down syndrome critical region gene 3) is one of the proteins that is lost in this syndrome.

DSCR3 has 10–17% identity and 17–27% homology with the α-/β-arrestins, the closest proteins being VPS26A (14.8% identity) and ARRDC5 (16.9% identity). DSCR3 has been predicted to harbor a VPS26 PFAM domain.[125] For the DSCR3 paralogue of some species such as *Dictyostelium*, it is possible to detect both a VPS26 domain (expect value = 2.91×10^{-43}) and an overlapping arrestin-N domain (expect value = 4.66×10^{-4}), a property also observed for human DSCR3 (Fig. 2.3A). The modeled structure of DSCR3 is indeed very close to that of VPS26 (Fig. 2.3B).[9] In a phylogenetic tree (Fig. 2.2), the DSCR3 clade is the first group to split off the arrestin branch that gives rise to the VPS26 branch, placing DSCR3 as a possible "missing link" between arrestins and VPS26.

DSCR3 has been suggested to function as a signaling platform reminiscent of arrestin in a novel MAPK signaling pathway, by acting as a scaffold for DYRK1A as MAPK, SNF1LK as MKK, and RIP4 as MKKK.[125]

4.3. RGP1, yet another arrestin-fold protein

It was of obvious interest to screen the whole human genome for still unknown candidate proteins with an arrestin fold. A recent tool, BackPhyre developed by Phyre2 (http://www.sbg.bio.ic.ac.uk/phyre2), allows one to search a known structure against a wide range of genomes. When this mode was used on the human genome, retrograde Golgi transport protein (RGP1) emerged as a potential arrestin-fold protein. Blast searches identified RGP1 in protists, plants, fungi, yeast, and metazoan.

A phylogenetic tree (Fig. 2.2), built on the basis of a Clustal Omega alignment (http://www.ebi.ac.uk/Tools/msa/clustalo/), revealed a new RGP1 sector branching off the ART proteins, and close to the VPS26 clade. Structure threading and modeling qualify RGP1 as a novel arrestin-fold protein. Indeed, the secondary structure of RGP1 as predicted by PsiPred is very close to that of VPS26 and consists essentially of a succession of positionally conserved β-strands (Fig. 2.6). RGP1 could be modeled with a confidence value above 99% on several templates, VPS26, visual- and β-arrestins, and visually, the structure of RGP1 unambiguously looks alike the characteristic structure of VPS26.

RGP1 forms a heterodimer with Ric1p to function as a nucleotide exchange factor (GEF) for Ypt6, the yeast homolog of mammalian Rab6. Activated Ypt6 then recruits the Golgi-associated retrograde protein tethering complex GARP and controls, together with Arl1, the tethering events at the trans-Golgi cisternae of incoming vesicles carrying over Golgi membrane proteins from endosomes.[126,127]

Two arrestin-fold proteins seem thus to be involved, although with very different functions, at the two ends of the retrograde transport pathway

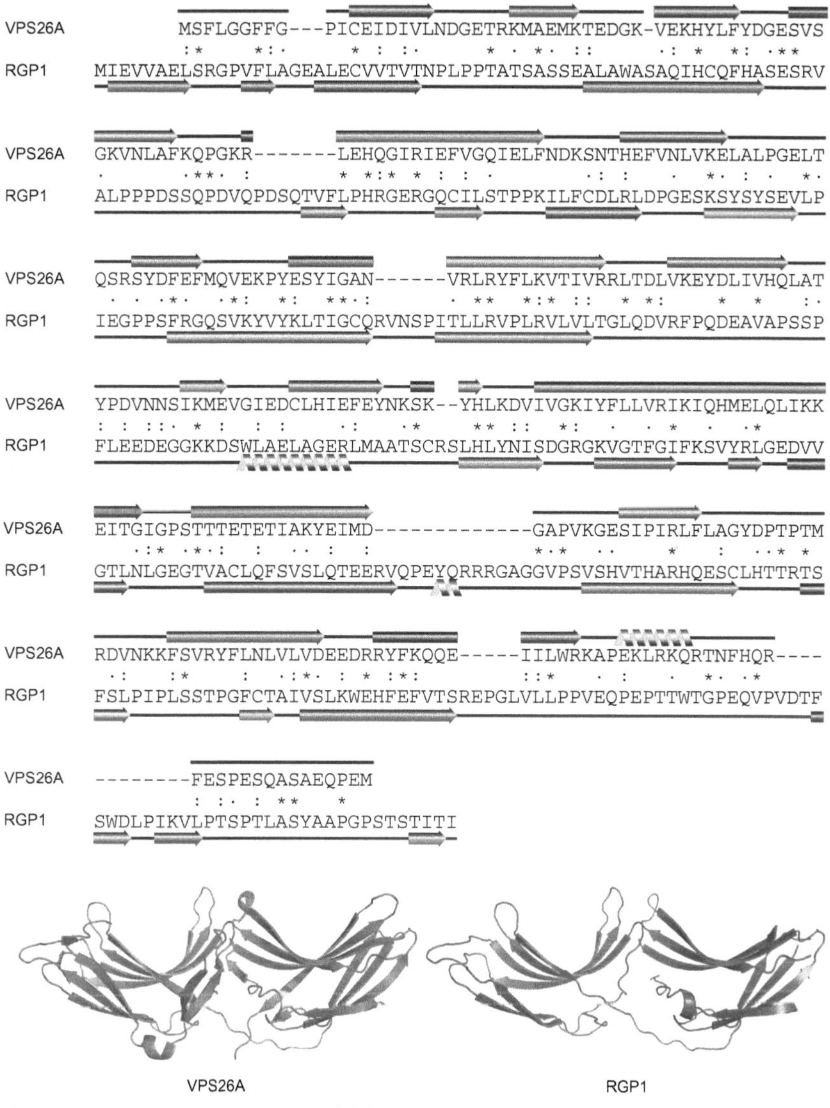

Figure 2.6 RGP1 as a new arrestin-fold protein. (A) The amino acid sequences of human RGP1 and VPS26A were aligned with Clustal Omega.[45] Identity of a given position is indicated by (*) and homology by either (:) [strong homology] or (.) [weak homology]. The secondary structure of each protein as predicted by PsiPred is indicated above or below the sequences. The structure of RGP1 was modeled on Phyre 2 in the intensive mode. It was modeled on four templates, 2FAU (human VPS26A), 1SUJ (*Ambystona* cone arrestin), 1AYR (bovine rod arrestin), and 1G4M (bovine β-arrestin 1) with a confidence value of 99.7–99.9%. (B) The modeled structure of human RGP1 is shown side by side with that of human VPS26A (2FAU). The characteristic double β-sheet sandwich is found in both structures and qualifies RGP1 as a novel arrestin-fold protein to be added to the arrestin clan. (See Color Insert.)

leading from the endosomes to the TGN; VPS26 within the retromer complex in the retrieval of Golgi proteins in endosome-derived vesicles, and RGP1 more downstream when these vesicles fuse with the TGN.

5. ARE ARRESTINS AND ARRESTIN-FOLD PROTEINS RELATED BY A SHARED MECHANISM FOR THEIR FUNCTION?

Recent findings have pinpointed several protein subfamilies with an arrestin fold. True arrestins, ARTs, ARRDCs, VPS26, and DSCR3 are now clustered in the arrestin clan.

The crucial question raised by the similar structure of arrestin-fold proteins only distantly related to arrestins in terms of sequence conservation is whether they are functionally related. Such a question seems all the more pertinent as we know that evolution has produced preserved basic folds in proteins with little conservation in their amino acid sequences or their biological functions.[128,129] Although proteins with a similar tertiary structure in the absence of sequence conservation may have evolved by processes of convergent evolution, it is in fact highly probable that they adopt one among a small number of thermodynamically preferred conformations, because of either the facility of the folding process or the stability of the obtained structure.[128]

The first discovered function of arrestins, and the origin of their name, is the arrest of the G-protein signaling triggered by the activation of 7TM-receptors. This desensitizing function was extended by their ability, once bound to an activated GPCR, to recruit components of the endocytosis machinery and to promote internalization of the GPCR. The interaction of arrestins with the phosphorylated tail of activated GPCRs exploits a subtle mechanism involving a polar core and the unfolding of their C-terminal tail, unmasking adaptin and clathrin-binding sites. More recently, arrestins were described as G-protein independent signaling platforms recruited to activated GPCRs.

We have examined the functions of the arrestin-fold proteins in the light of the above functions of arrestins to determine the extent to which they conform to this paradigm. We followed Alvarez's suggestion to divide arrestin fold proteins in two classes: α-arrestins (ARTs, ARRDCs) and β-arrestins that include visual arrestins.[130] VPS26 and DSCR3 proteins form a third subfamily. Similarities and differences between the functional mechanisms of diverse arrestin-fold proteins are summarized in Table 2.1.

Table 2.1 Similarities and differences between members of the arrestin clan

	β-Arrestins	α-Arrestins		Retromer	
	Visual/β-arrestins	ARRDCs	Yeast ARTs	VPS26	Other retromer subunits
Arrestin fold	1–3,5,6,8[a]	Predicted (9, b)	Predicted (9, 10, b)	11,12	
Polar core	3,5,20,31	n.d.	n.d.	11,12,95	
Cargo binding	36,131	81	86,89	132	VPS3597
Cargo phosphorylation	18,22,23	81	86		133
Cargo ubiquitination	134,135	81,82	89		
PxY sites	None	82,130	10,85,86,130,136		
Ubiquitination	41,68,69,137		10,85,87		
Phosphorylation	18,60,61,65		85,87		
Inositol phosphates binding	4,5,74,76,138				SNX139
Clathrin binding	18,49,55				114,121
AP adaptor protein binding	18,66				CI-MPR96;
Actin binding	40,140				110,114
Microtubule binding					118,119
ESCRT binding	28	79	88,141		
Regulation by small G proteins	18,40				107,109,110,142
Secondary signaling platform	19,24,30–35,41,143				

[a]Numbers correspond to references in text.
[b]This work.
Functions carried out by α-/β-arrestins, absent in VPS26, but found in other subunits of the retromer, are summarized in the last column.
n.d., not determined.

The initial function described for arrestins, their steric interference with G-proteins activated by their cognate GPCRs and attenuation of their signaling, is a characteristic that will not be found for the other members of the arrestin clan that regulate other classes of cargoes. It is now known that β-arrestins also interact with and desensitize PM transporters and receptors not coupled to G proteins, for example, NHE5 and insulin-like growth factor I receptor.[36,38,144]

The second hallmark of arrestin-mediated receptor desensitization is the arrestin-mediated internalization of cargoes on endocytic vesicles. Down-regulation of stimulated PM proteins is also observed for some ARTS and ARRDCs and clearly constitutes a point of convergence between β-arrestins and α-arrestins. However, the intimate interaction process based on the recognition by β-arrestins of phosphorylated serines/threonines of the GPCR, the disruption of a polar core, and the unfolding of the C-terminal tail preliminary to the scaffolding of the endocytosis machinery is not described for ARTs and ARRDCs. How cargoes regulated by α-arrestins are internalized still awaits extensive characterization.

The Alvarez split into α-arrestins and visual/β-arrestins is based on the existence of [L,P]PxY motifs within the C-terminal tail of α-arrestins that are able to bind the WW domains of HECT-type E3 ubiquitin ligases of the Nedd4 family, motifs that are absent in β-arrestins.[130] ARTs are used to scaffold E3 ubiquitin ligases for the ubiquitination of their cognate partner. In the process, they are themselves ubiquitinated and they regulate the endocytosis of ubiquitinated PM receptors.[10,85,87,89] Many yeast transporters of diverse types are delivered to the vacuole once internalized in an ART-dependent process.[89] Human ARRDC3 is similarly involved in the degradation pathway of β-integrins, with ubiquitination governing the commitment to this degradation itinerary.[81]

Interestingly, members of the β-arrestin subfamily also bind ubiquitin E3 ligases including Nedd4. β-Arrestin 2 is first ubiquitinated by the RING-type E3 ligase Mdm2, a modification that increases its affinity for the β_2-adrenergic receptor. Ubiquitinated β-arrestin 2 acts then as a scaffold for the binding of Nedd4, which mediates receptor ubiquitination. The interaction between β-arrestin 2 and Nedd4 occurs in the absence of PY motifs on β-arrestin 2 and does not require the WW domains of Nedd4.[145] The difference between α- and β-arrestins does not lie in the interaction with E3 ubiquitin ligases, but rather in the mechanism that governs this interaction, mediated by PY motifs in α-arrestins binding WW domains and yet unknown in the case of β-arrestins.[146]

α-Arrestins and visual/β-arrestins share another trait required to trigger endocytosis of their cargo, a dephosphorylation step. The bulk (70%) of β-arrestin is cytosolic, phosphorylated, and "inactive." To bind an activated GPCR, they are dephosphorylated at the PM. Once the receptor is internalized, β-arrestins can be rephosphorylated on the endosome.[59–62] Visual arrestin is phosphorylated upon light activation[63] and its dephosphorylation is required for the arrestin–clathrin interaction to proceed.[64] Inactive, phosphorylated Rod1/ART4 is mostly cytosolic and shielded by an interaction with 14-3-3. Its activation releases 14-3-3 and allows its dephosphorylation by PP1 phosphatase and its subsequent ubiquitination by Rsp5.[85] Upon activation, Ldb9(Cvs7)/ART1, but not Rod1/ART4, redistributes at the PM and their cargoes (Can1, Jen1) are endocytosed.[10,85] The example of PalF introduces some degree of complexity, as this paralogue of Rim8/ART9 in *Aspergillus* is phosphorylated and ubiquitinated in a signal-dependent manner before interacting with the signal sensor PalH.[87]

A clue to understanding the degradation mechanism of the cargoes comes from the demonstration that most ARRDCs interact with subunits of the ESCRT machinery.[79] The ESCRT complex is the sorting machinery scaffolded at the membrane of the multivesicular body (MVB) that recognizes and sorts cargoes destined for degradation by their ubiquitin label and generates intraluminal vesicles enriched with cargoes. After fusion of the MVB with lysosomes, these intraluminal vesicles are degraded by the lysosomal lytic enzymes.[147–149] ARTs or ARRDCs interact with ESCRT subunits not only at the endosome but also at the PM, as shown for the fungal signaling complex with PalF/Rim8[150] and with diverse ARRDCs in the budding process of viruses.[79] Recent results with the β_2-adrenergic receptor shed new light on the choice between recycling and degradation pathways. As discussed above, the class A β_2-adrenergic receptor normally recruits β-arrestin 2 upon activation and is only transiently downregulated.[151] This same receptor interacts also with ARRDC3 as the initial step of its ubiquitin-dependent degradation in the lysosome.[82] Interaction with ESCRT-0 has also been demonstrated for β-arrestin 1 in the degradation pathway of CXCR4.[28] If the choice between alternate interactors (arrestins or ARRDCs) is verified for other receptors and the full repertoire of α-/β-arrestins, it would provide a means to control the diversity of the followed itineraries, toward either recycling or degradation.

Apart from the involvement of both protein subfamilies in the trafficking of membrane-embedded receptors, the functional similarities between VPS26 and the α-/β-arrestins are rather weak. Despite its arrestin fold and polar core, VPS26 is generally not the cargo-recognition subunit in the retromer. Most examples of retromer–cargo interactions use the recognition by VPS35 of a specific [FW]L[MV] sequence carried by the cargoes.[96] VPS26 functions rather as an adaptor linking VPS35 and SNXs. There is, however, a recent example in which VPS26 directly binds a cargo. SorlA, a member of the VPS10 receptor family, transports the amyloid precursor protein to the Golgi for its processing by α-secretase. In the absence of retrograde transport, the amyloid precursor protein is transported to the lysosome where it is processed by β- and γ-secretases into the neurotoxic amyloid β peptide agent of Alzheimer's disease. VPS26 is the retromer subunit that recognizes SorlA's sorting motif FANSHY.[132] Interestingly, FANSHY is an extension of the FLV sequence previously identified on sortilin and CI-MPR and the same distance between two aromatic residues can be found in the extended FLV motif.[132] Substantial evidence demonstrates a role for clathrin and its adaptor epsinR in retromer trafficking.[120,152] The role of the adaptor AP-1 is still debated.[114,152] Tubulation of cargo-enriched endosomal patches and formation of the recycling endosomal vesicles proceed by a retromer-specific mechanism that takes advantage of the SNX's membrane-bending property. There is to our knowledge no evidence for posttranslational modifications of the retromer subunits or the transported cargoes, besides the CI-MPR's palmitoylation.[153]

The interaction of the retromer component SNX1 with Hrs may provide a molecular switch rerouting retromer cargoes to late endosomes/lysosomes for degradation.[120] This articulation between two trafficking routes observed for retromer cargoes is reminiscent of the dual fate of β_2-adrenergic receptor depending on the recruited scaffold, β-arrestin or ARRDC.

A functional kinship of RGP1 with the members of the arrestin clan described above is even more elusive as RGP1 is a component of a dimeric GEF with a regulatory function on the GARP multimeric complex involved in vesicle tethering at the TGN.

6. CONCLUSION AND PERSPECTIVES

Arrestin-fold proteins (true arrestins, human ARRDCs, ARTs, amoebal ADCs, VPS26, RGP1) share a convergent structure with two domains of antiparallel β-sheets oriented to optimally scaffold proteins in sorting/targeting events during membrane trafficking.

Members of the arrestin clan are clearly related, from a structural point of view and because of their functions in trafficking steps of membrane-embedded proteins. The mechanisms of cargo-recognition used by the diverse members of the arrestin clan seem clearly different, and so far only β-arrestins are involved in the interference with G-protein signaling and in the scaffolding of secondary signaling platforms on the endosome.

Visual/β-arrestins, ARRDCs and ARTs are all involved in the down-regulation of PM proteins. The initial steps of their cargo recognition are similar (Fig. 2.7A and C). Both proteins are cytosolic in their phosphorylated inactive state. Their dephosphorylation and ubiquitination are temporally correlated with their recruitment on their cargo(es). There is, however, no biochemical evidence that this recruitment proceeds through the same mechanism or that it involves a polar core and the unveiling of sites for the endocytic machinery for both adaptor proteins. The next step in the receptor/cargo desensitization process is the recruitment by arrestins/ARTs of an E3 ubiquitin ligase and the subsequent ubiquitination of the cargo. Data are lacking as to the detailed internalization mechanism of PM transporters by ARTs. The fate of internalized cargoes is either recycling to the PM and resensitization or degradation in the lysosome. For some human receptors, the choice between these two fates is dependent on the recruited adaptor, β-arrestins or ARRDCs. In yeast, the role of ARTs is related to the vacuolar degradation of internalized PM transporters. No data on the possible recycling of these transporters are currently available.

The retromer represents a very different intercompartmental trafficking machinery (Fig. 2.7B). It is a hetero-oligomer involved in retrograde transport from the endosome to the TGN of CI-MPR and other Golgi proteins. The trigger of this back transport is, in the case of the CI-MPR, related to the dissociation of lysosomal enzymes carried over from the TGN, but the detailed mechanism is not yet known. The arrestin-fold protein VPS26, along with VPS35 and VPS29, make up the trimeric cargo-recognition subcomplex, in which the primary Golgi protein recognition subunit is VPS35. The retromer complex is mechanistically very different from an arrestin scaffold in terms of direction and regulation of the transport, recognized cargoes, and recruited partners.

The last arrestin-fold protein described here is RGP1. It is part of a heterodimeric Rab6 GEF at the TGN that regulates the tethering of incoming endosomal vesicles involved in the retrograde transport of Golgi proteins (Fig. 2.7D). No functional kinship relates it to an arrestin, an ART, or even VPS26. Its exact function awaits biochemical characterization.

A challenge for the future is the attainment of full-length crystal structures for ARTs, ARRDCs, ADCs, and RGP1. Most of these arrestin-fold

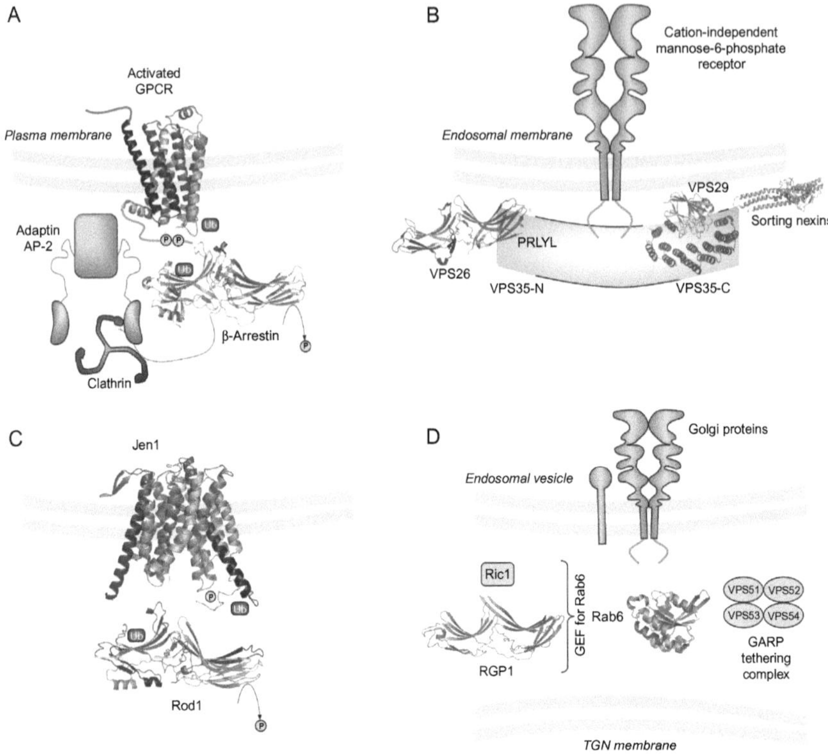

Figure 2.7 Functional context of arrestin-fold proteins. (A) β-Arrestins interact with activated, phosphorylated GPCRs to interfere with G-protein signaling and to recruit the endocytic machinery (adaptor protein AP-2, clathrin) to binding sites located on the C-terminal tail which is unfolded as a result of the receptor binding. Ubiquitination sites on the GPCR and the β-arrestin are only indicative. (B) VPS26 is a subunit of the pentameric retromer (consisting of, besides VPS26, the cargo-loading subunit VPS35, VPS29, and a pair of sorting nexins harboring PX- and BAR domains) involved in the retrograde transport of the mannose-6-phosphate receptor from the endosome to the TGN. The cation-independent mannose-6-phosphate receptor is outlined after a published tentative view.[154] (C) Rod1-ART4 is dephosphorylated and ubiquitinated by the ubiquitin ligase Nedd4. It mediates the ubiquitination and the surface removal of the MFS family protein Jen1. Rod 1 and Jen1 were modeled with Phyre2. It is uncertain where Rod1 acts. Phosphorylation and ubiquitination sites on Jen1 and Rod4 are only indicative. (D) RGP1 forms together with Ric1 a heterodimeric GEF for Rab6. They regulate the tethering complex GARP-dependent fusion in a Golgi retrograde transport mechanism.

proteins have important extensions outside the arrestin core, organized or not as known domains. The structure of the arrestin core, although modeled with a high confidence level ($\geq$95%) represents only part of the whole structure and the function of these arrestin-related proteins undoubtedly relies on the integration of the various structural portions.

Members of the arrestin clan described here share a stable tertiary topology essentially made of β-strands. All of them are involved in various intracellular trafficking steps and can thus be considered as relatives. However, no transversal functional unity emerges from their role in trafficking. In the arrestin clan, even the closest members are, at most, first cousins, clearly not twins.

Note added in proof: The structure of the N-terminal domain of human TXNIP has now been determined [Polekhina, G., Ascher, B. D., Kok, S. F., Beckham, S., Wilce, M. and Waltham, M. (2013) Structure of the N-terminal domain of human thioredoxin-interacting protein. Acta Cryst. D69, 333-344]. TXNIP adopts an arrestin fold as predicted.

REFERENCES

1. Granzin J, Wilden U, Choe HW, Labahn J, Krafft B, Buldt G. X-ray crystal structure of arrestin from bovine rod outer segments. *Nature*. 1998;391:918–921.
2. Han M, Gurevich VV, Vishnivetskiy SA, Sigler PB, Schubert C. Crystal structure of beta-arrestin at 1.9 Å: possible mechanism of receptor binding and membrane translocation. *Structure*. 2001;9:869–880.
3. Hirsch JA, Schubert C, Gurevich VV, Sigler PB. The 2.8 Å crystal structure of visual arrestin: a model for arrestin's regulation. *Cell*. 1999;97:257–269.
4. Milano SK, Kim YM, Stefano FP, Benovic JL, Brenner C. Nonvisual arrestin oligomerization and cellular localization are regulated by inositol hexakisphosphate binding. *J Biol Chem*. 2006;281:9812–9823.
5. Milano SK, Pace HC, Kim YM, Brenner C, Benovic JL. Scaffolding functions of arrestin-2 revealed by crystal structure and mutagenesis. *Biochemistry*. 2002;41:3321–3328.
6. Sutton RB, Vishnivetskiy SA, Robert J, et al. Crystal structure of cone arrestin at 2.3 Å: evolution of receptor specificity. *J Mol Biol*. 2005;354:1069–1080.
7. Granzin J, Cousin A, Weirauch M, Schlesinger R, Buldt G, Batra-Safferling R. Crystal structure of p44, a constitutively active splice variant of visual arrestin. *J Mol Biol*. 2012;416:611–618.
8. Zhan X, Gimenez LE, Gurevich VV, Spiller BW. Crystal structure of arrestin-3 reveals the basis of the difference in receptor binding between two non-visual subtypes. *J Mol Biol*. 2011;406:467–478.
9. Aubry L, Guetta D, Klein G. The arrestin fold: variations on a theme. *Curr Genomics*. 2009;10:133–142.
10. Lin CH, MacGurn JA, Chu T, Stefan CJ, Emr SD. Arrestin-related ubiquitin-ligase adaptors regulate endocytosis and protein turnover at the cell surface. *Cell*. 2008;135:714–725.
11. Shi H, Rojas R, Bonifacino JS, Hurley JH. The retromer subunit Vps26 has an arrestin fold and binds Vps35 through its C-terminal domain. *Nat Struct Mol Biol*. 2006;13:540–548.
12. Collins BM, Norwood SJ, Kerr MC, et al. Structure of Vps26B and mapping of its interaction with the retromer protein complex. *Traffic*. 2008;9:366–379.
13. Attramadal H, Arriza JL, Aoki C, et al. Beta-arrestin2, a novel member of the arrestin/beta-arrestin gene family. *J Biol Chem*. 1992;267:17882–17890.
14. Craft CM, Whitmore DH, Wiechmann AF. Cone arrestin identified by targeting expression of a functional family. *J Biol Chem*. 1994;269:4613–4619.

15. Lohse MJ, Benovic JL, Codina J, Caron MG, Lefkowitz RJ. Beta-arrestin: a protein that regulates beta-adrenergic receptor function. *Science*. 1990;248:1547–1550.
16. Wilden U, Hall SW, Kuhn H. Phosphodiesterase activation by photoexcited rhodopsin is quenched when rhodopsin is phosphorylated and binds the intrinsic 48-kDa protein of rod outer segments. *Proc Natl Acad Sci USA*. 1986;83:1174–1178.
17. Zuckerman R, Buzdygon B, Philp N, Liebman P, Sitaramayya A. Arrestin: an ATP/ADP exchange protein that regulates cGMP phosphodiesterase activity in retinal rod disk membranes (RDM). *Biophys J*. 1985;47:37a.
18. Claing A, Laporte SA, Caron MG, Lefkowitz RJ. Endocytosis of G protein-coupled receptors: roles of G protein-coupled receptor kinases and beta-arrestin proteins. *Prog Neurobiol*. 2002;66:61–79.
19. Ferguson SS. Evolving concepts in G protein-coupled receptor endocytosis: the role in receptor desensitization and signaling. *Pharmacol Rev*. 2001;53:1–24.
20. Gurevich VV, Gurevich EV. The structural basis of arrestin-mediated regulation of G-protein-coupled receptors. *Pharmacol Ther*. 2006;110:465–502.
21. Luttrell LM, Lefkowitz RJ. The role of beta-arrestins in the termination and transduction of G-protein-coupled receptor signals. *J Cell Sci*. 2002;115:455–465.
22. Moore CA, Milano SK, Benovic JL. Regulation of receptor trafficking by GRKs and arrestins. *Annu Rev Physiol*. 2007;69:451–482.
23. Premont RT, Gainetdinov RR. Physiological roles of G protein-coupled receptor kinases and arrestins. *Annu Rev Physiol*. 2007;69:511–534.
24. DeWire SM, Ahn S, Lefkowitz RJ, Shenoy SK. Beta-arrestins and cell signaling. *Annu Rev Physiol*. 2007;69:483–510.
25. Shenoy SK. Seven-transmembrane receptors and ubiquitination. *Circ Res*. 2007;100: 1142–1154.
26. Bhandari D, Trejo J, Benovic JL, Marchese A. Arrestin-2 interacts with the ubiquitin-protein isopeptide ligase atrophin-interacting protein 4 and mediates endosomal sorting of the chemokine receptor CXCR4. *J Biol Chem*. 2007;282:36971–36979.
27. Bhandari D, Robia SL, Marchese A. The E3 ubiquitin ligase atrophin interacting protein 4 binds directly to the chemokine receptor CXCR4 via a novel WW domain-mediated interaction. *Mol Biol Cell*. 2009;20:1324–1339.
28. Malik R, Marchese A. Arrestin-2 interacts with the endosomal sorting complex required for transport machinery to modulate endosomal sorting of CXCR4. *Mol Biol Cell*. 2010;21:2529–2541.
29. Barki-Harrington L, Rockman HA. Beta-arrestins: multifunctional cellular mediators. *Physiology (Bethesda)*. 2008;23:17–22.
30. DeFea KA. Beta-arrestins as regulators of signal termination and transduction: how do they determine what to scaffold? *Cell Signal*. 2011;23:621–629.
31. Gurevich EV, Gurevich VV. Arrestins: ubiquitous regulators of cellular signaling pathways. *Genome Biol*. 2006;7:236.
32. Luttrell LM, Gesty-Palmer D. Beyond desensitization: physiological relevance of arrestin-dependent signaling. *Pharmacol Rev*. 2010;62:305–330.
33. Ma L, Pei G. Beta-arrestin signaling and regulation of transcription. *J Cell Sci*. 2007;120:213–218.
34. Shenoy SK, Lefkowitz RJ. Beta-Arrestin-mediated receptor trafficking and signal transduction. *Trends Pharmacol Sci*. 2011;32:521–533.
35. Shukla AK, Xiao K, Lefkowitz RJ. Emerging paradigms of beta-arrestin-dependent seven transmembrane receptor signaling. *Trends Biochem Sci*. 2011;36:457–469.
36. Lefkowitz RJ, Rajagopal K, Whalen EJ. New roles for beta-arrestins in cell signaling: not just for seven-transmembrane receptors. *Mol Cell*. 2006;24:643–652.
37. Hupfeld CJ, Olefsky JM. Regulation of receptor tyrosine kinase signaling by GRKs and beta-arrestins. *Annu Rev Physiol*. 2007;69:561–577.

38. Szabo EZ, Numata M, Lukashova V, Iannuzzi P, Orlowski J. Beta-arrestins bind and decrease cell-surface abundance of the Na^+/H^+ exchanger NHE5 isoform. *Proc Natl Acad Sci USA*. 2005;102:2790–2795.
39. Lipsky R, Potts EM, Tarzami ST, et al. Beta-adrenergic receptor activation induces internalization of cardiac Cav1.2 channel complexes through a beta-Arrestin 1-mediated pathway. *J Biol Chem*. 2008;283:17221–17226.
40. DeFea KA. Stop that cell! Beta-arrestin-dependent chemotaxis: a tale of localized actin assembly and receptor desensitization. *Annu Rev Physiol*. 2007;69:535–560.
41. Kovacs JJ, Hara MR, Davenport CL, Kim J, Lefkowitz RJ. Arrestin development: emerging roles for beta-arrestins in developmental signaling pathways. *Dev Cell*. 2009;17:443–458.
42. Molla-Herman A, Boularan C, Ghossoub R, et al. Targeting of beta-arrestin2 to the centrosome and primary cilium: role in cell proliferation control. *PLoS One*. 2008;3: e3728.
43. Reinders Y, Schulz I, Graf R, Sickmann A. Identification of novel centrosomal proteins in *Dictyostelium discoideum* by comparative proteomic approaches. *J Proteome Res*. 2006;5:589–598.
44. Shankar H, Michal A, Kern RC, Kang DS, Gurevich VV, Benovic JL. Non-visual arrestins are constitutively associated with the centrosome and regulate centrosome function. *J Biol Chem*. 2010;285:8316–8329.
45. Sievers F, Wilm A, Dineen D, et al. Fast, scalable generation of high-quality protein multiple sequence alignments using Clustal Omega. *Mol Syst Biol*. 2011;7:539.
46. Ashkenazy H, Erez E, Martz E, Pupko T, Ben-Tal N. ConSurf 2010: calculating evolutionary conservation in sequence and structure of proteins and nucleic acids. *Nucleic Acids Res*. 2010;38:W529–W533.
47. Landau M, Mayrose I, Rosenberg Y, et al. ConSurf 2005: the projection of evolutionary conservation scores of residues on protein structures. *Nucleic Acids Res*. 2005;33: W299–W302.
48. Campbell ID, Spitzfaden C. Building proteins with fibronectin type III modules. *Structure*. 1994;2:333–337.
49. Kern RC, Kang DS, Benovic JL. Arrestin2/clathrin interaction is regulated by key N- and C-terminal regions in arrestin2. *Biochemistry*. 2009;48:7190–7200.
50. Kovoor A, Celver J, Abdryashitov RI, Chavkin C, Gurevich VV. Targeted construction of phosphorylation-independent beta-arrestin mutants with constitutive activity in cells. *J Biol Chem*. 1999;274:6831–6834.
51. Vishnivetskiy SA, Hosey MM, Benovic JL, Gurevich VV. Mapping the arrestin–receptor interface. Structural elements responsible for receptor specificity of arrestin proteins. *J Biol Chem*. 2004;279:1262–1268.
52. Goodman Jr OB, Krupnick JG, Santini F, et al. Beta-arrestin acts as a clathrin adaptor in endocytosis of the beta2-adrenergic receptor. *Nature*. 1996;383:447–450.
53. Gurevich VV, Gurevich EV. The new face of active receptor bound arrestin attracts new partners. *Structure*. 2003;11:1037–1042.
54. Vishnivetskiy SA, Hirsch JA, Velez MG, Gurevich YV, Gurevich VV. Transition of arrestin into the active receptor-binding state requires an extended interdomain hinge. *J Biol Chem*. 2002;277:43961–43967.
55. Kang DS, Kern RC, Puthenveedu MA, von Zastrow M, Williams JC, Benovic JL. Structure of an arrestin2-clathrin complex reveals a novel clathrin binding domain that modulates receptor trafficking. *J Biol Chem*. 2009;284:29860–29872.
56. Orem NR, Xia L, Dolph PJ. An essential role for endocytosis of rhodopsin through interaction of visual arrestin with the AP-2 adaptor. *J Cell Sci*. 2006;119:3141–3148.
57. Wang T, Montell C. Phototransduction and retinal degeneration in Drosophila. *Pflugers Arch*. 2007;454:821–847.

58. Satoh AK, Ready DF. Arrestin1 mediates light-dependent rhodopsin endocytosis and cell survival. *Curr Biol.* 2005;15:1722–1733.
59. Lin FT, Miller WE, Luttrell LM, Lefkowitz RJ. Feedback regulation of beta-arrestin1 function by extracellular signal-regulated kinases. *J Biol Chem.* 1999;274: 15971–15974.
60. Lin FT, Krueger KM, Kendall HE, et al. Clathrin-mediated endocytosis of the beta-adrenergic receptor is regulated by phosphorylation/dephosphorylation of beta-arrestin1. *J Biol Chem.* 1997;272:31051–31057.
61. Lin FT, Chen W, Shenoy S, Cong M, Exum ST, Lefkowitz RJ. Phosphorylation of beta-arrestin2 regulates its function in internalization of beta(2)-adrenergic receptors. *Biochemistry.* 2002;41:10692–10699.
62. Kim YM, Barak LS, Caron MG, Benovic JL. Regulation of arrestin-3 phosphorylation by casein kinase II. *J Biol Chem.* 2002;277:16837–16846.
63. Matsumoto H, Kurien BT, Takagi Y, et al. Phosrestin I undergoes the earliest light-induced phosphorylation by a calcium/calmodulin-dependent protein kinase in *Drosophila* photoreceptors. *Neuron.* 1994;12:997–1010.
64. Kiselev A, Socolich M, Vinos J, Hardy RW, Zuker CS, Ranganathan R. A molecular pathway for light-dependent photoreceptor apoptosis in *Drosophila. Neuron.* 2000;28: 139–152.
65. Barthet G, Carrat G, Cassier E, et al. Beta-arrestin1 phosphorylation by GRK5 regulates G protein-independent 5-HT4 receptor signalling. *EMBO J.* 2009;28:2706–2718.
66. Marion S, Fralish GB, Laporte S, Caron MG, Barak LS. N-terminal tyrosine modulation of the endocytic adaptor function of the beta-arrestins. *J Biol Chem.* 2007;282:18937–18944.
67. Ozawa K, Whalen EJ, Nelson CD, et al. S-Nitrosylation of beta-arrestin regulates beta-adrenergic receptor trafficking. *Mol Cell.* 2008;31:395–405.
68. Shenoy SK, Modi AS, Shukla AK, et al. Beta-arrestin-dependent signaling and trafficking of 7-transmembrane receptors is reciprocally regulated by the deubiquitinase USP33 and the E3 ligase Mdm2. *Proc Natl Acad Sci USA.* 2009;106:6650–6655.
69. Shenoy SK, Lefkowitz RJ. Receptor-specific ubiquitination of beta-arrestin directs assembly and targeting of seven-transmembrane receptor signalosomes. *J Biol Chem.* 2005;280:15315–15324.
70. Wyatt D, Malik R, Vesecky AC, Marchese A. Small ubiquitin-like modifier modification of arrestin-3 regulates receptor trafficking. *J Biol Chem.* 2011;286:3884–3893.
71. Kelley LA, Sternberg MJ. Protein structure prediction on the Web: a case study using the Phyre server. *Nat Protoc.* 2009;4:363–371.
72. Huson DH, Bryant D. Application and phylogenetic networks in evolutionary studies. *Mol Biol Evol.* 2006;23:254–267.
73. Nei M, Kumar S. *Molecular Evolution and Phylogenetics.* New York: Oxford University Press; 2000.
74. Zhuang T, Vishnivetskiy SA, Gurevich VV, Sanders CR. Elucidation of inositol hexaphosphate and heparin interaction sites and conformational changes in arrestin-1 by solution nuclear magnetic resonance. *Biochemistry.* 2010;49:10473–10485.
75. Pittet D, Schlegel W, Lew DP, Monod A, Mayr GW. Mass changes in inositol tetrakis- and pentakisphosphate isomers induced by chemotactic peptide stimulation in HL-60 cells. *J Biol Chem.* 1989;264:18489–18493.
76. Gaidarov I, Krupnick JG, Falck JR, Benovic JL, Keen JH. Arrestin function in G protein-coupled receptor endocytosis requires phosphoinositide binding. *EMBO J.* 1999;18:871–881.
77. Polekhina G, Ascher DB, Kok SF, Waltham M. Crystallization and preliminary X-ray analysis of the N-terminal domain of human thioredoxin-interacting protein. *Acta Crystallogr Sect F Struct Biol Cryst Commun.* 2011;67:613–617.

78. Patwari P, Lee RT. An expanded family of arrestins regulate metabolism. *Trends Endocrinol Metab.* 2012;23:216–222.
79. Rauch S, Martin-Serrano J. Multiple interactions between the ESCRT machinery and arrestin-related proteins: implications for PPXY-dependent budding. *J Virol.* 2011;85:3546–3556.
80. Nabhan JF, Hu R, Oh RS, Cohen SN, Lu Q. Formation and release of arrestin domain-containing protein 1-mediated microvesicles (ARMMs) at plasma membrane by recruitment of TSG101 protein. *Proc Natl Acad Sci USA.* 2012;109:4146–4151.
81. Draheim KM, Chen HB, Tao Q, Moore N, Roche M, Lyle S. ARRDC3 suppresses breast cancer progression by negatively regulating integrin beta4. *Oncogene.* 2010;29:5032–5047.
82. Nabhan JF, Pan H, Lu Q. Arrestin domain-containing protein 3 recruits the NEDD4 E3 ligase to mediate ubiquitination of the beta2-adrenergic receptor. *EMBO Rep.* 2010;11:605–611.
83. Patwari P, Chutkow WA, Cummings K, et al. Thioredoxin-independent regulation of metabolism by the alpha-arrestin proteins. *J Biol Chem.* 2009;284:24996–25003.
84. Oka S, Masutani H, Liu W, et al. Thioredoxin-binding protein-2-like inducible membrane protein is a novel vitamin D3 and peroxisome proliferator-activated receptor (PPAR)gamma ligand target protein that regulates PPARgamma signaling. *Endocrinology.* 2006;147:733–743.
85. Becuwe M, Vieira N, Lara D, et al. A molecular switch on an arrestin-like protein relays glucose signaling to transporter endocytosis. *J Cell Biol.* 2012;196:247–259.
86. Nikko E, Sullivan JA, Pelham HR. Arrestin-like proteins mediate ubiquitination and endocytosis of the yeast metal transporter Smf1. *EMBO Rep.* 2008;9:1216–1221.
87. Herranz S, Rodriguez JM, Bussink HJ, et al. Arrestin-related proteins mediate pH signaling in fungi. *Proc Natl Acad Sci USA.* 2005;102:12141–12146.
88. Herrador A, Herranz S, Lara D, Vincent O. Recruitment of the ESCRT machinery to a putative seven-transmembrane-domain receptor is mediated by an arrestin-related protein. *Mol Cell Biol.* 2010;30:897–907.
89. Nikko E, Pelham HR. Arrestin-mediated endocytosis of yeast plasma membrane transporters. *Traffic.* 2009;10:1856–1867.
90. Prabhu Y, Eichinger L. The Dictyostelium repertoire of seven transmembrane domain receptors. *Eur J Cell Biol.* 2006;85:937–946.
91. Guetta D, Langou K, Grunwald D, Klein G, Aubry L. FYVE-dependent endosomal targeting of an arrestin-related protein in amoeba. *PLoS One.* 2010;5:e15249.
92. Journet A, Klein G, Brugière S, et al. Investigating the macropinocytic proteome of *Dictyostelium* amoebae by high resolution mass spectrometry. *Proteomics.* 2012;12:241–245.
93. Seaman MN, McCaffery JM, Emr SD. A membrane coat complex essential for endosome-to-Golgi retrograde transport in yeast. *J Cell Biol.* 1998;142:665–681.
94. Arighi CN, Hartnell LM, Aguilar RC, Haft CR, Bonifacino JS. Role of the mammalian retromer in sorting of the cation-independent mannose 6-phosphate receptor. *J Cell Biol.* 2004;165:123–133.
95. Norwood SJ, Shaw DJ, Cowieson NP, Owen DJ, Teasdale RD, Collins BM. Assembly and solution structure of the core retromer protein complex. *Traffic.* 2011; 12:56–71.
96. Seaman MN. Identification of a novel conserved sorting motif required for retromer-mediated endosome-to-TGN retrieval. *J Cell Sci.* 2007;120:2378–2389.
97. Collins BM. The structure and function of the retromer protein complex. *Traffic.* 2008;9:1811–1822.
98. Hierro A, Rojas AL, Rojas R, et al. Functional architecture of the retromer cargo-recognition complex. *Nature.* 2007;449:1063–1067.

99. Wang D, Guo M, Liang Z, et al. Crystal structure of human vacuolar protein sorting protein 29 reveals a phosphodiesterase/nuclease-like fold and two protein–protein interaction sites. *J Biol Chem.* 2005;280:22962–22967.
100. Collins BM, Skinner CF, Watson PJ, Seaman MN, Owen DJ. Vps29 has a phosphoesterase fold that acts as a protein interaction scaffold for retromer assembly. *Nat Struct Mol Biol.* 2005;12:594–602.
101. Swarbrick JD, Shaw DJ, Chhabra S, et al. VPS29 is not an active metallo-phosphatase but is a rigid scaffold required for retromer interaction with accessory proteins. *PLoS One.* 2011;6:e20420.
102. Gullapalli A, Garrett TA, Paing MM, Griffin CT, Yang Y, Trejo J. A role for sorting nexin 2 in epidermal growth factor receptor down-regulation: evidence for distinct functions of sorting nexin 1 and 2 in protein trafficking. *Mol Biol Cell.* 2004;15:2143–2155.
103. Haft CR, de la Luz Sierra M, Bafford R, Lesniak MA, Barr VA, Taylor SI. Human orthologs of yeast vacuolar protein sorting proteins Vps26, 29, and 35: assembly into multimeric complexes. *Mol Biol Cell.* 2000;11:4105–4116.
104. Merino-Trigo A, Kerr MC, Houghton F, et al. Sorting nexin 5 is localized to a subdomain of the early endosomes and is recruited to the plasma membrane following EGF stimulation. *J Cell Sci.* 2004;117:6413–6424.
105. Ellson CD, Andrews S, Stephens LR, Hawkins PT. The PX domain: a new phosphoinositide-binding module. *J Cell Sci.* 2002;115:1099–1105.
106. Attar N, Cullen PJ. The retromer complex. *Adv Enzyme Regul.* 2010;50:216–236.
107. Cullen PJ, Korswagen HC. Sorting nexins provide diversity for retromer-dependent trafficking events. *Nat Cell Biol.* 2012;14:29–37.
108. Koumandou VL, Klute MJ, Herman EK, Nunez-Miguel R, Dacks JB, Field MC. Evolutionary reconstruction of the retromer complex and its function in *Trypanosoma brucei*. *J Cell Sci.* 2011;124:1496–1509.
109. Rojas R, van Vlijmen T, Mardones GA, et al. Regulation of retromer recruitment to endosomes by sequential action of Rab5 and Rab7. *J Cell Biol.* 2008;183:513–526.
110. Harbour ME, Breusegem SY, Antrobus R, Freeman C, Reid E, Seaman MN. The cargo-selective retromer complex is a recruiting hub for protein complexes that regulate endosomal tubule dynamics. *J Cell Sci.* 2010;123:3703–3717.
111. Bugarcic A, Zhe Y, Kerr MC, Griffin J, Collins BM, Teasdale RD. Vps26A and Vps26B subunits define distinct retromer complexes. *Traffic.* 2011;12:1759–1773.
112. Seaman MN, Harbour ME, Tattersall D, Read E, Bright N. Membrane recruitment of the cargo-selective retromer subcomplex is catalysed by the small GTPase Rab7 and inhibited by the Rab-GAP TBC1D5. *J Cell Sci.* 2009;122:2371–2382.
113. Johannes L, Wunder C. The SNXy flavours of endosomal sorting. *Nat Cell Biol.* 2011;13:884–886.
114. McGough IJ, Cullen PJ. Recent advances in retromer biology. *Traffic.* 2011;12: 963–971.
115. Harterink M, Port F, Lorenowicz MJ, et al. A SNX3-dependent retromer pathway mediates retrograde transport of the Wnt sorting receptor Wntless and is required for Wnt secretion. *Nat Cell Biol.* 2011;13:914–923.
116. Derivery E, Sousa C, Gautier JJ, Lombard B, Loew D, Gautreau A. The Arp2/3 activator WASH controls the fission of endosomes through a large multiprotein complex. *Dev Cell.* 2009;17:712–723.
117. Gomez TS, Billadeau DD. A FAM21-containing WASH complex regulates retromer-dependent sorting. *Dev Cell.* 2009;17:699–711.
118. Hong Z, Yang Y, Zhang C, et al. The retromer component SNX6 interacts with dynactin p150(Glued) and mediates endosome-to-TGN transport. *Cell Res.* 2009;19:1334–1349.

119. Wassmer T, Attar N, Harterink M, et al. The retromer coat complex coordinates endosomal sorting and dynein-mediated transport, with carrier recognition by the trans-Golgi network. *Dev Cell.* 2009;17:110–122.
120. Popoff V, Mardones GA, Bai SK, et al. Analysis of articulation between clathrin and retromer in retrograde sorting on early endosomes. *Traffic.* 2009;10:1868–1880.
121. Popoff V, Mardones GA, Tenza D, et al. The retromer complex and clathrin define an early endosomal retrograde exit site. *J Cell Sci.* 2007;120:2022–2031.
122. Shi A, Sun L, Banerjee R, Tobin M, Zhang Y, Grant BD. Regulation of endosomal clathrin and retromer-mediated endosome to Golgi retrograde transport by the J-domain protein RME-8. *EMBO J.* 2009;28:3290–3302.
123. Chin LS, Raynor MC, Wei X, Chen HQ, Li L. Hrs interacts with sorting nexin 1 and regulates degradation of epidermal growth factor receptor. *J Biol Chem.* 2001;276: 7069–7078.
124. Perez-Victoria FJ, Mardones GA, Bonifacino JS. Requirement of the human GARP complex for mannose 6-phosphate-receptor-dependent sorting of cathepsin D to lysosomes. *Mol Biol Cell.* 2008;19:2350–2362.
125. Pellegrini-Calace M, Tramontano A. Identification of a novel putative mitogen-activated kinase cascade on human chromosome 21 by computational approaches. *Bioinformatics (Oxford, England).* 2006;22:775–778.
126. Benjamin JJ, Poon PP, Drysdale JD, Wang X, Singer RA, Johnston GC. Dysregulated Arl1, a regulator of post-Golgi vesicle tethering, can inhibit endosomal transport and cell proliferation in yeast. *Mol Biol Cell.* 2011;22:2337–2347.
127. Siniossoglou S, Peak-Chew SY, Pelham HR. Ric1p and Rgp1p form a complex that catalyses nucleotide exchange on Ypt6p. *EMBO J.* 2000;19:4885–4894.
128. Salemme FR, Miller MD, Jordan SR. Structural convergence during protein evolution. *Proc Natl Acad Sci USA.* 1977;74:2820–2824.
129. Anantharaman V, Aravind L, Koonin EV. Emergence of diverse biochemical activities in evolutionarily conserved structural scaffolds of proteins. *Curr Opin Chem Biol.* 2003;7:12–20.
130. Alvarez CE. On the origins of arrestin and rhodopsin. *BMC Evol Biol.* 2008;8:222.
131. Skegro D, Pulvermuller A, Krafft B, et al. N-terminal and C-terminal domains of arrestin both contribute in binding to rhodopsin. *Photochem Photobiol.* 2007;83: 385–392.
132. Fjorback AW, Seaman M, Gustafsen C, et al. Retromer binds the FANSHY sorting motif in SorLA to regulate amyloid precursor protein sorting and processing. *J Neurosci.* 2012;32:1467–1480.
133. Vieira SI, Rebelo S, Esselmann H, et al. Retrieval of the Alzheimer's amyloid precursor protein from the endosome to the TGN is S655 phosphorylation state-dependent and retromer-mediated. *Mol Neurodegener.* 2010;5:40.
134. Shukla AK, Kim J, Ahn S, et al. Arresting a transient receptor potential (TRP) channel: beta-arrestin 1 mediates ubiquitination and functional down-regulation of TRPV4. *J Biol Chem.* 2010;285:30115–30125.
135. Simonin A, Fuster D. Nedd4-1 and beta-arrestin-1 are key regulators of Na^+/H^+ exchanger 1 ubiquitylation, endocytosis, and function. *J Biol Chem.* 2010;285: 38293–38303.
136. Andoh T, Hirata Y, Kikuchi A. PY motifs of Rod1 are required for binding to Rsp5 and for drug resistance. *FEBS Lett.* 2002;525:131–134.
137. Shenoy SK, Barak LS, Xiao K, et al. Ubiquitination of beta-arrestin links seven-transmembrane receptor endocytosis and ERK activation. *J Biol Chem.* 2007;282: 29549–29562.
138. Palczewski K, Pulvermuller A, Buczylko J, Gutmann C, Hofmann KP. Binding of inositol phosphates to arrestin. *FEBS Lett.* 1991;295:195–199.

139. van Weering JR, Verkade P, Cullen PJ. SNX-BAR proteins in phosphoinositide-mediated, tubular-based endosomal sorting. *Semin Cell Dev Biol.* 2010;21:371–380.
140. Min J, Defea K. Beta-arrestin-dependent actin reorganization: bringing the right players together at the leading edge. *Mol Pharmacol.* 2011;80:760–768.
141. Blanchin-Roland S. Identical domains of *Yarrowia lipolytica* Vps23 are required for both ESCRT and Rim pathways, but the latter needs an interaction between the Vps23 UEV domain and Rim8/PalF. *FEMS Yeast Res.* 2011;11:473–486.
142. Claing A, Chen W, Miller WE, et al. Beta-arrestin-mediated ADP-ribosylation factor 6 activation and beta 2-adrenergic receptor endocytosis. *J Biol Chem.* 2001;276: 42509–42513.
143. Defea K. Beta-arrestins and heterotrimeric G-proteins: collaborators and competitors in signal transduction. *Br J Pharmacol.* 2008;153(Suppl. 1):S298–S309.
144. Lin FT, Daaka Y, Lefkowitz RJ. Beta-arrestins regulate mitogenic signaling and clathrin-mediated endocytosis of the insulin-like growth factor I receptor. *J Biol Chem.* 1998;273:31640–31643.
145. Shenoy SK, Xiao K, Venkataramanan V, Snyder PM, Freedman NJ, Weissman AM. Nedd4 mediates agonist-dependent ubiquitination, lysosomal targeting, and degradation of the beta2-adrenergic receptor. *J Biol Chem.* 2008;283:22166–22176.
146. Mittal R, McMahon HT. Arrestins as adaptors for ubiquitination in endocytosis and sorting. *EMBO Rep.* 2009;10:41–43.
147. Hurley JH. The ESCRT complexes. *Crit Rev Biochem Mol Biol.* 2010;45:463–487.
148. Hurley JH, Emr SD. The ESCRT complexes: structure and mechanism of a membrane-trafficking network. *Annu Rev Biophys Biomol Struct.* 2006;35:277–298.
149. Schmidt O, Teis D. The ESCRT machinery. *Curr Biol.* 2012;22:R116–R120.
150. Galindo A, Calcagno-Pizarelli AM, Arst HN, Penalva MA. An ordered pathway for the assembly of ESCRT-containing fungal ambient pH signalling complexes at the plasma membrane. *J Cell Sci.* 2012;125:1784–1795.
151. Kendall RT, Luttrell LM. Diversity in arrestin function. *Cell Mol Life Sci.* 2009;66:2953–2973.
152. Saint-Pol A, Yelamos B, Amessou M, et al. Clathrin adaptor epsinR is required for retrograde sorting on early endosomal membranes. *Dev Cell.* 2004;6:525–538.
153. McCormick PJ, Dumaresq-Doiron K, Pluviose AS, Pichette V, Tosato G, Lefrancois S. Palmitoylation controls recycling in lysosomal sorting and trafficking. *Traffic.* 2008;9:1984–1997.
154. Brown J, Delaine C, Zaccheo OJ, et al. Structure and functional analysis of the IGF-II/IGF2R interaction. *EMBO J.* 2008;27:265–276.

CHAPTER THREE

Structural Determinants of Arrestin Functions

Vsevolod V. Gurevich, Eugenia V. Gurevich
Department of Pharmacology, Vanderbilt University, Nashville, Tennessee, USA

Contents

1. Introduction 58
2. What the Crystal Structure Reveals, and What It Does Not 59
3. How Do Arrestins Fit Receptors? 60
   3.1 Receptor-binding surface of arrestins 61
   3.2 The stoichiometry of the complex 63
   3.3 Phosphate-binding residues and the phosphate sensor 65
   3.4 The conformation of the receptor-bound arrestin 67
   3.5 Key determinants of receptor preference 69
4. Interactions with Other Signaling Proteins 72
   4.1 Where do the other partners bind? 72
   4.2 The shape of the arrestin–MAPK signaling complex 73
   4.3 Conformational preferences determine outcomes 73
   4.4 Binding and activation do not go hand-in-hand 76
5. Designing Signaling-Biased Arrestin Mutants 77
   5.1 Enhanced phosphorylation-independent mutants 78
   5.2 Constitutively monomeric arrestins 79
   5.3 Manipulating MAPK signaling 80
6. Conclusions: Where Do We Go from Here? 83
Acknowledgments 83
References 83

Abstract

Arrestins are a small protein family with only four members in mammals. Arrestins demonstrate an amazing versatility, interacting with hundreds of different G protein-coupled receptor (GPCR) subtypes, numerous nonreceptor signaling proteins, and components of the internalization machinery, as well as cytoskeletal elements, including regular microtubules and centrosomes. Here, we focus on the structural determinants that mediate various arrestin functions. The receptor-binding elements in arrestins were mapped fairly comprehensively, which set the stage for the construction of mutants targeting particular GPCRs. The elements engaged by other binding partners are only now being elucidated and in most cases we have more questions than answers.

Progress in Molecular Biology and Translational Science, Volume 118
ISSN 1877-1173
http://dx.doi.org/10.1016/B978-0-12-394440-5.00003-6

Interestingly, even very limited and imprecise identification of structural requirements for the interaction with very few other proteins has enabled the development of signaling-biased arrestin mutants. More comprehensive understanding of the structural underpinning of different arrestin functions will pave the way for the construction of arrestins that can link the receptor we want to the signaling pathway of our choosing.

1. INTRODUCTION

As far as size is concerned, arrestins are quite average, 44–48 kDa soluble proteins. Functionally, however, arrestins are far from being average in many ways, demonstrating that evolution can pack incredible versatility into ~400 amino acids.

The discovery of the first member of the arrestin family (modern systematic name arrestin-1[a]) was rather unremarkable, except that it was discovered twice: first as S-antigen, the target of auto-antibodies in uveitis,[1] then as a 48-kDa protein that binds light-activated rhodopsin,[2] preferring the phosphorylated form.[3] Eventually it became clear that both are the same protein, which prevents G protein activation by light-activated phosphorylated rhodopsin,[4] thereby blocking (arresting) further signaling.

All this happened before the seminal discovery of striking similarity in sequence and topology between the β2-adrenergic receptor (β2AR) and rhodopsin,[5] which led to the concept that there is a large family of G-protein-coupled receptors (GPCRs; also known as seven transmembrane domain receptors or 7TMRs) and fruitful ideas regarding the similarity of signaling and regulatory mechanisms in this family. The first nonvisual arrestin, termed β-arrestin because of its preference for the β2AR over rhodopsin, was cloned soon thereafter,[6] followed by another nonvisual subtype (termed β-arrestin2,[7] arrestin-3,[8] and hTHY-ARRX,[9] respectively) and cone-specific arrestin.[10,11] Considering that different vertebrate species express from 800 to >3400 distinct GPCRs (SEVENS database; http://sevens.cbrc.jp/), the fact that we only have four arrestin subtypes[12] is rather remarkable. Moreover, arrestin-1 and -4 are largely restricted to photoreceptors,[13] whereas the two nonvisual subtypes are ubiquitously expressed

[a] We use systematic names of arrestin proteins: arrestin-1 (historic names S-antigen, 48-kDa protein, visual or rod arrestin); arrestin-2 (β-arrestin or β-arrestin1); arrestin-3 (β-arrestin2 or hTHY-ARRX); and arrestin-4 (cone or X-arrestin; for unclear reasons, its gene is called *arrestin 3* in the HUGO database).

and interact with hundreds of different GPCRs.[12,14] Even this striking versatility is only half of the story—in addition to receptors, arrestins bind dozens,[12] and possibly hundreds,[15] of amazingly diverse proteins, serving as multifunctional signaling organizers in the cell (see Chapter 1).

2. WHAT THE CRYSTAL STRUCTURE REVEALS, AND WHAT IT DOES NOT

Visual arrestin-1 was the first subtype discovered,[2] functionally characterized,[4] cloned,[16] and crystallized.[17,18] The structure revealed a unique fold: an elongated molecule consisting of two cup-like domains with similar cores, each organized as a seven-strand β-sandwich (Fig. 3.1). Subsequently solved structures of arrestin-2,[19,21] arrestin-3,[20] and arrestin-4[22] and even the short splice variant of arrestin-1[23] showed rather disappointing similarity, offering surprisingly few clues regarding the structural underpinnings of the functional diversity of these proteins.

Several features revealed by X-ray crystal structure matched the predictions of previous mutagenesis studies surprisingly well. The N-terminal half of the molecule that was predicted to be a separable independently folding unit[24,25] turned out to be the N-domain[17,18] (Fig. 3.1). Predicted interaction between the N- and C-termini[26] was also revealed.[18] A molecule consisting of two domains with relatively few contacts between them appeared poised for a global conformational rearrangement predicted by the model of sequential multisite mechanism of receptor binding.[25] Finally, Arg175, predicted to interact with a negatively charged partner in arrestin and function as the phosphate sensor,[27,28] was found in an unusual (for a soluble protein) arrangement of solvent-excluded charged residues in the interface between the two domains, which was termed the "polar core".[18,29]

However, the structures solved thus far reveal only the basal conformation of all arrestin subtypes. Several lines of evidence suggest that the structure of "active" receptor-bound arrestin is likely to be quite different (reviewed in Refs. 30,31) and indicate that arrestins can assume yet another distinct conformation favorable for the binding to microtubules.[32–35] Both still remain to be elucidated.

Crystal structure of rhodopsin[36] and its subsequent refinements,[37–39] followed by a flurry of remarkably similar structures of nonvisual GPCRs,[40–54] raised another interesting question: how does arrestin, with the long axis of ~75 Å, fit GPCRs with a diameter of ~40 Å?

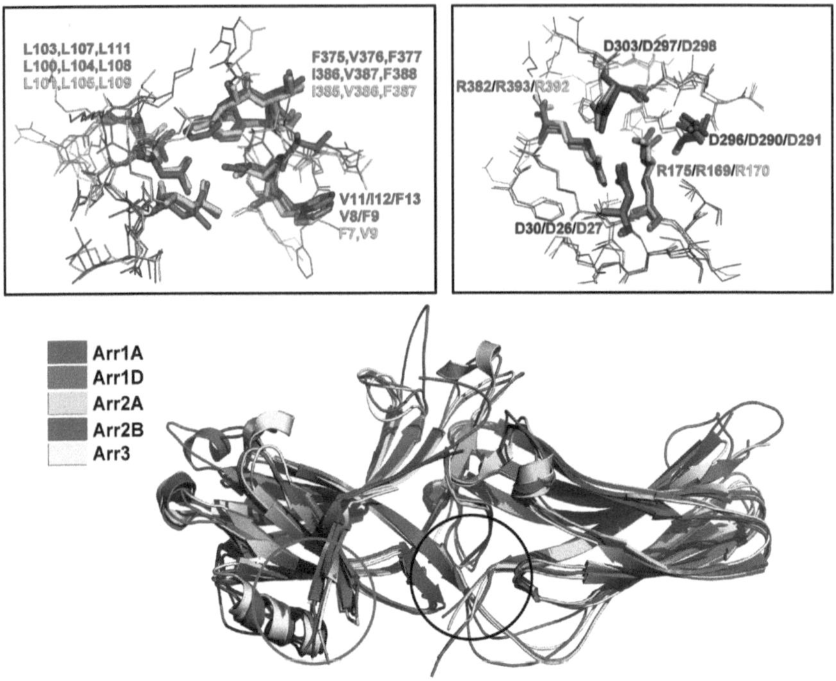

Figure 3.1 Basal conformation of different arrestin subtypes. Superimposition of the crystal structures of the two monomers in the arrestin-1 tetramer (Arr1A, red and Arr1D, pink),[18] two monomers of the arrestin-2 dimer (Arr2A, light blue and Arr2B, dark blue),[19] and arrestin-3 (Arr3, yellow)[20] shows remarkably similar cores of both domains and variable structure of the loops and inter-domain hinge. Importantly, the variability of these elements in different monomers of the same arrestin (compare Arr1A and Arr1D, as well as Arr2A and Arr2B) is essentially as great as between arrestin subtypes, suggesting that it reflects the flexibility of these loops, rather than their subtype-specific conformations. Black and red circles show the location of the two key intramolecular interactions that hold arrestins in their basal state, the polar core in the inter-domain interface, and the three-element interaction between β-strand I, α-helix I, and β-strand XX in the C-tail, respectively. The panels above show detailed structure of these elements revealing their extremely high conservation, down to the orientation of the side chains. *Right*: The polar core, main phosphate sensor. *Left*: The three-element interaction. In both panels, residue numbers of bovine proteins are indicated as follows: arrestin-1, red; arrestin-2, blue; arrestin-3, green. (See Color Insert.)

3. HOW DO ARRESTINS FIT RECEPTORS?

There is an obvious caveat in fitting known arrestin and receptor structures: for the complex to form, both arrestin and receptor must be in an active conformation, and the receptor also must be phosphorylated.[14,30]

While the effect of receptor-attached phosphates on its conformational state is completely unknown, activation-induced changes were well characterized, first by a series of site-directed spin-labeling studies of rhodopsin[55–59] then by the solution of several crystal structures of active forms of rhodopsin,[60–63] β2AR,[42] adenosine A2A receptor,[49,64] agonist-activated β2AR stabilized by nanobody,[65] and β2AR in a physiologically relevant complex with the G protein.[66,67] The common theme in GPCR activation is the outward movement of helices 5 and 6, and less dramatic rearrangements of other elements, which collectively open the cavity in the middle of the cytoplasmic side of the receptor, where the C-terminus of the α-subunit of their cognate G protein binds.[68–70] These movements increase the diameter of the receptor, making it larger than the short axis of arrestin, although still much smaller than its long axis (Fig. 3.2). To even begin meaningful fitting of an arrestin to a receptor, one needs to identify interacting residues in both proteins. While the arrestin-binding parts of the receptor have been identified imprecisely and only in a few GPCRs,[74–76] the receptor "footprint" on different arrestins was mapped fairly comprehensively by several groups using a variety of methods (Fig. 3.2).

3.1. Receptor-binding surface of arrestins

Receptor-binding residues in arrestins were identified using mutagenesis,[22,25–28,71,73,77] H/D exchange,[78] element swapping,[79,80] peptide competition,[81] epitope insertion,[82] solution NMR,[83–85] and site-directed spin-labeling EPR.[73,86] All these experiments yielded essentially the same answer: an extensive surface on the concave sides of both arrestin domains is involved (reviewed in Ref. 14) (Fig. 3.2). The distance between the extreme positions of arrestin residues implicated in receptor binding appears to be larger than even the expanded active receptor, at least the parts visible in crystal structures. However, we should not forget that spatial localization of the phosphates necessary for high-affinity arrestin binding *in vitro*,[25,80,87,88] in cells,[89] and in living animals[90,91] remains completely unknown. This is true for rhodopsin and the β2AR, where phosphorylation sites are localized in the C-terminus,[92–100] which was not resolved in any of the structures. This is equally true for M2 muscarinic cholinergic receptor, where phosphorylation sites are localized on the large third cytoplasmic loop,[101–103] which was deleted to obtain well-diffracting crystals.[45] Thus, we do not really know the size of the arrestin-binding surface of any GPCR.

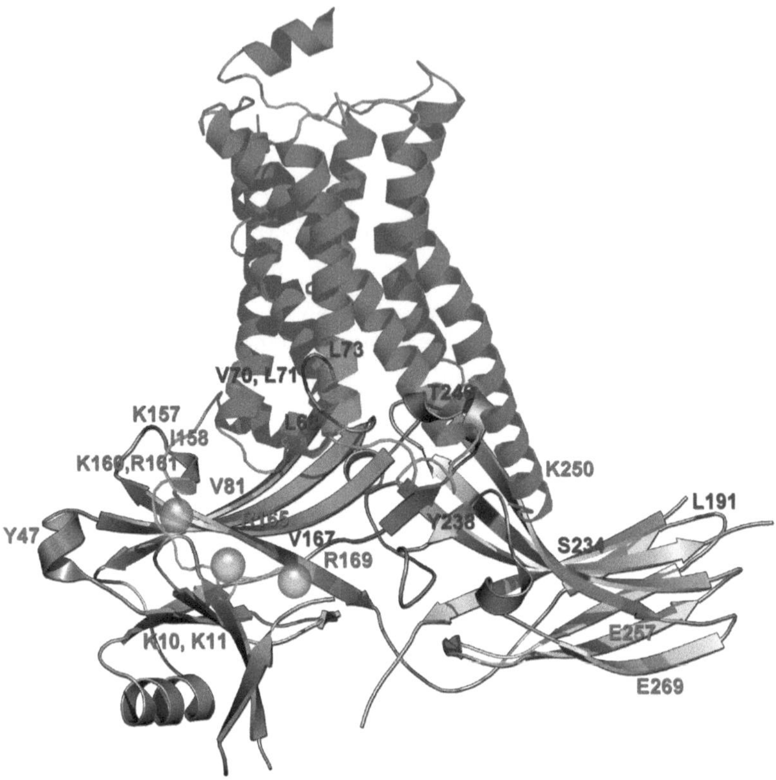

Figure 3.2 Comparison of the cytoplasmic side of active receptor and arrestin. Putative complex assembled from crystal structure of active β_2-adrenergic receptor (from the complex with G_s heterotrimer[44]) and arrestin-2.[19] Phosphate-binding residues and other elements that likely come into direct contact with receptor are shown in red and blue, respectively. Darker blue shows residues in positions where the mobility of the engineered spin label is dramatically decreased upon receptor binding, whereas lighter blue denotes smaller decreases in spin label mobility (based on Ref. 71). Residue numbers correspond to bovine arrestin-2 used in.[71] The comparison of these structures suggests that the receptor-binding surface of inactive arrestin-2 is greater than the cytoplasmic part of the receptor resolved in crystal. The receptor C-terminus (not resolved in any crystal structure) with attached phosphates (yellow spheres) was added manually to position the phosphates near known phosphate-binding positive charges in arrestin. The analysis of receptor-binding-induced conformational changes in arrestin-1[72] revealed very small shifts in relative positions of the two arrestin domains, moderate movement of the "finger loop" toward the receptor, large movement of the neighboring "139 loop" toward the N-domain and to the side (out of the way of incoming receptor), as well as the movement of two loops at the distal tips of the N- and C-domain toward the receptor. Collectively, these rearrangements would allow the finger loop to insert itself deep into the cavity between receptor helices that opens upon activation, and move the tips of the arrestin domains closer to the receptor. However, all contacts expected based on EPR studies of binding-induced changes of spin label mobility in arrestin-1[73] and -2[71] can only be readily achieved if the receptor helices move even further apart than they do in complex with G_s.[44] (See Color Insert.)

We also do not know the structure of the active receptor-bound arrestin so that the notion that the receptor and arrestin do not fit will be purely theoretical until the first structure of the arrestin–receptor complex is solved. Nonetheless, the idea that arrestins and the cytoplasmic tips of GPCRs do not fit has certain merits and remains quite popular. So far two elegant models have been proposed as possible solutions for this problem.

3.2. The stoichiometry of the complex

One model is based on the high Arrhenius activation energy of the arrestin–rhodopsin interaction[104] and subsequent finding that elements of arrestin-1 that do not interact either with light-activated (Rh*) or inactive phosphorylated rhodopsin (P-Rh) become engaged by the active phosphorylated form (P-Rh*).[25,27] Collectively, these data suggested that a substantial conformational change in arrestin, which brings additional parts into contact with the receptor, is necessary for high-affinity binding to P-Rh*.[25] The evidence that shortening of the inter-domain hinge impedes the binding of all arrestins to their cognate receptors,[35,105] suggested that this conformational change could be a clam-like movement of the two arrestin domains.[30] This type of movement could solve the problem of the misfit between the large receptor-binding surface of arrestin and the expected size of the arrestin-binding cytoplasmic side of GPCRs. Although this idea could still be considered plausible simply because it was not unambiguously refuted, there is no direct evidence for the large movement of the two domains relative to each other. Existing evidence suggests that while domain movement actually accompanies receptor binding, it is rather small,[72,106] certainly not big enough to significantly reduce the receptor-binding "face" of arrestin. However, detected movement of the loops on both distal tips of arrestin-1 toward the center of the receptor-binding side of the molecule likely contributes to the reduction of the surface of arrestin-1 that binds rhodopsin,[72] improving the fit in a different way (Fig. 3.2).

An alternative model attacks the problem even more radically, proposing that a single arrestin binds two GPCRs in a dimer.[107] Indeed, one can simultaneously fit two cytoplasmic tips of smaller GPCRs, such as rhodopsin, into the cavities of the two arrestin domains.[107] Unfortunately, to achieve this, one needs to disregard the receptor C-terminus, which is not visible in crystals, even though it is certainly involved in arrestin binding,[26,28,90,108] and ignore the fact that inactive rhodopsin does not bind arrestin with high affinity.[25,80,109] However, this model was proposed without mentioning these obvious caveats and has gained certain popularity, despite complete lack of supporting evidence. Unfortunately, it also did not survive experimental testing. The model predicts that the saturation point would be achieved upon binding of one arrestin molecule to two molecules of rhodopsin. This prediction was tested in mice *in vivo*.[110] In rod photoreceptors in the dark, arrestin-1 is largely localized away from the outer segment (OS), where all rhodopsin resides. In contrast, in bright light, the bulk of arrestin-1 moves to this compartment,[32,111–114] where it remains due to high-affinity binding to rhodopsin.[32] Considering that the expression ratio of arrestin-1 to rhodopsin in mouse rods is ~0.8:1,[110,115–117] if this model were true, one would never expect virtually quantitative translocation of arrestin-1 to the OS, which was reproducibly observed by many labs.[32,111–114] Genetic manipulation of arrestin-1 and rhodopsin expression levels in mouse rods revealed that the amount of arrestin-1 that can move to the OS in the light is, indeed, limited by the amount of rhodopsin there, but that saturation is achieved at the ratio of translocated arrestin-1 to rhodopsin that is greater than 0.8:1, which is consistent only with 1:1 binding model.[110] Obviously, in the photoreceptors of live mice, one cannot exclude the role of other proteins in arrestin-1 translocation. However, a variety of arrestin-1:rhodopsin ratios were tested *in vitro* using two carefully quantified pure proteins, which again yielded saturation at ~1:1.[110] Since arrestin-1 readily self-associates, forming dimers and tetramers,[118–121] this binding ratio could have been explained by an interaction of an arrestin-1 dimer with a rhodopsin dimer. However, it was shown that only monomeric arrestin-1 binds rhodopsin[120] because its receptor-binding surface is shielded by sister subunits in the arrestin-1 tetramer and both possible dimers.[122] Finally, monomeric rhodopsin was reconstituted into nanodiscs (HDL particles containing membrane-like lipid bilayer) and shown to bind arestin-1 not just efficiently,[123] but with physiologically relevant high affinity ($K_D \sim 3$–4 nM) and 1:1 stoichiometry.[124]

Thus, while the problem exists, neither model proposed so far has withstood the rigors of experimental testing. It appears very likely that in real life a single molecule of activated phosphorylated receptor fits arrestin in its active conformation well enough, but we do not know how exactly this fit is achieved. Two new ideas, which are not mutually exclusive, are suggested by the available evidence. On the arrestin side, unexpectedly large movement of the "139 loop" in the central crest, apparently out of the way of the incoming receptor,[72] likely allows the "finger loop" to insert itself fairly deeply into the cavity opening in the middle of the active receptor[69] (Fig. 3.2). This would result in extensive contacts between the cytoplasmic loops and extended helices of the receptor and the cavities of both arrestin domains. In addition, the loops at the tips of both arrestin-1 domains move toward the receptor,[72] possibly far enough to achieve direct contact with it, as suggested by reduced mobility of several residues in these loops in the complex.[71,73] As far as the receptor is concerned, the comparison of the structures of the same active β2AR in complex with an agonist,[42] nanobody mimicking G protein,[65] and cognate heterotrimeric G protein[66] shows that, with activation, the receptor helices on the cytoplasmic side progressively move further apart. Thus, it is entirely possible that in complex with arrestin the helices move out even more, thereby increasing the size of the cytoplasmic tip of the receptor to better accommodate arrestin. Naturally, these are no more than plausible speculations and should be viewed as such. Ultimately, the issue of the arrestin–receptor fit can only be definitively resolved by the elucidation of the structure of the complex.

3.3. Phosphate-binding residues and the phosphate sensor

The fact that arrestins preferentially bind phosphorylated forms of their cognate receptors was established early on, but the first model explaining how arrestin "selects" active phosphorylated receptors from among at least four coexisting forms (active and inactive, both of which can be unphosphorylated or phosphorylated) was proposed in 1993.[25] Specific binding to inactive phosphorylated (P-Rh) and light-activated unphosphorylated rhodopsin (Rh*) showed that arrestin-1 has interaction sites that recognize rhodopsin-attached phosphates and the active state of rhodopsin independently of each other.[25] However, the binding to light-activated phosphorhodopsin (P-Rh*) was many times greater than either to P-Rh or Rh*,[24,25] which cannot be explained by a simple cooperative two-site interaction. This led to the idea that primary binding sites engaged by Rh* and inactive P-Rh also serve as sensors. Only P-Rh* can engage both at the same time, suggesting that arrestin-1 acts as a molecular coincidence detector, where simultaneous activation of these two sensors triggers a global conformational change, which brings additional arrestin-1 elements in contact with rhodopsin, greatly increasing the energy of the interaction and therefore observed binding (reviewed in Ref. 30). This model predicts that among phosphate-binding residues in arrestin-1 and other family members, there must be at least one that not only contributes to the interaction but also serves as a sensor. The elimination of positive charges that simply bind phosphates was expected to reduce arrestin binding to P-Rh*, and even to a greater extent to inactive P-Rh, where the phosphates must be driving the interaction. In contrast, the neutralization or reversal of the charge of the putative phosphate sensor was expected to turn it on, enhancing the binding to unphosphorylated Rh*.

Since high-affinity arrestin binding was shown to require multiple phosphates on rhodopsin,[25] one would expect the multiphosphorylated and therefore highly negatively charged rhodopsin C-terminus to interact with several closely spaced positively charged residues in arrestin. The first scanning mutagenesis performed targeted a 14-amino acid stretch of the primary sequence carrying six positive charges (residues 163–176).[27] Considering that it was done long before crystal structure became available, this attempt proved amazingly well targeted: five out of six mutations reduced P-Rh and P-Rh* binding. The sixth, Arg175Asn, showed even more interesting phenotype: somewhat enhanced binding to P-Rh* and a dramatic ~fivefold increase in Rh* binding.[27] This is consistent with the idea that Arg175 serves as a phosphate sensor, which was artificially activated by charge neutralization, "fooling" arrestin-1 into perceiving any active form of rhodopsin as phosphorylated. However, the data with full-length protein did not prove one key point, that Arg175 actually interacts with phosphates, because the binding of Arg175Asn mutant to inactive P-Rh and P-Rh* was also increased.[27] Luckily, the N-terminal half of arrestin-1, residues 1-191 containing Arg175, was previously shown to be functional and act as an unsophisticated version of arrestin; it interacts P-Rh and Rh*, and its binding to P-Rh* is essentially the sum of the two.[24,25] In the context of this mini-arrestin, the Arg175Asn mutation reduced the binding to P-Rh and P-Rh* just like the other mutations, proving that Arg175 actually engages rhodopsin-attached phosphates.[27] Replacement of Arg175 with all 19 alternative residues showed that positive charge is the key: Arg175Lys retained high selectivity for P-Rh*, whereas all other substitutions showed enhanced binding to Rh*, with the charge reversal mutation Arg175Glu having the most potent effect.[28] The simplest interpretation of these data was that Arg175 interacts with a negatively charged partner within arrestin, and the breaking of this salt bridge by negatively charged phosphates turns the sensor on, allowing arrestin transition into high-affinity receptor-binding state.[27,28] Later, the crystal structure revealed that Arg175 is part of solvent-excluded arrangement of five charged residues between the two arrestin domains (Fig. 3.1), identifying three possible negatively charged partners.[17,18] Further mutagenesis proved that the salt bridge between Arg175 and Asp296 is the key phosphate sensor: charge reversal mutations of either greatly increase Rh* binding, whereas simultaneous reversal of both, restoring the salt bridge, also restores normal arrestin-1 selectivity for P-Rh*.[29] Interestingly, these five charges are conserved in arrestin evolution from *C. elegans* to mammals,[12] attesting to their important role in arrestin function. Virtually identical positions of all five side chains in different subtypes of mammalian arrestins (Fig. 3.1) support this notion. Charge reversals of homologous arginines in arrestin-2 and -3 also yield enhanced phosphorylation-independent mutants,[125–130] demonstrating that this residue plays the same role in all arrestins.

Further mutagenesis identified additional phosphate-binding elements in arrestins. Arg18 in the loop between β-strands I and II is unique for arrestin-1,[22] likely making is much more dependent on receptor-attached phosphates than the other subtypes.[89] In contrast, two lysines in β-strand I are present in all arrestins.[12] These charged residues appear to be critical for the "delivery" of phosphates to the shielded polar core.[77] Importantly, they are adjacent to the bulky hydrophobic residues in β-strand I that participate in its interactions with the arrestin C-tail and α-helix I (Fig. 3.1).[18] This suggested that their interaction with phosphates likely disrupts this three-element interaction, which would destabilize the basal arrestin conformation, similar to the effect of the disruption of the polar core.[77] These data support the main premise of the sequential multisite binding model of the arrestin–receptor interaction[25] that receptor binding is accompanied by a global conformational change in arrestin.

The action of the phosphate sensor is based on pure electrostatics; all that receptor-attached phosphates need to do to activate the sensor is to break the key salt bridge.[30] This makes it essentially insensitive to the sequence context of the phosphorylated residues. This mechanism explains how just two nonvisual arrestins in mammals and only one in *Drosophila* can interact with hundreds of different GPCRs, in which serines and threonines phosphorylated by GRKs are found within diverse sequences that can be localized in the receptor C-terminus, or any of the intracellular loops (reviewed in Ref. 14).

3.4. The conformation of the receptor-bound arrestin

Several lines of indirect evidence suggested that the conformation of receptor-bound arrestins is likely quite different from their basal state revealed by crystal structures. The first indication that this must be the case was unusually high-energy barrier of arrestin-1 binding to rhodopsin.[104] Receptor-binding-induced release of the arrestin C-tail has been well documented for more than 20 years.[73,86,109] In addition, the movement and/or structural rearrangement of the "finger loop" in the central crest of the receptor-binding side was also reported.[83,131] Both polar core and the three-element interaction clearly support the basal conformation in all arrestin structures,[18–20,22] so their destabilization by the phosphates was consistent with the idea that global rearrangement is necessary for receptor binding. Partial destabilization of the interface between the two domains enhanced arrestin binding to inactive receptor,[132] again suggesting that arrestin conformation must change upon receptor binding.

Arrestin interaction with nonreceptor partners also appears to be consistent with the idea that the conformations of free and receptor-bound arrestins must be different.[31] C-Raf1 and especially ERK1/2 preferentially interact with receptor-associated arrestins.[133] In contrast, the ubiquitin ligases Mdm2 and parkin[134,135] strongly prefer arrestins in a conformation induced by hinge deletions that impairs receptor binding.[35,105,136] JNK3 also appears to prefer this conformation, although the difference in binding is less dramatic.[134,136]

Thus, it appeared almost certain that the conformation of receptor-bound arrestin is significantly different from the basal one observed in crystal structures, but direct evidence was missing. A recent study employing site-directed spin-labeling and long-range distance measurements using pulse EPR technique double electron–electron resonance (DEER) yielded the first experimental data on conformational rearrangements in arrestin-1 beyond the release of the C-tail.[72] More than 25 distances between different residues in free and rhodopsin-associated arrestin-1 were measured. Significant changes in multiple distances combined with molecular modeling revealed binding-induced movements of several arrestin-1 elements. Some of the findings supported earlier predictions, whereas others were rather unexpected. Flexible "finger loop" (residues 67–79)[73] (Fig. 3.1) in the central crest of arrestins exist in fully extended or bent conformation in different protomers in crystal oligomers.[18,19] Multiple residues in this loop were shown to be immobilized upon receptor binding in both arrestin-1[73] and -2.[71] Previous studies using fluorescent labels[131] and NMR[83] suggested that this loop extends and forms an α-helix upon receptor binding. Indeed, this loop was found to move in the direction of the receptor, although not as much as previously proposed,[131] and the data were consistent with its helical conformation in receptor-associated arrestin-1.[72] However, hypothetical movement of the two arrestin domains relative to each other, which was proposed to improve the fit between arrestins and GPCRs,[30] turned out to be very small, clearly insufficient to significantly reduce the size of the receptor-binding arrestin surface.[72] Two other plastic loops containing residues 157 and 344, localized at the tips of the N- and C-domain, respectively, that had different conformations in crystal strcutures,[18] were found to move "inward," in the direction of the cavities of their respective domains. These movements also slightly reduce the long axis of the receptor-binding surface, but only by a few angstroms. The most unexpected finding was a dramatic movement of the loop with residue 139 at its tip, which is adjacent to the finger loop in the basal state. This element shifts by more than 10 Å, moving in the direction of the N-domain and to the side of the molecule.[72] This movement would take it out of the way of incoming receptor. Consistent with this model, spin label in position 139 was immobilized by inactive P-Rh, but in high-affinity complex with P-Rh*, its mobility increased to the level observed in free arrestin.[73] Moreover, deletions in this loop, taking it out of the way without movement, increased arrestin-1 binding to rhodopsin, most dramatically to the nonpreferred forms Rh* and inactive P-Rh.[72] Interestingly, the same deletions reduced the thermal stability of arrestin-1.[72] Thus, it appears that the 139-loop structurally stabilizes the basal arrestin conformation and serves as a "brake", precluding its binding to any form of rhodopsin except its preferred target, P-Rh*.

While this study clearly revealed multiple receptor-binding-induced rearrangements in arrestin-1, which are likely similar to those in nonvisual arrestins, biophysical methods cannot yield detailed atomic resolution structure of the receptor-bound arrestin. Crystal structure of the arrestin–receptor complex is necessary to obtain this information and clearly reveal the changes in arrestin molecule that underlie conformational preferences of nonreceptor binding partners.

3.5. Key determinants of receptor preference

As a general rule, arrestins preferentially bind active phosphorylated forms of their cognate receptors. Obviously, receptor-attached phosphates that activate the phosphate sensor are the common theme, so arrestin elements that bind other nonphosphorylated parts of the receptor in response to activation-induced conformational change must be responsible for receptor specificity of arrestin proteins. Two out of four arrestin subtypes in mammals are specialized and expressed primarily in photoreceptors. Interestingly, the specificity of cone arrestin-4 for cone opsins appears to be largely ensured by its selective expression in cone photoreceptors.[137] *In vitro*, it is quite promiscuous, binding other GPCRs almost as well as nonvisual arrestins,[22] both of which readily interact with dozens, if not hundreds, of different receptors.[80,138,139] Thus, evolution created only one receptor-specific subtype, arrestin-1, which demonstrates clear preference for rhodopsin over other receptors.[80,140]

Arrestin-1 binds rhodopsin efficiently and demonstrates fairly low binding to M2 muscarinic receptor, whereas arrestin-2 has the opposite preference.[80,140] Therefore, the first attempt to identify arrestin elements that determine its receptor specificity involved the construction of a series of arrestin-1/2 chimeras and testing their ability to bind these two receptors.[79] The premise of these experiments was that if an element is important for receptor preference, the introduction of that part from arrestin-2 into arrestin-1 would decrease rhodopsin and increase M2 binding, whereas the introduction of the corresponding arrestin-1 part into arrestin-2 would decrease M2 and increase rhodopsin binding. N-domain residues 49–90 (β-strands V and VI with adjacent loops) and C-domain residues 237–268 (β-strands XV and XVI) of arrestin-1 and homologous parts of arrestin-2 were found to play key role in receptor preference. The exchange of these two elements between arrestin-1 and -2 completely reversed receptor specificity of both, creating one chimera that was >90% arrestin-1, yet bound M2 muscarinic receptor much better than rhodopsin, and a symmetrical chimera with >90% of arrestin-2 residues that clearly preferred rhodopsin to M2.[79]

Fewer than 25 residues in these two elements are nonconservative substitutions. Their individual contributions to receptor preference were tested by introducing point mutations.[71] This approach led to the identification of 10 exposed side chains that collectively determine which receptor an arrestin protein prefers. Interestingly, the replacement of all 10 with alanines in arrestins-1, -2, and -3 yielded mutants that virtually lost the ability to bind any GPCRs.[71] One of the nonexposed residues in the N-domain element, Val90, was found to play an unexpectedly important role; its substitution with serine (Ser86 is the homologous arrestin-2 residue) increased arrestin-1 binding to M2 muscarinic receptor more than any other point mutation.[19] Arrestin-3, with similarly broad receptor specificity, has Ala87 in the same position,[20] which suggests that a small side chain is important. Crystal structure shows that arrestin-1 Val90, localized between the two β-sheets, makes contacts with several other hydrophobic residues, apparently stabilizing the core of the N-domain.[18] All potential partners are present in arrestin-2[19] and -3,[20] but they do not contact the smaller side chain of Ser86 or Ala 87. This analysis indicates that the presence of the bulky Val90 increases the rigidity of the N-domain, apparently enhancing receptor specificity of arrestin-1, whereas the smaller side chains in arrestin-2 and -3, as well as in the arrestin-1-V90S mutant, make the N-domain more flexible, which appears to correlate with broad receptor specificity. These results suggest that to construct a nonvisual arrestin with narrow receptor preference, one needs to place valine in this position.

The fact that the same 10 exposed residues determine receptor specificity and largely drive the arrestin–receptor interaction[71,89] suggests that the targeted manipulation of these side chains has the potential to yield nonvisual arrestin mutants with much narrower receptor specificity than their parental proteins. Since arrestins that can selectively target particular GPCRs would be very useful research tools and have clear therapeutic potential,[141,142] this idea was recently tested experimentally.[143] Ten different amino acid substitutions affecting 8 out of 10 identified receptor discriminator residues were introduced on the background of an arrestin-3-A87V base mutant. The binding of WT arrestin-3, A87V, and other mutant arrestins to β2AR, M2 muscarinic, as well as D1 and D2 dopamine receptors was tested in cells using a BRET-based assay.[143] Seven out of 10 mutations resulted in differential changes in the binding to these receptors, yielding up to fourfold increase in GPCR selectivity over WT protein.[143] This unexpectedly high success rate clearly indicates that correct targets were chosen to change the receptor preference of arrestin-3. The combination of two mutations that individually reduced β2AR binding, but did not affect the interactions with M2 or D2 receptors, yielded a variant with more than 50-fold preference for these receptors over β2AR. Similarly, two other mutations were shown to act additively; their combination yielded a version of arestin-3 with >fivefold preference for D1 over D2 receptor.[143]

Numerous GPCR mutations have been shown to underlie a variety of human disorders (reviewed in Ref. 144). Currently, there are no viable approaches to counteract the effects of gain-of-function mutations that result in excessive receptor signaling. Arrestins with enhanced ability to quench this signaling, such as the phosphorylation-independent mutants described above, have a potential to become a solution to this problem. However, WT nonvisual arrestins have very broad receptor specificity, so the introduction of an enhanced arrestin-2 or -3 will likely simultaneously blunt the signaling by the "bad" mutant receptor and numerous other perfectly normal GPCRs expressed in the same cell. Therefore, arrestin variants selectively targeting the overactive receptor are needed for therapeutic use in combination with other mutations changing its functional capabilities. This study was the first attempt to create something that evolution did not; a nonvisual arrestin with high receptor selectivity. Both the 70% success rate in changing receptor preference and additive effects of individual mutations show that it opened a promising new direction. However, a lot of additional experimentation is necessary to construct the arrestins that fit every GPCR whose signaling needs to be corrected to cure a particular human disorder. The same approach can be used to rein in excessive signaling by normal GPCRs that develops for various reasons in numerous human disorders.

4. INTERACTIONS WITH OTHER SIGNALING PROTEINS

Clathrin was the first nonreceptor-binding partner of arrestin proteins identified.[145] Since then arrestins were shown to interact with an amazing variety of trafficking and signaling proteins (see Chapter 1). The molecular mechanisms of most of these interactions remain to be elucidated. By comparison, it appears that we know a lot about the mechanics of receptor binding, although even in this area there are more questions than answers.

4.1. Where do the other partners bind?

Only two proteins have been shown to interact with the receptor-binding surface of arrestins: microtubules[35] and Ca^{2+}-liganded calmodulin.[146] In both cases, receptors bind arrestin with much higher affinity, easily winning the competition with these partners.[32,146] Most other proteins interact with receptor-bound arrestins,[145,147–150] which suggests that they engage nonreceptor-binding elements localized either on the convex sides of the two domains, or on the arrestin C-tail that is released upon receptor binding.

Considering the potential biological importance of these interactions, surprisingly few binding sites of nonreceptor partners have been identified, and most of those very imprecisely. The best characterized among these are clathrin and AP2 interaction sites, both of which are localized in the C-tail of arrestin-2 and -3. The main clathrin site is a short LIELD or LIEFE motif in the part of the C-tail upstream of its contact with the N-domain,[151] which is not visible in any of the crystal structures of free arrestins, likely because it is inherently disordered.[19–21] This interaction was resolved in the cocrystal of arrestin-2 with the clathrin β-propeller domain, where this arrestin element binds between blades 1 and 2.[152] Interestingly, in the prevalent long splice variant of arrestin-2,[8] another C-tail element also interacts with clathrin, binding between blades 4 and 5.[152] The AP2 binding site was localized downstream of the N-domain contact in the distal C-tail, by extensive mutagenesis[147,153] and cocrystallization of an arrestin peptide with the AP2 beta-appendage.[154]

No other partners have so far been cocrystallized with arrestins or even their elements. Only the binding sites of microtubules,[35] calmodulin,[146] and cAMP phosphodiesterase PDE4D5[155] were mapped with any degree of precision. Particular residues in arrestin-2 were implicated in binding of MEK1[156] and c-Raf1,[157] and several residues in arrestin-3 were found to be critical for its ability to promote JNK3 activation.[158] Thus, the structural basis of arrestin interactions with the great majority of putative partners still remains to be elucidated.

4.2. The shape of the arrestin–MAPK signaling complex

The first reports of arrestin-mediated activation of MAP kinases suggested that in each MAPKKK–MAPKK–MAPK cascade only MAPKKKs (ASK1 and c-Raf1) and MAPKs (JNK3 and ERK1/2) directly bind arrestins, whereas MAPKKs (MKK4 and MEK1) are recruited via interactions with corresponding MAPKKKs and/or MAPKs.[148,149] This model further suggested that MAPKKKs bind to the N-domain of arrestins, whereas MAPKs interact with the C-domain.[159] Subsequently, two MAPKKs, MEK1[156] and MKK4,[160] were shown to bind arrestins directly in the absence of MAPKKKs or MAPKs. Moreover, separately expressed arrestin N- and C-domains were shown to interact with all three kinases in c-Raf1–MEK1–ERK1/2 and ASK1–MKK4–JNK3 cascades.[161,162] This led to an alternative model of the arrestin complex with MAP kinases, where MAPKKK, MAPKK, and MAPK have bipartite binding sites, each interacting with both arrestin domains,[162] so that the three kinases are arranged on arrestin-like three hotdogs on a single bun.[141] Considering that receptor binding induces a conformational change in arrestins, the dependence of arrestin scaffolding function on GPCR activation can be easily explained by this arrangement.

4.3. Conformational preferences determine outcomes

The first reports of arrestin binding to nonreceptor partners focused on GPCR-associated arrestins,[145,147–149] creating an impression that these interactions are contingent on arrestin binding to receptors. This implied that only "active" receptor-bound arrestin can engage other proteins, which turned out not to be the case. In fact, accumulating evidence suggests that most arrestin-binding partners interact with arrestins in all three conformations, basal, receptor-bound, and microtubule-associated, but their conformational preferences vary widely.

ERK1/2 appears to be an extreme case; it interacts with reasonable affinity only with receptor-bound arrestins, whereas its transient interaction with free arrestins is not even detectable by coimmunoprecipitation without cross-linking.[133,149] C-Raf1 also prefers receptor-bound form, although the difference in binding is not as dramatic.[133] In contrast, MEK1 not only binds free arrestins[156] but also does not discriminate among the three conformations.[133] Unexpectedly, it was shown that both ERK1/2 and c-Raf1 bind arrestins in a conformation preferred by microtubules (which is mimicked by a 7-residue deletion in the inter-domain hinge, termed D7[35]) better than the basal state of free WT arrestins.[133] In cells, arrestins-1, -2, and -3 all bind ERK1/2 and recruit it to microtubules, reducing the overall ERK1/2 activation level, whereas arrestin-4 does not have this effect.[35] Moreover, arrestin-3-D7 effectively suppresses—ERK1/2 phosphorylation induced by receptor activation, which makes it a dominant-negative form in this regard.[136] The facilitation of signaling in c-Raf1–MEK1–ERK1/2 cascade appears to be strictly dependent on arrestin binding to GPCRs,[133,136,157] as originally hypothesized.[149] Therefore, mutant arrestins with impaired receptor binding that do interact with ERK1/2 normally recruit it away from places where it can be activated, blunting ERK1/2 activity.[35,136]

Arrestin-3, but not the other subtypes, promotes signaling in the ASK1–MKK4–JNK3 cascade, and this function was originally ascribed to the receptor-bound form.[148] However, subsequent studies showed that free arrestin-3 effectively facilitates JNK3 phosphorylation[136,158,159,162] and even localizes JNK3 to the cytoplasm independent of GPCR binding.[134,136,163] An assay based on the ability of arrestins to move JNK3 from the nucleus to the cytoplasm did not detect any differences among WT arrestins, their preactivated phosphorylation-independent forms where the C-tail is forcibly detached by a triple alanine substitution (3A), and D7 mutants with impaired receptor binding.[134] However, more a quantitative BRET-based assay in cells showed that arrestin-3–D7 actually binds JNK3 better.[136]

The ubiquitin ligase Mdm2 was first reported to interact with receptor-bound arrestins.[150] However, the comparison of the ability of different forms of arrestins to remove Mdm2 from the nucleus showed that D7 mutants do it more efficiently than WT, whereas preactivated 3A mutants are the least effective,[134,161] indicating that Mdm2 prefers free over receptor-bound arrestins. Yet arrestin ubiquitination by Mdm2 is clearly stimulated by receptor binding,[135,150,164] suggesting that receptor-bound form is a better substrate for Mdm2. Collectively, these data suggest that Mdm2 binds free arrestins and is recruited to the receptor, where it has limited time to ubiquitinate arrestin before falling off due to reduced binding affinity. This model explains limited and fairly selective ubiquitination of arrestins by Mdm2 in response to receptor activation.[164] Another E3 ubiquitin ligase, parkin, also shows strict preference for the basal and D7 forms of arrestins over 3A mutants partially mimicking receptor-bound state.[135] Interestingly, in contrast to Mdm2, parkin appears to be able to shift the conformational equilibrium of arrestin to the form it prefers.[135] Parkin greatly increases the binding of Mdm2 to WT arrestins, but not to D7 mutants that show enhanced Mdm2 interaction, suggesting that parkin acts by stabilizing the D7-like conformation.[135] We found that parkin suppresses Mdm2-dependent arrestin ubiquitination.[135] This is consistent with the idea that parkin stabilizes a basal-like arrestin conformation, whereas receptor-bound arrestins are better Mdm2 substrates.

Even though they appear almost inseparable in evolution (see Chapter 2),[12] the two arrestin domains are independent folding units that can be expressed separately and retain certain functions.[24–26,80] Therefore, the expression of separated domains has been repeatedly used to determine which part of arrestin contains the binding sites for different partners. Interestingly, whenever the localization of binding sites was tested by this method or using peptide arrays, it was found that the protein of interest binds both arrestin domains. This was shown for microtubules,[35] the MAP kinases JNK3,[161] ASK1,[162] MKK4,[162] c-Raf1,[162] MEK1,[162] ERK2,[162] the cAMP phosphodiesterase PDE4D5,[155] as well as the ubiquitin ligases Mdm2[135,161] and parkin.[135] Thus, with very few possible exceptions, the interaction with both arrestin domains appears to be a general rule. Even though receptor binding does not induce a large movement of the two arrestin domains, the domains actually shift relative to each other.[72] Therefore, this rearrangement can explain conformational preference of certain partners; the relative positions of the two parts of the binding site localized on different domains actually change. The distance between these two parts and/or their relative orientation can become favorable or unfavorable for the interaction in one of the conformations.[31]

It is harder to reconcile the bipartite binding site with the observations that some proteins, such as MEK1, appear to bind arrestins in all conformations equally well,[133] unless the interaction with one of the domains predominates, as some data suggest.[156] It is even harder to explain mechanistically how ERK1/2 can prefer both receptor-bound and D7-like conformation over basal; existing data suggest that the arrestin domains shift in opposite directions in these two states.[34,35] It is becoming increasingly clear that without direct structural information obtained by cocrystallization of arrestins with different partners these questions will remain unanswered.

4.4. Binding and activation do not go hand-in-hand

The original explanation for why arrestin-3 promotes JNK3 phosphorylation while the closely related arrestin-2 does not was beautifully simple; it was suggested that arrestin-2 does not bind this MAP kinase,[148] and this idea appeared to be supported by the identification of putative JNK3-binding sequence RRS present in the C-domain of arrestin-3, whereas arrestin-2 has KP in homologous position.[159] However, RRS is only found in rodent arrestin-3, while the prevalent sequence in mammalian species is RS.[12] Subsequent studies showed that arrestin-3 from other species also binds JNK3[134,163] and effectively facilitates its activation.[136,158,162] In fact, all four mammalian arrestins were found to bind JNK3 comparably and remove it from the nucleus, where it spontaneously localizes, to the cytoplasm.[134,161] This was the first indication that JNK3 binding and its activation are two distinct functions of arrestins. The idea that arrestin-2 might not bind the upstream kinases ASK1 and MKK4, and therefore fails to assemble the complete MAPK module necessary for JNK3 activation, also was not supported by the evidence; both nonvisual arrestins appeared to interact with all three kinases comparably in cells,[162] although subsequent more precise direct binding experiments with purified proteins suggested that the affinity of arestin-3 for JNK3 and MKK4 is higher than that of arrestin-2.[160] Since to assemble a productive complex arrestin must hold all three kinases simultaneously, even a subtle reduction in affinity for two of them could make a qualitative difference in the cellular environment, where the concentrations of all proteins involved are fairly low.

Arrestins-2 and -3 are 78% identical (and 88% similar),[8] both bind ASK1, MKK4, and JNK3,[134,162] yet only arrestin-3 facilitates JNK3 phosphorylation in cells.[148,158,162] Thus, arrestin-3-specific residues must be responsible for its unique ability to promote JNK3 activation. This idea was tested by the construction of arrestin-2/3 chimeras.[158] It turned out that virtually every chimera was less efficient than WT arrestin-3,[158] indicating that it is much easier to destroy a function than to build it. Although these data were consistent with earlier conclusion that each kinase in the ASK1–MKK4–JNK3 module has multipartite site, interacting with both arrestin domains,[162] this approach identified the C-domain as the key element in JNK3 activation.[158] Interestingly, this region included residues 182–376, encompassing the Arg196–Ser197 sequence in bovine arrestin-3 that is replaced with Lys195–Pro196 in arrestin-2. To further dissect the key arrestin-3 elements necessary for JNK3 activation, the residues unique for this subtype were replaced with their arrestin-2 homologues.[158] Quite a few mutations, including RS→KP, had no detectable effect, ruling out the role of those residues in JNK3 activation. Multiple other substitutions reduced arrestin-3 potency in JNK3 activation assay.[158] This study identified Val343 as the most important residue, as a Val343Thr point mutation reduced the ability of arrestin-3 to activate JNK3 by two-thirds. Residues Leu278, Ser280, His350, Asp351, His352, and Ile353 were found to play supporting roles, as their substitution with arrestin-2 homologues also reduced the ability of arrestin-3 to facilitate JNK3 phosphorylation.[158] Interestingly, virtually all arrestin-3 mutants with reduced or even abolished ability to activate JNK3 demonstrated essentially normal binding to all three kinases in this module, supporting the idea that the interaction with these kinases and the ability to promote JNK3 activation are different arrestin-3 functions that can be separated by targeted mutagenesis. Evolution actually separated these functions in arrestin-2, which binds the same kinases but does not facilitate JNK3 phosphorylation.[162]

5. DESIGNING SIGNALING-BIASED ARRESTIN MUTANTS

To a certain extent, the natural structural organization of arrestins makes targeting different aspects of their function easier. The concave sides of both domains contain allknown receptor-binding residues (Section 3.3), most other partners interact with elements on the other side of the molecule that remain accessible in the arrestin–receptor complex, whereas clathrin and AP2 engage distinct sites on the arrestin C-tail (Section 4.1) that do not appear to overlap with the binding sites of any other known partner. These sites were the first to be eliminated by targeted mutagenesis.[145,147,153] Predictably, the disruption of either site precluded arrestin interactions with its respective partners without affecting GPCR binding. Moreover, arrestin mutants with disabled binding sites for clathrin, AP2, and particularly both, acted as dominant-negative, selectively suppressing arrestin-dependent GPCR internalization via coated pits.[145,147,153] These findings firmly established that arrestin functions can be manipulated independently of each other. In addition, these studies showed that an arrestin can act as adominant negative in a particular regard only when a single function is destroyed in a way that preserves all other functional capabilities. Similarly, to generate signaling-biased arrestins, a single function should be enhanced or disabled by the mutations that do not significantly affect others. In reality, it turned out to be extremely difficult to change anything within the arrestin molecule without affecting more than one function. However, as long as a single aspect of arrestin functions is affected to a much greater extent than the others, the mutant can serve a particular purpose.

5.1. Enhanced phosphorylation-independent mutants

The first arrestin that binds with high affinity to any active form of its cognate receptor regardless of phosphorylation, arrestin-1–Arg175Asn, was created in the process of elucidation of the mechanism arrestins use to respond to the receptor-attached phosphates in 1995.[27] This happened when arrestins were still believed to interact only with GPCRs, before arrestin binding to the first nonreceptor partner, clathrin, was discovered in 1996.[145] Arg175 was later substituted by every residue in the book, and charge reversal mutation Arg175Glu was found to be the most potent.[28] Later, a whole family of structurally distinct phosphorylation-independent mutants of arrestin-1 was constructed.[77,165] The Arg175Glu mutant was shown to effectively suppress transducin activation by unphosphorylated light-activated rhodopsin (Rh*).[166] Several mutations homologous to those that enable arrestin-1 binding to unphosphorylated Rh* were shown to yield similar phosphorylation-independent binding of both nonvisual arrestins to their cognate receptors,[125–128] and to block the coupling of unphosphorylated WT and mutant GPCRs to their cognate G proteins.[125,126,129]

One of the phosphorylation-independent arrestin-1 mutants was recently tested for its ability to compensate for defects of rhodopsin phosphorylation *in vivo*.[91] This enhanced arrestin-1 was shown to significantly improve survival and functional performance of rod photoreceptors lacking rhodopsin kinase, and to facilitate the rate of rod recovery ~threefold, as compared to parental WT arrestin-1.[91] This proof-of-concept study showed that one mutant protein can be used to compensate for a molecular defect in another.[141] However, photoresponse recovery in "compensated" rods was much slower than in WT photoreceptors, suggesting that a significant further redesign of arrestin-1 is necessary to achieve a perfect fit and high-affinity binding to unphosphorylated Rh*.[91] These results clearly showed that, even though we know about the molecular mechanism of the arrestin–receptor interaction much more than about any other arrestin function, we still need a lot of additional information to construct a perfect tool for therapeutic use.

Enhanced phosphorylation-independent versions of all arrestins have been constructed[125–128] and shown to effectively desensitize several GPCRs without receptor phosphorylation.[125,126,129] These mutants proved to be useful in the studies of GPCR phosphorylation[129,130] and its role in particular biological processes.[130,167,168] However, wider practical use of enhanced nonvisual arrestins, particularly for therapeutic purposes, is contingent on our ability to make them target the receptors of our choosing, rather than all GPCRs indiscriminately, like parental WT arrestin-2 and -3.

5.2. Constitutively monomeric arrestins

Self-association of arrestin-1 (known as S-antigen at the time) was discovered even before its role in quenching rhodopsin signaling.[1] Arrestin-1 was subsequently crystallized as a tetramer (dimer of dimers) by two independent groups under different conditions.[17,18] It was shown to form dimers and tetramers in solution by a variety of methods: analytical ultracentrifugation,[118] small angle X-ray scattering,[119] and multiangle light scattering.[120] The study of arrestin-1 self-association using long-range measurements of intersubunit distances by DEER yielded an unexpected result; the solution tetramer was found to be strikingly different from that in crystal form.[120] Elucidation of its structure[122] confirmed the earlier idea that arrestin-1 oligomers are storage forms.[118] It also explained the earlier finding that only monomeric arrestin-1 can bind rhodopsin[120]; in the physiologically relevant solution tetramer and both possible dimers the rhodopsin-binding surfaces of all arestin-1 molecules are shielded by sister protomers.[122]

Bovine, mouse, and human arrestin-1 were shown to self-associate via the same monomer–dimer–tetramer equilibrium, although dimerization and tetramerization constants were very different in these species.[121] Model-based targeted elimination of the same two phenylalanines generated self-association-deficient bovine and mouse arrestin-1.[121,122] These results showed that the structure of the solution tetramer of all mammalian arrestin-1 is likely very similar, despite striking differences in dimerization and tetramerization constants. This study also generated a constitutively monomeric form of arrestin-1 that retained perfectly normal binding to rhodopsin and microtubules.[121] This mutant is the molecular tool necessary to determine the biological role of robust arrestin-1 self-association at concentrations normally found in photoreceptors.[110,115–117] This is not a trivial question, as only monomeric arrestin binds receptor,[120] and no other signaling protein in photoreceptors has an oligomeric storage form. Interestingly, cone-specific arrestin-4 is a constitutive monomer,[169] but it represents only ~2% of total arrestin complement in cones,[170] the rest being arrestin-1 that readily self-associates. This finding, along with apparent conservation of self-association among mammalian arrestin-1 homologues,[121] strongly suggests that oligomerization of this subtype has an important biological function that needs to be elucidated experimentally.

Both nonvisual arrestins were also found to self-associate[171–173] and even form heterodimers.[171,172] The mechanism of self-association of arrestin-1 and nonvisual subtypes appears to be different, as IP6 inhibits oligomerization of arrestin-1, while promoting self-association of nonvisual arrestins.[169,172] Two IP6 binding sites in arrestin-2 were identified by crystallography and confirmed by structure-based mutagenesis.[172] Self-association was shown to regulate the subcellular distribution of arrestin-2; oligomers were predominantly cytoplasmic, whereas monomers showed increased nuclear localization.[172] Oligomerization of arrestin-3 was reported to be required for its interaction with the E3 ubiquitin ligase Mdm2, and elimination of IP6 binding residues inhibited Mdm2 binding and p53-dependent antiproliferative effects of arrestin-3.[173] It should be noted that while self-association of arrestin-1 certainly stops at tetramers that have the shape of closed diamonds, where all potential interaction surfaces are fully engaged by sister subunits,[122] it is not clear whether oligomerization of nonvisual arrestins has a natural limit. Crystallography suggests that they might form chains of any length[172]; biochemical data obtained with pure protein indicate that arrestin-2 can form at least tetramers but do not provide evidence that the process stops at that stage.[169] Further experimentation with arrestin-2, arrestin-3, and a mixed population of the two nonvisual arrestins in the presence and absence of IP6 is necessary to resolve these issues and attempt construction of constitutively monomeric forms of these proteins with all other functions intact. These tools are necessary to determine the biological role of the oligomerization of nonvisual arrestins.

5.3. Manipulating MAPK signaling

MAP kinases regulate many vital cellular functions, from proliferation to apoptotic death.[174] MAPK cascades consist of three kinases that sequentially activate each other by phosphorylation: upstream MAPKKKs (e.g., ASK1 and cRaf1), MAPKKs (e.g., MEK1/2, MKK4/7), and MAPKs (e.g.,

ERK1/2, JNK1/2/3, p38 MAPK). All MAP kinases are soluble proteins, but phosphatase activity in the cytoplasm makes it highly likely that an upstream kinase would be dephosphorylated before it finds its target by random diffusion. In addition, there are many different MAPKs, MAPKKs, and especially MAPKKKs in the cell.[174] Therefore, productive complexes containing matching components are usually assembled by scaffolding proteins that bring the three kinases together, thereby facilitating signaling.[175]

Arrestins were shown to promote the activation of JNK3,[148] ERK1/2,[149] and p38 MAPK[168] by scaffolding the relevant three-kinase modules. The first reports suggested that receptor activation is necessary for arrestin-dependent scaffolding of MAPK cascades.[148,149,168] Subsequent studies confirmed that receptor binding is a prerequisite for arrestin-dependent activation of ERK1/2.[133,136] However, arrestin-3-mediated facilitation of JNK3 phosphorylation was shown to be receptor-independent,[159,162] indicating that free arrestin-3 performs this function.[136,158] This was unambiguously confirmed by reconstitution of an arrestin-3–MKK4–JNK3 signaling module *in vitro* from purified proteins in the absence of the receptor.[160]

Regardless of whether free, receptor-bound, or both forms of arrestins scaffold MAPK cascades, several ways of manipulating MAPK signaling by mutant arrestins are conceivable. First, targeted mutations disrupting an interaction with one of the kinases would yield arrestins that can only form incomplete complexes, thereby sequestering the other two kinases and suppressing the activation of a particular MAPK. Second, a mutant that binds all kinases but holds them in a "wrong" configuration not conducive to signal transduction would serve this purpose even better. Third, mutations enhancing the interactions with one or more of the kinases in a particular cascade would channel signaling in that direction. Finally, mutations that create preference for ERK1/2 over JNK3 would facilitate prosurvival and suppress proapoptotic signaling, while mutations biasing arrestin toward JNK3 would have the opposite effect.

Even though the binding sites of MAP kinases remain unknown or poorly defined, several signaling-biased forms of arrestins that significantly affect MAPK signaling were created. The first in this line was arrestin-2-R307A mutant, which was shown to have impaired c-Raf1 binding, while interacting with ERK1/2 and MEK1 normally.[157] In contrast to WT arrestin-2, the expression of this mutant did not rescue β2AR-dependent arrestin-mediated ERK1/2 activation in arrestin-2/3 double knockout cells.[157] Whether this mutant can suppress ERK1/2 activation via other scaffolds still remains to be elucidated. Interestingly, a homologous K308A mutation in arrestin-3 did not affect c-Raf1 binding or its ability to promote ERK1/2 phosphorylation upon receptor stimulation.[157] These results suggest that despite very high homology between the two nonvisual subtypes, the molecular mechanisms of scaffolding c-Raf1–MEK1–ERK1/2 cascade by arrestin-2 and -3 are subtly different.

Two signaling-biased mutants of arrestin-3 have been characterized more comprehensively.[136] The ability of WT arrestin-3, its "preactivated" 3A mutant that binds GPCRs even more readily,[125] and two receptor-binding-impaired forms; one with 7-residue deletion in the inter-domain hinge (D7) and the other where key receptor-binding residues were mutated to alanines (KNC),[71,89] to bind JNK3 and promote its activation in the cell was compared.[136] It turned out that the ability of these proteins to bind GPCRs and activate JNK3 do not correlate at all: WT arrestin-3 that binds receptors and D7 mutant that does not promoted JNK3 phosphorylation equally well, whereas both 3A and KNC did not.[136] This was the most convincing evidence that arrestin-3 scaffolds ASK1–MKK4–JNK3 module independently of receptor binding. Moreover, the inactive KNC mutant was shown to interact with ASK1 and MKK4 as well as WT, and bind JNK3 even better,[136] providing the strongest proof so far that recruitment of MAP kinases and the ability to promote signaling in the cascade are also separable arrestin functions. Indeed, the KNC mutant was shown to act in a dominant-negative manner, suppressing JNK3 activation in the cell. Thus, arrestin-3–KNC is a silent scaffold, inhibiting JNK3 activation by sequestering ASK1, MKK4, and JNK3 away from productive scaffolds. As could be expected, both receptor-binding-impaired forms of arrestin-3, D7 and KNC, were found to suppress ERK1/2 activation.[136] Apparently, these mutants retain the ability to recruit kinases of c-Raf1–MEK1–ERK1/2 module but cannot assume the conformation conducive to ERK1/2 activation due to the defect in GPCR binding.

These studies generated tools that selectively suppress either ERK1/2 or JNK3 activity, which is believed to send prosurvival or proapoptotic signal to the cell, respectively. The ability of the arrestin-3–KNC mutant to enhance cell survival and that of arrestin-2–R307A to suppress proliferation and possibly facilitate cell death needs to be tested. Mutants where individual functions are enhanced, rather than suppressed, also must be developed. Targeted construction of distinct signaling-biased forms of arrestins would be greatly facilitated by better understanding of the molecular mechanisms involved in their binding to nonreceptor partners and the ability to scaffold MAP kinase cascades productively.

6. CONCLUSIONS: WHERE DO WE GO FROM HERE?

It might appear that as proteins go, arrestins are fairly well studied structurally and functionally. It is clear that members of this small protein family, which apparently emerged relatively late in evolution,[12] likely after GPCR kinases,[176] serve as multifunctional signaling organizers in the cell. However, while known phenomenology is rich and pretty well described,[177] the structural basis of most arrestin functions remains obscure. The molecular mechanisms underlying arrestin interactions with most partners, as well as expected inter-dependence of these interactions, must be elucidated. This will greatly improve our understanding of cell signaling, particularly the integration of different inputs into coherent cell behavior, which is arguably the greatest challenge in modern cell biology.[142] As an added bonus, elucidation of the molecular mechanisms underlying individual arrestin functions and identification of the residues involved will pave the way to intelligent design of mutant arrestins with desired signaling bias. The mutants of this type are not just valuable research tools but also hold promise of enormous therapeutic potential.

ACKNOWLEDGMENTS

The authors are grateful to our collaborators, whose expertise and efforts made many of the studies discussed here possible. Supported by NIH Grants EY011500, GM077561, GM081756 (VVG), NS065868, and DA030103 (EVG).

REFERENCES

1. Wacker WB, Donoso LA, Kalsow CM, Yankeelov Jr JA, Organisciak DT. Experimental allergic uveitis. Isolation, characterization, and localization of a soluble uveitopathogenic antigen from bovine retina. *J Immunol.* 1977;119:1949–1958.
2. Kuhn H. Light-regulated binding of rhodopsin kinase and other proteins to cattle photoreceptor membranes. *Biochemistry.* 1978;17:4389–4395.
3. Kuhn H, Hall SW, Wilden U. Light-induced binding of 48-kDa protein to photoreceptor membranes is highly enhanced by phosphorylation of rhodopsin. *FEBS Lett.* 1984;176:473–478.
4. Wilden U, Hall SW, Kühn H. Phosphodiesterase activation by photoexcited rhodopsin is quenched when rhodopsin is phosphorylated and binds the intrinsic 48-kDa protein of rod outer segments. *Proc Natl Acad Sci USA.* 1986;83:1174–1178.
5. Dixon RA, Kobilka BK, Strader DJ, et al. Cloning of the gene and cDNA for mammalian beta-adrenergic receptor and homology with rhodopsin. *Nature.* 1986;321:75–79.
6. Lohse MJ, Benovic JL, Codina J, Caron MG, Lefkowitz RJ. Beta-arrestin: a protein that regulates beta-adrenergic receptor function. *Science.* 1990;248:1547–1550.
7. Attramadal H, Arriza JL, Aoki C, et al. Beta-arrestin2, a novel member of the arrestin/beta-arrestin gene family. *J Biol Chem.* 1992;267:17882–17890.

8. Sterne-Marr R, Gurevich VV, Goldsmith P, et al. Polypeptide variants of beta-arrestin and arrestin3. *J Biol Chem*. 1993;268:15640–15648.
9. Rapoport B, Kaufman KD, Chazenbalk GD. Cloning of a member of the arrestin family from a human thyroid cDNA library. *Mol Cell Endocrinol*. 1992;84:R39–R43.
10. Craft CM, Whitmore DH, Wiechmann AF. Cone arrestin identified by targeting expression of a functional family. *J Biol Chem*. 1994;269:4613–4619.
11. Sakuma H, Murakami A, Fujimaki T, Inana G. Isolation and characterization of the human X-arrestin gene. *Gene*. 1998;224:87–95.
12. Gurevich EV, Gurevich VV. Arrestins: ubiquitous regulators of cellular signaling pathways. *Genome Biol*. 2006;7:236.
13. Gurevich VV, Hanson SM, Song X, Vishnivetskiy SA, Gurevich EV. The functional cycle of visual arrestins in photoreceptor cells. *Prog Retin Eye Res*. 2011;30:405–430.
14. Gurevich VV, Gurevich EV. The structural basis of arrestin-mediated regulation of G-protein-coupled receptors. *Pharm Ther*. 2006;110:465–502.
15. Xiao K, McClatchy DB, Shukla AK, et al. Functional specialization of beta-arrestin interactions revealed by proteomic analysis. *Proc Natl Acad Sci USA*. 2007;104:12011–12016.
16. Shinohara T, Dietzschold B, Craft CM, et al. Primary and secondary structure of bovine retinal S antigen (48-kDa protein). *Proc Natl Acad Sci USA*. 1987;84:6975–6979.
17. Granzin J, Wilden U, Choe HW, Labahn J, Krafft B, Buldt G. X-ray crystal structure of arrestin from bovine rod outer segments. *Nature*. 1998;391:918–921.
18. Hirsch JA, Schubert C, Gurevich VV, Sigler PB. The 2.8 A crystal structure of visual arrestin: a model for arrestin's regulation. *Cell*. 1999;97:257–269.
19. Han M, Gurevich VV, Vishnivetskiy SA, Sigler PB, Schubert C. Crystal structure of beta-arrestin at 1.9 A: possible mechanism of receptor binding and membrane Translocation. *Structure*. 2001;9:869–880.
20. Zhan X, Gimenez LE, Gurevich VV, Spiller BW. Crystal structure of arrestin-3 reveals the basis of the difference in receptor binding between two non-visual subtypes. *J Mol Biol*. 2011;406:467–478.
21. Milano SK, Pace HC, Kim YM, Brenner C, Benovic JL. Scaffolding functions of arrestin-2 revealed by crystal structure and mutagenesis. *Biochemistry*. 2002; 41:3321–3328.
22. Sutton RB, Vishnivetskiy SA, Robert J, et al. Crystal structure of cone arrestin at 2.3A: evolution of receptor specificity. *J Mol Biol*. 2005;354:1069–1080.
23. Granzin J, Cousin A, Weirauch M, Schlesinger R, Büldt G, Batra-Safferling R. Crystal structure of p44, a constitutively active splice variant of visual arrestin. *J Mol Biol*. 2012;416:611–618.
24. Gurevich VV, Benovic JL. Cell-free expression of visual arrestin. Truncation mutagenesis identifies multiple domains involved in rhodopsin interaction. *J Biol Chem*. 1992;267:21919–21923.
25. Gurevich VV, Benovic JL. Visual arrestin interaction with rhodopsin. Sequential multisite binding ensures strict selectivity toward light-activated phosphorylated rhodopsin. *J Biol Chem*. 1993;268:11628–11638.
26. Gurevich VV, Chen C-Y, Kim CM, Benovic JL. Visual arrestin binding to rhodopsin. Intramolecular interaction between the basic N terminus and acidic C terminus of arrestin may regulate binding selectivity. *J Biol Chem*. 1994;269:8721–8727.
27. Gurevich VV, Benovic JL. Visual arrestin binding to rhodopsin. Diverse functional roles of positively charged residues within the phosphorylation-recognition region of arrestin. *J Biol Chem*. 1995;270:6010–6016.
28. Gurevich VV, Benovic JL. Mechanism of phosphorylation-recognition by visual arrestin and the transition of arrestin into a high affinity binding state. *Mol Pharmacol*. 1997;51:161–169.

29. Vishnivetskiy SA, Paz CL, Schubert C, Hirsch JA, Sigler PB, Gurevich VV. How does arrestin respond to the phosphorylated state of rhodopsin? *J Biol Chem.* 1999;274:11451–11454.
30. Gurevich VV, Gurevich EV. The molecular acrobatics of arrestin activation. *Trends Pharmacol Sci.* 2004;25:105–111.
31. Gurevich VV, Gurevich EV. The new face of active receptor bound arrestin attracts new partners. *Structure.* 2003;11:1037–1042.
32. Nair KS, Hanson SM, Mendez A, et al. Light-dependent redistribution of arrestin in vertebrate rods is an energy-independent process governed by protein-protein interactions. *Neuron.* 2005;46:555–567.
33. Nair KS, Hanson SM, Kennedy MJ, Hurley JB, Gurevich VV, Slepak VZ. Direct binding of visual arrestin to microtubules determines the differential subcellular localization of its splice variants in rod photoreceptors. *J Biol Chem.* 2004;279:41240–41248.
34. Hanson SM, Francis DJ, Vishnivetskiy SA, Klug CS, Gurevich VV. Visual arrestin binding to microtubules involves a distinct conformational change. *J Biol Chem.* 2006;281:9765–9772.
35. Hanson SM, Cleghorn WM, Francis DJ, et al. Arrestin mobilizes signaling proteins to the cytoskeleton and redirects their activity. *J Mol Biol.* 2007;368:375–387.
36. Palczewski K, Kumasaka T, Hori T, et al. Crystal structure of rhodopsin: A G protein-coupled receptor. *Science.* 2000;289:739–745.
37. Okada T, Fujiyoshi Y, Silow M, Navarro J, Landau EM, Shichida Y. Functional role of internal water molecules in rhodopsin revealed by X-ray crystallography. *Proc Natl Acad Sci USA.* 2002;99:5982–5987.
38. Li J, Edwards PC, Burghammer M, Villa C, Schertler GF. Structure of bovine rhodopsin in a trigonal crystal form. *J Mol Biol.* 2004;343:1409–1438.
39. Standfuss J, Xie G, Edwards PC, Burghammer M, Oprian DD, Schertler GF. Crystal structure of a thermally stable rhodopsin mutant. *J Mol Biol.* 2007;372:1179–1188.
40. Cherezov V, Rosenbaum DM, Hanson MA, et al. High-resolution crystal structure of an engineered human beta2-adrenergic G protein-coupled receptor. *Science.* 2007;318:1258–1265.
41. Chien EYT, Liu W, Zhao Q, et al. Structure of the human dopamine D3 receptor in complex with a D2/D3 selective antagonist. *Science.* 2010;330:1091–1095.
42. Rosenbaum DM, Zhang C, Lyons JA, et al. Structure and function of an irreversible agonist-β(2) adrenoceptor complex. *Nature.* 2011;469:236–240.
43. Wu B, Chien EYT, Mol CD, et al. Structures of the CXCR4 chemokine GPCR with small-molecule and cyclic peptide antagonists. *Science.* 2010;330:1066–1071.
44. Rasmussen SG, Choi HJ, Rosenbaum DM, et al. Crystal structure of the human beta2 adrenergic G-protein-coupled receptor. *Nature.* 2007;450:383–387.
45. Haga K, Kruse AC, Asada H, et al. Structure of the human M2 muscarinic acetylcholine receptor bound to an antagonist. *Nature.* 2012;482:547–551.
46. Kruse AC, Hu J, Pan AC, et al. Structure and dynamics of the M3 muscarinic acetylcholine receptor. *Nature.* 2012;482:552–556.
47. Manglik A, Kruse AC, Kobilka TS, et al. Crystal structure of the μ-opioid receptor bound to a morphinan antagonist. *Nature.* 2012;485:321–326.
48. Jaakola VP, Griffith MT, Hanson MA, et al. The 2.6 angstrom crystal structure of a human A2A adenosine receptor bound to an antagonist. *Science.* 2008;322:1211–1217.
49. Xu F, Wu H, Katritch V, et al. Structure of an agonist-bound human A2A adenosine receptor. *Science.* 2011;332:322–327.
50. Shimamura T, Shiroishi M, Weyand S, et al. Structure of the human histamine H1 receptor complex with doxepin. *Nature.* 2011;475:65–70.
51. Wu H, Wacker D, Mileni M, et al. Structure of the human κ-opioid receptor in complex with JDTic. *Nature.* 2012;485:327–332.

52. Moukhametzianov R, Warne T, Edwards PC, et al. Two distinct conformations of helix 6 observed in antagonist-bound structures of a beta1-adrenergic receptor. *Proc Natl Acad Sci USA*. 2011;108:8228–8232.
53. Warne T, Serrano-Vega MJ, Baker JG, et al. Structure of a beta1-adrenergic G-protein-coupled receptor. *Nature*. 2008;454:486–491.
54. Doré AS, Robertson N, Errey JC, et al. Structure of the adenosine A(2A) receptor in complex with ZM241385 and the xanthines XAC and caffeine. *Structure*. 2011;19:1283–1293.
55. Hubbell WL, Altenbach C, Hubbell CM, Khorana HG. Rhodopsin structure, dynamics, and activation: a perspective from crystallography, site-directed spin labeling, sulfhydryl reactivity, and disulfide cross-linking. *Adv Protein Chem*. 2003;63:243–290.
56. Altenbach C, Kusnetzow AK, Ernst OP, Hofmann KP, Hubbell WL. High-resolution distance mapping in rhodopsin reveals the pattern of helix movement due to activation. *Proc Natl Acad Sci USA*. 2008;105:7439–7444.
57. Farrens DL, Altenbach C, Yang K, Hubbell WL, Khorana HG. Requirement of rigid-body motion of transmembrane helices for light activation of rhodopsin. *Science*. 1996;274:768–770.
58. Resek JF, Farahbakhsh ZT, Hubbell WL, Khorana HG. Formation of the meta II photointermediate is accompanied by conformational changes in the cytoplasmic surface of rhodopsin. *Biochemistry*. 1993;32:12025–12032.
59. Farahbakhsh ZT, Ridge KD, Khorana HG, Hubbell WL. Mapping light-dependent structural changes in the cytoplasmic loop connecting helices C and D in rhodopsin: a site-directed spin labeling study. *Biochemistry*. 1995;34:8812–8819.
60. Scheerer P, Park JH, Hildebrand PW, et al. Crystal structure of opsin in its G-protein-interacting conformation. *Nature*. 2008;455:497–502.
61. Park JH, Scheerer P, Hofmann KP, Choe HW, Ernst OP. Crystal structure of the ligand-free G-protein-coupled receptor opsin. *Nature*. 2008;454:183–187.
62. Choe HW, Kim YJ, Park JH, et al. Crystal structure of metarhodopsin II. *Nature*. 2011;471:651–655.
63. Deupi X, Edwards P, Singhal A, et al. Stabilized G protein binding site in the structure of constitutively active metarhodopsin-II. *Proc Natl Acad Sci USA*. 2012;109:119–124.
64. Lebon G, Warne T, Edwards PC, et al. Agonist-bound adenosine A2A receptor structures reveal common features of GPCR activation. *Nature*. 2011;474:521–525.
65. Rasmussen SG, Choi HJ, Fung JJ, et al. Structure of a nanobody-stabilized active state of the β(2) adrenoceptor. *Nature*. 2011;469:175–180.
66. Rasmussen SG, DeVree BT, Zou Y, et al. Crystal structure of the β2 adrenergic receptor-Gs protein complex. *Nature*. 2011;477:549–555.
67. Chung KY, Rasmussen SG, Liu T, et al. Conformational changes in the G protein Gs induced by the β2 adrenergic receptor. *Nature*. 2011;477:611–615.
68. Kobilka BK. G protein coupled receptor structure and activation. *Biochim Biophys Acta*. 2007;1768:794–807.
69. Kobilka BK. Cold Spring Harbor Asia membrane protein meeting, May 16–20, 2011; 79.
70. Warne T, Moukhametzianov R, Baker JG, et al. The structural basis for agonist and partial agonist action on a β(1)-adrenergic receptor. *Nature*. 2011;469:241–244.
71. Vishnivetskiy SA, Gimenez LE, Francis DJ, et al. Few residues within an extensive binding interface drive receptor interaction and determine the specificity of arrestin proteins. *J Biol Chem*. 2011;286:24288–24299.
72. Kim M, Vishnivetskiy SA, Van Eps N, et al. Conformation of receptor-bound visual arrestin. *Proc Natl Acad Sci USA*. 2012;109:18407–18412.
73. Hanson SM, Francis DJ, Vishnivetskiy SA, et al. Differential interaction of spin-labeled arrestin with inactive and active phosphorhodopsin. *Proc Natl Acad Sci USA*. 2006;103:4900–4905.

74. DeGraff JL, Gurevich VV, Benovic JL. The third intracellular loop of alpha 2-adrenergic receptors determines subtype specificity of arrestin interaction. *J Biol Chem*. 2002;277:43247–43252.
75. Gelber EI, Kroeze WK, Willins DL, et al. Structure and function of the third intracellular loop of the 5-hydroxytryptamine2A receptor: the third intracellular loop is alpha-helical and binds purified arrestins. *J Neurochem*. 1999;72:2206–2214.
76. Raman D, Osawa S, Gurevich VV, Weiss ER. The interaction with the cytoplasmic loops of rhodopsin plays a crucial role in arrestin activation and binding. *J Neurochem*. 2003;84:1040–1050.
77. Vishnivetskiy SA, Schubert C, Climaco GC, Gurevich YV, Velez M-G, Gurevich VV. An additional phosphate-binding element in arrestin molecule. Implications for the mechanism of arrestin activation. *J Biol Chem*. 2000;275:41049–41057.
78. Ohguro H, Palczewski K, Walsh KA, Johnson RS. Topographic study of arrestin using differential chemical modifications and hydrogen/deuterium exchange. *Protein Sci*. 1994;3:2428–2434.
79. Vishnivetskiy SA, Hosey MM, Benovic JL, Gurevich VV. Mapping the arrestin-receptor interface. Structural elements responsible for receptor specificity of arrestin proteins. *J Biol Chem*. 2004;279:1262–1268.
80. Gurevich VV, Dion SB, Onorato JJ, et al. Arrestin interactions with G protein-coupled receptors. Direct binding studies of wild type and mutant arrestins with rhodopsin, beta 2-adrenergic, and m2 muscarinic cholinergic receptors. *J Biol Chem*. 1995;270:720–731.
81. Pulvermuller A, Schroder K, Fischer T, Hofmann KP. Interactions of metarhodopsin II. Arrestin peptides compete with arrestin and transducin. *J Biol Chem*. 2000;275: 37679–37685.
82. Dinculescu A, McDowell JH, Amici SA, et al. Insertional mutagenesis and immunochemical analysis of visual arrestin interaction with rhodopsin. *J Biol Chem*. 2002;277:11703–11708.
83. Feuerstein SE, Pulvermüller A, Hartmann R, et al. Helix formation in arrestin accompanies recognition of photoactivated rhodopsin. *Biochemistry*. 2009;48: 10733–10742.
84. Zhuang T, Vishnivetskiy SA, Gurevich VV, Sanders CR. Elucidation of inositol hexaphosphate and heparin interaction sites and conformational changes in arrestin-1 by solution nuclear magnetic resonance. *Biochemistry*. 2010;49:10473–10485.
85. Zhuang T, Chen Q, Cho M-K, et al. Involvement of distinct arrestin-1 elements in binding to different functional forms of rhodopsin. *Proc Natl Acad Sci USA*. 2013;110:942–947.
86. Vishnivetskiy SA, Francis DJ, Van Eps N, et al. The role of arrestin alpha-helix I in receptor binding. *J Mol Biol*. 2010;395:42–54.
87. Vishnivetskiy SA, Raman D, Wei J, Kennedy MJ, Hurley JB, Gurevich VV. Regulation of arrestin binding by rhodopsin phosphorylation level. *J Biol Chem*. 2007;282:32075–32083.
88. Krasel C, Bunemann M, Lorenz K, Lohse MJ. Beta-arrestin binding to the beta2-adrenergic receptor requires both receptor phosphorylation and receptor activation. *J Biol Chem*. 2005;280:9528–9535.
89. Gimenez LE, Kook S, Vishnivetskiy SA, Ahmed MR, Gurevich EV, Gurevich VV. Role of receptor-attached phosphates in binding of visual and non-visual arrestins to G protein-coupled receptors. *J Biol Chem*. 2012;287:9028–9040.
90. Mendez A, Burns ME, Roca A, et al. Rapid and reproducible deactivation of rhodopsin requires multiple phosphorylation sites. *Neuron*. 2000;28:153–164.
91. Song X, Vishnivetskiy SA, Gross OP, et al. Enhanced arrestin facilitates recovery and protects rods lacking rhodopsin phosphorylation. *Curr Biol*. 2009;19:700–705.

92. Wilden U, Kuhn H. Light-dependent phosphorylation of rhodopsin: number of phosphorylation sites. *Biochemistry*. 1982;21:3014–3022.
93. Bouvier M, Hausdorff WP, De Blasi A, et al. Removal of phosphorylation sites from the beta 2-adrenergic receptor delays onset of agonist-promoted desensitization. *Nature*. 1988;333:370–373.
94. Fredericks ZL, Pitcher JA, Lefkowitz RJ. Identification of the G protein-coupled receptor kinase phosphorylation sites in the human beta2-adrenergic receptor. *J Biol Chem*. 1996;271:13796–13803.
95. Hausdorff WP, Bouvier M, O'Dowd BF, Irons GP, Caron MG, Lefkowitz RJ. Phosphorylation sites on two domains of the beta 2-adrenergic receptor are involved in distinct pathways of receptor desensitization. *J Biol Chem*. 1989;264:12657–12665.
96. Krasel C, Zabel U, Lorenz K, Reiner S, Al-Sabah S, Lohse MJ. Dual role of the beta2-adrenergic receptor C terminus for the binding of beta-arrestin and receptor internalization. *J Biol Chem*. 2008;283:31840–31848.
97. Seibold A, January BG, Friedman J, Hipkin RW, Clark RB. Desensitization of beta2-adrenergic receptors with mutations of the proposed G protein-coupled receptor kinase phosphorylation sites. *J Biol Chem*. 1998;273:7637–7642.
98. Seibold A, Williams B, Huang ZF, et al. Localization of the sites mediating desensitization of the beta(2)-adrenergic receptor by the GRK pathway. *Mol Pharmacol*. 2000;58:1162–1173.
99. Trester-Zedlitz M, Burlingame A, Kobilka B, von Zastrow M. Mass spectrometric analysis of agonist effects on posttranslational modifications of the beta-2 adrenoceptor in mammalian cells. *Biochemistry*. 2005;44:6133–6143.
100. Vaughan DJ, Millman EE, Godines V, et al. Role of the G protein-coupled receptor kinase site serine cluster in beta2-adrenergic receptor internalization, desensitization, and beta-arrestin translocation. *J Biol Chem*. 2006;281:7684–7692.
101. Lee KB, Ptasienski JA, Pals-Rylaarsdam R, Gurevich VV, Hosey MM. Arrestin binding to the M(2) muscarinic acetylcholine receptor is precluded by an inhibitory element in the third intracellular loop of the receptor. *J Biol Chem*. 2000;275:9284–9289.
102. Pals-Rylaarsdam R, Hosey MM. Two homologous phosphorylation domains differentially contribute to desensitization and internalization of the m2 muscarinic acetylcholine receptor. *J Biol Chem*. 1997;272:14152–14158.
103. Pals-Rylaarsdam R, Gurevich VV, Lee KB, Ptasienski J, Benovic JL, Hosey MM. Internalization of the m2 muscarinic acetylcholine receptor. Arrestin-independent and -dependent pathways. *J Biol Chem*. 1997;272:23682–23689.
104. Schleicher A, Kuhn H, Hofmann KP. Kinetics, binding constant, and activation energy of the 48-kDa protein-rhodopsin complex by extra-metarhodopsin II. *Biochemistry*. 1989;28:1770–1775.
105. Vishnivetskiy SA, Hirsch JA, Velez M-G, Gurevich YV, Gurevich VV. Transition of arrestin into the active receptor-binding state requires an extended interdomain hinge. *J Biol Chem*. 2002;277:43961–43968.
106. Kim M, Vishnivetskiy SA, Van Eps N, et al. ASBMB meeting, 2011; 750–759.
107. Fotiadis D, Jastrzebska B, Philippsen A, Muller DJ, Palczewski K, Engel A. Structure of the rhodopsin dimer: a working model for G-protein-coupled receptors. *Curr Opin Struct Biol*. 2006;16:252–259.
108. Chen J, Makino CL, Peachey NS, Baylor DA, Simon MI. Mechanisms of rhodopsin inactivation in vivo as revealed by a COOH-terminal truncation mutant. *Science*. 1995;267:374–377.
109. Palczewski K, Pulvermuller A, Buczylko J, Hofmann KP. Phosphorylated rhodopsin and heparin induce similar conformational changes in arrestin. *J Biol Chem*. 1991;266:18649–18654.

110. Hanson SM, Gurevich EV, Vishnivetskiy SA, Ahmed MR, Song X, Gurevich VV. Each rhodopsin molecule binds its own arrestin. *Proc Natl Acad Sci USA*. 2007;104:3125–3128.
111. Broekhuyse RM, Tolhuizen EF, Janssen AP, Winkens HJ. Photoreceptor cell-specific localization of S-antigen in retina. *Curr Eye Res*. 1985;4:613–618.
112. Whelan JP, McGinnis JF. Light-dependent subcellular movement of photoreceptor proteins. *J Neurosci Res*. 1988;20:263–270.
113. Elias RV, Sezate SS, Cao W, McGinnis JF. Temporal kinetics of the light/dark translocation and compartmentation of arrestin and alpha-transducin in mouse photoreceptor cells. *Mol Vis*. 2004;10:672–681.
114. Mendez A, Lem J, Simon M, Chen J. Light-dependent translocation of arrestin in the absence of rhodopsin phosphorylation and transducin signaling. *J Neurosci*. 2003;23:3124–3129.
115. Strissel KJ, Sokolov M, Trieu LH, Arshavsky VY. Arrestin translocation is induced at a critical threshold of visual signaling and is superstoichiometric to bleached rhodopsin. *J Neurosci*. 2006;26:1146–1153.
116. Song X, Vishnivetskiy SA, Seo J, Chen J, Gurevich EV, Gurevich VV. Arrestin-1 expression level in rods: balancing functional performance and photoreceptor health. *Neuroscience*. 2011;174:37–49.
117. Cleghorn WM, Tsakem EL, Song X, et al. Progressive reduction of its expression in rods reveals two pools of arrestin-1 in the outer segment with different roles in photoresponse recovery. *PLoS One*. 2011;6:e22797.
118. Schubert C, Hirsch JA, Gurevich VV, Engelman DM, Sigler PB, Fleming KG. How does arrestin respond to the phosphorylated state of rhodopsin? *J Biol Chem*. 1999;274:21186–21190.
119. Imamoto Y, Tamura C, Kamikubo H, Kataoka M. Concentration-dependent tetramerization of bovine visual arrestin. *Biophys J*. 2003;85:1186–1195.
120. Hanson SM, Van Eps N, Francis DJ, et al. Structure and function of the visual arrestin oligomer. *EMBO J*. 2007;26:1726–1736.
121. Kim M, Hanson SM, Vishnivetskiy SA, et al. Robust self-association is a common feature of mammalian visual arrestin-1. *Biochemistry*. 2011;50:2235–2242.
122. Hanson SM, Dawson ES, Francis DJ, et al. A model for the solution structure of the rod arrestin tetramer. *Structure*. 2008;16:924–934.
123. Tsukamoto H, Sinha A, Dewitt M, Farrens DL. Monomeric rhodopsin is the minimal functional unit required for arrestin binding. *J Mol Biol*. 2010;399:501–511.
124. Bayburt TH, Vishnivetskiy SA, McLean M, et al. Monomeric rhodopsin is sufficient for normal rhodopsin kinase (GRK1) phosphorylation and arrestin-1 binding. *J Biol Chem*. 2011;286:1420–1428.
125. Celver J, Vishnivetskiy SA, Chavkin C, Gurevich VV. Conservation of the phosphate-sensitive elements in the arrestin family of proteins. *J Biol Chem*. 2002;277:9043–9048.
126. Kovoor A, Celver J, Abdryashitov RI, Chavkin C, Gurevich VV. Targeted construction of phosphorylation-independent beta-arrestin mutants with constitutive activity in cells. *J Biol Chem*. 1999;274:6831–6834.
127. Gurevich VV, Pals-Rylaarsdam R, Benovic JL, Hosey MM, Onorato JJ. Agonist-receptor-arrestin, an alternative ternary complex with high agonist affinity. *J Biol Chem*. 1997;272:28849–28852.
128. Pan L, Gurevich EV, Gurevich VV. The nature of the arrestin x receptor complex determines the ultimate fate of the internalized receptor. *J Biol Chem*. 2003;278: 11623–11632.
129. Celver J, Lowe J, Kovoor A, Gurevich VV, Chavkin C. Threonine 180 is required for G-protein-coupled receptor kinase 3- and beta-arrestin 2-mediated

desensitization of the mu-opioid receptor in Xenopus oocytes. *J Biol Chem*. 2001; 276:4894–4900.

130. Lowe JD, Celver JP, Gurevich VV, Chavkin C. mu-Opioid receptors desensitize less rapidly than delta-opioid receptors due to less efficient activation of arrestin. *J Biol Chem*. 2002;277:15729–15735.

131. Sommer ME, Farrens DL, McDowell JH, Weber LA, Smith WC. Dynamics of arrestin-rhodopsin interactions: loop movement is involved in arrestin activation and receptor binding. *J Biol Chem*. 2007;282:25560–25568.

132. Hanson SM, Gurevich VV. The differential engagement of arrestin surface charges by the various functional forms of the receptor. *J Biol Chem*. 2006;281:3458–3462.

133. Coffa S, Breitman M, Hanson SM, et al. The effect of arrestin conformation on the recruitment of c-Raf1, MEK1, and ERK1/2 activation. *PLoS One*. 2011;6:e28723.

134. Song X, Raman D, Gurevich EV, Vishnivetskiy SA, Gurevich VV. Visual and both non-visual arrestins in their "inactive" conformation bind JNK3 and Mdm2 and relocalize them from the nucleus to the cytoplasm. *J Biol Chem*. 2006;281:21491–21499.

135. Ahmed MR, Zhan X, Song X, Kook S, Gurevich VV, Gurevich EV. Ubiquitin ligase parkin promotes Mdm2-arrestin interaction but inhibits arrestin ubiquitination. *Biochemistry*. 2011;50:3749–3763.

136. Breitman M, Kook S, Gimenez LE, et al. Silent scaffolds: inhibition of c-Jun N-terminal kinase 3 activity in cell by dominant-negative arrestin-3 mutant. *J Biol Chem*. 2012;287:19653–19664.

137. Smith WC, Gurevich EV, Dugger DR, et al. Cloning and functional characterization of salamander rod and cone arrestins. *Invest Ophthalmol Vis Sci*. 2000;41:2445–2455.

138. Barak LS, Ferguson SS, Zhang J, Caron MG. A beta-arrestin/green fluorescent protein biosensor for detecting G protein-coupled receptor activation. *J Biol Chem*. 1997;272:27497–27500.

139. Walther C, Nagel S, Gimenez LE, Mörl K, Gurevich VV, Beck-Sickinger AG. Ligand-induced internalization and recycling of the human neuropeptide Y2 receptor is regulated by its carboxyl-terminal tail. *J Biol Chem*. 2010;285:41578–41590.

140. Gurevich VV, Richardson RM, Kim CM, Hosey MM, Benovic JL. Binding of wild type and chimeric arrestins to the m2 muscarinic cholinergic receptor. *J Biol Chem*. 1993;268:16879–16882.

141. Gurevich VV, Gurevich EV. Custom-designed proteins as novel therapeutic tools? The case of arrestins. *Expert Rev Mol Med*. 2010;12:e13.

142. Gurevich VV, Gurevich EV. Synthetic biology with surgical precision: targeted reengineering of signaling proteins. *Cell Signal*. 2012;24:899–1908.

143. Gimenez LE, Vishnivetskiy SA, Baameur F, Gurevich VV. Manipulation of very few receptor discriminator residues greatly enhances receptor specificity of non-visual arrestins. *J Biol Chem*. 2012;287:29495–29505.

144. Schöneberg T, Schulz A, Biebermann H, Hermsdorf T, Römpler H, Sangkuhl K. Mutant G-protein-coupled receptors as a cause of human diseases. *Pharmacol Ther*. 2004;104:173–206.

145. Goodman Jr OB, Krupnick JG, Santini F, et al. Beta-arrestin acts as a clathrin adaptor in endocytosis of the beta2-adrenergic receptor. *Nature*. 1996;383:447–450.

146. Wu N, Hanson SM, Francis DJ, et al. Arrestin binding to calmodulin: a direct interaction between two ubiquitous signaling proteins. *J Mol Biol*. 2006;364:955–963.

147. Laporte SA, Oakley RH, Zhang J, et al. The beta2-adrenergic receptor/betaarrestin complex recruits the clathrin adaptor AP-2 during endocytosis. *Proc Natl Acad Sci USA*. 1999;96:3712–3717.

148. McDonald PH, Chow CW, Miller WE, et al. Beta-arrestin 2: a receptor-regulated MAPK scaffold for the activation of JNK3. *Science*. 2000;290:1574–1577.

149. Luttrell LM, Roudabush FL, Choy EW, et al. Activation and targeting of extracellular signal-regulated kinases by beta-arrestin scaffolds. *Proc Natl Acad Sci USA*. 2001;98:2449–2454.
150. Shenoy SK, McDonald PH, Kohout TA, Lefkowitz RJ. Regulation of receptor fate by ubiquitination of activated beta 2-adrenergic receptor and beta-arrestin. *Science*. 2001;294:1307–1313.
151. Krupnick JG, Goodman Jr OB, Keen JH, Benovic JL. Arrestin/clathrin interaction. Localization of the clathrin binding domain of nonvisual arrestins to the carboxy terminus. *J Biol Chem*. 1997;272:15011–15016.
152. Kang DS, Kern RC, Puthenveedu MA, von Zastrow M, Williams JC, Benovic JL. Structure of an arrestin2-clathrin complex reveals a novel clathrin binding domain that modulates receptor trafficking. *J Biol Chem*. 2009;284:29860–29872.
153. Kim YM, Benovic JL. Differential roles of arrestin-2 interaction with clathrin and adaptor protein 2 in G protein-coupled receptor trafficking. *J Biol Chem*. 2002;277: 30760–30768.
154. Schmid EM, Ford MG, Burtey A, et al. Role of the AP2 beta-appendage hub in recruiting partners for clathrin-coated vesicle assembly. *PLoS Biol*. 2006;4:e262.
155. Baillie GS, Adams DR, Bhari N, et al. Mapping binding sites for the PDE4D5 cAMP-specific phosphodiesterase to the N- and C-domains of beta-arrestin using spot-immobilized peptide arrays. *Biochem J*. 2007;404:71–80.
156. Meng D, Lynch MJ, Huston E, et al. MEK1 binds directly to betaarrestin1, influencing both its phosphorylation by ERK and the timing of its isoprenaline-stimulated internalization. *J Biol Chem*. 2009;284:11425–11435.
157. Coffa S, Breitman M, Spiller BW, Gurevich VV. A single mutation in arrestin-2 prevents ERK1/2 activation by reducing c-Raf1 binding. *Biochemistry*. 2011;50: 6951–6958.
158. Seo J, Tsakem EL, Breitman M, Gurevich VV. Identification of arrestin-3-specific residues necessary for JNK3 kinase activation. *J Biol Chem*. 2011;286:27894–27901.
159. Miller WE, McDonald PH, Cai SF, Field ME, Davis RJ, Lefkowitz RJ. Identification of a motif in the carboxyl terminus of beta-arrestin2 responsible for activation of JNK3. *J Biol Chem*. 2001;276:27770–27777.
160. Zhan X, Kaoud TS, Dalby KN, Gurevich VV. Nonvisual arrestins function as simple scaffolds assembling the MKK4-JNK3⟨2 signaling complex. *Biochemistry*. 2011;50:10520–10529.
161. Song X, Gurevich EV, Gurevich VV. Cone arrestin binding to JNK3 and Mdm2: conformational preference and localization of interaction sites. *J Neurochem*. 2007;103: 1053–1062.
162. Song X, Coffa S, Fu H, Gurevich VV. How does arrestin assemble MAPKs into a signaling complex? *J Biol Chem*. 2009;284:685–695.
163. Scott MG, Le Rouzic E, Perianin A, et al. Differential nucleocytoplasmic shuttling of beta-arrestins. Characterization of a leucine-rich nuclear export signal in beta-arrestin2. *J Biol Chem*. 2002;277:37693–37701.
164. Shenoy SK, Lefkowitz RJ. Receptor-specific ubiquitination of beta-arrestin directs assembly and targeting of seven-transmembrane receptor signalosomes. *J Biol Chem*. 2005;280:15315–15324.
165. Gurevich VV. The selectivity of visual arrestin for light-activated phosphorhodopsin is controlled by multiple nonredundant mechanisms. *J Biol Chem*. 1998;273: 15501–15506.
166. Gray-Keller MP, Detwiler PB, Benovic JL, Gurevich VV. Arrestin with a single amino acid substitution quenches light-activated rhodopsin in a phosphorylation-independent fashion. *Biochemistry*. 1997;36:7058–7063.

167. Macey TA, Lowe JD, Chavkin C. Mu opioid receptor activation of ERK1/2 is GRK3 and arrestin dependent in striatal neurons. *J Biol Chem.* 2006;281:34515–34524.
168. Bruchas MR, Macey TA, Lowe JD, Chavkin C. Kappa opioid receptor activation of p38 MAPK is GRK3- and arrestin-dependent in neurons and astrocytes. *J Biol Chem.* 2006;281:18081–18089.
169. Hanson SM, Vishnivetskiy SA, Hubbell WL, Gurevich VV. Opposing effects of inositol hexakisphosphate on rod arrestin and arrestin2 self-association. *Biochemistry.* 2008;47:1070–1075.
170. Nikonov SS, Brown BM, Davis JA, et al. Mouse cones require an arrestin for normal inactivation of phototransduction. *Neuron.* 2008;59:462–474.
171. Storez H, Scott MG, Issafras H, et al. Homo- and hetero-oligomerization of beta-arrestins in living cells. *J Biol Chem.* 2005;280:40210–40215.
172. Milano SK, Kim YM, Stefano FP, Benovic JL, Brenner C. Nonvisual arrestin oligomerization and cellular localization are regulated by inositol hexakisphosphate binding. *J Biol Chem.* 2006;281:9812–9823.
173. Boularan C, Scott MG, Bourougaa K, et al. Beta-arrestin 2 oligomerization controls the Mdm2-dependent inhibition of p53. *Proc Natl Acad Sci USA.* 2007;104:18061–18066.
174. Johnson GL, Dohlman HG, Graves LM. MAPK kinase kinases (MKKKs) as a target class for small-molecule inhibition to modulate signaling networks and gene expression. *Curr Opin Chem Biol.* 2005;9:325–331.
175. Pearson G, Robinson F, Beers Gibson T, et al. Mitogen-activated protein (MAP) kinase pathways: regulation and physiological functions. *Endocr Rev.* 2001;22:153–183.
176. Mushegian A, Gurevich VV, Gurevich EV. The origin and evolution of G protein-coupled receptor kinases. *PLoS One.* 2012;7:e33806.
177. DeWire SM, Ahn S, Lefkowitz RJ, Shenoy SK. Beta-arrestins and cell signaling. *Annu Rev Physiol.* 2007;69:483–510.

CHAPTER FOUR

Arrestins: Role in the Desensitization, Sequestration, and Vesicular Trafficking of G Protein-Coupled Receptors

Cornelia Walther[*,†], **Stephen S.G. Ferguson**[*,†]

[*]J. Allyn Taylor Centre for Cell Biology, Robarts Research Institute, Western University Canada, London, Ontario, Canada

[†]Department of Physiology and Pharmacology, Western University Canada, London, Ontario, Canada

Contents

1. Introduction 93
2. Arrestins in GPCR Desensitization 95
3. Arrestins in GPCR Trafficking 99
   3.1 Sequestration 100
   3.2 Postendocytic vesicular trafficking of GPCRs 104
4. Conclusions 107
Acknowledgments 107
References 107

Abstract

Over the years, β-arrestins have emerged as multifunctional molecular scaffolding proteins regulating almost every imaginable G protein-coupled receptor (GPCR) function. Originally discovered as GPCR-desensitizing molecules, they have been shown to also serve as important regulators of GPCR signaling, sequestration, and vesicular trafficking. This broad functional role implicates β-arrestins as key regulatory proteins for cellular function. Hence, this chapter summarizes the current understanding of the β-arrestin family's unique ability to control the kinetics as well as the extent of GPCR activity at the level of desensitization, sequestration, and subsequent intracellular trafficking.

1. INTRODUCTION

G protein-coupled receptors (GPCRs) are integral membrane proteins that represent the largest and functionally most diverse family of cell-surface receptor proteins. Members of this family share a common

Progress in Molecular Biology and Translational Science, Volume 118
ISSN 1877-1173
http://dx.doi.org/10.1016/B978-0-12-394440-5.00004-8

topographical organization with seven transmembrane spanning α-helices connected by three intracellular and three extracellular loops with extracellular amino and intracellular carboxyl termini. GPCRs transduce the information provided by a multitude of physical and chemical extracellular stimuli into intracellular second messengers and thereby have the ability to mediate and modulate essential biological functions. However, besides their numerous important roles in basic cellular physiology, dysregulation of GPCR function eventually results in pathophysiologic conditions. As a consequence, understanding the molecular mechanisms underlying GPCR function is fundamental to the development of new drugs for a host of human diseases.

The classical paradigm for transduction of external signals across the plasma membrane into the cell following ligand binding to GPCRs involves the coupling of GPCRs to heterotrimeric guanine nucleotide-binding proteins (G proteins), which promotes the exchange of GDP for GTP on the G protein α subunit. This, in turn, leads to the dissociation of Gα and Gβγ subunits.[1] The activated subunits subsequently regulate the activity of a wide variety of effectors, thus regulating intracellular second messenger levels that mediate the cellular response to receptor activation. Besides the G protein-dependent activation of downstream effectors such as ion channels, phospholipases, and adenylyl cyclases, agonist-mediated GPCR activation also results in multiple molecular protein interactions that initiate (1) receptor desensitization, (2) receptor endocytosis, (3) intracellular trafficking between intracellular vesicular compartments, (4) the activation of G protein-independent signaling pathways, and (5) either receptor resensitization or downregulation.[2]

These cellular processes display remarkable kinetic differences. Whereas desensitization occurs within seconds, endocytosis takes place over minutes and resensitization ensues within minutes to hours.[3] All of these processes are governed by a myriad of intracellular accessory proteins, generally termed GPCR-interacting proteins (GIPs). Within the last two decades of research, the list of GIPs has expanded rapidly and continues to grow.[4–8] Notably, one of the first GIPs identified is the cytosolic protein arrestin, which was first shown to bind to GRK1-phosphorylated rhodopsin.[9] The arrestin family in vertebrates is now known to consist of four members: two visual arrestins, arrestin-1 and arrestin-4, that are limited in their expression to the phototransduction pathway (retinal rods and cones), and the two nonvisual arrestins, β-arrestin-1 and β-arrestin-2 (alternatively known as arrestin-2 and arrestin-3). β-Arrestin-1 and β-arrestin-2 are ubiquitously expressed in mammalian tissue and are known to contribute to the uncoupling of many GPCRs from heterotrimeric G proteins.[10,11] For many years, β-arrestin actions were thought to be limited to the desensitization and internalization processes of GPCRs.[12–15] However, over the past several years, β-arrestins have emerged as one of the key players involved in the regulation of multiple facets of GPCR function; hence, they have evolved as multifunctional cellular mediators.[16] Besides the well-known participation of β-arrestins in receptor desensitization, due to their ability to uncouple GPCR/G protein complexes, β-arrestins have also been identified as signal scaffolding proteins as they contain specific interaction sites for various signaling as well as accessory molecules.[12,16,17] Consequently, β-arrestins not only have the capacity to regulate GPCR desensitization, they also control a vast array of additional cellular functions including cell signaling as well as membrane, cytosolic, and nuclear-associated trafficking of GPCRs.[16] Intense research has been conducted to identify the kinetics of β-arrestin binding, as well as the molecular determinants that dictate whether β-arrestins bind to a given GPCR and/or subsequently regulate GPCR activity at multiple steps within the GPCR life cycle. This chapter will focus on the current understanding of how and to what extent β-arrestins control GPCR desensitization, sequestration, and vesicular trafficking.

2. ARRESTINS IN GPCR DESENSITIZATION

Receptor desensitization represents an important physiological process that prevents GPCRs from overstimulation due to prolonged agonist exposure by signal attenuation or termination.[2,3] The classical model of GPCR desensitization involves three processes: (1) receptor phosphorylation and subsequent uncoupling of the receptor from its cognate G protein, (2) receptor sequestration (internalization) to intracellular compartments, and (3) downregulation. Desensitization is typically elicited by three classes of regulatory molecules: (1) second messenger-dependent protein kinases, (2) G protein-coupled receptor kinases (GRKs), and (3) β-arrestins.[3,11] Second messenger-dependent protein kinases, such as protein kinase A and C, that mediate heterologous desensitization, phosphorylate GPCRs at specific consensus sequences within intracellular loops or carboxyl terminal domains that prevent G protein coupling. However, second messenger-dependent protein kinases do not discriminate between inactive and agonist-activated GPCRs. Hence, they prevent the activation of GPCRs that have never been exposed to agonists as well as receptors that have been agonist activated. There is no evidence that second messenger-dependent protein kinase phosphorylation-dependent uncoupling of GPCRs from heterotrimeric G proteins requires the association of β-arrestin with GPCRs.[3] In contrast, phosphorylation by GRKs (homologous desensitization) occurs only following the agonist-dependent isomerization of GPCRs to an activated state that results in GRK-mediated phosphorylation of distinct serine and threonine residues within intracellular GPCR domains.[18] Specifically, GRKs phosphorylate agonist-activated GPCRs because the activated receptor itself activates the kinase.[19] The GRK family consists of seven members (GRK1–7) that display different distribution patterns and show receptor-specific preferences.[20–22] Whereas GRK1 and 7 expression is restricted to retinal rods and cones and GRK4 displays limited expression in the kidneys, testis, and cerebellum, GRK2, -3, -5, and -6 are ubiquitously expressed in mammalian tissues.[22,23] However, GRK-mediated phosphorylation is not sufficient to cause desensitization of most GPCRs, but rather requires the recruitment of the cytosolic cofactor protein arrestin. β-Arrestins bind to agonist-activated and phosphorylated GPCRs and serve to sterically uncouple receptors from heterotrimeric G proteins resulting in the termination of G protein-dependent signal transduction[10,11,24] (Fig. 4.1). Accordingly, the first function of β-arrestins is to

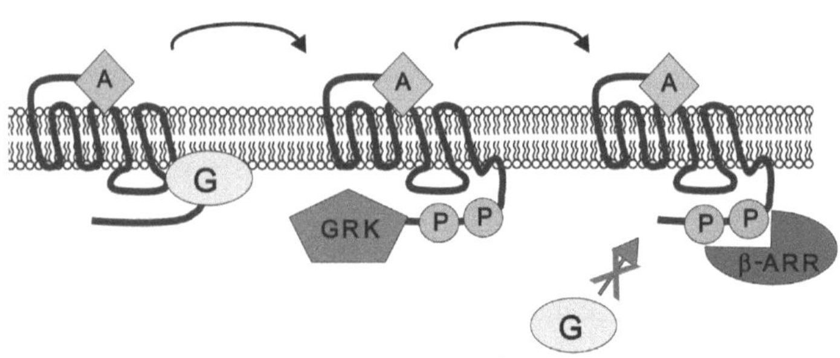

Figure 4.1 Classical model of G protein-coupled receptor kinase (GRK)-mediated G protein-coupled receptor (GPCR) desensitization. Agonist binding leads to conformational changes allowing for coupling and activation of heterotrimeric G proteins. This leads to the specific phosphorylation of agonist-activated receptors by GRKs at various intracellular domains, primarily intracellular loop three and the carboxyl terminus. Subsequently, arrestin is recruited to the phosphorylated receptor, where it sterically uncouples the receptor from its cognate G protein and, in turn, leads to receptor desensitization. A, agonist; G, G protein; GRK, G protein-coupled receptor kinase; P, phosphate moiety; β-ARR, β-arrestin. (For color version of this figure, the reader is referred to the online version of this chapter.)

"arrest" receptor signaling via G proteins. However, there is clearly a tight interplay of GRKs and β-arrestins to control the desensitization process. It has been found for numerous receptors that predominantly GRK2 and GRK3 over GRK5 and GRK6 are responsible for phosphorylation and subsequent β-arrestin recruitment.[25,26] However, several studies exist, where primarily GRK5 and GRK6 have been implicated in desensitization, for example, calcitonin gene-related peptide receptors[27] and dopamine D1A receptors.[28] A proposed GRK-specific "bar code"[29,30] dictates subsequent downstream effects and whether or not a specific β-arrestin isoform binds to the GPCR. For example, GRK2- and GRK3-mediated phosphorylation of CXCR4 receptors leads to β-arrestin-1 recruitment, whereas GRK2-/6-dependent events involve β-arrestin-2 binding.[31] Comparably, the stimulation of the CCR7 receptor by different ligands either leads to the activation of both GRK3 and GRK6 or just GRK6 alone. Different phosphorylation patterns subsequently lead to distinct β-arrestin recruitment: either to the membrane or to endocytic vesicles.[32] Thus, it has become evident that the cellular GRK repertoire, depending on the respective receptor and tissue studied, may dictate the characteristics and scope of receptor responsiveness and desensitization.[28]

Generally, β-arrestin binding to GRK-phosphorylated receptors is determined by the absence or presence of conserved clusters of Ser/Thr residues within the carboxyl terminal tails of the receptor. GPCRs can be broadly divided into two classes, A and B, depending on the stability of β-arrestin binding to the receptor.[33,34] The carboxyl terminal tails of class A receptors, such as the β2-adrenergic receptor (β2AR), consist of a diffusely Ser/Thr-rich cluster and only transiently recruit β-arrestin-2 upon phosphorylation. In contrast, class B receptors like the V2 vasopressin receptor and the angiotensin II type 1A receptor recruit β-arrestin-1 as well as β-arrestin-2 with high affinity and allow for the stable association of β-arrestin with the receptors.

How is this β-arrestin selectivity mechanistically achieved? A "multisite arrestin-receptor interaction" model has been proposed to account for the differences.[35] This model hypothesizes that β-arrestins have two binding sites: one that binds distinct receptor elements which are subject to conformational change upon receptor activation, and a second one that binds receptor-attached phosphates. When a GPCR is activated and subsequently phosphorylated and β-arrestin binds via both interaction sites, it is subject to transition into its active high-affinity receptor-binding state. This mechanism most likely contributes to the preferential desensitization of certain

GPCRs by β-arrestins.[22] To bind arrestins with high affinity, many GPCRs must be phosphorylated. This was first shown for rhodopsin[36] and later for numerous other GPCRs such as the β2AR,[24,37–40] angiotensin II type 1 receptor,[41] α2-adrenergic receptor,[42] m2 muscarinic receptor,[38] and neuropeptide Y1 receptor.[43] The majority of relevant phosphorylation sites are localized to the carboxyl terminal tails of most GPCRs, but relevant phosphorylation sites have been mapped to any intracellular domain including the intracellular loops (as reviewed in Ref. 44). Nonetheless, β-arrestins were also found to bind to several unphosphorylated receptors.[45–47] Phosphorylation-independent β-arrestin binding can be explained by acidic amino acid stretches within intracellular domains. Acidic amino acids can successfully mimic phosphate groups[48,49]; thus, the density of negative charges is sufficient to activate the arrestin phospho-binding site which leads to receptor interaction, such has been demonstrated for the D6 chemokine receptor.[50]

Due to the vast complexity of phosphorylation events that occur on different GPCRs, it is still a matter of debate how only seven GRKs and four arrestins can specifically regulate several hundred different GPCRs. For many GPCRs, it appears that all GRKs or arrestins (except retinal arrestins) contribute to the regulation of their desensitization. However, many GPCRs display a distinct preference for a specific GRK and arrestin isoform to mediate their desensitization. For example, GRK6 and β-arrestin-2 regulate μ-opioid receptors and D2-like dopamine receptors in striatum.[21] It has also become evident that the regulation of a given GPCR depends on cell background, as its regulation can be affected by differences in the expression levels of distinct GRKs and β-arrestins between different cell types and tissues. It remains an open question whether certain GPCRs are regulated either randomly by various GRK/β-arrestin combinations or by specific GRK and β-arrestin pairs, and whether a particular GRK/β-arrestin pairing takes precedence over other pairings.[21] To add an additional level of complexity to GRK and β-arrestin regulation of GPCR activity, Schulz *et al.*[51] recently identified a novel mechanism whereby distinct opioid agonists stimulated site-specific μ-opioid receptor phosphorylation patterns. Thus, morphine binding results in a μ-opioid receptor conformation with next to no affinity for β-arrestins, whereas enkephalin analogs induce a rapid high-affinity binding of both β-arrestin-1 and -2.[51,52] Similarly, the chemokine receptor CCR2 displays different affinities for both β-arrestin-1 and -2, depending on whether the receptor was stimulated with different CCR2 ligands. Furthermore, while CCL7 induces transient binding of β-arrestins, CCL8/13 leads to the stable formation of a CCLR7/β-arrestin complex. Consequently, stimulation with different ligands stabilizes different receptor conformations, leading to qualitative differences in arrestin responses.[53]

The importance of arrestins in GPCR desensitization has been clearly demonstrated in studies using *Drosophila* and mice. Drosophila strains that display mutations in the arrestin gene show impaired inactivation of metarhodopsin.[54] Similarly, β-arrestin-2 knock-out mice exhibit a significant potentiation and prolongation of morphine-induced analgesia indicating altered μ-opioid receptor desensitization.[55] In addition, and apart from its well-established and extensively studied function in uncoupling the GPCR from its cognate G protein, more recent studies discovered novel roles for arrestins in the regulation of GPCR desensitization. β-Arrestins have been identified to serve as shuttle proteins that localize phosphodiesterases within the vicinity of agonist-activated βARs.[56] In cardiac myocytes, β-arrestin-mediated phosphodiesterase recruitment to activated βARs promotes the switch from G_s to G_i coupling and subsequently results in βAR signaling shifting toward a pathway that limits cAMP production.[57] In addition, β-arrestins have been shown to reduce the level of the second messenger diacylglycerol upon agonist stimulation of the M1 cholinergic receptor.[58] These results suggest novel impacts of β-arrestins on GPCR desensitization, as they presumably function to limit the generation of second messengers, but are also capable of enhancing the rate of second messenger degradation.[16] These findings clearly expand the functional relevance of arrestin in GPCR desensitization.

3. ARRESTINS IN GPCR TRAFFICKING

While desensitization serves to protect receptors from overstimulation, receptor sequestration (internalization) is required to prevent prolonged desensitization, as well as to enable GPCRs to either resensitize or become downregulated upon agonist removal.[2,3] In addition to the well-established importance of β-arrestins in desensitization, they also play a central role in mediating GPCR endocytosis. β-Arrestins function as endocytic adaptor proteins, recruiting membrane-bound activated receptors to the internalization machinery.[59] The process of agonist-induced sequestration leads to the functional removal of activated receptors from the plasma membrane and targets them to intracellular compartments for either cell surface recycling and resensitization or degradation. But the question arises as to how this is accomplished mechanistically. It turns out that the underlying mechanism(s) for β-arrestin-dependent regulation of GPCR trafficking is far more complex than was initially assumed. This is because β-arrestins display varying binding affinities to different GPCRs depending on the cell type studied.

3.1. Sequestration

The sequestration of GPCRs is essential to the maintenance and regulation of agonist responsiveness, as it can result in not only dephosphorylation, recycling, and resensitization of the receptor but also in receptor degradation and activation of G protein-independent intracellular signaling pathways.[22] The first evidence that β-arrestins have functions beyond arresting G protein-mediated signaling was reported in 1996.[60] In this pioneering study, endocytosis-deficient β2AR mutants were rescued by the over-expression of wild-type β-arrestins and β2AR internalization was prevented by the expression of dominant-negative β-arrestin mutants. The sequestration of most GPCRs requires GRK phosphorylation and β-arrestin binding such that these events are indispensible prerequisites. However, depending on the GRK involved in receptor phosphorylation and subsequent β-arrestin binding, distinct downstream processes are favored. For example, it has been shown that only GRK2-mediated phosphorylation initiates β-arrestin-dependent endocytosis of the V2 vasopressin and angiotensin II type 1A receptors.[25,26] In contrast, vasopressin and angiotensin receptor phosphorylation by GRK5 and GRK6 triggers β-arrestin-dependent activation of ERK pathways, rather than internalization. Similar effects have been described for β2AR, as well as for the follicle-stimulating hormone receptor.[25,26,31]

In general, three common mechanisms have been implicated in GPCR internalization. These mechanisms either involve (1) clathrin-coated pits, (2) caveolae, or (3) other yet to be characterized uncoated vesicles.[61,62] Whereas GRK-mediated phosphorylation and β-arrestin binding are critical to clathrin-dependent sequestration,[14] receptors that lack GRK phosphorylation sites are sequestered via caveolae in a β-arrestin-independent manner, for example, the β1AR.[63] β-Arrestins function as adaptor molecules to link GPCRs to the clathrin-coated pit machinery through their ability to bind directly and stoichiometrically to the major structural components of the endocytic machinery.[64–66] Such components include the heavy chain of clathrin,[67] as well as β-adaptin, a component of the clathrin-associated adaptor protein complex AP-2.[68,69] Clathrin was discovered as the first nonreceptor β-arrestin-interacting protein in 1996.[67] AP-2 is involved in the recruitment of clathrin and the subsequent assembly of clathrin lattices, the major components of the coat of the internalized membrane. Binding of clathrin and AP-2 is facilitated by β-arrestin binding to the receptor, which leads to the release of the β-arrestin carboxy-terminal tail, the region where both interaction partners bind. Thus, clathrin and AP-2 binding to β-arrestins during endocytosis has turned out to be a critical step in internalization.[12]

Whether GPCR sequestration involves clathrin-coated pits, caveolae, and/or other uncoated vesicles, internalization rates are highly receptor and cell-type dependent. Almost every possible internalization mechanism for different GPCRs has been described in the literature: β-arrestin- and clathrin-dependent, β-arrestin- and clathrin-independent, β-arrestin-independent and clathrin-dependent, and also β-arrestin-dependent and clathrin-independent.[62] Thus, it seems that β-arrestins are sometimes redundant for the internalization of some GPCRs.[70] There is also evidence that certain receptors can choose different internalization pathways and can also dictate whether they internalize via either β-arrestin-dependent or independent pathways. For example, the chemokine receptor CCR5 and m2 muscarinic receptors internalize in a β-arrestin-dependent manner and, although bound to β-arrestin, can also internalize in a β-arrestin-independent manner.[71] Other examples include the full length A2B adenosine receptor and the neuropeptide Y2 receptor, which typically internalize via a β-arrestin-dependent pathway. However, carboxyl terminal truncations of both receptors serve to redirect them to a β-arrestin-independent internalization pathway.[72,73] Thus, β-arrestins provide GPCRs with the ability to interact with the clathrin-coated pit internalization machinery, but this interaction does not necessarily predetermine whether GPCRs are endocytosed in clathrin-coated vesicles. However, it is now clear that the majority of GPCRs utilize the β-arrestin-dependent, clathrin-mediated internalization pathway (Fig. 4.2).

The two nonvisual mammalian β-arrestins display different characteristics with respect to cellular distribution patterns, and binding properties to their GPCR substrates. While β-arrestin-1, comparable to visual arrestin, is localized in the cytoplasm and the nucleus, β-arrestin-2 is limited to the cytoplasm.[34,74] They also have different abilities to mediate internalization. Evidence to support this has come from studies with knockout mice either lacking β-arrestin-1 or -2. Although both β-arrestin-1 and -2 contribute to the desensitization of the β2AR, β-arrestin-2 appears to be significantly

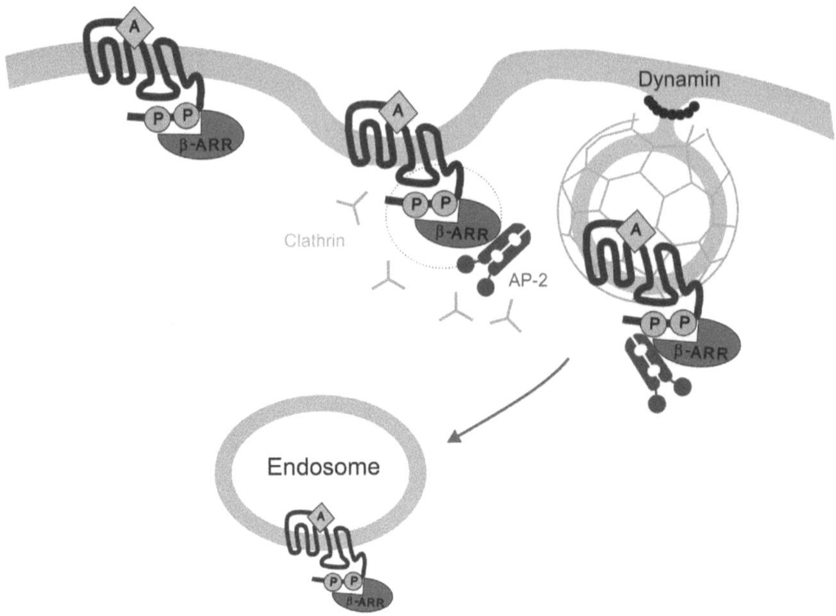

Figure 4.2 The prototypical model of β-arrestin-promoted sequestration of GPCRs via the clathrin-mediated internalization pathway. Agonist-activated phosphorylated receptors are guided to clathrin-coated pits by the recruitment of β-arrestin, which binds to AP-2 and clathrin. The vesicles are subsequently pinched off by the GTPase dynamin, which leads to receptor internalization into early endosomes. (For color version of this figure, the reader is referred to the online version of this chapter.)

more efficient at promoting its internalization.[75] In contrast, activation of the angiotensin II type 1 receptor results in the recruitment of both β-arrestin-1 and -2, which display comparable efficiency in desensitizing and internalizing the receptor.[34,75] These findings have led to the classification of GPCRs into two subgroups that reflect the stability and outcome of GPCR/arrestin interactions. Class A receptors, including β2AR, μ-opioid receptors, endothelin type A receptors, and dopamine D1A receptors, have a higher affinity for β-arrestin-2 and form transient GPCR/arrestin complexes that dissociate at/near the cell surface. In contrast, class B receptors, such as V2 vasopressin, neurotensin 1, and angiotensin II type 1A receptors, bind β-arrestin-1 and -2 with equivalent affinity and form stable GPCR/β-arrestin complexes that remain intact as the receptors endocytose.[34,76,77] Chimeric receptors (β2AR/angiotensin II type 1 receptors) have provided strong evidence that the stability of GPCR/β-arrestin complexes are controlled by determinants within the GPCR carboxyl terminal tail domains. Consequently, we now know that the receptor carboxyl terminus determines the stability of the GPCR/β-arrestin complex and subsequent cellular distribution of β-arrestin.[76] This is likely achieved by the distinct phosphorylation patterns induced by different GRKs, the density of phosphorylation sites within a given GPCR C-tail or intracellular loop domain, and the induction of conformational changes in receptor-bound β-arrestins, as these conformational states may vary and will likely reflect dynamic changes in the binding of downstream accessory proteins.[33,44]

The list of GPCRs that have been shown to bind to β-arrestins via the phosphorylated Ser/Thr residues within their carboxyl terminal tails initially consisted only of the rhodopsin receptor and the β2AR. However, due to intense investigation in this field over the past two decades worldwide, this list is continuously expanding. To date, evidence for receptor/β-arrestin interactions involved in internalization is provided for β2AR,[60,78,79] M2-muscarinic acetylcholine receptors,[70,80,81] δ-opioid receptors,[82] angiotensin II type 1 receptors,[83] human P2Y1 receptors,[84] dopamine D1 receptors,[85] neuropeptide Y1, Y2, and Y4 receptors,[43,73,86,87] and many more. The contribution of specific intracellular receptor domains to β-arrestin binding and subsequent sequestration has been studied intensively. Most commonly, the carboxyl terminus (prototypical model) and, to a lesser extent, the third intracellular loop contribute to endocytic trafficking. However, even within the class of adrenergic receptor subtypes, specific differences occur. For example, β-arrestins bind to the carboxyl terminus of the β2AR, but they bind to the third intracellular loop of α-adrenergic receptors.[88] Similar characteristics have been documented for the muscarinic acetylcholine receptors 2 and 3.[88,89] Clearly, receptor domains are significantly involved in the regulation of sequestration.

Another important regulator of β-arrestin-promoted clathrin-dependent receptor internalizations is the posttranslational modification of β-arrestin itself. β-Arrestin-1 is regulated by phosphorylation/dephosphorylation. Ser412 within the carboxyl termini of β-arrestin-1 is constitutively phosphorylated and recruited to the plasma membrane as soon as the receptor is activated in an agonist-dependent manner. Subsequently, β-arrestin-1 becomes dephosphorylated, which is required for clathrin binding and targeting to clathrin-coated pits.[49] Besides phosphorylation, ubiquitination of β-arrestins is another required process for effective internalization. β-Arrestin-2 binds the mouse double minute 2 (mdm2) protein, an E3 ubiquitin ligase, and becomes ubiquitinated, a process shown to regulate the stability of the β2AR/β-arrestin interaction internalization.[90] Likewise, β-arrestin ubiquitinylation is fundamental to subsequent clathrin-mediated endocytosis.[15]

Recently, a new interaction partner for β-arrestin, phosphatidylinositol 4-phosphate 5-kinase, has been shown to be involved in the agonist-mediated sequestration of the β2AR receptor. The complex formation of β-arrestin-2, phosphatidylinositol 4-phosphate 5-kinase, and the agonist-bound β2AR appears to be essential to regulate receptor internalization.[91] Thus, strong evidence is provided that the function of β-arrestins at the level of GPCR sequestration goes far beyond the fact that β-arrestin binds agonist-activated and GRK-phosphorylated receptors.

3.2. Postendocytic vesicular trafficking of GPCRs

Once GPCRs are sequestered and transported into intracellular compartments, β-arrestins also have the ability to influence the intracellular fate of GPCRs by regulating their postendocytic trafficking. The majority of GPCRs are either recycled or degradated. It is still largely unknown how cells decide which pathway a particular receptor will take. There is a growing body of evidence suggesting that cells decide on the ultimate fate of certain receptors based on the properties of the receptor/β-arrestin complex. The division of GPCRs into class A and B reflects the apparent affinities for β-arrestin. While class A receptors, which show higher affinity for β-arrestin-2 than β-arrestin-1, traffic to endosomes and recycle directly back afterward, class B receptors, which bind both β-arrestins equally well, form a stable complex with arrestin and are either retained in endosomes for hours before being recycled or are targeted to lysosomes for final degradation.[20,33,34,76,92] Apparently, β-arrestin and its receptor affinity represent a critical determinant for GPCR postendocytic sorting as it determines the rate of receptor resensitization (Fig. 4.3). Evidence is provided by studies with chimeric β2AR and V2 vasopressin receptors: exchange of the carboxyl terminal tails of the rapidly recycling β2AR and the extremely slow recycling V2 vasopressin receptor completely reverses the pattern of β-arrestin binding and the propensity of these receptors to dephosphorylate, recycle, and resensitize.[92] The same effect was observed for the exchange of the C-termini of two vasopressin receptor subtypes: V1A, which recycles rapidly, and V2, which moves slowly. Exchange of the carboxyl termini between these two receptors also switches the trafficking pattern of the receptors.[93] As the stable association of β-arrestin is promoted by a specific cluster of phosphorylated Ser/Thr residues within the carboxyl terminus, the interaction of β-arrestins with such a specific phospho-cluster, a

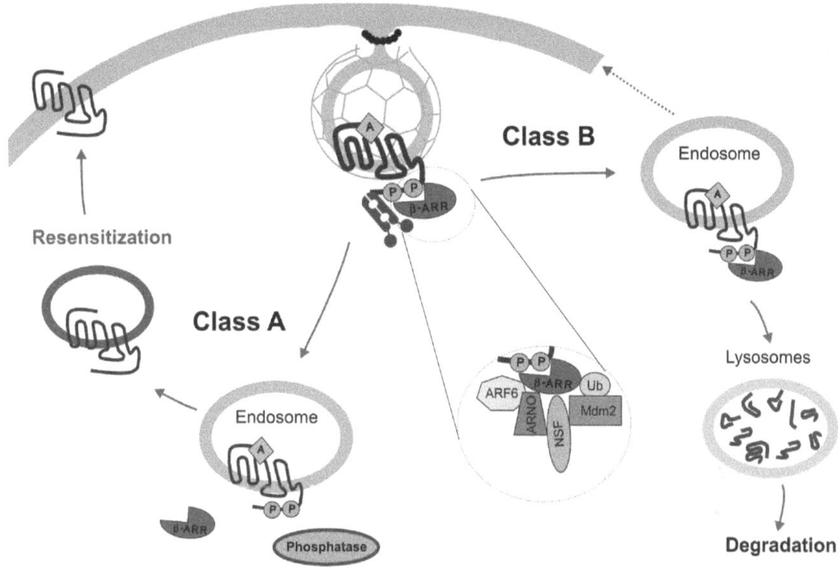

Figure 4.3 Postendocytic trafficking of GPCRs strongly depends on the stability of receptor/arrestin complexes as well as the interaction with various other intracellular proteins that interact with the receptor-bound β-arrestin during the internalization process. The magnified region depicts those arrestin-interacting proteins influencing internalization and thus subsequent trafficking to intracellular compartments. As internalization is the prerequisite for subcellular trafficking, these arrestin-interacting proteins are involved in both mechanisms and boundaries are blurred. Once internalized, GPCRs can be divided into two classes, depending on the properties of the receptor–arrestin complex. Class A receptors, which preferentially bind β-arrestin-2, dissociate rapidly from the bound arrestin, become dephosphorylated, and transit through recycling endosomes before reaching the plasma membrane for efficient resensitization. In contrast, class B receptors, which recruit both arrestins with equal affinity, remain associated with arrestin during internalization and are targeted to lysosomes for final degradation. Occasionally, class B receptors are retained in early endosomes for a prolonged period before undergoing recycling. A, agonist; P, phosphate moiety; β-ARR, β-arrestin; ARF6, G protein ADP-ribosylation factor 6; ARNO, ARF nucleotide-binding site opener; Mdm2, mouse double minute protein 2, an E3 ubiquitin ligase; NSF, *N*-ethylmaleimide-sensitive fusion protein; Ub, Ubiquitin. (For color version of this figure, the reader is referred to the online version of this chapter.)

predetermined distance from the seventh transmembrane domain, appears to dictate the rate of dephosphorylation, recycling, and resensitization.[92] Studies examining β2AR endocytosis in the presence of β-arrestin mutants that have either decreased or increased affinities for GPCRs and endocytic adaptor proteins revealed that arrestin directs the receptor to recycling or degradation pathways, respectively.[94,95] Therefore, it can be concluded that β-arrestins regulate a GPCR's ability to resensitize and ultimately reestablish responsiveness.[96] Both β-arrestin expression levels and the rate of β-arrestin release dictate fate.[44,96]

Not surprisingly, exceptions to the rule exist as not all GPCRs fit into the class A/B dichotomy. Somatostatin (sst) receptors, for example, do not follow the classical postendocytic trafficking patterns for class A and B receptors as described earlier. While the sst2A somatostatin receptor exhibits a typical class B receptor β-arrestin-dependent trafficking pattern (stable β-arrestin–receptor complexes, cointernalization), this receptor is rapidly recycled and resensitized rather than targeted for degradation. In contrast, the somatostatin receptor subtypes sst3 and sst5 display typical class A receptor properties (transient β-arrestin–receptor complexes, receptor internalization without β-arrestin), but unlike other class A receptors, are targeted to lysosomes for degradation.[97] Similar observations were made for the spontaneous internalization of the α1A-adrenergic receptor. Although this receptor internalizes in a β-arrestin-dependent manner, it travels with β-arrestin bound to recycling endosomes.[98] A third example is the metabotropic glutamate receptor 1, which interacts selectively with β-arrestin-1, not β-arrestin-2.[99] Even so, for most GPCRs, the stability of the receptor-β-arrestin complex strongly affects the postendocytic trafficking pattern, and thereby determines the ultimate subcellular fate.

In the past few years, it has become evident that the capacity of β-arrestins to engage in protein–protein interactions is not limited to receptors and components of the clathrin-mediated endocytosis machinery, or dedicated to the control of receptor desensitization and internalization. The number of identified β-arrestin-interacting partners is growing rapidly. β-Arrestins are capable of forming an interface between intracellular components of the vesicular trafficking and sorting machinery and internalized receptors. Recently, novel intracellular interaction partners involved in arrestin-promoted clathrin-mediated trafficking have been identified. Among them are the small G protein ADP-ribosylation factor 6 and its guanine nucleotide exchange factor ADP-ribosylation factor nucleotide-binding site opener,[100] and the *N*-ethylmaleimide-sensitive fusion protein[101] (Fig. 4.3). These proteins subsequently act as adaptors to facilitate the clathrin-mediated endocytosis of GPCRs.

Another regulatory factor in β-arrestin-mediated postendocytic sorting is the ubiquitination status of β-arrestin, which dramatically influences the subcellular fate of GPCRs. For example, a constitutively ubiquitinated β-arrestin-ubiquitin chimera is unable to dissociate from the β2AR, and its expression in cells directs the receptor toward degradative pathways rather than the naturally occurring recycling and resensitization.[15]

4. CONCLUSIONS

Arrestin function in GPCR desensitization, sequestration, and vesicular trafficking has been studied intensively over decades. Thereby, researchers worldwide were able to establish a prototypical mechanism on how arrestin regulates each of these steps individually. Nonetheless, exceptions to the predominant rules governing the regulation of GPCRs will always exist and may provide advantages for the development of new therapeutics to treat disease. It is increasingly becoming evident that GPCRs are not the only substrates for β-arrestin. The β-arrestin interaction network is far more complex than assumed 20 years ago. Arrestins are now recognized as key scaffolding molecules that connect membrane receptors to multiple cytosolic proteins that use the β-arrestin-bound receptor as the platform for assembling multiple protein signaling complexes. Thus, β-arrestins have the remarkable ability both to regulate the kinetics and the extent of GPCR cellular activity at the level of desensitization, internalization, and subsequent intracellular fate, and control the activation and compartmentation of a wide variety of G protein-independent signal transduction cascades.

ACKNOWLEDGMENTS

This work was supported by an operating grant to S. S. G. F. from the Canadian Institutes of Health Research (CIHR) (MOP-119437, MOP-62738, and MOP-111093) and a Navigator Grant from the Huntington's Society of Canada. S. S. G. F. also holds a Tier I Canada Research Chair in Molecular Neurobiology and is a Career Investigator of the Heart and Stroke Foundation of Ontario.

REFERENCES

1. Neer EJ. Heterotrimeric G proteins: organizers of transmembrane signals. *Cell.* 1995;80:249–257.
2. Seachrist JL, Ferguson SS. Regulation of G protein-coupled receptor endocytosis and trafficking by Rab GTPases. *Life Sci.* 2003;74:225–235.
3. Ferguson SS. Evolving concepts in G protein-coupled receptor endocytosis: the role in receptor desensitization and signaling. *Pharmacol Rev.* 2001;53:1–24.
4. Bockaert J, Fagni L, Dumuis A, Marin P. GPCR interacting proteins (GIP). *Pharmacol Ther.* 2004;103:203–221.
5. Magalhaes AC, Dunn H, Ferguson SS. Regulation of GPCR activity, trafficking and localization by GPCR-interacting proteins. *Br J Pharmacol.* 2012;165:1717–1736.
6. Bockaert J, Roussignol G, Becamel C, et al. GPCR-interacting proteins (GIPs): nature and functions. *Biochem Soc Trans.* 2004;32:851–855.

7. Ritter SL, Hall RA. Fine-tuning of GPCR activity by receptor-interacting proteins. *Nat Rev Mol Cell Biol*. 2009;10:819–830.
8. Maurice P, Guillaume JL, Benleulmi-Chaachoua A, Daulat AM, Kamal M, Jockers R. GPCR-interacting proteins, major players of GPCR function. *Adv Pharmacol*. 2011;62:349–380.
9. Pfister C, Chabre M, Plouet J, et al. Retinal S antigen identified as the 48 K protein regulating light-dependent phosphodiesterase in rods. *Science*. 1985;228:891–893.
10. Ferguson SS, Barak LS, Zhang J, Caron MG. G-protein-coupled receptor regulation: role of G-protein-coupled receptor kinases and arrestins. *Can J Physiol Pharmacol*. 1996;74:1095–1110.
11. Krupnick JG, Benovic JL. The role of receptor kinases and arrestins in G protein-coupled receptor regulation. *Annu Rev Pharmacol Toxicol*. 1998;38:289–319.
12. McDonald PH, Lefkowitz RJ. Beta-arrestins: new roles in regulating heptahelical receptors' functions. *Cell Signal*. 2001;13:683–689.
13. Pierce KL, Premont RT, Lefkowitz RJ. Seven-transmembrane receptors. *Nat Rev Mol Cell Biol*. 2002;3:639–650.
14. Shenoy SK, Lefkowitz RJ. Multifaceted roles of beta-arrestins in the regulation of seven-membrane-spanning receptor trafficking and signalling. *Biochem J*. 2003;375: 503–515.
15. Shenoy SK, Lefkowitz RJ. Trafficking patterns of beta-arrestin and G protein-coupled receptors determined by the kinetics of beta-arrestin deubiquitination. *J Biol Chem*. 2003;278:14498–14506.
16. Barki-Harrington L, Rockman HA. Beta-arrestins: multifunctional cellular mediators. *Physiology (Bethesda)*. 2008;23:17–22.
17. Milano SK, Pace HC, Kim YM, Brenner C, Benovic JL. Scaffolding functions of arrestin-2 revealed by crystal structure and mutagenesis. *Biochemistry*. 2002;41:3321–3328.
18. Kohout TA, Lefkowitz RJ. Regulation of G protein-coupled receptor kinases and arrestins during receptor desensitization. *Mol Pharmacol*. 2003;63:9–18.
19. Palczewski K, Buczylko J, Kaplan MW, Polans AS, Crabb JW. Mechanism of rhodopsin kinase activation. *J Biol Chem*. 1991;266:12949–12955.
20. Moore CA, Milano SK, Benovic JL. Regulation of receptor trafficking by GRKs and arrestins. *Annu Rev Physiol*. 2007;69:451–482.
21. Premont RT, Gainetdinov RR. Physiological roles of G protein-coupled receptor kinases and arrestins. *Annu Rev Physiol*. 2007;69:511–534.
22. Reiter E, Lefkowitz RJ. GRKs and beta-arrestins: roles in receptor silencing, trafficking and signaling. *Trends Endocrinol Metab*. 2006;17:159–165.
23. Pitcher JA, Freedman NJ, Lefkowitz RJ. G protein-coupled receptor kinases. *Annu Rev Biochem*. 1998;67:653–692.
24. Benovic JL, Kuhn H, Weyand I, Codina J, Caron MG, Lefkowitz RJ. Functional desensitization of the isolated beta-adrenergic receptor by the beta-adrenergic receptor kinase: potential role of an analog of the retinal protein arrestin (48-kDa protein). *Proc Natl Acad Sci USA*. 1987;84:8879–8882.
25. Kim J, Ahn S, Ren XR, et al. Functional antagonism of different G protein-coupled receptor kinases for beta-arrestin-mediated angiotensin II receptor signaling. *Proc Natl Acad Sci USA*. 2005;102:1442–1447.
26. Ren XR, Reiter E, Ahn S, Kim J, Chen W, Lefkowitz RJ. Different G protein-coupled receptor kinases govern G protein and beta-arrestin-mediated signaling of V2 vasopressin receptor. *Proc Natl Acad Sci USA*. 2005;102:1448–1453.
27. Aiyar N, Disa J, Dang K, Pronin AN, Benovic JL, Nambi P. Involvement of G protein-coupled receptor kinase-6 in desensitization of CGRP receptors. *Eur J Pharmacol*. 2000;403:1–7.

28. Tiberi M, Nash SR, Bertrand L, Lefkowitz RJ, Caron MG. Differential regulation of dopamine D1A receptor responsiveness by various G protein-coupled receptor kinases. *J Biol Chem*. 1996;271:3771–3778.
29. Tobin AB, Butcher AJ, Kong KC. Location, location, location...site-specific GPCR phosphorylation offers a mechanism for cell-type-specific signalling. *Trends Pharmacol Sci*. 2008;29:413–420.
30. Nobles KN, Xiao K, Ahn S, et al. Distinct phosphorylation sites on the beta(2)-adrenergic receptor establish a barcode that encodes differential functions of beta-arrestin. *Sci Signal*. 2011;4:ra51.
31. Busillo JM, Armando S, Sengupta R, Meucci O, Bouvier M, Benovic JL. Site-specific phosphorylation of CXCR4 is dynamically regulated by multiple kinases and results in differential modulation of CXCR4 signaling. *J Biol Chem*. 2010;285:7805–7817.
32. Zidar DA, Violin JD, Whalen EJ, Lefkowitz RJ. Selective engagement of G protein coupled receptor kinases (GRKs) encodes distinct functions of biased ligands. *Proc Natl Acad Sci USA*. 2009;106:9649–9654.
33. Oakley RH, Laporte SA, Holt JA, Barak LS, Caron MG. Molecular determinants underlying the formation of stable intracellular G protein-coupled receptor-beta-arrestin complexes after receptor endocytosis*. *J Biol Chem*. 2001;276: 19452–19460.
34. Oakley RH, Laporte SA, Holt JA, Caron MG, Barak LS. Differential affinities of visual arrestin, beta arrestin1, and beta arrestin2 for G protein-coupled receptors delineate two major classes of receptors. *J Biol Chem*. 2000;275:17201–17210.
35. Gurevich VV, Benovic JL. Visual arrestin interaction with rhodopsin. Sequential multisite binding ensures strict selectivity toward light-activated phosphorylated rhodopsin. *J Biol Chem*. 1993;268:11628–11638.
36. Kuhn H, Hall SW, Wilden U. Light-induced binding of 48-kDa protein to photoreceptor membranes is highly enhanced by phosphorylation of rhodopsin. *FEBS Lett*. 1984;176:473–478.
37. Attramadal H, Arriza JL, Aoki C, et al. Beta-arrestin2, a novel member of the arrestin/beta-arrestin gene family. *J Biol Chem*. 1992;267:17882–17890.
38. Gurevich VV, Dion SB, Onorato JJ, et al. Arrestin interactions with G protein-coupled receptors. Direct binding studies of wild type and mutant arrestins with rhodopsin, beta 2-adrenergic, and m2 muscarinic cholinergic receptors. *J Biol Chem*. 1995;270: 720–731.
39. Lohse MJ, Andexinger S, Pitcher J, et al. Receptor-specific desensitization with purified proteins. Kinase dependence and receptor specificity of beta-arrestin and arrestin in the beta 2-adrenergic receptor and rhodopsin systems. *J Biol Chem*. 1992;267: 8558–8564.
40. Lohse MJ, Benovic JL, Codina J, Caron MG, Lefkowitz RJ. Beta-arrestin: a protein that regulates beta-adrenergic receptor function. *Science*. 1990;248:1547–1550.
41. Qian H, Pipolo L, Thomas WG. Association of beta-arrestin 1 with the type 1A angiotensin II receptor involves phosphorylation of the receptor carboxyl terminus and correlates with receptor internalization. *Mol Endocrinol*. 2001;15:1706–1719.
42. Wang Q, Limbird LE. Regulated interactions of the alpha 2A adrenergic receptor with spinophilin, 14-3-3zeta, and arrestin 3. *J Biol Chem*. 2002;277:50589–50596.
43. Holliday ND, Lam CW, Tough IR, Cox HM. Role of the C terminus in neuropeptide Y Y1 receptor desensitization and internalization. *Mol Pharmacol*. 2005;67:655–664.
44. Gurevich VV, Gurevich EV. The structural basis of arrestin-mediated regulation of G-protein-coupled receptors. *Pharmacol Ther*. 2006;110:465–502.
45. Milasta S, Evans NA, Ormiston L, Wilson S, Lefkowitz RJ, Milligan G. The sustainability of interactions between the orexin-1 receptor and beta-arrestin-2 is defined by a

single C-terminal cluster of hydroxy amino acids and modulates the kinetics of ERK MAPK regulation. *Biochem J.* 2005;387:573–584.
46. Min L, Galet C, Ascoli M. The association of arrestin-3 with the human lutropin/choriogonadotropin receptor depends mostly on receptor activation rather than on receptor phosphorylation. *J Biol Chem.* 2002;277:702–710.
47. Mukherjee S, Gurevich VV, Preninger A, et al. Aspartic acid 564 in the third cytoplasmic loop of the luteinizing hormone/choriogonadotropin receptor is crucial for phosphorylation-independent interaction with arrestin2. *J Biol Chem.* 2002;277:17916–17927.
48. Lin FT, Chen W, Shenoy S, Cong M, Exum ST, Lefkowitz RJ. Phosphorylation of beta-arrestin2 regulates its function in internalization of beta(2)-adrenergic receptors. *Biochemistry.* 2002;41:10692–10699.
49. Lin FT, Krueger KM, Kendall HE, et al. Clathrin-mediated endocytosis of the beta-adrenergic receptor is regulated by phosphorylation/dephosphorylation of beta-arrestin1. *J Biol Chem.* 1997;272:31051–31057.
50. Galliera E, Jala VR, Trent JO, et al. Beta-arrestin-dependent constitutive internalization of the human chemokine decoy receptor D6. *J Biol Chem.* 2004;279:25590–25597.
51. Doll C, Konietzko J, Poll F, Koch T, Hollt V, Schulz S. Agonist-selective patterns of micro-opioid receptor phosphorylation revealed by phosphosite-specific antibodies. *Br J Pharmacol.* 2011;164:298–307.
52. Groer CE, Tidgewell K, Moyer RA, et al. An opioid agonist that does not induce mu-opioid receptor-arrestin interactions or receptor internalization. *Mol Pharmacol.* 2007;71:549–557.
53. Berchiche YA, Gravel S, Pelletier ME, St-Onge G, Heveker N. Different effects of the different natural CC chemokine receptor 2b ligands on beta-arrestin recruitment, G alpha i signaling, and receptor internalization. *Mol Pharmacol.* 2011;79:488–498.
54. Dolph PJ, Ranganathan R, Colley NJ, Hardy RW, Socolich M, Zuker CS. Arrestin function in inactivation of G protein-coupled receptor rhodopsin in vivo. *Science.* 1993;260:1910–1916.
55. Bohn LM, Lefkowitz RJ, Gainetdinov RR, Peppel K, Caron MG, Lin FT. Enhanced morphine analgesia in mice lacking beta-arrestin 2. *Science.* 1999;286:2495–2498.
56. Perry SJ, Baillie GS, Kohout TA, et al. Targeting of cyclic AMP degradation to beta 2-adrenergic receptors by beta-arrestins. *Science.* 2002;298:834–836.
57. Baillie GS, Sood A, McPhee I, et al. Beta-arrestin-mediated PDE4 cAMP phosphodiesterase recruitment regulates beta-adrenoceptor switching from Gs to Gi. *Proc Natl Acad Sci USA.* 2003;100:940–945.
58. Nelson CD, Perry SJ, Regier DS, Prescott SM, Topham MK, Lefkowitz RJ. Targeting of diacylglycerol degradation to M1 muscarinic receptors by beta-arrestins. *Science.* 2007;315:663–666.
59. Ferguson SS. Phosphorylation-independent attenuation of GPCR signalling. *Trends Pharmacol Sci.* 2007;28:173–179.
60. Ferguson SS, Downey 3rd WE, Colapietro AM, Barak LS, Menard L, Caron MG. Role of beta-arrestin in mediating agonist-promoted G protein-coupled receptor internalization. *Science.* 1996;271:363–366.
61. Claing A, Laporte SA, Caron MG, Lefkowitz RJ. Endocytosis of G protein-coupled receptors: roles of G protein-coupled receptor kinases and beta-arrestin proteins. *Prog Neurobiol.* 2002;66:61–79.
62. Marchese A, Chen C, Kim YM, Benovic JL. The ins and outs of G protein-coupled receptor trafficking. *Trends Biochem Sci.* 2003;28:369–376.
63. Rapacciuolo A, Suvarna S, Barki-Harrington L, et al. Protein kinase A and G protein-coupled receptor kinase phosphorylation mediates beta-1 adrenergic receptor endocytosis through different pathways. *J Biol Chem.* 2003;278:35403–35411.

64. Goodman Jr OB, Krupnick JG, Gurevich VV, Benovic JL, Keen JH. Arrestin/clathrin interaction. Localization of the arrestin binding locus to the clathrin terminal domain. *J Biol Chem.* 1997;272:15017–15022.
65. Krupnick JG, Goodman Jr OB, Keen JH, Benovic JL. Arrestin/clathrin interaction. Localization of the clathrin binding domain of nonvisual arrestins to the carboxy terminus. *J Biol Chem.* 1997;272:15011–15016.
66. Krupnick JG, Santini F, Gagnon AW, Keen JH, Benovic JL. Modulation of the arrestin-clathrin interaction in cells. Characterization of beta-arrestin dominant-negative mutants. *J Biol Chem.* 1997;272:32507–32512.
67. Goodman Jr OB, Krupnick JG, Santini F, et al. Beta-arrestin acts as a clathrin adaptor in endocytosis of the beta2-adrenergic receptor. *Nature.* 1996;383:447–450.
68. Laporte SA, Oakley RH, Holt JA, Barak LS, Caron MG. The interaction of beta-arrestin with the AP-2 adaptor is required for the clustering of beta 2-adrenergic receptor into clathrin-coated pits. *J Biol Chem.* 2000;275:23120–23126.
69. Laporte SA, Oakley RH, Zhang J, et al. The beta2-adrenergic receptor/betaarrestin complex recruits the clathrin adaptor AP-2 during endocytosis. *Proc Natl Acad Sci USA.* 1999;96:3712–3717.
70. Pals-Rylaarsdam R, Gurevich VV, Lee KB, Ptasienski JA, Benovic JL, Hosey MM. Internalization of the m2 muscarinic acetylcholine receptor. Arrestin-independent and -dependent pathways. *J Biol Chem.* 1997;272:23682–23689.
71. Kraft K, Olbrich H, Majoul I, Mack M, Proudfoot A, Oppermann M. Characterization of sequence determinants within the carboxyl-terminal domain of chemokine receptor CCR5 that regulate signaling and receptor internalization. *J Biol Chem.* 2001;276: 34408–34418.
72. Matharu AL, Mundell SJ, Benovic JL, Kelly E. Rapid agonist-induced desensitization and internalization of the A(2B) adenosine receptor is mediated by a serine residue close to the COOH terminus. *J Biol Chem.* 2001;276:30199–30207.
73. Walther C, Nagel S, Gimenez LE, Morl K, Gurevich VV, Beck-Sickinger AG. Ligand-induced internalization and recycling of the human neuropeptide Y2 receptor is regulated by its carboxyl-terminal tail. *J Biol Chem.* 2010;285:41578–41590.
74. Scott MG, Le Rouzic E, Perianin A, et al. Differential nucleocytoplasmic shuttling of beta-arrestins. Characterization of a leucine-rich nuclear export signal in beta-arrestin2. *J Biol Chem.* 2002;277:37693–37701.
75. Kohout TA, Lin FS, Perry SJ, Conner DA, Lefkowitz RJ. Beta-arrestin 1 and 2 differentially regulate heptahelical receptor signaling and trafficking. *Proc Natl Acad Sci USA.* 2001;98:1601–1606.
76. Zhang J, Barak LS, Anborgh PH, Laporte SA, Caron MG, Ferguson SS. Cellular trafficking of G protein-coupled receptor/beta-arrestin endocytic complexes. *J Biol Chem.* 1999;274:10999–11006.
77. Anborgh PH, Seachrist JL, Dale LB, Ferguson SS. Receptor/beta-arrestin complex formation and the differential trafficking and resensitization of beta2-adrenergic and angiotensin II type 1A receptors. *Mol Endocrinol.* 2000;14:2040–2053.
78. Krasel C, Zabel U, Lorenz K, Reiner S, Al-Sabah S, Lohse MJ. Dual role of the beta2-adrenergic receptor C terminus for the binding of beta-arrestin and receptor internalization. *J Biol Chem.* 2008;283:31840–31848.
79. Vaughan DJ, Millman EE, Godines V, et al. Role of the G protein-coupled receptor kinase site serine cluster in beta2-adrenergic receptor internalization, desensitization, and beta-arrestin translocation. *J Biol Chem.* 2006;281:7684–7692.
80. Lee KB, Ptasienski JA, Pals-Rylaarsdam R, Gurevich VV, Hosey MM. Arrestin binding to the M(2) muscarinic acetylcholine receptor is precluded by an inhibitory element in the third intracellular loop of the receptor. *J Biol Chem.* 2000;275: 9284–9289.

81. Pals-Rylaarsdam R, Hosey MM. Two homologous phosphorylation domains differentially contribute to desensitization and internalization of the m2 muscarinic acetylcholine receptor. *J Biol Chem.* 1997;272:14152–14158.
82. Qiu Y, Loh HH, Law PY. Phosphorylation of the delta-opioid receptor regulates its beta-arrestins selectivity and subsequent receptor internalization and adenylyl cyclase desensitization. *J Biol Chem.* 2007;282:22315–22323.
83. Kule CE, Karoor V, Day JN, et al. Agonist-dependent internalization of the angiotensin II type one receptor (AT1): role of C-terminus phosphorylation in recruitment of beta-arrestins. *Regul Pept.* 2004;120:141–148.
84. Reiner S, Ziegler N, Leon C, et al. Beta-arrestin-2 interaction and internalization of the human P2Y1 receptor are dependent on C-terminal phosphorylation sites. *Mol Pharmacol.* 2009;76:1162–1171.
85. Kim OJ, Gardner BR, Williams DB, et al. The role of phosphorylation in D1 dopamine receptor desensitization: evidence for a novel mechanism of arrestin association. *J Biol Chem.* 2004;279:7999–8010.
86. Kilpatrick LE, Briddon SJ, Hill SJ, Holliday ND. Quantitative analysis of neuropeptide Y receptor association with beta-arrestin2 measured by bimolecular fluorescence complementation. *Br J Pharmacol.* 2010;160:892–906.
87. Kilpatrick LE, Briddon SJ, Holliday ND. Fluorescence correlation spectroscopy, combined with bimolecular fluorescence complementation, reveals the effects of beta-arrestin complexes and endocytic targeting on the membrane mobility of neuropeptide Y receptors. *Biochim Biophys Acta.* 2012;1823:1068–1081.
88. DeGraff JL, Gurevich VV, Benovic JL. The third intracellular loop of alpha 2-adrenergic receptors determines subtype specificity of arrestin interaction. *J Biol Chem.* 2002;277:43247–43252.
89. Wu G, Krupnick JG, Benovic JL, Lanier SM. Interaction of arrestins with intracellular domains of muscarinic and alpha2-adrenergic receptors. *J Biol Chem.* 1997;272: 17836–17842.
90. Shenoy SK, McDonald PH, Kohout TA, Lefkowitz RJ. Regulation of receptor fate by ubiquitination of activated beta 2-adrenergic receptor and beta-arrestin. *Science.* 2001;294:1307–1313.
91. Nelson CD, Kovacs JJ, Nobles KN, Whalen EJ, Lefkowitz RJ. Beta-arrestin scaffolding of phosphatidylinositol 4-phosphate 5-kinase Ialpha promotes agonist-stimulated sequestration of the beta2-adrenergic receptor. *J Biol Chem.* 2008;283:21093–21101.
92. Oakley RH, Laporte SA, Holt JA, Barak LS, Caron MG. Association of beta-arrestin with G protein-coupled receptors during clathrin-mediated endocytosis dictates the profile of receptor resensitization. *J Biol Chem.* 1999;274:32248–32257.
93. Innamorati G, Le Gouill C, Balamotis M, Birnbaumer M. The long and the short cycle. Alternative intracellular routes for trafficking of G-protein-coupled receptors. *J Biol Chem.* 2001;276:13096–13103.
94. Pan L, Gurevich EV, Gurevich VV. The nature of the arrestin x receptor complex determines the ultimate fate of the internalized receptor. *J Biol Chem.* 2003;278: 11623–11632.
95. Pierce KL, Lefkowitz RJ. Classical and new roles of beta-arrestins in the regulation of G-protein-coupled receptors. *Nat Rev Neurosci.* 2001;2:727–733.
96. Zhang J, Barak LS, Winkler KE, Caron MG, Ferguson SS. A central role for beta-arrestins and clathrin-coated vesicle-mediated endocytosis in beta2-adrenergic receptor resensitization. Differential regulation of receptor resensitization in two distinct cell types. *J Biol Chem.* 1997;272:27005–27014.
97. Tulipano G, Stumm R, Pfeiffer M, Kreienkamp HJ, Hollt V, Schulz S. Differential beta-arrestin trafficking and endosomal sorting of somatostatin receptor subtypes. *J Biol Chem.* 2004;279:21374–21382.

98. Pediani JD, Colston JF, Caldwell D, Milligan G, Daly CJ, McGrath JC. Beta-arrestin-dependent spontaneous alpha1a-adrenoceptor endocytosis causes intracellular transportation of alpha-blockers via recycling compartments. *Mol Pharmacol.* 2005;67:992–1004.
99. Dale LB, Bhattacharya M, Seachrist JL, Anborgh PH, Ferguson SS. Agonist-stimulated and tonic internalization of metabotropic glutamate receptor 1a in human embryonic kidney 293 cells: agonist-stimulated endocytosis is beta-arrestin1 isoform-specific. *Mol Pharmacol.* 2001;60:1243–1253.
100. Claing A, Chen W, Miller WE, et al. Beta-arrestin-mediated ADP-ribosylation factor 6 activation and beta 2-adrenergic receptor endocytosis. *J Biol Chem.* 2001;276: 42509–42513.
101. McDonald PH, Cote NL, Lin FT, Premont RT, Pitcher JA, Lefkowitz RJ. Identification of NSF as a beta-arrestin1-binding protein. Implications for beta2-adrenergic receptor regulation. *J Biol Chem.* 1999;274:10677–10680.

CHAPTER FIVE

Arrestins as Regulators of Kinases and Phosphatases

Louis M. Luttrell*,†,‡, William E. Miller§
*Department of Medicine, Medical University of South Carolina, Charleston, South Carolina, USA
†Department of Biochemistry & Molecular Biology, Medical University of South Carolina, Charleston, South Carolina, USA
‡The Research Service of the Ralph H. Johnson Veterans Affairs Medical Center, Charleston, South Carolina, USA
§Department of Molecular Genetics, Biochemistry & Microbiology, University of Cincinnati College of Medicine, Cincinnati, Ohio, USA

Contents

1. Introduction 116
2. Arrestins as GPCR Effectors 118
    2.1 Adaptors or scaffolds? 118
    2.2 Mechanism of arrestin-signaling scaffolds 121
    2.3 Spatial and temporal control of kinase activity 122
    2.4 The arrestin-regulated kinome 123
3. Positive and Negative Regulation of Kinase Pathways 124
    3.1 Negative regulation of second messenger-dependent protein kinases 124
    3.2 Positive and negative regulation of arrestin-associated kinases 125
4. Arrestin-Regulated Kinase and Phosphatase Pathways 126
    4.1 Tyrosine protein kinases 126
    4.2 Mitogen-activated protein kinases 130
    4.3 Other SER/THR kinases 135
    4.4 Protein phosphatases 136
    4.5 Lipid kinases 139
5. Conclusions 140
Acknowledgments 141
References 141

Abstract

The discovery that, in addition to mediating G protein-coupled receptor (GPCR) desensitization and endocytosis, arrestins bind to diverse catalytically active nonreceptor proteins and act as ligand-regulated signaling scaffolds led to a paradigm shift in the study of GPCR signal transduction. Research over the past decade has solidified the concept that arrestins confer novel GPCR-signaling capacity by recruiting protein and lipid kinase, phosphatase, phosphodiesterase, and ubiquitin ligase activity into receptor-based multiprotein "signalsome" complexes. Signalsomes regulate downstream pathways controlled by Src

ISSN 1877-1173
http://dx.doi.org/10.1016/B978-0-12-394440-5.00005-X

family nonreceptor tyrosine kinases, mitogen-activated protein kinases, protein kinase B (AKT), glycogen synthase kinase 3, protein phosphatase 2A, nuclear factor-κB, and several others, imposing spatial and temporal control on their function. While many arrestin-bound kinases and phosphatases are involved in the control of cytoskeletal rearrangement, vesicle endocytosis, exocytosis, and cell migration, other signals reach into the nucleus, affecting cell proliferation, apoptosis, and survival. Indeed, the kinase/phosphatase network regulated by arrestins may be fully as diverse as that regulated by heterotrimeric G proteins.

1. INTRODUCTION

By the mid-1990s, the central roles of visual and nonvisual arrestins in G protein-coupled receptor (GPCR) desensitization were well understood.[1] The at-the-time startling discovery that arrestins, like the heterotrimeric G proteins, functioned as discrete GPCR effectors arose from two lines of research. The first was the study of the mechanisms of GPCR endocytosis, which led to the discovery that upon binding to activated receptors, the two nonvisual arrestins, arrestin2 and 3 (β-arrestin1 and 2), act as adapter proteins linking the receptor to components of the clathrin-dependent endocytic machinery.[2,3] The second line of research involved the study of the mechanisms by which GPCRs regulated the nonreceptor tyrosine protein kinase c-Src and the small GTPase Ras pathway.[4,5] The realization that nonvisual arrestins form relatively stable arrestin–GPCR complexes that persist over a timescale of minutes to hours as receptors transit the endocytic vesicle compartment, combined with the observation that arrestin2 bound directly to activated c-Src and recruited it into a multiprotein "signalsome" complex with the receptor,[6] evolved into what has amounted to a reenvisioning of GPCR signal transduction. Rather than functioning solely as ligand-activated guanine nucleotide exchange factors for heterotrimeric G protein, GPCRs have become multifunctional signaling platforms that transmit "pluridimensional" intracellular signals via both G protein and non-G protein effectors.[7] And rather than serving simply as terminators of receptor–G protein coupling, arrestins have become versatile signaling platforms whose binding confers additional GPCR-signaling capability.[8–10]

It is now known that arrestins function as ligand-regulated scaffolds, bringing a host of arrestin-bound proteins to agonist-occupied GPCRs, among them Src family nonreceptor tyrosine kinases[4,11,12]; components of the extracellular signal-regulated kinase (ERK)1/2 and c-Jun N-terminal kinase (JNK)3 mitogen-activated protein kinase (MAPK) cascades[13–15]; the E3 ubiquitin ligase; Mdm2[16]; the cAMP phosphodiesterases, PDE4D3/5[17]; diacylglycerol kinase (DGK)[18]; the inhibitor of nuclear factor-κB, IκBα[19]; the Ral-GDP dissociation stimulator, Ral-GDS[20]; the actin filament-severing protein, cofilin[21]; and the Ser/Thr protein phosphatase 2A (PP2A).[22,23] Indeed, the signaling repertoire accessible to GPCRs through arrestin "coupling" may be as diverse as that mediated by heterotrimeric G -protein-regulated effectors (Fig. 5.1). Moreover, since arrestins uncouple

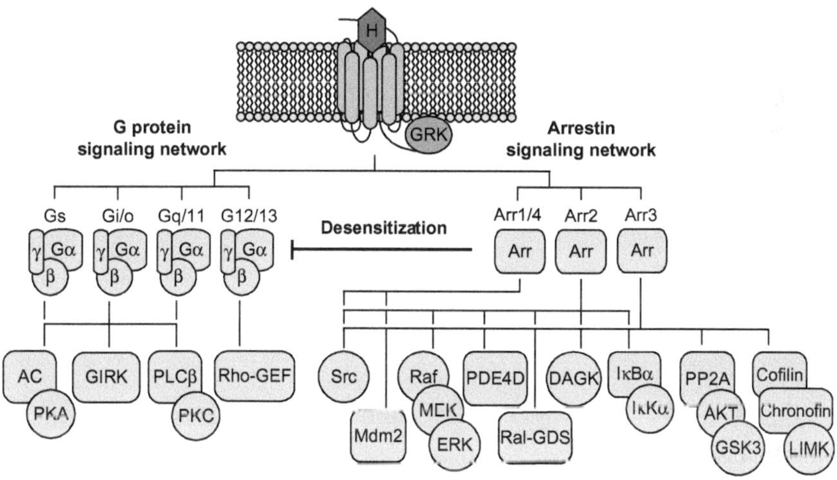

Figure 5.1 Pluridimensional GPCR-signaling networks. Agonist binding to a GPCR stimulates the intrinsic guanine nucleotide exchange factor activity of the receptor, activating heterotrimeric G proteins and initiating second messenger-dependent pathways, including Gs-adenylyl cyclase (AC)–PKA and phospholipase Cβ (PLCβ)–PKC. Numerous other effectors, including gated inwardly rectifying K+ channels (GIRK) and small GTPases like Rho-guanine nucleotide exchange factor (Rho-GEF) are regulated by activated Gα or Gβγ subunits. Agonist-occupied GPCRs are phosphorylated by GRKs, promoting arrestin binding. Arrestins uncouple the receptor and G protein, leading to desensitization of G protein signaling and internalization of the receptor. Arrestins also function as ligand-regulated scaffolds, recruiting catalytic proteins to initiate a second wave of signaling events. Arrestin-coupled effectors include Src family tyrosine kinases (Src), E3 ubiquitin ligases (Mdm2), components of the ERK1/2 mitogen-activated protein kinase cascade (Raf–MEK–ERK1/2), cAMP phosphodiesterases (PDE4D), the Ral-GDP dissociation stimulator (Ral-GDS), diacylglycerol kinases (DAGK), regulators of nuclear factor-κB signaling (IκBα–IκKα), the glycogen synthase kinase 3 regulatory complex (PP2A–Akt–GSK3), and the actin filament-severing complex (cofilin–chronofin–LIMK). Evidence suggests that the signaling network mediated by arrestins may rival the G protein-signaling network in diversity. *Reproduced from Ref. 24.*

the receptor and G protein, arrestin binding can be viewed as the dividing line between two mutually exclusive and temporally discrete GPCR-signaling states.

The first discovered, and still the most diverse, of the arrestin-signaling functions is the regulation of protein and lipid phosphorylation. This chapter discusses the characteristics of arrestins that enable them to function as adapters and scaffolds for protein kinases, their roles as both positive and negative regulators of kinase activity, and the numerous protein and lipid kinase, and protein phosphatase effectors controlled by arrestins. Subsequent chapters in this volume cover the other major classes of arrestin-regulated effectors, including small G proteins, ubiquitin ligases, and regulators of cytoskeletal dynamics. How these signals affect physiological function in embryologic development, intermediary metabolism, vision, cancer, bone, and the cardiovascular, immune, and central nervous systems are then discussed separately.

2. ARRESTINS AS GPCR EFFECTORS

The fundamental duality of arrestin function is well illustrated by a simple experiment. When angiotensin II-stimulated phosphatidylinositol production was assayed in COS-7 cells expressing the angiotensin AT_{1A} receptor, overexpression of either arrestin2 or 3 led to a reduction in the maximal response, an entirely predictable effect of a protein known to promote uncoupling of AT_{1A} receptors from the Gq/11–phospholipase Cβ effector pathway. Yet when angiotensin II-stimulated ERK1/2 activation was assayed under identical conditions, arrestin overexpression led to a paradoxical increase in the response.[25] Such a result could not be a reflection of arrestin-dependent termination of G protein signaling, but rather suggested that the arrestin itself was somehow coupling the AT_{1A} receptor to the ERK1/2 cascade.

2.1. Adaptors or scaffolds?

The nonvisual arrestins clearly function as adapter proteins in the context of clathrin-dependent GPCR endocytosis, binding to agonist-bound GPCRs while at the same time engaging clathrin and the β2-adaptin subunit of the AP-2 complex.[2,3] To transmit signals, in many cases they also function as scaffolds. Scaffolding proteins perform at least three functions in cells: to increase the efficiency of signaling between successive components of an enzymatic cascade, ensure signaling fidelity by dampening cross talk between parallel cascades, and localize signaling to specific subcellular locations.[26]

The prototypic MAP kinase scaffold is the *Saccharomyces cerevisiae* protein Ste5p.[27] In the yeast pheromone mating pathway, Ste5p binds to each of the three components of the yeast MAP kinase pathway, Ste11p, Ste7p, and either Fus3 or Kss1. Binding of mating factor to the pheromone receptor, a GPCR, leads to translocation of Ste5 to the plasma membrane and activation of the Fus3/Kss1 cascade. Deletion of the *Ste5* gene disrupts the yeast mating process, indicating the essential role of this scaffold in MAP kinase regulation. While there are no structural homologues of Ste5 in mammalian cells, arrestins perform an analogous function, organizing individual pathway components to increase efficiency and fidelity of signaling, imposing spatial constraints, and, importantly, providing for pathway regulation by extracellular signals detected via GPCRs on the plasma membrane.

This scaffolding property is illustrated by the effect of arrestin3 on the activation state of c-Jun N-terminal kinase 3 (JNK3). JNK3 is a neuronal MAPK that is regulated by a phosphorylation cascade composed of the upstream MAPK kinase kinase, Ask1; the MAPK kinase, MKK4; and JNK3. Arrestin3 binds the three component kinases, increasing the efficiency of phosphorylation. Ectopic expression of Ask1 and JNK3 in COS-7 cells is not sufficient to activate JNK3, but simultaneous expression of arrestin3 with Ask1 and JNK3 results in dramatic activation.[13] Arrestin3 is excluded from the nucleus by the presence of a classical leucine-rich nuclear export signal in its C-terminus that is absent in arrestin2.[28,29] Consequently, expression of arrestin3 also leads to a dramatic redistribution of JNK3 from the nucleus into an arrestin-bound cytosolic pool.[13,30]

Arrestin-dependent regulation of the ERK1/2 MAPK cascade provided another early example of arrestin scaffolding. Both arrestin2 and 3 can assemble a complex composed of c-Raf1, MEK1/2, and ERK1/2, and enhance ERK1/2 activation in response to stimulation of protease-activated receptor 2 (PAR2) and angiotensin II AT_{1A} receptors.[14,15,25,31] Agonist binding induces the assembly of a protein complex containing the internalized receptor, arrestin, and activated ERK1/2. This arrestin-mediated signal is independent of heterotrimeric G protein activation, as it can be produced by mutated AT_{1A} receptors that lack G protein-coupling efficacy and by "biased" AT_{1A} receptor agonists that promote arrestin recruitment and receptor internalization without G protein activation.[32] Because the complex formed between PAR2 and AT_{1A} receptors and arrestins is stable, these signalsomes localize activated ERK1/2 to early endosomes, where it remains

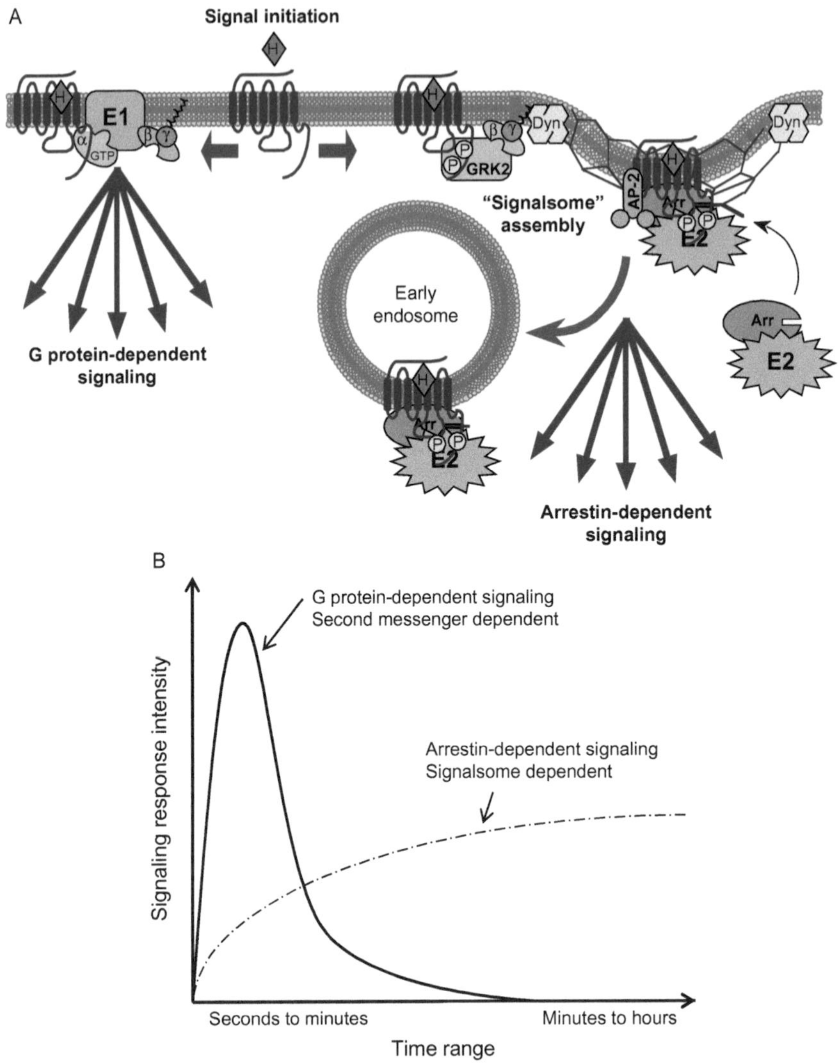

Figure 5.2 Arrestin scaffolds impose spatial and temporal regulation of signaling pathways. (A) Upon agonist (H) binding, GPCRs engage heterotrimeric G proteins, activating G protein-regulated effectors (E1) at the plasma membrane. Within seconds, GRK phosphorylation of the activated receptor creates high-affinity arrestin-binding sites. Arrestin binding uncouples the receptor from heterotrimeric G proteins while targeting it for endocytosis. As arrestins translocate to the receptor, they recruit additional catalytically active proteins (E2) into receptor–arrestin signalsomes. These stoichiometric signaling complexes transmit a distinct set of signals as the receptor internalizes and transits the intracellular compartment. (B) G protein-dependent signaling is characterized by rapid onset followed by waning intensity, reflecting desensitization due to receptor phosphorylation by second messenger-dependent protein kinases and GRKs, and arrestin binding. In contrast, arrestin-mediated signals are of slower onset and often sustained in duration. *(A) Reproduced from Ref. 33.* (See Color Insert.)

associated with the receptor–arrestin complex[14,15] (Fig. 5.2A). In fact, an estimated 75–80% of the active ERK1/2 produced in response to short-term stimulation of the PAR2 receptors is associated with the GPCR–arrestin signalsome.[14] As a result, nuclear translocation of active ERK1/2 is retarded and its kinase activity is directed away from nuclear, and toward cytosolic, targets.[31,34,35]

2.2. Mechanism of arrestin-signaling scaffolds

In contrast to the clathrin and AP-2 binding sites in the C-terminus, and receptor recognition motifs within the N- and C-terminal globular domains of arrestin (see Chapter 3), the binding sites for most arrestin-regulated signaling partners have not been precisely mapped. In fact, all three components of the c-Raf1–MEK1–ERK1/2 and ASK1–MKK4–JNK3 cascades can bind to separately expressed arrestin N- and C-domains, suggesting that they each make multiple contacts across the exposed cytosolic face of the receptor-bound arrestin.[36,37]

Since arrestins undergo a conformational change when they bind to GPCRs, activation of arrestin-bound effectors might then proceed by either of two mechanisms. Receptor binding might change the affinity of arrestin for its nonreceptor partners, allowing the signalsome complex to form only on the receptor. Alternatively, the arrestin–effector complex may be preformed, with activation resulting either from conformational changes in the arrestin that provide the proper orientation of pathway components or from the receptor-dependent translocation of components to the plasma membrane where they gain access to otherwise unavailable upstream pathway activators. Evidence suggests that both mechanisms play a role in arrestin signaling.

ERK1/2 and, to a lesser extent, c-Raf1 bind with high affinity only to the receptor-bound conformation of arrestin.[38] Although MEK1 binds equivalently to both receptor-bound and free arrestin, assembly of a productive c-Raf1–MEK1–ERK1/2 scaffolding complex requires an active GPCR docking site. Membrane translocation may also play a role. Expression of a G protein-uncoupled neurokinin NK1-arrestin2 chimera leads to constitutive activation of a pool of ERK1/2 that remains bound, along with c-Raf1 and MEK1/2 to the endosomal membrane-delimited receptor–arrestin chimera.[39] Since membrane targeting of c-Raf1 is itself sufficient to activate ERK1/2,[40] one possibility is that the arrestin functions simply to move cytosolic c-Raf1 to the membrane for activation.

The finding that plasma membrane recruitment of arrestin3 independent of receptor binding is sufficient to activate ERK1/2, albeit inefficiently, is consistent with this model.[41] ERK1/2 bound to the signalsome complex is also relatively protected from dephosphorylation by MAP kinase phosphatases, suggesting that a slower rate of inactivation also promotes sustained activity.[39] The converse situation, where receptor binding releases an arrestin-bound effector allowing it to become active, occurs during arrestin-dependent regulation of the small G protein, Ral, by Ral-GDP dissociation simulator (Ral-GDS). Ral-GDS constitutively interacts with cytoplasmic arrestin2 and 3. Upon activation of the formyl-Met-Leu-Phe receptor and arrestin recruitment, Ral-GDS is released from the arrestin complex, whereupon it regulates cytoskeleton rearrangement and exocytic granule release in polymorphonuclear neutrophilic leukocytes.[20]

In contrast to ERK1/2, both the receptor-bound and free conformations of arrestin3 are able to bind and activate the ASK1–MKK4–JNK3 cascade equivalently, indicating that the scaffold complex is preformed in cells and an activated GPCR is not required for signaling.[37] In fact all four mammalian arrestins bind JNK3 comparably and can remove it from the nucleus, where it otherwise spontaneously resides.[42] Yet only arrestin3 is able to support JNK3 activation in cells.[30,43,44] Targeted mutagenesis studies identified several residues in arrestin3, most notably Val343, that enable it alone to activate JNK3.[43] The finding that the ability of arrestins to bind pathway components is not synonymous with the ability to activate them has significant implications for the ability of arrestin scaffolds to positively or negatively regulate kinase activity in cells.

2.3. Spatial and temporal control of kinase activity

Because arrestin-dependent ERK1/2 activation occurs within relatively stable receptor–arrestin signalsomes, the arrestin-bound pool of activated ERK1/2 remains with the receptor, first appearing at the plasma membrane and later within the endosomal compartment, but not transiting to the nucleus.[14,15,45] The kinetics of pathway activation differ markedly as well. Whereas G protein activation leads to a rapid and transient rise in ERK1/2 activity, arrestin-dependent ERK1/2 activation is slow and sustained, likely because ERK1/2 activated via the arrestin pathway is protected from rapid dephosphorylation by nuclear MAPK phosphatases[34,37] (Fig. 5.2B).

These differences have profound effects on ERK1/2 function. Whereas ERK1/2 activated by classical receptor tyrosine kinase growth factor receptors or G protein-dependent GPCR signaling is able to translocate to the nucleus and elicit a transcriptional response, ERK1/2 activated through the arrestin pathway is confined to the cytosol and silent in Elk-1 reporter assays.[25,46] Arrestin-bound ERK1/2 performs other functions, for example, regulating arrestin–clathrin interaction during GPCR endocytosis[47,48] and localized actin cytoskeletal reorganization during chemotaxis.[49] Similarly, arrestin-bound ERK1/2 mediates AngII-stimulated phosphorylation of the cytosolic targets, Mnk1 and eIF4E, leading to increased rates of mRNA translation.[35] Thus, by compartmentalizing signaling, arrestin scaffolding can change the functional consequences of pathway activation, even when the pathway is subject to convergent regulation by multiple mechanisms.

2.4. The arrestin-regulated kinome

Before considering the multitude of individual kinase and phosphatase pathways regulated by arrestins, it is worth examining the scope of arrestin-dependent effects on protein phosphorylation globally. Taking advantage of the ability of the biased angiotensin AT_{1A} receptor agonist, [Sar^1,Ile^4, Ile^8]-AngII (SII), to promote arrestin recruitment and arrestin-dependent signaling independent of significant G protein activation,[32,50] two studies have surveyed the arrestin-dependent "kinome" using whole-cell quantitative phosphoproteomic approaches. These studies point to the existence of a robust arrestin-dependent signaling network with far-reaching regulatory functions.

In AT_{1A} receptor-expressing HEK293 cells, Xiao *et al.*[51] identified 171 unique proteins whose phosphorylation increased, and 53 whose phosphorylation decreased, upon stimulation with SII, including 38 protein kinases and 3 phosphatases. A subsequent bioinformatic network analysis based on these results suggested that much of the arrestin-dependent signaling network was focused on regulation of cytoskeletal rearrangement. Using a similar strategy, Christensen *et al.*[52] performed a side-by-side global phosphoproteomic comparison of angiotensin II and SII, thus revealing the extent to which arrestin-dependent kinase regulation contributes to the overall response. These investigators detected over 1183 regulated protein phosphorylation sites out of 10,000 sites surveyed using high-resolution LTQ-Orbitrap mass spectrometry. Of these, 756 (64%) were unique to angiotensin II, 369 (34%) were regulated by both angiotensin II and SII, and 58 (5%) were unique to SII. Analysis of consensus phosphorylation sites indicated a striking difference between the kinases regulated by

G protein-dependent and arrestin-dependent signals, supporting the concept that these two networks regulate largely distinct subsets of downstream effectors. The arrestin-dependent kinase network included several kinases previously reported to be regulated by G protein-independent mechanisms, including A-Raf, Rsk-1, p70S6 kinase, and calmodulin-dependent protein kinase 2, and also pointed to protein kinase D as a downstream target of arrestin signaling.

3. POSITIVE AND NEGATIVE REGULATION OF KINASE PATHWAYS

The duality of arrestin function that characterizes its actions with respect to G protein-mediated and arrestin-mediated signaling processes has curious parallels in its kinase and phosphatase scaffolding role. The binding of pathway components to arrestins not only facilitates ligand-dependent arrestin signaling but also allows arrestins to sequester kinases and regulatory proteins in a manner that dampens basal pathway activity.

3.1. Negative regulation of second messenger-dependent protein kinases

The simplest negative regulatory role of arrestin scaffolds is one that complements its roles in GPCR desensitization and sequestration. Two enzymes involved in second messenger breakdown have been reported to bind arrestins: type 4D cAMP phosphodiesterases (PDE4D) and DGK. Arrestin-dependent recruitment of these enzymes to the locus of second messenger production serves as an additional means of limiting the magnitude and duration of G protein-mediated activation of the second messenger-dependent protein kinases, protein kinase (PK)A and PKC. Arrestins 2 and 3 interact with all five PDE4D isoforms, PDE4D1–5.[17] The Gs-coupled β2-adrenergic receptor forms a signaling complex with arrestin3 and PDE4D3 and PDE4D5, leading to accelerated cAMP degradation. Recruitment of PDE4D into the signalsome appears to be highly receptor specific, since the closely related β1-adrenergic receptor was shown to recruit a different alternative-spliced isoform, PDE4D8, and to do so without the aid of arrestin.[53] Arrestin-dependent recruitment of DGK appears to dampen M1 muscarinic receptor-mediated PKC activity.[18] DGK converts diacylglycerol produced by PLCβ to phosphatidic acid. Besides terminating PKC activity, this mechanism may generate physiologically relevant concentrations of phosphatidic acid, a precursor of the multifunctional lipid second messenger, lysophosphatidic acid (LPA).

3.2. Positive and negative regulation of arrestin-associated kinases

The complexity of arrestin-dependent regulation of kinase activity can be illustrated by two examples: its opposing effects on ERK1/2 and NF-κB signaling and its bidirectional regulation of the PP2A–AKT–GSK3 cascade. As previously discussed, ERK1/2 is recruited preferentially to the receptor-bound conformation of arrestin2 and 3,[38] permitting GPCRs to selectively activate ERK1/2 within receptor–arrestin signalsomes.[14,15] The resulting spatial compartmentalization of ERK1/2 leads to increased phosphorylation of cytosolic targets,[35,47–49] while inhibiting its ability to stimulate Elk-1-dependent transcription.[25,46] At the same, the dominant effect of arrestins on nuclear factor (NF)-κB signaling is to dampen pathway activity. In NF-κB signaling, phosphorylation of the inhibitory protein, IκBα, by IκBα kinases accelerates its proteosomal degradation, permitting nuclear translocation of NF-κB and increased NF-κB transcriptional activity. IκBα binds arrestin3, while both arrestin2 and 3 interact with IκB kinase α/β and NR-IκB-inducing kinase.[54,55] Downregulating arrestin2 expression increases NF-κB activation by tumor necrosis factor α (TNFα), consistent with the hypothesis that arrestins tonically inhibit NF-κB signaling by protecting IκBα from degradation.[19] In HEK293 cells, downregulating arrestin expression attenuates toll-like receptor (TLR)4-mediated ERK1/2 activation while simultaneously enhancing NF-κB reporter activity, suggesting that arrestins exert opposing effects on the ERK1/2 and NF-κB pathways.[56]

The situation with respect to PP2A–AKT–GSK3 signaling is similarly complex. β-Catenin and Akt signaling is regulated by an arrestin3 signalsome complex composed of the catalytic subunit of PP2A, Akt, and glycogen synthase kinase 3β (GSK3β).[23] As with NF-κB, the dominant arrestin effect is to dampen β-catenin signaling. Within the arrestin3 complex, PP2A maintains Akt in an inactive state by dephosphorylating Thr[308]. Since Akt phosphorylation of GSK3β inhibits its activity, keeping Akt inactive increases GSK3α/β activity. GSK3β, in turn, phosphorylates β-catenin, accelerating its degradation and inhibiting β-catenin-dependent transcription. Predictably, brain extracts from arrestin3 null mice show higher levels of β-catenin expression, presumably reflecting the loss of tonic arrestin-mediated Akt inhibition.[57] On the other hand, a phosphoproteomic screen of SII-mediated protein phosphorylation identified two peptide inhibitors of PP2A (I1PP2A and I2PP2A) as targets for increased phosphorylation.[58] Agonist-stimulated phosphorylation of I2PP2A was associated with rapid and transient inhibition of the arrestin3-associated pool of PP2A, leading to activation of Akt and increased phosphorylation of GSK3β in an arrestin signalsome complex. Thus, while arrestins maintain tonic inhibition of GSK3β under homeostatic conditions *in vivo*,[57] mechanisms may exist to release that inhibition in response to GPCR stimulation.

4. ARRESTIN-REGULATED KINASE AND PHOSPHATASE PATHWAYS

Table 5.1 summarizes many of the experimentally validated arrestin-dependent GPCR-signaling pathways reported to date.[10,72] The evidence supporting their existence ranges from coprecipitation studies using overexpressed pathway components to isolation of endogenous arrestin–effector complexes from native tissues and from loss of function studies using arrestin dominant-negative mutants and isoform-selective RNA interference to rescue studies performed using arrestin2/3 null murine embryo fibroblasts (MEFs). Viewed as a whole, arrestin-regulated kinase and phosphatase signaling appear to encompass a fairly discrete set of functions, linking GPCRs to receptor and nonreceptor tyrosine kinases, MAPKs, regulators of NF-κB and β-catenin signaling, and a few protein phosphatases and lipid kinases. Many of these putative effectors are not known to be regulated by heterotrimeric G protein subunits, suggesting that these GPCR–arrestin–effector pathways function in parallel with GPCR–G protein–effector pathways to add additional dimensions to GPCR signaling.

4.1. Tyrosine protein kinases

4.1.1 Src family nonreceptor tyrosine kinases

Arrestins bind to several members of the Src family of nonreceptor tyrosine kinases and recruit them to activated GPCRs. Arrestin-dependent recruitment of c-Src to β2-adrenergic receptors on the plasma membrane can be visualized after isoproterenol stimulation.[6] As with several other nonreceptor arrestin-binding partners, Src kinases appear to make contact with arrestins at several points. The N-terminus of arrestin2 is proline rich and contains three PXXP motifs that interact with the Src homology (SH)3 domain of c-Src.[6] Additional contacts involving the c-Src SH1 (catalytic) domain confer added binding affinity.[59] In contrast, arrestin1 has only a

Table 5.1 Arrestin-regulated kinase and phosphatase pathways

Effector	Arrestin	Reported functions	References
Tyrosine protein kinases			
Src family tyrosine kinases c-Src, c-Yes, c-Hck, c-Fgr, c-Fyn	Arrestin1 Arrestin2 Arrestin3	ERK1/2 activation	6,11
		Dynamin 1 phosphorylation	59
		Exocytosis/granule release	12
		Phosphorylation/destabilization of GRK2	60
		FAK phosphorylation	61
		EGF receptor transactivation	62
		Phosphorylation of β2-adaptin subunit of AP-2 complex	48
SER/THR protein kinases			
c-Raf1–MEK1/2–ERK1/2	Arrestin2 Arrestin3	Activation of cytosolic ERK1/2	14,15
		p90RSK phosphorylation	63
		Actin cytoskeletal reorganization/chemotaxis	49
		ERK1/2-dependent transcription	64
		Mnk1/eIF4E phosphorylation/protein translation	35
ASK1–MKK4–JNK3	Arrestin1	Inhibition of cytosolic JNK3	36
	Arrestin2 Arrestin3	Activation of cytosolic JNK3	13
ASK1–MKK3–p38 MAPK	Arrestin2	Activation of p38 MAPK	65,66
	Arrestin3	Inhibition of p38 MAPK	67
IκBα–IκB kinase α/β	Arrestin2 Arrestin3	Attenuation of NF-κB signaling	19,55,56
Casein kinase 2	Arrestin3	Activation of CK2	58
Protein phosphatases			
PP2A–Akt–GSK3β	Arrestin3	Activation of Akt	58,68
		Inactivation of Akt-GSK3β/Activation of β-catenin signaling	23
Cofilin, chronophin, LIM kinase	Arrestin3	Actin cytoskeletal reorganization/chemotaxis	21
SHP-1, SHP-2	Arrestin3	Inhibition of NK cell cytotoxicity	69
Lipid kinases			
Diacylglycerol kinase	Arrestin2 Arrestin3	Attenuation of PKC signaling	18
Phosphatidylinositol 3-kinase	Arrestin2	Inhibition of PI3K	70
PI 4-phosphate 5-kinase Iα	Arrestin2 Arrestin3	Control of GPCR internalization	71

single PXXP motif and binding appears to involve the c-Src SH2 domain.[73] It is unclear whether c-Src exhibits a binding preference for the receptor-bound conformation of arrestins, but immunostaining for the active (Y^{530} dephosphorylated) form of c-Src suggests that c-Src in the receptor–arrestin complex is active.[6] It is also unclear whether the association of arrestin and c-Src is regulated, although phosphorylation of Ser^{412} in the C-terminus of arrestin2, which destabilizes the receptor–arrestin complex, has been reported to disrupt the arrestin–Src interaction.[47]

Arrestin-dependent recruitment of Src family kinases appears to regulate clathrin-dependent GPCR endocytosis. Arrestin-scaffolded c-Src phosphorylates GRK2, providing negative feedback on receptor desensitization by destabilizing GRK2 and promoting its entry into the proteosome pathway where it is rapidly degraded.[60] Arrestin–Src binding is required for phosphorylation of Tyr^{497} of dynamin 1, which regulates dynamin self-assembly.[59,74,75] Expression of a Y497F mutant of dynamin 1 impairs the internalization of both the β2-adrenergic[74] and the M2 muscarinic acetylcholine receptor.[76] The β2-adaptin subunit of AP-2 is another endocytic protein whose regulation by Src is arrestin dependent.[48,77,78] c-Src stabilizes the constitutive association between arrestin3 and β2-adaptin independent of its kinase activity. Src-mediated phosphorylation of β2-adaptin Tyr^{737} occurs in clathrin-coated pits in response to AT_{1A}, β2 adrenergic, V2 vasopressin, or B2 bradykinin receptor activation, leading to dissociation of AP-2 from the complex. If β2-adaptin phosphorylation is blocked, receptor–arrestin complexes are retained at the membrane.

The involvement of arrestin–Src complexes in vesicle trafficking extends to exocytosis. Arrestin-dependent activation of the Src family kinases, c-Hck and c-Fgr, by the interleukin 8 receptor (CXCR1) appears to be important for granule release, since expression of P91G/P121G arrestin2 with mutations in the PXXP motifs antagonizes CXCR1-induced exocytosis in granulocyte cells.[12] Similarly, the endothelin type A receptor assembles an arrestin2-dependent complex with c-Yes that positively regulates endothelin-1-stimulated translocation of the glucose transporter Glut4 to the plasma membrane.[79]

Arrestin-dependent Src signaling has been implicated in several other GPCR-regulated processes, notably activation of ERK1/2. Src recruitment to β2-adrenergic receptors contributes to ERK1/2 MAP activation.[6] The neurokinin NK1 receptor forms a receptor–arrestin–c-Src signalsome in response to substance P stimulation.[11] Assembly of this signalsome has been implicated in NK1 receptor-mediated ERK1/2 activation, cell survival, and proliferation. D2 dopamine receptor activation of the NF-κB pathway reportedly requires c-Src as well as Gαi proteins, and this response is potentiated by overexpression of arrestin2.[80] c-Src is also recruited to rhodopsin–arrestin1 complexes, which concentrate in the rod outer segment compartment upon light exposure.[73] It has been proposed that c-Src binding to arrestin1 promotes the formation of high-affinity phospho-Tyr binding sites in the rod outer segment that lead to translocation of SH2-containing proteins.

4.1.2 Receptor tyrosine kinases

Members of EGF receptor family of receptor tyrosine kinases (HER1–4) are key convergence points for mitogenic stimuli. A well-characterized mechanism whereby GPCRs affect cell proliferation and survival is by stimulating the matrix metalloprotease (MMP)-dependent shedding of preformed EGF-family growth factors, leading to paracrine transactivation of EGF receptors (reviewed in Ref. 81). In some settings, arrestin–Src complexes play a role in GPCR-mediated EGF receptor transactivation, and arrestin-dependent activation of the ERK1/2 cascade downstream of transactivated EGF receptors can stimulate transcription and promote cell proliferation. This contrasts with direct arrestin-dependent scaffolding mechanism of ERK1/2 activation, wherein the signalsome-bound active ERK1/2 is usually transcriptionally repressed.[25,46]

The luteinizing hormone (LH) receptor activates c-Fyn in an arrestin3-dependent manner.[61] Downregulating arrestin expression reduces the rate of internalization of hCG by 50% and inhibits LH receptor-mediated activation of c-Fyn, phosphorylation of the antiapoptotic focal adhesion kinase (FAK), and the release of EGF-like growth factors. In HEK293 cells expressing β1-adrenergic receptors, EGF receptor transactivation and ERK1/2 activation are inhibited by downregulation of arrestin2 or 3, or GRK5 or 6, inhibiting Src or MMP activity, or exposure to a heparin-binding-EGF neutralizing antibody, suggesting that β1-receptor-mediated EGF receptor transactivation is arrestin dependent.[62] Consistent with this, a mutant β1-receptor lacking 14 GRK phosphorylation sites in its C-terminal tail (-GRKβ1), which cannot undergo arrestin-dependent desensitization, fails to transactivate EGF receptors despite exaggerated G protein activation. Transgenic mice expressing the -GRKβ1 receptor in cardiomyocytes develop more severe dilated cardiomyopathy in response to chronic isoproterenol stimulation, with significantly increased LV end-diastolic dimension, decreased fractional shortening, and increased myocardial apoptosis. In this model, inhibiting EGF receptors worsens the dilated cardiomyopathy, suggesting a protective role for transactivated EGF receptors in the heart.[62]

Arrestin signaling is not the only mechanism of GPCR-stimulated ectodomain shedding, and, in some cases, arrestin-mediated GPCR desensitization antagonizes signaling by transactivated EGF receptors. In arrestin2/3 null MEFs, ERK1/2 activation by LPA receptors reflects primarily G protein-dependent transactivation of EGF receptors.[64] Because LPA receptor desensitization is impaired, the EGF receptor-dependent ERK1/2 signal is persistent, lasting for several hours in the continued presence of LPA. Reintroducing arrestin3, which restores desensitization, makes the transactivation-dependent signal transient, such that it contributes significantly to ERK1/2 activation only during the first few minutes of stimulation. At the same time, arrestin3 confers a long-lasting EGF receptor-independent ERK1/2 signal that presumably reflects activation of the arrestin-scaffolding pathway. Whereas most of the early LPA-stimulated transcriptional responses in arrestin2/3 null MEFs are EGF receptor-dependent, expression of arrestin3 attenuates EGF receptor-dependent transcription and enables LPA to elicit EGF receptor-independent transcription.

4.2. Mitogen-activated protein kinases

4.2.1 Extracellular signal-regulated kinases

ERK1/2 activity is required for G0–G1 cell-cycle transition and the passage of cells through mitosis or meiosis.[82] Heptahelical receptors employ multiple mechanisms to activate ERK1/2, often simultaneously, from PKA- and PKC-dependent signals downstream of heterotrimeric G proteins, to transactivation of EGF receptors, to signals transmitted via arrestins.[81,83] As a result, GPCR-mediated ERK1/2 activation in any given setting is complex, as are the functional consequences. Regarding arrestin-dependent ERK1/2 activation, several factors influence the spatial distribution and duration of ERK1/2 activity including the specific GRK and arrestin isoforms acting on the receptor, and the stability of the receptor–arrestin interaction.

Although arrestins 1, 2, and 3, but not arrestin4, are able to bind ERK1/2,[84] evidence suggests that, at least with some receptors, the arrestin2 and 3 isoforms play opposing roles in arrestin-mediated signaling and desensitization. In HEK293 cells expressing AT_{1A} receptors, isoform-selective downregulation of endogenous arrestin3 reduces angiotensin II-stimulated ERK1/2 activation by about 50% and abrogates activation the arrestin pathway-selective ligand SII. Paradoxically, silencing arrestin2 expression has the opposite effect, enhancing the arrestin-dependent component of the ERK1/2 signal,[34,46] suggesting that, with respect to ERK1/2 activation, arrestin3 is the signaling arrestin isoform, while arrestin2 functions only in desensitization. Arrestin3 is certainly the key isoform for Class A receptors like the β2 adrenergic, which do not bind arrestin2 efficiently,[85] but the dichotomy of isoform-specific arrestin function does not appear to hold for all Class B receptors. Arrestin-dependent ERK1/2 activation by the type 1 parathyroid hormone receptor, for example, is inhibited when either isoform is downregulated, suggesting that both are required to assemble functional signalsomes.[86]

Additional evidence suggests that arrestins adopt different "active" conformations depending on which GPCR they bind and which GRK phosphorylated the receptor. Evidence of the former comes from characterization of arrestin ubiquitination. Ubiquitination of lysines 11 and 12 of arrestin3 is necessary for it to remain stably bound to the AT_{1A} receptor, yet an arrestin3 (K11,12R) mutant is still ubiquitinated and fully functional when recruited to the vasopressin V2 receptor.[45] All 31 lysines must be mutated before arrestin3 ubiquitination is lost upon β2-adrenergic receptor binding.[87] The variability suggests that either the conformation or the accessibility of surface epitopes on arrestin3 differs depending on the GPCR-binding partner. Additional support for different arrestin conformations comes from data obtained using isoform-selective silencing of GRKs that suggest GRK2 and GRK3 phosphorylation of the AT_{1A} and V2 receptors promotes arrestin-dependent desensitization, while GRK5 and GRK6 appear to be exclusively responsible for initiating arrestin-dependent ERK1/2 activation.[88,89] Similar work with the β2-adrenergic receptor suggests that GRK2 and GRK6 phosphorylate the receptor C-terminus at different sites, and that only the GRK6-induced pattern of phosphorylation supports arrestin-dependent ERK1/2 activation.[90] This has led to the hypothesis that different GRKs establish a phosphorylation "barcode" that imparts distinct arrestin3 conformations to regulate its functional activity.

Signalsomes regulate the spatial, temporal, and functional characteristics of ERK1/2. When recruited to a Class B receptor, active ERK1/2 accumulates in early endosomes, failing to translocate to the cell nucleus. Wild-type PAR2 receptors predominantly utilize the arrestin-dependent pathway to activate ERK1/2.[14] As a result, ERK1/2 is excluded from the nucleus and does not stimulate proliferation. In contrast, a C-terminal phosphorylation site mutant PAR2, that does not bind arrestins or internalize, activates ERK1/2 via a G protein-dependent pathway that promotes its nuclear translocation and stimulates proliferation. Similarly, the wild-type AT_{1A} receptor activates ERK1/2 using both G protein- and arrestin-dependent pathways, increasing both cytosolic and nuclear ERK1/2,[25,34] whereas a G protein-uncoupled DRY–AAY AT_{1A} receptor mutant, which utilizes only the arrestin pathway, only activates cytosolic ERK1/2 and fails to elicit a detectable transcriptional response.[46] Native V2 receptors similarly engage both pathways. Exchanging the V2 receptor C-terminus for that of the Class A β2-adrenergic receptor, which converts the receptor from stable to transient arrestin binding, increases the proportion of ERK1/2 that enters the cell nucleus and permits the chimeric receptor to stimulate cell proliferation. The opposite effect is obtained when the V2 receptor tail is appended to the β2 receptor.[31]

Whether ERK1/2 activated by Class A receptors using the arrestin pathway is transcriptionally competent is less clear. Class A receptors, like the β2-adrenergic and LPA receptors, also use arrestin scaffolds to activate ERK1/2, but the transient nature of the receptor–arrestin interaction does not support endosomal targeting.[64,85] Reintroduction of arrestin3 into arrestin2/3 double null MEFs restores arrestin-dependent ERK1/2 activation, and while it represses transcription mediated through G protein-dependent transactivation of EGF receptors, it enables LPA to elicit ERK1/2 dependent, but EGF receptor-independent transcription.[64] Such findings suggest that dissociation of the LPA receptor–arrestin complex upon internalization may permit ERK1/2 activated by the arrestin pathway to enter the cell nucleus.

The spatial constraint imposed by the assembly of stable signalsomes appears to direct ERK1/2 kinase activity toward membrane or cytosolic substrates. ERK1/2 phosphorylates Ser^{412} in the C-terminus of arrestin2, limiting its ability to bind clathrin.[47] Arrestin2 in the cytosol is almost stoichiometrically phosphorylated on Ser^{412}, and dephosphorylation of Ser^{412} upon receptor binding promotes receptor internalization and ERK1/2 activation. Rephosphorylation by ERK1/2 in the signalsome complex probably provides either negative feedback regulation of receptor endocytosis or facilitates receptor internalization by promoting dissociation of arrestin and clathrin, allowing the receptor to exit clathrin-coated vesicles. Arrestin-dependent targeting of ERK1/2 to the plasma membrane also appears to play a role in chemotaxis. During PAR2-induced chemotaxis, PAR2 receptor–arrestin–ERK1/2 complexes localize to the leading edge of the cell where ERK1/2 activity is required for actin cytoskeletal reorganization.[49] Other cytosolic ERK1/2 substrates include the ribosomal S6 kinase, p90RSK. ERK1/2 phosphorylation of p90RSK is activated by a mutant AT_{1A} receptor with a deletion in its second intracellular loop that inhibits G protein coupling.[63] This arrestin-dependent activation of the ERK1/2 substrate p90RSK acts in concert with another arrestin-mediated signal, phosphatidylinositol 3-kinase (PI3K)–AKT to downregulate phospho-BAD, inducing antiapoptotic cytoprotective effects in rat vascular smooth muscle.[91] Using RNA interference to downregulate arrestin3, it has also been possible to show that arrestin-dependent ERK1/2 activation by the AT_{1A} receptor mediates phosphorylation of Mnk1 and eIF4E, increasing rates of mRNA translation.[35]

4.2.2 c-Jun N-terminal kinase 3

JNK1–3 are stress-activated kinases that regulate apoptosis by stimulating cytochrome *C* release from the mitochondria during cellular stress and control transcription by phosphorylating the transcription factor c-Jun.[92] There are three JNK isoforms, of which JNK1/2 are widely expressed, while JNK3 is highly expressed only in brain, heart, and testes.[93] JNK2 and JNK3 were originally found to interact with arrestin3 in yeast two-hybrid screens, but only JNK3 interacts with arrestins in mammalian cells.[13] The three components of the JNK3 cascade, Ask1, MKK4, and JNK3, are able to bind all four arrestin isoforms. While all four arrestins can cause JNK3 to redistribute from its normal nuclear location into the cytosol, only arrestin3 is able to potentiate JNK3 activation.[36,42]

It is unclear whether GPCRs exert any control over arrestin3-dependent activation of JNK3 or indeed whether activation of arrestin3–JNK3 has a physiologic role.[94] Interestingly, JNK3 exhibits higher affinity for the "inactive" conformation of arrestins, suggesting that its dominant role is to repress basal JNK3 signaling.[36,42] Even though the JNK3 bound to arrestin3 is active, it is sequestered in the cytosol away from its nuclear transcription factor targets. The original study, performed in transfected cells, reported that stimulation of AT_{1A} receptors activated JNK3 and caused it to colocalize with arrestin3 in endosomal vesicles,[13] but later work performed with the β2-adrenergic receptor found no evidence of receptor-mediated JNK3 activation under conditions where ERK1/2 was being robustly activated via the arrestin3 pathway. To the contrary, expression of inactive arrestin3 mutants that do not bind GPCRs increased basal JNK3 phosphorylation while simultaneously decreasing receptor-catalyzed ERK1/2 activation, confirming that while arrestin3-dependent ERK1/2 activation is receptor dependent, JNK3 activation is not.[30] In a more physiologic animal model of cerebral ischemia, the angiotensin receptor blocker, losartan, has been reported to attenuate neuronal damage by inhibiting the assembly of an AT_{1A} receptor–arrestin3–Ask1–MKK4-signaling module and repressing the activation of JNK3, c-jun, and caspase-3, and the release of cytochrome *C*.[95]

4.2.3 p38 MAP kinase

Although the molecular mechanisms of activation have not been characterized in any detail, several studies have reported arrestin-dependent regulation of the p38 MAP kinase cascade. In HeLa and HEK293 cells, overexpression of arrestin3 enhanced, and downregulation inhibited, activation of both p38 MAPK and ERK1/2 following stimulation of the chemokine receptor CXCR4 by its ligand, stromal cell-derived factor 1α.[65] Inhibiting p38 MAPK, but not ERK1/2, blocked arrestin3-dependent chemotaxis, suggesting a specific role for arrestin-dependent p38 MAPK in CXCR4 signaling. In primary cultured astrocytes, κ-opioid receptor-stimulated activation of p38 MAPK appears to involve GRK3 and arrestin3.[66] p38 activation does not occur in astrocytes derived from κ-opioid receptor or GRK3 null mice, or following downregulation of arrestin3 expression.

As with ERK1/2, however, the dominant effect of arrestins in other systems appears to attenuate p38 MAPK activation via G protein-dependent pathways. Arrestin2/3 null MEFs exhibit greatly enhanced activation of ERK1/2, JNK1/2, and p38 MAPK in response to the CXCR2 agonist, interleukin 8.[67] Activation of the stress kinases, JNK1/2 and p38 MAPK, was dependent on reactive oxygen species generated by NADPH oxidase, and arrestin expression conferred protection from oxidative burst-induced cell death resulting from sustained CXCR2 activation. Similarly, β2-adrenergic receptor-mediated regulation of immunoglobulin IgE expression on CD40L/interleukin-4-activated B lymphocytes involves PKA and p38 MAPK, but not arrestins. β2-Adrenergic receptor stimulation leads to Gs/cAMP/PKA-dependent phosphorylation and inactivation of the hematopoietic protein tyrosine phosphatase, HePTP, which releases bound p38 MAPK, making more available for phosphorylation and subsequent IgE regulation.[96] In this setting, arrestin-dependent β2-adrenergic receptor desensitization would attenuate, not enhance p38 MAPK activity.

Besides CXCR4-mediated chemotaxis,[65] arrestin-dependent regulation of p38 MAPK has been implicated in several cellular processes. In polymorphonuclear neutrophils, clathrin-mediated endocytosis of the platelet-activating receptor (PAF) requires p38 MAPK-dependent rearrangement of actin bundles. Activation of PAF receptors leads to recruitment and activation of an arrestin2–ASK1–MKK3–p38 MAPK signalsome, which is required for subsequent cell polarization, actin bundle formation, and receptor endocytosis.[97] While the antiapoptotic effects of arrestins are mediated primarily through activation of the prosurvival kinase AKT (*vide infra*), such that genetic deletion of either arrestin increases apoptosis after serum withdrawal, deletion of both arrestin2 and 3 is paradoxically protective.[98] Individual deletion of either arrestin2 or 3 enhances starvation-induced activation of ERK1/2 and p38 MAPK and caspase 3 cleavage, while decreasing AKT activity. When both are deleted, starvation-induced ERK1/2 and p38 MAPK activation returns to wild-type levels. This suggests that both arrestins are required to maintain a balance between proapoptotic p38 MAPK and antiapoptotic AKT-signaling pathways.

4.3. Other SER/THR kinases

4.3.1 Regulators of nuclear factor-κB

As previously discussed, arrestins2 and 3 appear to act as negative regulators of NF-κB transcription by sequestering the NF-κB inhibitor, IκBα, and protecting it from phosphorylation by IκBα kinases, which promotes its degradation. β2-Adrenergic receptor stimulation increases arrestin3 binding to IκBα, preventing its phosphorylation-dependent degradation and inhibiting IL-8-stimulated NF-κB activity.[54,55] Consistent with this, TLR4-mediated activation of an NF-κB reporter is enhanced by downregulating arrestin expression.[56] Arrestin binding to the E3 ubiquitin ligase, TRAF6, probably augments the inhibition of TLR signaling.[99] TRAF6 is normally recruited to TLR/IL-1 family receptors where it facilitates IκB kinase and NF-κB activation. Binding of TRAF6 to arrestin2 and 3 in response to lipopolysaccharide or IL-1 stimulation prevents TRAF6 oligomerization and autoubiquitination, negatively regulating LPS and IL-1 signaling.

The capacity to negatively regulate NF-κB transcriptional pathways involved in cytokine signaling suggests that arrestins may dampen immune responses by inhibiting the production of proinflammatory cytokines. Consistent with this, isolated peritoneal neutrophils from arrestin3 null animals exhibit increased basal and LPS-stimulated TNFα and IL-6 production.[100] Arrestin effects in cancer may also derive in part from modulation of NF-κB signaling. Loss of expression of the type III transforming growth factor-β (TβRIII) occurs in a variety of human malignancies. TβRIII is thought to function as a tumor suppressor by reducing cell motility. Although not a GPCR, clathrin-dependent endocytosis of TβRIII and downregulation of TGFβ signaling depends on the interaction with arrestin3.[101] *In vitro* evidence suggests that arrestin3 plays a key role in the ability of TβRIII to inhibit cell migration. In breast and ovarian cancer cell lines, activation of the small GTPase Cdc42 by TβRIII alters actin cytoskeletal rearrangement and reduces random cell migration.[102] A TβRIII mutant unable to interact with arrestins fails to inhibit migration and the wild-type receptor effect is blocked by downregulating arrestin3. In addition, the interaction between TβRIII and arrestin3 negatively regulates NF-κB transcriptional activity, further inhibiting cell migration.[103]

4.3.2 Casein kinase 2

Casein kinase 2 (CK2) is a ubiquitously expressed, constitutively active Ser/Thr protein kinase that performs diverse functions related to cell survival and tumorigenesis. The catalytic subunit of CK2 was identified in a proteomic screen of arrestin3 binding proteins as well as a phosphoproteomic screen of AT_{1A} receptor-mediated phosphorylation following stimulation with the arrestin-selective biased agonist, SII.[51,58] Since CK2 has been implicated in phosphorylation of Thr^{383} of arrestin3, which destabilizes the interaction between arrestin3 and β2-adrenergic receptors,[104] it is possible that arrestin recruitment of CK2 plays a role similar to that proposed for ERK1/2-mediated phosphorylation of arrestin2 Ser^{412}.

4.4. Protein phosphatases

4.4.1 PP2A–AKT–GSK3β

The Ser/Thr phosphatase PP2A is ubiquitously expressed and has broad substrate specificity. The PP2A holoenzyme is heterotrimeric, composed of regulatory A and B subunits that target the catalytic C subunit to specific intracellular locations. The PP2A catalytic subunit was identified as an arrestin3-interacting protein in a proteomic screen,[51] and a native arrestin3–PP2A–AKT–GSK3β complex has been purified from the D2 dopamine receptor-rich striatum of mice.[23] The regulation of PP2A–AKT–GSK3 signaling by arrestins is probably one of the most physiologically important arrestin-regulated signaling processes, being critically involved in both AKT-dependent survival signaling and regulation of β-catenin-mediated transcription. As previously discussed, the major homeostatic function of the complex in the striatum is to modulate D2 dopamine receptor-mediated behaviors by tonic repression of β-catenin signaling.[23,57] At the same time, AT_{1A} receptor-mediated, G protein-independent, phosphorylation of the PP2A inhibitor, I2PP2A, may provide a mechanism for acute modulation of AKT and GSK3β kinase activity by transiently inhibiting PP2A in the signalsome complex.[58] PAR1 receptor activation also reportedly causes rapid activation of AKT through an unknown arrestin2-dependent mechanism.[68]

Dopaminergic neurotransmission in the brain regulates behavioral responses such as locomotor activity and neural reward mechanisms. Loss of dopaminergic cells in the substantia nigra leads to a loss of locomotor control in Parkinson's disease. Conversely, D2 dopamine receptor antagonists are effective neuroleptic drugs used in the treatment of schizophrenia and attention-deficit hyperactivity disorder, which are thought to result from excess dopaminergic neurotransmission. Several lines of evidence suggest that arrestin-signaling complexes regulate dopamine-dependent behaviors. Locomotor hyperactivity induced by the dopaminergic drug apomorphine, a D2 receptor agonist, is reduced in arrestin3 knockout mice.[57] Similarly, the hyperactivity displayed by dopamine-transporter knockout mice, which results from increased synaptic dopamine concentration, is reduced when DAT knockout mice are crossbred with arrestin3 knockouts, a paradoxical result, since G protein-mediated responses would be enhanced by the loss of arrestin-dependent desensitization. The molecular basis of these effects may lie in the scaffolding of PP2A–AKT–GSK3 by arrestin3. Amphetamine treatment, which increases synaptic dopamine release, increases the PP2A–AKT association in wild type, but not arrestin3 knockout mice, suggesting that arrestins mediate the interaction.[57] Directly inhibiting PP2A or GSK3β also mimics the effect of arrestin3 knockout on hyperactive locomotor activity in DAT knockout mice, suggesting that dopamine-mediated activation of GSK3β results from PP2A-dependent AKT inhibition that is scaffolded by arrestin3.[23]

Arrestin3–PP2A–AKT scaffolds may underlie the mechanism of action of many antipsychotic drugs. Lithium, a mood stabilizer used in the treatment of schizophrenia, modulates dopamine-dependent behavior in mice such as horizontal activity. GSK3β haploinsufficient mice exhibit augmented lithium-induced antidepressant and anxiolytic effects compared to wild-type animals, suggesting that lithium acts by inhibiting GSK3β. Therapeutic concentrations of lithium disrupt the interaction between arrestin3, AKT, and PP2A, relieving PP2A-mediated negative regulation of AKT, allowing it to phosphorylate and inactivate GSK3β.[57] The arrestin3–PP2A–AKT complex requires magnesium, and lithium is thought to destabilize the complex by competing for magnesium binding. The clinical efficacy of essentially all other classes of antipsychotic drug correlates directly to their dopamine

D2-receptor binding affinity and ability to antagonize the receptor. In an *in vitro* screen using fluorescence-based reporters, it was found that while different classes of antipsychotics exhibit complex efficacy profiles with respect to D2 receptor–G protein coupling, they share the property of antagonizing the D2 receptor–arrestin3 interaction.[105]

PP2A also plays an important role in receptor trafficking and resensitization, and some of these effects may be arrestin dependent. Arrestin2-bound PP2A reportedly dephosphorylates Ser^{412} on arrestin2, a step that regulates the interaction between arrestin and the clathrin-coated pit and subsequent internalization.[106] Dephosphorylation of GRK-phosphorylated receptors, a prerequisite for receptor resensitization, also involves PP2A. A 150-kDa oligomeric form of PP2A catalyzes the dephosphorylation of β2- and α2-adrenergic receptors.[107] The acidic microenvironment of endosomes is important for PP2A association with internalized β2-adrenergic receptors, their dephosphorylation, and subsequent recycling to the plasma membrane.[108] At present, it is not clear whether the PP2A pool involved in receptor dephosphorylation is targeted by the arrestin. Finally, PP2A is known to promote ERK1/2 activation by acting on c-Raf1 Ser^{259}, an inhibitory site that must be dephosphorylated for Raf activation.[109] Since PP2A both positively regulates c-Raf[109,110] and negatively regulates ERK1/2,[111,112] its presence in the signalsome may play a role in arrestin-dependent ERK1/2 regulation.

4.4.2 Cofilin–chronophin–LIM kinase

Arrestin kinase/phosphatase scaffolds play a central role in the control of GPCR-mediated chemotaxis, the process whereby migrating cells follow a concentration gradient to its source. Chemoattractant receptor activation induces actin cytoskeletal rearrangement forming leading and trailing edges. A dominant pseudopodium forms at the leading edge that protrudes forward driven by F-actin polymerization and actin-myosin contraction forces.[113] Splenocytes derived from arrestin3 null mice exhibit strikingly impaired chemotactic responses to stromal cell-derived factor-1, CXCL12.[114] While impaired gradient sensing due to the loss of arrestin-mediated desensitization might be a contributing factor,[115] evidence suggests that arrestin-dependent regulation ofERK1/2 and cortical actin assembly at the leading edge is required for GPCR-mediated chemotaxis.[49,116–118] In addition, arrestins scaffold a complex containing the actin filament-severing protein, cofilin, LIM kinase, and the cofilin-specific phosphatase, chronophin, which is required for the dephosphorylation and activation of cophilin.[21] Arrestin-bound cophilin generates the free barbed ends on actin filaments that permit filament extension. Arrestins also affect cell shape change by interacting with the actin-bundling protein, filamin A. Assembly of an AT_{1A} receptor–arrestin–ERK1/2–filamin A complex is required for the formation of membrane ruffles in Hep2 cells.[119]

4.4.3 Protein tyrosine phosphatases

The Src homology 2-containing protein tyrosine phosphatase, SHP-1, modulates AKT activation by the ghrelin receptor, growth hormone secretagogue receptor type 1α (GHSR1α). In adipocytes, ghrelin activates AKT by both an early pertussis toxin-sensitive Gi/o-mediated pathway and a slower arrestin-scaffolded pathway.[120] AKT activation by GHSR1α involves tyrosine phosphorylation of the PI3K regulatory subunit p85, leading to an increase in c-Src and phosphoinositide-dependent protein kinase 1 (PDK1) phosphorylation and activation of AKT. SHP-1 is present in a receptor-associated arrestin-scaffold complex, where it attenuates ghrelin-induced c-Src and AKT activation. An arrestin3-dependent mechanism also appears to negatively regulate the activity of natural killer cells, a key component of the innate immune response. Arrestin3 mediates recruitment of SHP-1 and SHP-2 to KIR2DL1, an inhibitory receptor of natural killer cells, facilitating downstream inhibitory signaling.[69]

4.5. Lipid kinases

4.5.1 Diacylglycerol kinases

DGKs convert diacylglycerol produced by PLCβ to phosphatidic acid. As previously discussed, classical arrestin-dependent desensitization of M1 muscarinic acetylcholine receptor signaling via the Gq/11–PLCβ–PKC pathway is augmented by the arrestin-dependent recruitment of DGK to the receptor, which accelerates second messenger degradation.[18]

4.5.2 Phosphatidylinositol 3-kinase

As an upstream modulator of AKT, PI3K plays an integral role in several arrestin scaffold functions. PI3K activity is required to translocate PDK1 the plasma membrane. PDK1 is a constitutively active kinase and, upon translocation, phosphorylates AKT Thr^{308}, leading to AKT activation. As previously discussed, arrestin scaffolding of PP2A, which dephosphorylates AKT Thr^{308}, restrains AKT activity in brain.[22,23] While there is no known mechanism whereby arrestins promote PI3K activity, there are data suggesting that they inhibit it. PAR2 receptors stimulate PI3K activity through a Gq/11–Ca^{2+}-dependent pathway involving the nonreceptor tyrosine kinase Pyk2 and c-Src.[70] Arrestin2 can directly associate with the catalytic p110α subunit of PI3K and inhibit its activity. PI3K is recruited into a PAR2–arrestin1 complex, and arrestin2 colocalizes with the regulatory p85 PI3K subunit within the pseudopodia of cells during PAR2-mediated chemotaxis, suggesting that the arrestin2–PI3K association may spatially restrict PI3K activity and that this localized inhibition PI3K may be involved in PAR2-stimulated chemotaxis.

4.5.3 Phosphatidylinositol 4-phosphate 5-kinase

Another lipid kinase, phosphatidylinositol 4-phosphate 5-kinase (PIP5K) 1α, regulates clathrin and AP2 function during clathrin-dependent GPCR endocytosis.[71] PIP5K1α produces phosphatidylinositol 4,5-bisphosphate (PIP2) on the inner leaflet of the clathrin-coated pit, promoting polymerization of clathrin and AP-2 and assembly of the clathrin coat. Arrestin3 recruits PIP5K1α to activated β2-adrenergic receptors, increasing PIP2 formation and enhancing receptor endocytosis.

5. CONCLUSIONS

Arrestins regulate a robust network of protein and lipid kinases and phosphatases that allow GPCRs to modulate several downstream signaling cascades not available to heterotrimeric G protein-regulated effectors. In functioning as ligand-regulated scaffolds, they control the timing and location of signaling, directing multifunctional enzymes, like ERK1/2 and PP2A, toward some substrates and away from others. In a curious parallel to their role in terminating G protein signaling, arrestins appear to provide tonic inhibition of some kinase pathways, like JNK3 and NF-κB, while at the same time enabling GPCR-catalyzed activation of others. Arrestin kinase/phosphatase scaffolds regulate cytoskeletal rearrangement, vesicle endocytosis and exocytosis, cell migration and chemotaxis, and transmit signals to the cell nucleus to control cell proliferation, apoptosis, and survival. As subsequent chapters in this volume will attest, arrestin-dependent kinase/phosphatase signaling is integrated into GPCR-regulated physiologic processes at multiple levels, and improving our understanding of these multifunctional scaffolds may provide keys to novel treatments for diseases as seemingly diverse as cancer, inflammation, osteoporosis, and mental illness.

ACKNOWLEDGMENTS

Supported in part by National Institutes of Health Grant DK55524 (L. M. L.) and the Research Service of the Charleston, SC Veterans Affairs Medical Center. The contents of this chapter do not represent the views of the Department of Veterans Affairs or the United States Government.

REFERENCES

1. Ferguson SS. Evolving concepts in G protein-coupled receptor endocytosis: the role in receptor desensitization and signaling. *Pharmacol Rev*. 2001;53:1–24.
2. Goodman Jr OB, Krupnick JG, Santini F, et al. Beta-arrestin acts as a clathrin adaptor in endocytosis of the beta2-adrenergic receptor. *Nature*. 1996;383:447–450.
3. Laporte SA, Oakley RH, Zhang J, et al. The beta2-adrenergic receptor/beta arrestin complex recruits the clathrin adaptor AP-2 during endocytosis. *Proc Natl Acad Sci USA*. 1999;96:3712–3717.
4. Luttrell LM, Hawes BE, van Biesen T, Luttrell DK, Lansing TJ, Lefkowitz RJ. Role of c-Src in G protein-coupled receptor- and Gβγ subunit-mediated activation of mitogen activated protein kinases. *J Biol Chem*. 1996;271:19443–19450.
5. van Biesen T, Hawes BE, Luttrell DK, et al. Receptor-tyrosine-kinase- and Gβγ-mediated MAP kinase activation by a common signalling pathway. *Nature*. 1995;376:781–784.
6. Luttrell LM, Ferguson SSG, Daaka Y, et al. β-Arrestin-dependent formation of β2 adrenergic receptor/Src protein kinase complexes. *Science*. 1999;283:655–661.
7. Galandrin S, Bouvier M. Distinct signaling profiles of beta1 and beta2 adrenergic receptor ligands toward adenylyl cyclase and mitogen-activated protein kinase reveals the pluridimensionality of efficacy. *Mol Pharmacol*. 2006;70:1575–1584.
8. Maudsley S, Martin B, Luttrell LM. Perspectives in pharmacology: the origins of diversity and specificity in G protein-coupled receptor signaling. *J Pharmacol Exp Ther*. 2005;314:485–494.
9. Luttrell LM. Reviews in molecular biology and biotechnology: transmembrane signaling by G protein-coupled receptors. *Mol Biotechnol*. 2008;39:239–264.
10. Luttrell LM, Gesty-Palmer D. Beyond desensitization: physiological relevance of arrestin-dependent signaling. *Pharmacol Rev*. 2010;62:305–330.
11. DeFea KA, Vaughn ZD, O'Bryan EM, Nishijima D, Dery O, Bunnett NW. The proliferative and antiapoptotic effects of substance P are facilitated by formation of a beta-arrestin-dependent scaffolding complex. *Proc Natl Acad Sci USA*. 2000;97: 11086–11091.
12. Barlic J, Andrews JD, Kelvin AA, et al. Regulation of tyrosine kinase activation and granule release through beta-arrestin by CXCRI. *Nat Immunol*. 2000;1:227–233.
13. McDonald PH, Chow CW, Miller WE, et al. Beta-arrestin 2: a receptor-regulated MAPK scaffold for the activation of JNK3. *Science*. 2000;290:1574–1577.
14. DeFea KA, Zalevsky J, Thoma MS, Dery O, Mullins RD, Bunnett NW. Beta-arrestin-dependent endocytosis of proteinase-activated receptor 2 is required for intracellular targeting of activated ERK1/2. *J Cell Biol*. 2000;148:1267–1281.
15. Luttrell LM, Roudabush FL, Choy EW, et al. Activation and targeting of extracellular signal-regulated kinases by beta-arrestin scaffolds. *Proc Natl Acad Sci USA*. 2001;98:2449–2454.
16. Shenoy SK, McDonald PH, Kohout TA, Lefkowitz RJ. Regulation of receptor fate by ubiquitination of activated β2-adrenergic receptor and β-arrestin. *Science*. 2001;294:1307–1313.

17. Perry SJ, Baillie GS, Kohout TA, et al. Targeting of cyclic AMP degradation to beta 2-adrenergic receptors by beta-arrestins. *Science*. 2002;298:834–836.
18. Nelson CD, Perry SJ, Regier DS, Prescott SM, Topham MK, Lefkowitz RJ. Targeting of diacylglycerol degradation to M1 muscarinic receptors by beta-arrestins. *Science*. 2007;315:663–666.
19. Witherow DS, Garrison TR, Miller WE, Lefkowitz RJ. Beta-arrestin inhibits NF-kappaB activity by means of its interaction with the NF-kappaB inhibitor IkappaBalpha. *Proc Natl Acad Sci USA*. 2004;101:8603–8607.
20. Bhattacharya M, Anborgh PH, Babwah AV, et al. Beta-arrestins regulate a Ral-GDS Ral effector pathway that mediates cytoskeletal reorganization. *Nat Cell Biol*. 2002;4:547–555.
21. Zoudilova M, Kumar P, Ge L, Wang P, Bokoch GM, DeFea KA. Beta-arrestin-dependent regulation of the cofilin pathway downstream of protease-activated receptor-2. *J Biol Chem*. 2007;282:20634–20646.
22. Beaulieu JM, Sotnikova TD, Yao WD, et al. Lithium antagonizes dopamine-dependent behaviors mediated by an AKT/glycogen synthase kinase 3 signaling cascade. *Proc Natl Acad Sci USA*. 2004;101:5099–5104.
23. Beaulieu JM, Sotnikova TD, Marion S, Lefkowitz RJ, Gainetdinov RR, Caron MG. An Akt/beta-arrestin 2/PP2A signaling complex mediates dopaminergic neurotransmission and behavior. *Cell*. 2005;122:261–273.
24. Gesty-Palmer D, Luttrell LM. Refining efficacy: exploiting functional selectivity for drug discovery. *Adv Pharmacol*. 2011;62:79–107.
25. Tohgo A, Pierce KL, Choy EW, Lefkowitz RJ, Luttrell LM. β–Arrestin scaffolding of the ERK cascade enhances cytosolic ERK activity but inhibits ERK mediated transcription following angiotensin AT1a receptor stimulation. *J Biol Chem*. 2002;277:9429–9436.
26. Burack WR, Shaw AS. Signal transduction: hanging on a scaffold. *Curr Opin Cell Biol*. 2000;12:211–216.
27. Elion EA. The Ste5p scaffold. *J Cell Sci*. 2001;114:3967–3978.
28. Scott MG, Le Rouzic E, Perianin A, et al. Differential nucleocytoplasmic shuttling of beta-arrestins. Characterization of a leucine-rich nuclear export signal in beta-arrestin2. *J Biol Chem*. 2002;277:37693–37701.
29. Wang P, Wu Y, Ge X, Ma L, Pei G. Subcellular localization of beta-arrestins is determined by their intact N domain and the nuclear export signal at the C terminus. *J Biol Chem*. 2003;278:11648–11653.
30. Breitman M, Kook S, Gimenez LE, et al. Silent scaffolds: inhibition of c-Jun N-terminal kinase 3 activity in cell by dominant-negative arrestin-3 mutant. *J Biol Chem*. 2012;287:19653–19664.
31. Tohgo A, Choy EW, Gesty-Palmer D, et al. The stability of the G protein-coupled receptor-beta-arrestin interaction determines the mechanism and functional consequence of ERK activation. *J Biol Chem*. 2003;278:6258–6267.
32. Wei H, Ahn S, Shenoy SK, et al. Independent G protein and beta-Arrestin2 mediated activation of ERK by angiotensin. *Proc Natl Acad Sci USA*. 2003;100: 10782–10787.
33. Kendall RT, Luttrell LM. Diversity in β-Arrestin Function. *Cell Mol Life Sci*. 2009;66:2953–2973.
34. Ahn S, Shenoy SK, Wei H, Lefkowitz RJ. Differential kinetic and spatial patterns of beta-arrestin and G protein-mediated ERK activation by the angiotensin II receptor. *J Biol Chem*. 2004;279:35518–35525.
35. DeWire SM, Kim J, Whalen EJ, Ahn S, Chen M, Lefkowitz RJ. Beta-arrestin-mediated signaling regulates protein synthesis. *J Biol Chem*. 2008;283: 10611–10620.

36. Song X, Gurevich EV, Gurevich VV. Cone arrestin binding to JNK3 and Mdm2: conformational preference and localization of interaction sites. *J Neurochem.* 2007;103:1053–1062.
37. Song X, Coffa S, Fu H, Gurevich VV. How does arrestin assemble MAPKs into a signaling complex? *J Biol Chem.* 2009;284:685–695.
38. Coffa S, Breitman M, Hanson SM, et al. The effect of arrestin conformation on the recruitment of c-Raf1, MEK1, and ERK1/2 activation. *PLoS One.* 2011;6:e28723.
39. Jafri F, El-Shewy HM, Lee MH, Kelly M, Luttrell DK, Luttrell LM. Constitutive ERK1/2 activation by a chimeric neurokinin 1 receptor-beta-arrestin1 fusion protein. Probing the composition and function of the G protein-coupled receptor "signalsome" *J Biol Chem.* 2006;281:19346–19357.
40. Stokoe D, Macdonald SG, Cadwallader K, Symons M, Hancock JF. Activation of Raf as a result of recruitment to the plasma membrane. *Science.* 1994;264:1463–1467.
41. Terrillon S, Bouvier M. Receptor activity-independent recruitment of betaarrestin2 reveals specific signalling modes. *EMBO J.* 2004;23:3950–3961.
42. Song X, Raman D, Gurevich EV, Vishnivetskiy SA, Gurevich VV. Visual and both non-visual arrestins in their "inactive" conformation bind JNK3 and Mdm2 and relocalize them from the nucleus to the cytoplasm. *J Biol Chem.* 2006;281: 21491–21499.
43. Seo J, Tsakem EL, Breitman M, Gurevich VV. Identification of arrestin-3-specific residues necessary for JNK3 kinase activation. *J Biol Chem.* 2011;286:27894–27901.
44. Miller WE, McDonald PH, Cai SF, Field ME, Davis RJ, Lefkowitz RJ. Identification of a motif in the carboxyl terminus of beta-arrestin2 responsible for activation of JNK3. *J Biol Chem.* 2001;276:27770–27777.
45. Shenoy SK, Lefkowitz RJ. Receptor-specific ubiquitination of beta-arrestin directs assembly and targeting of seven-transmembrane receptor signalosomes. *J Biol Chem.* 2005;280:15315–15324.
46. Lee M-H, El-Shewy HM, Luttrell DK, Luttrell LM. Role of beta-arrestin-mediated desensitization and signaling in the control of angiotensin AT1a receptor-stimulated transcription. *J Biol Chem.* 2008;283:2088–2097.
47. Lin F-T, Miller WE, Luttrell LM, Lefkowitz RJ. Feedback regulation of beta-arrestin1 function by extracellular signal-regulated kinases. *J Biol Chem.* 1999;274:15971–15974.
48. Zimmerman B, Simaan M, Lee M-H, Luttrell LM, Laporte SA. c-Src-mediated phosphorylation of AP-2 reveals a general mechanism for receptors internalizing through the clathrin pathway. *Cell Signal.* 2009;21:103–110.
49. Ge L, Ly Y, Hollenberg M, DeFea K. A beta-arrestin-dependent scaffold is associated with prolonged MAPK activation in pseudopodia during protease-activated receptor-2-induced chemotaxis. *J Biol Chem.* 2003;278:34418–34426.
50. Holloway AC, Qian H, Pipolo L, et al. Side-chain substitutions within angiotensin II reveal different requirements for signaling, internalization, and phosphorylation of type 1A angiotensin receptors. *Mol Pharmacol.* 2002;61:768–777.
51. Xiao K, Sun J, Kim J, et al. Global phosphorylation analysis of beta-arrestin-mediated signaling downstream of a seven transmembrane receptor (7TMR). *Proc Natl Acad Sci USA.* 2010;107:15299–15304.
52. Christensen GL, Kelstrup CD, Lyngsø C, et al. Quantitative phosphoproteomics dissection of seven-transmembrane receptor signaling using full and biased agonists. *Mol Cell Proteomics.* 2010;9:1540–1553.
53. Richter W, Day P, Agrawal R, et al. Signaling from beta1- and beta2-adrenergic receptors is defined by differential interactions with PDE4. *EMBO J.* 2008;27:384–393.
54. Luan B, Zhang Z, Wu Y, Kang J, Pei G. Beta-arrestin2 functions as a phosphorylation-regulated suppressor of UV-induced NF-kappaB activation. *EMBO J.* 2005;24: 4237–4246.

55. Gao H, Sun Y, Wu Y, et al. Identification of beta-arrestin2 as a G protein-coupled receptor-stimulated regulator of NF-kappaB pathways. *Mol Cell.* 2004;14:303–317.
56. Fan H, Luttrell LM, Tempel GE, Senn JJ, Halushka PV, Cook JA. β-Arrestins 1 and 2 differentially regulate LPS-induced signaling and pro-inflammatory gene expression. *Mol Immunol.* 2007;44:3092–3099.
57. Beaulieu JM, Marion S, Rodriguiz RM, et al. A beta-arrestin 2 signaling complex mediates lithium action on behavior. *Cell.* 2008;132:125–136.
58. Kendall RT, Strungs EG, Rachidi SM, et al. The beta-arrestin pathway-selective type 1A angiotensin receptor (AT1A) agonist [Sar1, Ile4, Ile8]angiotensin II regulates a robust G protein-independent signaling network. *J Biol Chem.* 2011;286: 19880–19891.
59. Miller WE, Maudsley S, Ahn S, Khan KD, Luttrell LM, Lefkowitz RJ. Beta-arrestin1 interacts with the catalytic domain of the tyrosine kinase c-SRC. Role of beta-arrestin1-dependent targeting of c-SRC in receptor endocytosis. *J Biol Chem.* 2000;275:11312–11319.
60. Penela P, Elorza A, Sarnago S, Mayor Jr F. Beta-arrestin- and c-Src-dependent degradation of G-protein-coupled receptor kinase 2. *EMBO J.* 2001;20:5129–5138.
61. Galet C, Ascoli M. Arrestin-3 is essential for the activation of Fyn by the luteinizing hormone receptor (LHR) in MA-10 cells. *Cell Signal.* 2008;20:1822–2829.
62. Noma T, Lemaire A, Naga Prasad SV, et al. Beta-arrestin-mediated beta1-adrenergic receptor transactivation of the EGFR confers cardioprotection. *J Clin Invest.* 2007;117:2445–2458.
63. Seta K, Nanamori M, Modrall JG, Neubig RR, Sadoshima J. AT1 receptor mutant lacking heterotrimeric G protein coupling activates the Src-Ras-ERK pathway without nuclear translocation of ERKs. *J Biol Chem.* 2002;277:9268–9277.
64. Gesty-Palmer D, El Shewy H, Kohout TA, Luttrell LM. Beta-arrestin 2 expression determines the transcriptional response to lysophosphatidic acid stimulation in murine embryo fibroblasts. *J Biol Chem.* 2005;280:32157–32167.
65. Sun Y, Cheng Z, Ma L, Pei G. Beta-arrestin2 is critically involved in CXCR4-mediated chemotaxis, and this is mediated by its enhancement of p38 MAPK activation. *J Biol Chem.* 2002;277:49212–49219.
66. Bruchas MR, Macey TA, Lowe JD, Chavkin C. Kappa opioid receptor activation of p38 MAPK is GRK3- and arrestin-dependent in neurons and astrocytes. *J Biol Chem.* 2006;281:18081–18089.
67. Zhao M, Wimmer A, Trieu K, Discipio RG, Schraufstatter IU. Arrestin regulates MAPK activation and prevents NADPH oxidase-dependent death of cells expressing CXCR2. *J Biol Chem.* 2004;279:49259–49267.
68. Goel R, Phillips-Mason PJ, Raben DM, Baldassare JJ. Alpha-thrombin induces rapid and sustained Akt phosphorylation by beta-arrestin1-dependent and -independent mechanisms, and only the sustained Akt phosphorylation is essential for G1 phase progression. *J Biol Chem.* 2002;277:18640–18648.
69. Yu MC, Su LL, Zou L, et al. An essential function for beta-arrestin 2 in the inhibitory signaling of natural killer cells. *Nat Immunol.* 2008;9:898–907.
70. Wang P, DeFea KA. Protease-activated receptor-2 simultaneously directs beta-arrestin-1-dependent inhibition and Galphaq-dependent activation of phosphatidylinositol 3-kinase. *Biochemistry.* 2006;45:9374–9385.
71. Nelson CD, Kovacs JJ, Nobles KN, Whalen EJ, Lefkowitz RJ. Beta-arrestin scaffolding of phosphatidylinositol 4-phosphate 5-kinase Ialpha promotes agonist-stimulated sequestration of the beta2-adrenergic receptor. *J Biol Chem.* 2008;283:21093–21101.
72. Whalen EJ, Rajagopal S, Lefkowitz RJ. Therapeutic potential of β-arrestin- and G protein-biased agonists. Trends Mol Med. 2011;17:126–139.

73. Ghalayini AJ, Desai N, Smith KR, Holbrook RM, Elliott MH, Kawakatsu H. Light-dependent association of Src with photoreceptor rod outer segment membrane proteins in vivo. *J Biol Chem.* 2002;277:1469–1476.
74. Ahn S, Maudsley S, Luttrell LM, Lefkowitz RJ, Daaka Y. Src-mediated tyrosine phosphorylation of dynamin is required for beta2-adrenergic receptor internalization and mitogen-activated protein kinase signaling. *J Biol Chem.* 1999;274:1185–1188.
75. Ahn S, Kim J, Lucaveche CL, et al. Src-dependent tyrosine phosphorylation regulates dynamin self-assembly and ligand-induced endocytosis of the epidermal growth factor receptor. *J Biol Chem.* 2002;277:26642–26651.
76. Werbonat Y, Kleutges N, Jakobs KH, van Koppen CJ. Essential role of dynamin in internalization of M2 muscarinic acetylcholine and angiotensin AT1A receptors. *J Biol Chem.* 2000;275:21969–21974.
77. Fessart D, Simaan M, Laporte SA. c-Src regulates clathrin adapter protein 2 interaction with beta-arrestin and the angiotensin II type 1 receptor during clathrin-mediated internalization. *Mol Endocrinol.* 2005;19:491–503.
78. Fessart D, Simaan M, Zimmerman B, et al. Src-dependent phosphorylation of beta2-adaptin dissociates the beta-arrestin-AP-2 complex. *J Cell Sci.* 2007;120:1723–1732.
79. Imamura T, Huang J, Dalle S, et al. Beta-arrestin-mediated recruitment of the Src family kinase Yes mediates endothelin-1-stimulated glucose transport. *J Biol Chem.* 2001;276:43663–43667.
80. Yang M, Zhang H, Voyno-Yasenetskaya T, Ye RD. Requirement of Gbetagamma and c-Src in D2 dopamine receptor-mediated nuclear factor-kappaB activation. *Mol Pharmacol.* 2003;64:447–455.
81. Luttrell LM. Location, Location. Location. Spatial and temporal regulation of MAP kinases by G protein-coupled receptors. *J Mol Endo.* 2003;30:117–126.
82. Pearson G, Robinson F, Beers Gibson T, et al. Mitogen-activated protein (MAP) kinase pathways: regulation and physiologic functions. *Endocr Rev.* 2001; 22:153–183.
83. Gutkind JS. The pathways connecting G protein-coupled receptors to the nucleus through divergent mitogen-activated protein kinase cascades. *J Biol Chem.* 1998;273:1839–1842.
84. Hanson SM, Cleghorn WM, Francis DJ, et al. Arrestin mobilizes signaling proteins to the cytoskeleton and redirects their activity. *J Mol Biol.* 2007;368:375–387.
85. Shenoy SK, Drake MT, Nelson CD, et al. Beta-arrestin-dependent, G protein-independent ERK1/2 activation by the beta2 adrenergic receptor. *J Biol Chem.* 2006;281:1261–1273.
86. Gesty-Palmer D, Chen M, Reiter E, et al. Distinct beta-arrestin- and G protein-dependent pathways for parathyroid hormone receptor-stimulated ERK1/2 activation. *J Biol Chem.* 2006;281:10856–10864.
87. Shenoy SK, Barak LS, Xiao K, et al. Ubiquitination of beta-arrestin links seven-transmembrane receptor endocytosis and ERK activation. *J Biol Chem.* 2007; 282:29549–29562.
88. Kim J, Ahn S, Ren XR, et al. Functional antagonism of different G protein-coupled receptor kinases for beta-arrestin-mediated angiotensin II receptor signaling. *Proc Natl Acad Sci USA.* 2005;102:1442–1447.
89. Ren XR, Reiter E, Ahn S, Kim J, Chen W, Lefkowitz RJ. Different G protein-coupled receptor kinases govern G protein and beta-arrestin-mediated signaling of V2 vasopressin receptor. *Proc Natl Acad Sci USA.* 2005;102:1448–1453.
90. Nobles KN, Xiao K, Ahn S, et al. Distinct phosphorylation sites on the β(2)-adrenergic receptor establish a barcode that encodes differential functions of β-arrestin. *Sci Signal.* 2011;4:ra51.

91. Ahn S, Kim J, Hara MR, Ren XR, Lefkowitz RJ. Beta-arrestin-2 mediates anti-apoptotic signaling through regulation of BAD phosphorylation. *J Biol Chem.* 2009;284:8855–8865.
92. Tournier C, Hess P, Yang DD, et al. Requirement of JNK for stress-induced activation of the cytochrome c-mediated death pathway. *Science.* 2000;288:870–874.
93. Kuan CY, Whitmarsh AJ, Yang DD, et al. A critical role of neural-specific JNK3 for ischemic apoptosis. *Proc Natl Acad Sci USA.* 2003;100:15184–15189.
94. Lin A, Defea KA. β-Arrestin-kinase scaffolds: turn them on or turn them off? *Wiley Interdiscip Rev Syst Biol Med.* 2013;5:231–241.
95. Zhang TL, Fu JL, Geng Z, Yang JJ, Sun XJ. The neuroprotective effect of losartan through inhibiting AT1/ASK1/MKK4/JNK3 pathway following cerebral I/R in rat hippocampal CA1 region. *CNS Neurosci Ther.* 2012;18:981–987.
96. McAlees JW, Sanders VM. Hematopoietic protein tyrosine phosphatase mediates beta2-adrenergic receptor-induced regulation of p38 mitogen-activated protein kinase in B lymphocytes. *Mol Cell Biol.* 2009;29:675–686.
97. McLaughlin NJ, Banerjee A, Kelher MR, et al. Platelet-activating factor-induced clathrin-mediated endocytosis requires beta-arrestin-1 recruitment and activation of the p38 MAPK signalosome at the plasma membrane for actin bundle formation. *J Immunol.* 2006;176:7039–7050.
98. Yang X, Zhou G, Ren T, et al. β-Arrestin prevents cell apoptosis through pro-apoptotic ERK1/2 and p38 MAPKs and anti-apoptotic Akt pathways. *Apoptosis.* 2012;17:1019–1026.
99. Wang Y, Tang Y, Teng L, Wu Y, Zhao X, Pei G. Association of beta-arrestin and TRAF6 negatively regulates Toll-like receptor-interleukin 1 receptor signaling. *Nat Immunol.* 2006;7:139–147.
100. Basher F, Fan H, Zingarelli B, et al. Beta-arrestin 2: a negative regulator of inflammatory responses in polymorphonuclear leukocytes. *Int J Clin Exp Med.* 2008;1:32–41.
101. Finger EC, Lee NY, You HJ, Blobe GC. Endocytosis of the type III transforming growth factor-beta (TGF-beta) receptor through the clathrin-independent/lipid raft pathway regulates TGF-beta signaling and receptor down-regulation. *J Biol Chem.* 2008;283:34808–34818.
102. Mythreye K, Blobe GC. The type III TGF-beta receptor regulates epithelial and cancer cell migration through beta-arrestin2-mediated activation of Cdc42. *Proc Natl Acad Sci USA.* 2009;106:8221–8226.
103. You HJ, How T, Blobe GC. The type III transforming growth factor-beta receptor negatively regulates nuclear factor kappa B signaling through its interaction with beta-arrestin2. *Carcinogenesis.* 2009;30:1281–1287.
104. Lin FT, Chen W, Shenoy S, Cong M, Exum ST, Lefkowitz RJ. Phosphorylation of beta-arrestin2 regulates its function in internalization of beta(2)-adrenergic receptors. *Biochemistry.* 2002;41:10692–10699.
105. Masri B, Salahpour A, Didriksen M, et al. Antagonism of dopamine D2 receptor/beta-arrestin 2 interaction is a common property of clinically effective antipsychotics. *Proc Natl Acad Sci USA.* 2008;105:13656–13661.
106. Hupfeld CJ, Resnik JL, Ugi S, Olefsky JM. Insulin-induced beta-arrestin1 Ser-412 phosphorylation is a mechanism for desensitization of ERK activation by Galphai-coupled receptors. *J Biol Chem.* 2005;280:1016–1023.
107. Pitcher JA, Payne ES, Csortos C, DePaoli-Roach AA, Lefkowitz RJ. The G-protein-coupled receptor phosphatase: a protein phosphatase type 2A with a distinct subcellular distribution and substrate specificity. *Proc Natl Acad Sci USA.* 1995;92:8343–8347.
108. Krueger KM, Daaka Y, Pitcher JA, Lefkowitz RJ. The role of sequestration in G protein-coupled receptor resensitization. Regulation of beta2-adrenergic receptor dephosphorylation by vesicular acidification. *J Biol Chem.* 1997;272:5–8.

109. Abraham D, Podar K, Pacher M, et al. Raf-1-associated protein phosphatase 2A as a positive regulator of kinase activation. *J Biol Chem.* 2000;275:22300–22304.
110. Adams DG, Coffee Jr RL, Zhang H, Pelech S, Strack S, Wadzinski BE. Positive regulation of Raf1-MEK1/2-ERK1/2 signaling by protein serine/threonine phosphatase 2A holoenzymes. *J Biol Chem.* 2005;280:42644–42654.
111. Zhou B, Wang ZX, Zhao Y, Brautigan DL, Zhang ZY. The specificity of extracellular signal-regulated kinase 2 dephosphorylation by protein phosphatases. *J Biol Chem.* 2002;277:31818–31825.
112. Silverstein AM, Barrow CA, Davis AJ, Mumby MC. Actions of PP2A on the MAP kinase pathway and apoptosis are mediated by distinct regulatory subunits. *Proc Natl Acad Sci USA.* 2002;99:4221–4226.
113. Machesky LM. Cell motility: complex dynamics at the leading edge. *Curr Biol.* 1997;7: R164–R167.
114. Fong AM, Premont RT, Richardson RM, Yu YR, Lefkowitz RJ, Patel DD. Defective lymphocyte chemotaxis in beta-arrestin2- and GRK6-deficient mice. *Proc Natl Acad Sci USA.* 2002;99:7478–7483.
115. Aragay AM, Mellado M, Frade JM, et al. Monocyte chemoattractant protein-1-induced CCR2B receptor desensitization mediated by the G protein-coupled receptor kinase 2. *Proc Natl Acad Sci USA.* 1998;95:2985–2990.
116. Ge L, Shenoy SK, Lefkowitz RJ, DeFea K. Constitutive protease-activated receptor-2-mediated migration of MDA MB-231 breast cancer cells requires both beta-arrestin-1 and -2. *J Biol Chem.* 2004;279:55419–55424.
117. Barnes WG, Reiter E, Violin JD, Ren XR, Milligan G, Lefkowitz RJ. Beta-arrestin 1 and Galphaq/11 coordinately activate RhoA and stress fiber formation following receptor stimulation. *J Biol Chem.* 2005;280:8041–8050.
118. Hunton DL, Barnes WG, Kim J, et al. Beta-arrestin 2-dependent angiotensin II type 1A receptor-mediated pathway of chemotaxis. *Mol Pharmacol.* 2005;67:1229–1236.
119. Scott MG, Pierotti V, Storez H, et al. Cooperative regulation of extracellular signal-regulated kinase activation and cell shape change by filamin A and beta-arrestins. *Mol Cell Biol.* 2006;26:3432–3445.
120. Lodeiro M, Alén BO, Mosteiro CS, et al. The SHP-1 protein tyrosine phosphatase negatively modulates Akt signaling in the ghrelin/GHSR1a system. *Mol Biol Cell.* 2011;22:4182–4191.

CHAPTER SIX

β-Arrestins: Modulators of Small GTPase Activation and Function

Audrey Claing
Department of Pharmacology, Université de Montréal, Montréal, Quebec, Canada

Contents

1. Introduction 150
   1.1 The family of GTPases 150
   1.2 Modes of regulation of GTPase activity 152
2. Ras Family GTPases 154
   2.1 Ras and the MAPK pathway 154
   2.2 Ral—A multifunctional GTPase 155
3. Rho Family GTPases 157
   3.1 Rho and β-arrestin modulation of stress fiber formation 158
   3.2 Rac—The GTPase associated with membrane ruffling and protrusions 160
   3.3 Cdc42 and filopodia formation 161
4. Rab Family GTPases 162
   4.1 Colocalization of Rabs with receptors and β-arrestin 163
   4.2 Modulation of Rab function through interaction 164
5. ARF Family GTPases 164
   5.1 ARF6 and receptor endocytosis 165
   5.2 Remodeling of the actin cytoskeleton via β-arrestin and ARF6 166
6. Ran Family GTPases 166
7. Perspectives and Future Directions 167
   7.1 Defining the role of β-arrestin 1 and β-arrestin 2 167
   7.2 Identification of the mechanisms whereby β-arrestins regulate GTPase function 168
8. Conclusions 169
References 169

Abstract

Most cellular events responsible for accurate G protein-coupled receptor trafficking involve small GTP-binding proteins. For example, trafficking of receptors via the endocytic and exocytic pathways requires activation of ADP-ribosylation factors and Rab proteins, while receptor-mediated complex responses such as migration are well characterized to be dependent upon Rho family members. Because β-arrestin proteins are recruited to activated receptors and now considered as key signaling molecules,

ISSN 1877-1173
http://dx.doi.org/10.1016/B978-0-12-394440-5.00006-1

whether they act to control small GTPase activity remains a subject of great interest. Over the years, considerable evidence has suggested that β-arrestins and GTPases might be effectors of the same signaling pathways. One example is the roles of both β-arrestin and Ras, the prototypical GTPase, in coordinating activation of mitogen-activated protein kinase. Recently developed tools effective in suppressing the expression of β-arrestins will help define whether they are essential for small G protein activation. Furthermore, novel approaches to identify protein complexes will greatly advance our understanding of the possible cross talk between β-arrestin and small GTPases.

1. INTRODUCTION

Initially identified as key proteins regulating desensitization, internalization, and recycling of seven-transmembrane (7TM) receptors, β-arrestin proteins are now considered multifunctional scaffolds and central regulators of signal transduction events.[1,2] Through their ability to interact with numerous partners, they coordinate signaling from 7TM receptors and small GTP-binding proteins. The latter are characterized as molecular switches regulating a wide range of cellular responses. In this chapter, we review the evidence suggesting that β-arrestins are key regulators of GTPase activity and our current understanding of the mechanisms whereby β-arrestins regulate the function of monomeric G proteins and the pathways they activate. We focus on how β-arrestins modulate the regulation of GTPase activity to affect cellular processes like receptor trafficking and cytoskeletal rearrangement.

1.1. The family of GTPases

Small GTPases are single-chain polypeptides of 20–40 kDa. These proteins are considered molecular switches that determine the temporal aspects of a broad variety of signaling events involved in numerous cellular processes and responses. The G domain responsible for guanine nucleotide binding is highly conserved between the different families of GTPases. When bound to guanosine diphosphate (GDP), small GTPases are inactive. Some families are maintained inactive through binding guanosine nucleotide dissociation inhibitors (GDIs). These accessory proteins not only prevent nucleotide exchange but also restrict relocalization of GTPases to membranes. Loading of guanosine triphosphate (GTP) results in the stabilization of a new 3D conformation that enables them to interact with and modulate the activity of different and specific effector proteins. The GTP-bound state is therefore regarded as the active (ON) state. Typically, these proteins require guanine nucleotide exchange factors (GEFs) to exchange GTP for GDP. Some GTPases possess limited intrinsic GTPase activity, which depends on a few conserved amino acids in critical positions. GTPase-activating proteins (GAPs) activate the intrinsic GTPase activity, catalyzing the hydrolysis of GTP into GDP. It is the interplay between the GTPases, the GEFs, and the GAPs that coordinates signal transduction regulated by GTPases (Fig. 6.1).

More than 150 small G proteins have been identified in humans. These are traditionally classified into five families.[3] Ras proteins were the first members of the entire superfamily and were initially discovered on the basis of their homology to rat sarcoma virus genes.[4,5] The best-characterized members are H-Ras, N-Ras, and K-Ras, which have been implicated in many types of cancers and considered protooncogenes. However, the Ral and Rap families of small GTPases also belong to this prototypical class of more than 30 members. Members of the Ras homologous (Rho) family comprise a second group of more than 20 proteins. The three major members are Rho (A, B, C), Rac,[1–3] and Cdc42. This family of GTPase is well known for its role in remodeling of the actin cytoskeleton and gene

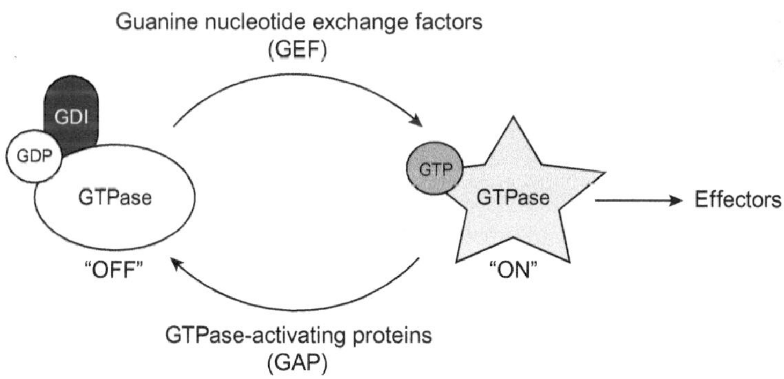

Figure 6.1 Mode of activation of small GTPases. Small GTPases are inactive (OFF) when bound to GDP and active when bound to GTP (ON). Cycling of these molecular switches is controlled by the combined activities of guanine nucleotide exchange factors (GEFs), which catalyze the exchange of GDP for GTP, and GTPase-activating proteins (GAPs), which increase the rate of GTP hydrolysis. For some GTPases, another level of regulation is provided by guanosine dissociation inhibitors (GDIs) that assure stability of the inactive GDP-bound state. Upon their activation, GTPases control a broad variety of effectors to regulate the timing, localization, and specificity of the cellular response.

expression. The third family, the Rabs, includes more than 60 isoforms. Their main function in cells is to finely tune many steps of vesicle trafficking (vesicle formation, vesicle movement, and membrane fusion). The fourth family is the ADP-ribosylation factors (ARFs), which also regulate vesicle budding through regulation of coat polymerization and disassembly. Some ARF isoforms are also involved in actin rearrangement and lipid remodeling. The last family of GTPases consists of only one member, Ran (Ras-related Nuclear protein), which regulates nucleoplasmic transport during interphase. This small G protein controls cell cycle progression. GTPases that have been linked directly or indirectly to β-arrestins are listed in Table 6.1.

1.2. Modes of regulation of GTPase activity

As mentioned previously, GEFs and GAPs tightly regulate the activity of monomeric G proteins. At any given time, the fraction of activated GTPases is dependent on the activity of their GEFs and GAPs. Cells express a large number of GEFs, GAPs, and GDIs, outnumbering GTPases themselves. GEFs possess specific domains responsible for their activity as well as protein- and lipid-binding domains that provide targeting. These allow GEFs to exert their effects in multimeric protein complexes linking signaling proteins to specific GTPases. GEFs are, therefore, themselves considered scaffolding proteins. Some GEFs are specific to a GTPase, while others can modulate the activation of different families. Their counterparts, the GAP proteins, are responsible for the hydrolysis of GTP into GDP. Their function is therefore monomeric G protein inactivation and downregulation of signaling events activated by GTPases. Certain GAPs are also considered effector proteins because they can link activated GTPases to new downstream signaling events. In addition to the GAP domain, these regulatory proteins also possess a large variety of domains that mediate their diverse interactions with binding partners. GAPs generally act specifically on a family of GTPase, but a few are known to exhibit dual specificity. Finally, as discussed earlier, the function of some small GTP-binding proteins, namely, Rho and Rab family members, is further regulated by GDIs. These preferentially interact with the GDP-bound form to prevent nucleotide exchange. By binding to inactive GTPases, it is believed that they serve to sequester and release GTPases in a regulated fashion.

In most signaling pathways, activation of small GTPases is basally regulated. However, in many cases, receptor stimulation promotes GTP loading

Table 6.1 Cross talk between β-arrestin and small GTPase family of proteins in mammals

Ras family		Rho family		Rab family		ARF family	
GTPase	**References**	**GTPase**	**References**	**GTPase**	**References**	**GTPase**	**References**
Ras	6,7	Rho	8–11	Rab4	12	ARF6	13–17
Ral	9,18,19	Rac	20–23	Rab5	12,24		
		Cdc42	25,26	Rab7	12,27		
Cellular function		**Cellular function**		**Cellular function**		**Cellular function**	
Proliferation Death/survival		Cell shape Migration Contraction		Intracellular trafficking		Vesicular trafficking Actin remodeling Lipid transformation	

of small G proteins leading to the modulation of agonist-driven cellular responses. As β-arrestins are also important signaling molecules for the regulation of receptor-mediated events, numerous reports have demonstrated that β-arrestins can facilitate small GTPase-mediated events. In some cases, β-arrestins were shown to be upstream regulators, while in others, they acted downstream. Over the years, an increasing body of evidence has demonstrated that β-arrestins interact with a variety of small GTP-binding proteins as well as their regulatory proteins. This chapter focuses on the cross talk between β-arrestins and the different families of small GTP-binding proteins.

2. RAS FAMILY GTPases

2.1. Ras and the MAPK pathway

Ras proteins are the founding and prototypical small GTPases. Their activation has been linked to gene expression leading to mitogenic and prosurvival responses. When mutated and constitutively activated, H-Ras, N-Ras, and K-Ras are capable of transforming mammalian cells causing diverse cancers.[28–31] In normal cells, Ras activity is controlled by a large variety of extracellular signals varying from growth factors that activate receptor tyrosine kinases to agonists that act via 7TM receptors. Activation of Ras is classically dependent upon son of sevenless, a GEF that is recruited to Grb2 upon receptor tyrosine kinase stimulation. The most prominent effector of activated Ras is Raf, a mitogen-activated protein kinase (MAPK) that centrally regulates activation of mitogenic cascades. A classical means to measure Ras activity is therefore to assess the phosphorylation level of ERK1/2.

The different pathways by which 7TM receptors activate mitogenic pathways have been extensively reviewed and β-arrestins have been characterized as key scaffolds in this process.[32] The use of dominant negative β-arrestin mutants (β-arrestin 1^{V53D}), knock-out mouse embryonic fibroblasts (MEFs), and small interfering RNA (siRNA) has revealed the importance of β-arrestins in regulating the ERK1/2 pathway.[33–35] The best understood role of β-arrestins in this process is at the level of the scaffolding of the MAPK pathway components, Src activation, and receptor tyrosine kinase receptor transactivation (see Chapter 5). Depletion of β-arrestin by siRNA has been shown to effectively reduce upregulation of the matrix metalloproteinase-1 gene upon exposure of human bronchial epithelium, the first line of protection against environmental insults, to diesel exhaust particles. This cellular response is known to require the Ras/Raf/MEK/ERK pathway, suggesting that in at least some cases Ras-dependent ERK activation by β-arrestins is physiologically relevant.[6]

The possibility of direct effects of β-arrestins on Ras activation had not been addressed until very recently. Using BRET-based biosensors, angiotensin II (Ang II) stimulation was shown to promote Ras activation in the Golgi and endoplasmic reticulum, but not in early endosomes, of HEK293 cells stably expressing the Ang II type 1 receptor (AT_1R). Plasma membrane Ras activation was also detected in this study, but at a more subtle level. In cells expressing an AT_1R mutant impaired in its ability to internalize ($AT_{1A}R^{\Delta 319}$), Ang II stimulation enhanced total Ras activation to a greater extent than in cells expressing wild-type AT_1Rs, suggesting that β-arrestin predominantly played a negative regulatory role in Ras activation. Instead, expression of a mutant receptor defective in $G\alpha_q$ coupling ($AT_1R^{DRY/AAY}$) inhibited Ras activation, suggesting that this process is predominantly G_q-mediated.[7] Such results underscore the multiplicity of different means, including heterotrimeric G protein activation, receptor tyrosine kinase transactivation, and β-arrestin scaffolding, that 7TM receptors use to signal to this key pathway, and it is clear that the role of β-arrestins in modulating Ras activity will need to be defined for a broad variety of receptors in different cellular models. BRET-based Ras probes, as well as more classical approaches such as pull down of activated Ras by Raf1–Ras-binding domain fusion proteins, should provide a means to directly test whether β-arrestins contribute to receptor dependent Ras activation. Knockdown of β-arrestin is likely to impair Ras activity if a specific receptor utilizes this scaffold protein to promote MAPK activation and enhance it if β-arrestin-mediated desensitization is providing negative regulation.

2.2. Ral—A multifunctional GTPase

The Ras-like GTPases RalA and RalB were identified based on sequence similarity to H-, K-, and N-Ras.[36] The two Ral isoforms, therefore, constitute a distinct subfamily within the Ras family. Among all GTPases, Rals are the most closely related small G proteins to Ras with more than 50% homology. RalA and RalB are ubiquitously expressed and highly homologous sharing close to 85% homology among themselves. They both participate in the signaling events leading to endocytosis and exocytosis, actin remodeling, and transcription.[37] However, they also display distinct functions. For example, RalA is important for anchorage-independent proliferation while RalB is critical for cell viability.[38,39] In some signaling pathways, Ral acts downstream of Ras. For example, Ras-mediated oncogenic transformation and propagation of mitogenic signals have been shown to require signaling via RalGEFs, considered direct effectors of Ras.[40] It is generally accepted that binding of RalGEFs to activated Ras is responsible for Ral activation.[41] Activation of Ral-dependent, but Ras-independent, signaling pathways has also been identified. For example, this family of small GTPases can be activated following calcium influx.[42] Furthermore, formyl-Met-Leu-Phe (fMLP) stimulation promotes Ral activation by a mechanism involving a GEF, RalGDS.[18]

RalGDS was first identified as a binding partner for H-Ras and R-Ras.[43,44] The CDC25 GEF domain and the Ras-binding domain are responsible for the ability of RalGDS proteins to mediate cross talk between Ras and Ral in the context of EGF receptor activation.[45] However, Ras-independent mechanisms of RalGDS activation have also been reported. In a yeast two-hybrid assay, this RalGEF was identified as a binding partner for β-arrestin. The interaction of RalGDS with β-arrestin 1 was subsequently confirmed by coimmunoprecipitation experiments.[18] The two proteins interact in the cytosol in the absence of stimulation, and it has been proposed that β-arrestin 1 serves to maintain RalGDS in an inactive state. Stimulation by fMLP destabilized the interaction leading to rapid dissociation of RalGDS and relocalization of the GEF to its GTPase at the plasma membrane. Interestingly, translocation of β-arrestin to the plasma membrane contributed, in part, to the redistribution of RalGDS to sites of membrane ruffling. Attenuation of the ruffling response correlated with the reassociation of β-arrestin 1 with RalGDS in the cytosol. The β-arrestin/RalGDS/Ral signaling axis, therefore, plays a key role in fMLP-stimulated plasma membrane ruffling in HEK293 cells as well as to chemoattractant receptor signaling in hematopoietic cell lines.[18] A role for RalA was also observed in the context of second messenger production following stimulation of the AT_1R. In HEK293 cells, Ang II stimulation is known to promote inositol phosphate (IP) synthesis following activation of $G\alpha_q$ and PLC. A careful examination of this signaling pathway has revealed that for this receptor, the β-arrestin-dependent translocation of RalGDS is required for the activation of RalA, which controls PLC-δ1 activity and generation of IP.[9] Ral GTPases were also reported to stimulate breast cancer cell proliferation and metastasis.[46–48] Rals, as well as β-arrestins, are aberrantly expressed in human breast tumors, and knockdown of either β-arrestin or Ral reduced the ability of invasive breast cancer cells to migrate and proliferate upon lysophosphatidic acid (LPA) treatment.[49] These observations provide evidence that misregulation of the β-arrestin/Ral pathway might contribute to the pathogenesis of diverse conditions.

Still other roles for the β-arrestin/RalGDS/Ral pathway have been reported. For example, this signaling axis was shown to regulate exocytosis of Weibel–Palade bodies (WPB) in endothelial cells. These intracellular granules contain a number of proteins and cofactors, ready to be released in the extracellular environment. Their fusion with the plasma membrane is tightly regulated and requires α-synuclein (α-Syn). Overexpression of α-Syn inhibited the activity of RalA through a mechanism involving potentiation of the β-arrestin/RalGDS interaction in human umbilical vein endothelial cells. Interestingly, α-Syn can interact with both β-arrestin and RalGDS. Together, these data suggest that this interaction might serve to restrict Ral activation and may function as a negative regulator of WPB exocytosis. In endothelial cells, the physiological function of this pathway remains to be defined. However, the release of WPB components such as von Willebrand factor is likely to impact vascular homeostasis.[19]

The other closely related Ras family GTPases are the Rap proteins. Although these small G proteins are known to be activated by 7TM receptors and mediate cellular proliferation through the MAPK pathway,[50,51] there is no evidence so far to suggest that their function is regulated by β-arrestins. Rather, reports have suggested that 7TM receptor activation of Raps occurs through heterotrimeric G proteins.[50,52]

3. RHO FAMILY GTPases

Members of the Rho family of small GTPases are primarily involved in cytoskeletal rearrangement, and they are broadly expressed in eukaryotes. Although the function of the three major members is interconnected in the regulation of cell migration and polarity, each GTPase has been associated with a particular function. RhoA is important for actin polymerization leading to stress fiber formation, Rac regulates formation of the branching actin chains of the lamellipodia, and Cdc42 controls filopodia formation. As signaling molecules, Rho GTPases activate a broad variety of downstream effectors that impact the function of various proteins. For example, Rac is well known for its role on NADPH oxidase and the production of superoxide. Increased Rho, Rac, and Cdc42 activity can be observed upon agonist stimulation of receptors. As described below, β-arrestins have been reported to regulate the function of the GTPases themselves as well as their regulatory factors, GEFs and GAPs.

3.1. Rho and β-arrestin modulation of stress fiber formation

Three Rho GTPases have been identified in higher vertebrates: RhoA, RhoB, and RhoC. Despite their 85% homology of sequence, they play distinct roles. Over the years, much attention has focused on RhoA in the context of receptor modulation of actin reorganization. The identification of Rho-dependent signaling pathways leading to specific cellular responses has largely relied on the use of dominant negative mutants and biochemical inhibitors such as the exoenzyme C3 transferase. LPA was the first agonist identified to activate Rho leading to stress fiber formation and cell migration.[53] Several other agonists acting via 7TM receptors, such as sphingosine 1-phosphate, bombesin, thrombin, endothelin, and Ang II, were also reported to promote Rho activation.[54–56] Efforts to identify the molecular mechanisms leading to activation of this GTPase have implicated heterotrimeric G proteins as key molecules. Specifically, $G\alpha_{12/13}$ were identified as heterotrimeric G protein isoforms mediating Rho activation,[57] and it was proposed that $G\alpha$ subunits directly bind RhoGEFs to modulate their activity.[58–60] The potential of $G\alpha_{q/11}$ to mediate Rho activation has also been reported[8,61] and the identity of the G protein implicated in Rho activation following 7TM receptor stimulation remains of great interest.

The role of β-arrestins in mediating RhoA activation has been studied in the context of Ang II stimulation of HEK293 cells stably expressing the $AT_{1A}R$.[8] Knockdown of β-arrestin 1, but not β-arrestin 2, by siRNA, impaired the ability of Ang II stimulation to promote GTP loading of RhoA by about 60% and reduced stress fiber formation. The involvement of β-arrestins was further supported by experiments where an AT_1R ligand known to mediate biased signaling via β-arrestin, [Sarcosine^1-Ile^4-Ile^8]-Ang II, was effective in promoting RhoA activation. The authors concluded that in concert with $G\alpha_{q/11}$, β-arrestin 1 was critical for RhoA activation and formation of stress fibers. The molecular mechanism by which G proteins and β-arrestin coordinate Rho activation remains to be defined. Whether β-arrestin 1 acts as a scaffold to promote the recruitment of GEFs to the receptor has not been investigated. Two Rho GEFs, Lbc and LARG, were previously reported to interact with $G\alpha_{q/11}$.[62,63] β-Arrestin 1 may play a role in translocation of these GEFs, to the receptor/G_q complex to promote RhoA activation.[8] The role of β-arrestins in Ang II-mediated membrane bleb formation, a cellular response dependent on remodeling of the actin cytoskeleton, was also studied. This agonist-mediated effect was shown to be dependent upon the activation of RhoA, since overexpression of the dominant negative $RhoT^{19}N$ mutant almost completely inhibited Ang II-mediated membrane blebbing. Inhibition of β-arrestin function by different molecular approaches, such as expression of the amino-terminal domain of β-arrestin or transfection of siRNA, revealed that β-arrestin 2-mediated AT_1R signaling to this RhoA-dependent pathway. However, the loading of GTP on RhoA in β-arrestin 2-depleted cells was not assessed. In contrast, β-arrestin 1 played no role in Ang II-stimulated membrane blebbing.[9]

The role of β-arrestins in RhoA activation does not appear to be limited to mammalians. In *Xenopus*, xβ-arrestin 2 was shown to interact with RhoA.[10] In this species, xβ-arrestin 2 is required for disheveled activation of the small G protein for convergence and extension movement. In contrast, there are also examples in the literature where RhoA signaling is not affected by the inhibition of β-arrestin expression. Signaling via RhoA is known to be critical for thrombin-induced endothelial barrier dysfunction. However, β-arrestin is not essential for PAR1 activation of this GTPase.[64]

Interestingly, RhoA itself has been shown to regulate the interactions of β-arrestin with its signaling partners, suggesting that RhoA–β-arrestin regulation is bidirectional. For example, RhoA is one of the few proteins known to enhance the phosphatase activity of PTEN.[65,66] Expression of constitutively active mutants of RhoA was demonstrated to be effective in potentiating the interaction between β-arrestin 1 and PTEN, while the C3 transferase Rho inhibitor prevented this interaction. These observations suggest that activation of RhoA plays a role in β-arrestin binding to PTEN. Conversely, inhibition of endogenous expression of β-arrestin 1/2 by siRNA limited the ability of RhoA to increase PTEN activity upon stimulation of the LPA_1 receptor.[67] Because other 7TM receptors also regulate PTEN activity through a RhoA-dependent mechanism, β-arrestin may be a key molecule generally regulating this pathway. Further studies are required to examine this hypothesis.

Another means by which β-arrestins may regulate Rho GTPases is through modulating the activity of their regulatory proteins. As scaffolds, β-arrestin may interact with GEFs and/or GAPs. One example that well depicts this concept is the binding of β-arrestin 1 to ARFGAP21, an effector and GAP for RhoA. Biochemical analysis has shown that β-arrestin binds directly ARFGAP21 through a region that transects the RhoA effector GAP domain. This interaction was enhanced by Ang II stimulation and as a consequence was shown to have great impact on the level of active RhoA. Disruption of the interaction led to increased activation of the GAP, attenuation of RhoA activity, and less efficient stress fiber formation.[11] Binding to GEF and/or GAPs might provide an additional way for β-arrestin proteins to regulate monomeric G protein function.

3.2. Rac—The GTPase associated with membrane ruffling and protrusions

Rac1 was first discovered as Ras-related C3 botulinium toxin substrate 1 and is the most studied member of the three Rac isoforms. Expression of Rac1 is ubiquitous, while Rac2 is present only in hematopoietic cells and Rac3 is most highly expressed in brain. All Rac isoforms are highly homologous, and their main role is to promote the actin assembly required for formation of lamellipodia and membrane ruffling. Activation of Rac is therefore often associated with cell migration. RhoG, the most divergent isoform, was suggested to be important for the cell cycle. However, other functions have been attributed to this isoform, namely, the activation of Rac1 through DOCK180.[68–70] Altogether, Rac GTPases play numerous roles in addition to cytoskeletal remodeling. These vary from cell growth and adhesion to axonal guidance and superoxide production.[71]

The role of β-arrestins in the control of Rac activation remains controversial since conflicting results have been reported over the years. First, enhanced recruitment of Rac1 to plasma membranes was observed in MEFs lacking the two β-arrestin isoforms, at basal levels and upon stimulation with IL-8, suggesting that the presence of β-arrestin limited the degree of Rac activation. This β-arrestin/Rac axis, when intact, would limit the NADPH oxidase-dependent oxidative burst and protect against cell death.[20] Alternatively, it was reported that activation of Rac1 by Wnt-5A was blocked in β-arrestin knock-out MEFs. Measurements of the basal activation level of Rac1 were similar in control and double β-arrestin 1/2 cells, but levels of GTP-bound Rac1 were enhanced upon Wnt-5A stimulation only in arrestin replete MEFs. Furthermore, overexpression of β-arrestin 2 increased activation of the GTPase in MEFs.[21] Similar results were observed in another system model, *Xenopus laevis*.[21] If one assumes that translocation of Rac1 reflects its activity, the observations reported by these two groups, who have used the same cells, appear contradictory.

In HEK 293 cells, isoproterenol treatment promotes the rapid and sustained activation of Rac1 almost to the degree obtained from GTPγS loading of the small G protein. Inhibition of β-arrestin 1 expression by RNA interference has been shown to abrogate the ability of this agonist to activate Rac1, a key event important for NADPH oxidase activation and p38 phosphorylation.[22] These data suggest that β-arrestin is required for β_2-adrenergic receptor signaling to this GTPase and that Rac1 acts downstream of β-arrestin. The involvement of β-arrestins in the activation of Rac proteins, therefore, remains a subject of great interest and requires further studies to define the roles of β-arrestin isoforms in different cellular contexts.

Interplay between arrestins and Rac2 has also been reported. One report has proposed that Rac2 acts upstream and regulates the translocation of arrestin 2, the major arrestin protein present in *Drosophila*, to activated rhodopsin.[23] Similar to mammalian photoreceptors, translocation of arrestin upon receptor activation resulted in desensitization of the photoresponse by preventing further heterotrimeric G protein coupling. In control experiments, arrestin was rapidly redistributed to the rhabdomeres upon light exposure. In Rac2 null mutant photoreceptor cells, arrestin 2 remained distributed throughout the cell bodies after light exposure. In *Drosophila*, relocalization of arrestin did not depend on any of the known components of the phototransduction cascade. Altogether, these findings suggest that light-dependent translocation of arrestin 2 occurs through a rhodopsin/Rac2-dependent pathway in flies.[23] Whether this is an evolutionary conserved mechanism remains to be addressed. In mammalian cells, our understanding of the mechanisms by which receptor activation results in arrestin translocation remains poorly understood.

Finally, other examples of β-arrestin signaling to Rac-dependent pathways have been reported. Complex responses such as vascular permeability, controlled by internalization of VE-cadherins, which like 7TM receptors are membrane proteins, were shown to be dependent upon a β-arrestin/Rac1 axis. In this context, it was proposed that the vascular endothelial growth factor receptor promoted endocytosis of VE-cadherin through a mechanism involving Src, β-arrestin, Vav2, Rac, and PAK.[72]

3.3. Cdc42 and filopodia formation

The human Cdc42 protein was initially purified from placental membranes and named G_p. This newly identified 21 kDa protein was considered an analog of G_s and G_i since it copurified with $G\beta\gamma$ subunits.[73] Upon cloning of the gene from a human placental library, it was renamed as the human homolog of the yeast cell division cycle protein Cdc42.[74] Similar to other GTPase of the Rho family, Cdc42 participates in actin remodeling. Activation of this isoform has been associated with filopodia formation as well as numerous other cellular processes including cell cycle progression. Stimulation of the serotonin 5-HT_7 receptor leads to activation of Cdc42 via a mechanism involving $G\alpha_{12}$ in NIH3T3 cells.[75]

The role of β-arrestins in the activation of this monomeric G protein was studied in MEFs. It was previously reported that stimulation of the PAR2 receptor with a synthetic ligand (2-furoyl-LIGRL-O-NH_2) promoted Cdc42 deactivation in this cell line. However, this dampening of Cdc42 activity was not observed in double β-arrestin1/2 knock-out MEFs treated with this agonist. Instead, a twofold increase in Cdc42 activation was reported. Transfection of either β-arrestin 1 or β-arrestin 2 inhibited Cdc42 activation, suggesting that both β-arrestin isoforms contributed to PAR2-mediated Cdc42 deactivation.[25] Stimulation of numerous 7TM receptors has been shown to promote Cdc42 activation. These include the endothelin, the Ang II, adenosine, and the bradykinin receptors.[76–79] Whether β-arrestin regulates Cdc42 activation in response to these stimuli remains to be addressed.

The involvement of β-arrestins in regulation of Cdc42 by non-7TM receptors has also been investigated. In an ovarian cancer cell line (Ovca429), simple expression of the type III TGFβ receptor (TβRIII) markedly enhanced basal activation levels of Cdc42. Inhibition of β-arrestin 2 expression by transfection of specific siRNA inhibited this TβRIII-mediated Cdc42 activation. Furthermore, expression of a mutant receptor impaired in its ability to bind β-arrestin 2 ($T\beta RIII^{T841A}$) failed to constitutively activate the GTPase. These data suggest that the TβRIII/β-arrestin 2 interaction is essential for the activation of Cdc42.[26]

4. RAB FAMILY GTPases

The Rab family of GTPases is the largest of the Ras superfamily of small G proteins with over 60 members identified. The first *rab* gene was discovered in yeast (SEC4/YPT1), and numerous mammalian homologs were subsequently cloned and named *rab* (*ras*-like in rat brain).[80,81] These small GTPases are ubiquitously expressed, and although their structure is highly homologous, their intracellular localization is quite diverse. Rab GTPases have been associated with many steps of membrane trafficking such as vesicle formation, docking, and fusion to membranes as well as vesicle movement along the actin and tubulin networks. Together, Rabs control almost all aspects of vesicular trafficking. The use of dominant negative forms and constitutively active mutants has greatly helped in delineating the roles of each isoform in trafficking of proteins in general but most importantly of 7TM receptors.

4.1. Colocalization of Rabs with receptors and β-arrestin

Over the years, Rab4, Rab5, Rab7, and Rab11 were shown to regulate receptor endocytosis, their trafficking through the endosomal pathway, as well as their targeting to lysosomes.[82] Several receptors were shown to colocalize with Rabs in different intracellular compartments. For example, internalized β_2-adrenergic receptors are present in Rab5a-positive endosomes after 15 min of agonist treatment.[83] Similar observations were reported for the endothelin receptors and the AT_1R.[24,84] It is through the use of Rab mutants mimicking the inactive or active form that elucidation of the endocytic and trafficking routes utilized by 7TM receptors were identified. For example, expression of a dominant negative Rab5a mutant was shown to inhibit trafficking of the AT_1R to enlarged vesicular structures, without altering its internalization from the plasma membrane.[24] This particular receptor directly bound this Rab isoform as well as Rab 4, 7, and 11.[85] Other receptors such as the β_2-adrenergic, the thromboxane A2, and the prostacyclin receptor were all shown to bind Rab11.[24,86–88]

It was reported that the AT_1R/β-arrestin complex localized to Rab5 positive endosomes.[24] However, β-arrestin overexpression was also reported to decrease colocalization of the platelet-activating factor receptor with Rab7 in endothelial cells.[27] As discussed in Chapter 4, some 7TM receptors (class B) have the ability to internalize along with β-arrestin while others (class A) dissociate from the β-arrestin at the plasma membrane.[89] Therefore, binding of β-arrestin to a receptor impacts the intracellular route an active receptor uses. One such example is the Apelin receptor. When stimulated with Apelin-13, this 7TM receptor acts as a class A receptor (internalization without β-arrestin) while Apelin-36 stimulation leads to trafficking reminiscent of the class B family (internalization with β-arrestin). Interestingly, both Apelin-13 and -36 stimulation promoted receptor internalization to Rab5 positive early endosomes. However, Apelin-13 internalized receptors rapidly recycled back through a Rab4-dependent pathway, while Apelin-36-stimulated receptors were targeted to lysosomes by Rab7.[12]

4.2. Modulation of Rab function through interaction

The question of interest for this chapter is whether β-arrestins directly modulate GTP loading of Rabs. Unlike many other GTPases, no evidence in the literature clearly demonstrates that they do. Rabs function to control trafficking of a vesicle from one organelle to another. They are highly specialized, and although they are key players dictating the route a 7TM receptor will utilize, their activation does not appear to depend on any extracellular stimuli. Like other small G proteins, activation is possible via GEFs and inactivation facilitated by GAPs. The molecular mechanisms triggering activation of Rab regulatory factors remains to be fully elucidated. It is, however, becoming more and more evident that β-arrestins and Rabs act in concert to dictate intracellular trafficking. For example, Rab5 and β-arrestin binding to the carboxyl-terminal tail of the AT_1R is an event important for retention of this receptor in early endosomes.[90] Whether as scaffold proteins, β-arrestins bind to Rabs or their regulatory proteins to modulate their activity remains to be explored.

5. ARF FAMILY GTPases

The ARFs were first identified as a cellular activity required for ADP-ribosylation of $G\alpha_s$ by cholera toxin, a process by which it exerts its toxic effect.[91,92] ARFs are part of a larger family of small Ras-related G proteins that include Sar and Arl (ARF-like), identified mainly by genome sequencing. The focus here will, however, be on ARFs since evidence shows that their activity is regulated by receptor stimulation. Although they are typically called small GTPases, ARFs have actually no intrinsic GTPase activity and require GAPs for GTP hydrolysis. The numerous roles of ARFs in cells are not related to ADP-ribosylation. Rather, these small G proteins are best known to participate in the formation of coated vesicles responsible for membrane traffic, lipid transformation, and reorganization of the actin cytoskeleton. Six isoforms of ARFs have been identified (ARF1–6), and they are ubiquitously expressed with the exception of ARF2 that is not present in humans. The best-characterized ARFs are ARF1 and ARF6. ARF1 is known for orchestrating the coating of budding vesicles at the Golgi.[93] However, some evidence suggests that this ARF isoform can localize to plasma membranes to mediate receptor-dependent events.[94–97] No data suggesting cross talk with β-arrestins have been reported in the literature. ARF6 is the small G protein classically present at the cell periphery, and its activation can be stimulated by agonist treatment of the cells. Evidence demonstrating regulation of its function by β-arrestins is described below.

5.1. ARF6 and receptor endocytosis

The first evidence that ARF6 could be a modulator of 7TM receptor function came from the early studies describing the function of GIT, a new ARF GAP and interacting partner of the G protein-coupled receptor kinases (GRKs).[98] The observation that an ARF GAP would modulate internalization of a receptor sparked some interest in addressing the role of ARFs themselves in this process. Expression of dominant negative and constitutively active mutants of ARF6 inhibited sequestration of the β_2-adrenergic receptor.[13] Later, it was shown that inhibition of ARF6 expression by RNA interference blocked internalization of this particular receptor, as well as the AT_1R, the V2R, the ET_BR, and the M2MR.[99] The proposed mechanisms by which ARF6 regulated receptor internalization involve β-arrestins. It was shown that both β-arrestin isoforms could be found in complex with the GTPase upon agonist stimulation of the β_2-adrenergic receptor in HEK293 cells. According to *in vitro* data, the interaction appeared to be direct and modulated by the nature of the nucleotide bound to the ARF, in that ARF6-GDP preferentially bound the C-terminal region of β-arrestin.[13,14] In this same context, β-arrestin was shown to be constitutively associated with ARNO, an ARF GEF that promoted loading of GTP on ARF6.[13] In addition, a second exchange factor was also reported to interact with β-arrestin, the protein EFA6.[14] Activated ARF6 was shown to mediate the recruitment of AP-2 and clathrin to the receptor/β-arrestin complex, a step essential for the formation of endocytic vesicles.[100,101] A role of β-arrestin in the activation process of ARF6 had been suggested after addition of purified recombinant β-arrestin proteins improved the ability of ARNO to promote loading of GTP on the small G in an *in vitro* setting.[13] The direct demonstration that β-arrestin is a key protein controlling β_2-adrenergic receptor-mediated ARF6 activation was recently reported in HEK293 cells. Depletion of β-arrestin by siRNA markedly inhibited isoproterenol-induced GTP loading of ARF6.[14] In addition to mediating the initial step of receptor endocytosis,[99] activation of ARF6 controlled recycling. In these conditions, the pool of EFA6-activated ARF6 inhibited fast recycling through the Rab4 pathway and relocalized receptors to late endosomes and lysosomes.[14] A role for the β-arrestin/ARF signaling axis has also been described in *Dictyostelium discoideum*. In this model organism, six members of the arrestin subfamily were identified and named arrestin domain-containing protein (Adc) A–F. It was reported that AdcA interacted with GDP-bound ARFα, an homologue of human ARF proteins, further demonstrating the importance of these two proteins across species.[15]

Interestingly, activation of ARF6 by ARNO was reported to cause β-arrestin release from its membrane docking site to promote luteinizing hormone/choriogonadotropin receptor desensitization using a cell-free plasma membrane model of ovarian cells.[16] Indeed, the addition of recombinant ARNO promoted binding of β-arrestin to the third intracellular loop of the active receptor thereby limiting signaling to heterotrimeric $G\alpha_s$ protein and adenylate cyclase. Further studies revealed that in this system, ARNO is readily detected in follicular membranes. Its activation to promote β-arrestin release from the plasma membrane pool was demonstrated to be independent of PI-3 kinase and Gβγ but required PIP2.[16]

5.2. Remodeling of the actin cytoskeleton via β-arrestin and ARF6

A role for β-arrestin and ARF6 in other receptor-mediated cellular responses was also observed. Stimulation of the calcium-sensing receptor leads to increased intracellular calcium levels, membrane ruffling and ultimately chemotaxis. These responses were blocked by expression of a dominant negative β-arrestin mutant or transfection of specific β-arrestin siRNA, expression of a catalytically inactive ARNO construct, and expression of a dominant negative form of ARF6 as well as a truncated form of ELMO. Furthermore, ARNO and β-arrestin were shown to coimmunoprecipitate independent of receptor stimulation and to colocalize in HEK293 cells.[17] This chapter was the first suggesting that the β-arrestin/ARNO/ARF6/ELMO signaling axis was involved in 7TM receptor-dependent cell shape changes. As many other receptors are known to promote actin remodeling in a β-arrestin and ARF6-dependent manner, the β-arrestin/GEF/ARF6 signaling axis may represent a general mechanism.

Together, these newly identified ARF-dependent signaling events contribute to the understanding of the molecular mechanisms by which β-arrestin regulates receptor endocytosis and trafficking.

6. RAN FAMILY GTPases

The last family of Ras proteins comprises the Ras-like nuclear protein (Ran). Only one isoform has been identified, and its main role has been associated with the regulation of nucleocytoplasmic transport during the cell cycle (interphase) and the organization of the spindle apparatus during mitosis. Evidence in the literature suggests a role of β-arrestin 1 in the nucleus.[102] Although no studies have tested whether β-arrestin 1 can modulate the activity of Ran, the possibility that it does remains to be examined.

7. PERSPECTIVES AND FUTURE DIRECTIONS

Although it is clear that 7TM receptors can activate small GTPases and that GTPases can in turn participate in defining the cellular response mediated by agonist stimulation, both at the levels of receptor trafficking and signaling, numerous questions remain unanswered. The studies described in this chapter provide a framework for future experimental work that will clarify the role of β-arrestins as modulators of small GTPase activity and function. For some small G proteins, great progress has been made, and we are now able to provide a model. However, for others, little is known. A comprehensive study of GTPase activation, in conditions of β-arrestin knockdown, could represent an initial step toward delineating possible involvement of these ubiquitous scaffold proteins.

7.1. Defining the role of β-arrestin 1 and β-arrestin 2

Very few studies have addressed the roles of both β arrestin isoforms in modulating small GTPase activity. Although the contribution of one isoform versus the other might depend upon the stimulus, the nature of receptor, and the cellular context, few studies have systematically reported the role of β-arrestin 1 and β-arrestin 2 simultaneously. Now that reagents such as si- and shRNAs are in common use, these experiments could easily be revisited. Although β-arrestins might have redundant effects, it is generally appreciated that they often play specific, as well as complementary, roles. A significant amount of evidence suggests that β-arrestin 1 is a key intermediate in Ral, Rho, and ARF6 activation.[8,14,18] However, some studies have proposed that both β-arrestins might contribute.[13] Differentially studying complex formation between the β-arrestins and the GTPase, and the consequence of one isoform versus the other in the activation process of a given small G, should give us better insight into the function of each isoform in mediating receptor activation of small GTPase-dependent cellular responses. This aspect has been well defined in the endocytic paradigm.

7.2. Identification of the mechanisms whereby β-arrestins regulate GTPase function

In order to better understand the biological significance of the role of β-arrestins in small GTPase activation, we must better define the molecular mechanisms underlying their actions. We now know that β-arrestins can colocalize and interact with some small G proteins. In addition, their depletion in cells impacts the activation of some small GTPases. Although for years investigators have been examining the role β-arrestin plays in GTPase-mediated events, we can only draw simple schematic diagrams to propose a model (Fig. 6.2). Sustained agonist stimulation leads to conformational changes of receptor structure and translocation of β-arrestins from the cytosol to the plasma membrane. As scaffold proteins, β-arrestins can interact with GEFs and GTPases to promote their activation. This simple snap shot of the signaling axis remains incomplete considering the numerous events occurring during agonist stimulation, including heterotrimeric G protein activation, plasma membrane lipid transformation, and posttranslational modification of proteins through enzymes, as well as cytoskeletal remodeling. Whether and how β-arrestin controls these biological events to impact small GTPase activation need to be addressed.

Studies aiming at identifying the binding site between β-arrestins, small GTPases, and their GEFs and GAPs might be useful to create new tools to study the relevance of a specific interaction. Instead of knocking down the expression of the β-arrestin to assess its role in the activation of a specific GTPase, one could express a mutant protein in a null background to only

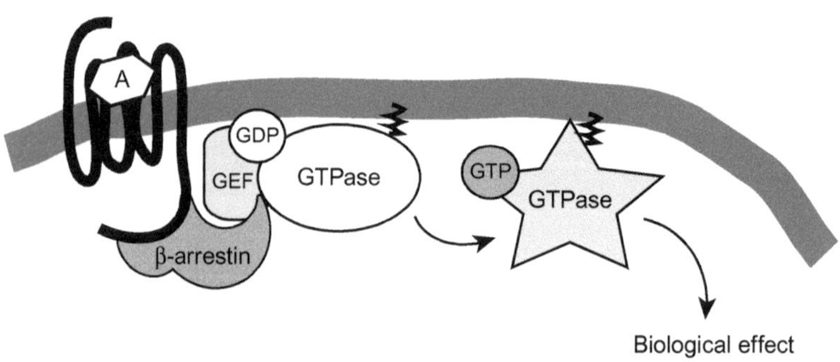

Figure 6.2 General mechanism of GTPase activation by 7TM receptors. Agonist stimulation of a receptor classically leads to conformational changes within the receptor that not only activates heterotrimeric G proteins but also promotes the recruitment of β-arrestin proteins. β-Arrestins have been shown to interact with several GEFs and small GTPases. In many cases, their presence is necessary for activation of the small G protein (loading of GTP) and its related downstream signaling events.

study that specific function of β-arrestin. In this context, β-arrestin would function to regulate all its other functions unrelated to the small GTPase. This approach could provide a clearer picture of the role protein complex formation plays in GTPase activation. Although for a specific GTPase, β-arrestin may interact with both the small G protein and its GEF, the contribution of each interaction is presently unknown.

Identification of new β-arrestin interacting partners may also help defining the molecular mechanism by which β-arrestins regulate small GTPase function. Different types of screens have helped find new β-arrestin interactors that are related to small GTPases. For example, a proteomic analysis of β-arrestin-binding partners revealed that many GAPs can associate with β-arrestin.[103] Demonstration that in cells, these different GAPs can interact with β-arrestins and impact GTPase activity would further contribute to our understanding of their regulatory mechanism.

8. CONCLUSIONS

Although small GTPases are involved in almost all cellular processes, their mode of activation remains simple and highly conserved. How 7TM receptor stimulation activates signaling to this family of proteins remains an intense and competitive domain of research. As most 7TM receptors only directly couple to a few effectors, the heterotrimeric G proteins, the GRKs, and the β-arrestins, the mechanisms by which they activate small G proteins must be limited. Different reports have implicated both heterotrimeric G proteins and β-arrestins as proximal effectors. Now that the latter are considered signaling molecules, we may find that they act to direct signaling to many GTPases. The challenge for the future is to define how they do this in a general way.

REFERENCES

1. Reiter E, Ahn S, Shukla AK, Lefkowitz RJ. Molecular mechanism of beta-arrestin-biased agonism at seven-transmembrane receptors. *Annu Rev Pharmacol Toxicol.* 2012;52:179–197.
2. Shenoy SK, Lefkowitz RJ. beta-Arrestin-mediated receptor trafficking and signal transduction. *Trends Pharmacol Sci.* 2011;32:521–533.
3. Takai Y, Sasaki T, Matozaki T. Small GTP-binding proteins. *Physiol Rev.* 2001;81:153–208.
4. DeFeo D, Gonda MA, Young HA, et al. Analysis of two divergent rat genomic clones homologous to the transforming gene of Harvey murine sarcoma virus. *Proc Natl Acad Sci USA.* 1981;78:3328–3332.

5. Chang EH, Gonda MA, Ellis RW, Scolnick EM, Lowy DR. Human genome contains four genes homologous to transforming genes of Harvey and Kirsten murine sarcoma viruses. *Proc Natl Acad Sci USA*. 1982;79:4848–4852.
6. Li J, Ghio AJ, Cho SH, Brinckerhoff CE, Simon SA, Liedtke W. Diesel exhaust particles activate the matrix-metalloproteinase-1 gene in human bronchial epithelia in a beta-arrestin-dependent manner via activation of RAS. *Environ Health Perspect*. 2009;117:400–409.
7. Balla A, Erdelyi LS, Soltesz-Katona E, Balla T, Varnai P, Hunyady L. Demonstration of angiotensin II-induced Ras activation in the trans-Golgi network and endoplasmic reticulum using bioluminescence resonance energy transfer-based biosensors. *J Biol Chem*. 2011;286:5319–5327.
8. Barnes WG, Reiter E, Violin JD, Ren XR, Milligan G, Lefkowitz RJ. beta-Arrestin 1 and Galpha q/11 coordinately activate RhoA and stress fiber formation following receptor stimulation. *J Biol Chem*. 2005;280:8041–8050.
9. Godin CM, Ferreira LT, Dale LB, Gros R, Cregan SP, Ferguson SS. The small GTPase Ral couples the angiotensin II type 1 receptor to the activation of phospholipase C-delta 1. *Mol Pharmacol*. 2010;77:388–395.
10. Kim GH, Han JK. Essential role for beta-arrestin 2 in the regulation of Xenopus convergent extension movements. *EMBO J*. 2007;26:2513–2526.
11. Anthony DF, Sin YY, Vadrevu S, et al. beta-Arrestin 1 inhibits the GTPase-activating protein function of ARHGAP21, promoting activation of RhoA following angiotensin II type 1A receptor stimulation. *Mol Cell Biol*. 2011;31:1066–1075.
12. Lee DK, Ferguson SS, George SR, O'Dowd BF. The fate of the internalized apelin receptor is determined by different isoforms of apelin mediating differential interaction with beta-arrestin. *Biochem Biophys Res Commun*. 2010;395:185–189.
13. Claing A, Chen W, Miller WE, et al. Beta arrestin-mediated ARF6 activation and beta2-adrenergic receptor endocytosis. *J Biol Chem*. 2001;276:42509–42513.
14. Macia E, Partisani M, Paleotti O, Luton F, Franco M. Arf6 negatively controls the rapid recycling of the beta2AR. *J Cell Sci*. 2012;125:4026–4035.
15. Guetta D, Langou K, Grunwald D, Klein G, Aubry L. FYVE-dependent endosomal targeting of an arrestin-related protein in amoeba. *PLoS One*. 2010;5:e15249.
16. Mukherjee S, Gurevich VV, Jones JC, et al. The ADP ribosylation factor nucleotide exchange factor ARNO promotes beta-arrestin release necessary for luteinizing hormone/choriogonadotropin receptor desensitization. *Proc Natl Acad Sci USA*. 2000; 97:5901–5906.
17. Bouschet T, Martin S, Kanamarlapudi V, Mundell S, Henley JM. The calcium-sensing receptor changes cell shape via a beta-arrestin-1 ARNO ARF6 ELMO protein network. *J Cell Sci*. 2007;120:2489–2497.
18. Bhattacharya M, Anborgh PH, Babwah AV, et al. Beta-arrestins regulate a Ral-GDS Ral effector pathway that mediates cytoskeletal reorganization. *Nat Cell Biol*. 2002;4:547–555.
19. Kim KS, Park JY, Jou I, Park SM. Regulation of Weibel-Palade body exocytosis by alpha-synuclein in endothelial cells. *J Biol Chem*. 2010;285:21416–21425.
20. Zhao M, Wimmer A, Trieu K, Discipio RG, Schraufstatter IU. Arrestin regulates MAPK activation and prevents NADPH oxidase-dependent death of cells expressing CXCR2. *J Biol Chem*. 2004;279:49259–49267.
21. Bryja V, Schambony A, Cajanek L, Dominguez I, Arenas E, Schulte G. Beta-arrestin and casein kinase 1/2 define distinct branches of non-canonical WNT signalling pathways. *EMBO Rep*. 2008;9:1244–1250.
22. Gong K, Li Z, Xu M, Du J, Lv Z, Zhang Y. A novel protein kinase A-independent, beta-arrestin-1-dependent signaling pathway for p38 mitogen-activated protein kinase activation by beta2-adrenergic receptors. *J Biol Chem*. 2008;283:29028–29036.

23. Elsaesser R, Kalra D, Li R, Montell C. Light-induced translocation of Drosophila visual arrestin2 depends on Rac2. *Proc Natl Acad Sci USA*. 2010;107:4740–4745.
24. Seachrist JL, Laporte SA, Dale LB, et al. Rab5 association with the angiotensin II type 1A receptor promotes Rab5 GTP binding and vesicular fusion. *J Biol Chem*. 2002;277:679–685.
25. Wang P, Kumar P, Wang C, Defea KA. Differential regulation of class IA phosphoinositide 3-kinase catalytic subunits p110 alpha and beta by protease-activated receptor 2 and beta-arrestins. *Biochem J*. 2007;408:221–230.
26. Mythreye K, Blobe GC. The type III TGF-beta receptor regulates epithelial and cancer cell migration through beta-arrestin2-mediated activation of Cdc42. *Proc Natl Acad Sci USA*. 2009;106:8221–8226.
27. Radin JN, Orihuela CJ, Murti G, Guglielmo C, Murray PJ, Tuomanen EI. beta-Arrestin 1 participates in platelet-activating factor receptor-mediated endocytosis of Streptococcus pneumoniae. *Infect Immun*. 2005;73:7827–7835.
28. Brown R, Marshall CJ, Pennie SG, Hall A. Mechanism of activation of an N-ras gene in the human fibrosarcoma cell line HT1080. *EMBO J*. 1984;3:1321–1326.
29. Capon DJ, Seeburg PH, McGrath JP, et al. Activation of Ki-ras2 gene in human colon and lung carcinomas by two different point mutations. *Nature*. 1983;304:507–513.
30. Feramisco JR, Gross M, Kamata T, Rosenberg M, Sweet RW. Microinjection of the oncogene form of the human H-ras (T-24) protein results in rapid proliferation of quiescent cells. *Cell*. 1984;38:109–117.
31. Stacey DW, Kung HF. Transformation of NIH 3 T3 cells by microinjection of Ha-ras p21 protein. *Nature*. 1984;310:508–511.
32. DeWire SM, Ahn S, Lefkowitz RJ, Shenoy SK. Beta-arrestins and cell signaling. *Annu Rev Physiol*. 2007;69:483–510.
33. Daaka Y, Luttrell LM, Ahn S, et al. Essential role for G protein-coupled receptor endocytosis in the activation of mitogen-activated protein kinase. *J Biol Chem*. 1998;273:685–688.
34. Ahn S, Nelson CD, Garrison TR, Miller WE, Lefkowitz RJ. Desensitization, internalization, and signaling functions of beta-arrestins demonstrated by RNA interference. *Proc Natl Acad Sci USA*. 2003;100:1740–1744.
35. Azzi M, Charest PG, Angers S, et al. Beta-arrestin-mediated activation of MAPK by inverse agonists reveals distinct active conformations for G protein-coupled receptors. *Proc Natl Acad Sci USA*. 2003;100:11406–11411.
36. Chardin P, Tavitian A. The ral gene: a new ras related gene isolated by the use of a synthetic probe. *EMBO J*. 1986;5:2203–2208.
37. Feig LA. Ral-GTPases: approaching their 15 minutes of fame. *Trends Cell Biol*. 2003;13:419–425.
38. Chien Y, White MA. RAL GTPases are linchpin modulators of human tumour-cell proliferation and survival. *EMBO Rep*. 2003;4:800–806.
39. Moskalenko S, Henry DO, Rosse C, Mirey G, Camonis JH, White MA. The exocyst is a Ral effector complex. *Nat Cell Biol*. 2002;4:66–72.
40. Wolthuis RM, Bos JL. Ras caught in another affair: the exchange factors for Ral. *Curr Opin Genet Dev*. 1999;9:112–117.
41. Kishida S, Koyama S, Matsubara K, Kishida M, Matsuura Y, Kikuchi A. Colocalization of Ras and Ral on the membrane is required for Ras-dependent Ral activation through Ral GDP dissociation stimulator. *Oncogene*. 1997;15:2899–2907.
42. Hofer F, Berdeaux R, Martin GS. Ras-independent activation of Ral by a Ca(2+)-dependent pathway. *Curr Biol*. 1998;8:839–842.
43. Hofer F, Fields S, Schneider C, Martin GS. Activated Ras interacts with the Ral guanine nucleotide dissociation stimulator. *Proc Natl Acad Sci USA*. 1994;91:11089–11093.

44. Spaargaren M, Bischoff JR. Identification of the guanine nucleotide dissociation stimulator for Ral as a putative effector molecule of R-ras, H-ras, K-ras, and Rap. *Proc Natl Acad Sci USA*. 1994;91:12609–12613.
45. Ferro E, Trabalzini L. RalGDS family members couple Ras to Ral signalling and that's not all. *Cell Signal*. 2010;22:1804–1810.
46. Rosario M, Paterson HF, Marshall CJ. Activation of the Ral and phosphatidylinositol 3′ kinase signaling pathways by the ras-related protein TC21. *Mol Cell Biol*. 2001;21:3750–3762.
47. Yu Y, Feig LA. Involvement of R-Ras and Ral GTPases in estrogen-independent proliferation of breast cancer cells. *Oncogene*. 2002;21:7557–7568.
48. Gildea JJ, Harding MA, Seraj MJ, Gulding KM, Theodorescu D. The role of Ral A in epidermal growth factor receptor-regulated cell motility. *Cancer Res*. 2002;62: 982–985.
49. Li TT, Alemayehu M, Aziziyeh AI, et al. Beta-arrestin/Ral signaling regulates lysophosphatidic acid-mediated migration and invasion of human breast tumor cells. *Mol Cancer Res*. 2009;7:1064–1077.
50. Weissman JT, Ma JN, Essex A, Gao Y, Burstein ES. G-protein-coupled receptor-mediated activation of rap GTPases: characterization of a novel Galphai regulated pathway. *Oncogene*. 2004;23:241–249.
51. Ohtsuka T, Shimizu K, Yamamori B, Kuroda S, Takai Y. Activation of brain B-Raf protein kinase by Rap1B small GTP-binding protein. *J Biol Chem*. 1996;271: 1258–1261.
52. Kelley GG, Reks SE, Smrcka AV. Hormonal regulation of phospholipase Cepsilon through distinct and overlapping pathways involving G12 and Ras family G-proteins. *Biochem J*. 2004;378:129–139.
53. Ridley AJ, Hall A. The small GTP-binding protein rho regulates the assembly of focal adhesions and actin stress fibers in response to growth factors. *Cell*. 1992;70:389–399.
54. Seasholtz TM, Majumdar M, Brown JH. Rho as a mediator of G protein-coupled receptor signaling. *Mol Pharmacol*. 1999;55:949–956.
55. Kjoller L, Hall A. Signaling to Rho GTPases. *Exp Cell Res*. 1999;253:166–179.
56. Aoki H, Izumo S, Sadoshima J. Angiotensin II activates RhoA in cardiac myocytes: a critical role of RhoA in angiotensin II-induced premyofibril formation. *Circ Res*. 1998;82:666–676.
57. Buhl AM, Johnson NL, Dhanasekaran N, Johnson GL. G alpha 12 and G alpha 13 stimulate Rho-dependent stress fiber formation and focal adhesion assembly. *J Biol Chem*. 1995;270:24631–24634.
58. Hart MJ, Jiang X, Kozasa T, et al. Direct stimulation of the guanine nucleotide exchange activity of p115 RhoGEF by Galpha13. *Science*. 1998;280:2112–2114.
59. Kozasa T, Jiang X, Hart MJ, et al. p115 RhoGEF, a GTPase activating protein for Galpha12 and Galpha13. *Science*. 1998;280:2109–2111.
60. Fukuhara S, Murga C, Zohar M, Igishi T, Gutkind JS. A novel PDZ domain containing guanine nucleotide exchange factor links heterotrimeric G proteins to Rho. *J Biol Chem*. 1999;274:5868–5879.
61. Katoh H, Aoki J, Yamaguchi Y, Kitano Y, Ichikawa A, Negishi M. Constitutively active Galpha12, Galpha13, and Galphaq induce Rho-dependent neurite retraction through different signaling pathways. *J Biol Chem*. 1998;273:28700–28707.
62. Sagi SA, Seasholtz TM, Kobiashvili M, Wilson BA, Toksoz D, Brown JH. Physical and functional interactions of Galphaq with Rho and its exchange factors. *J Biol Chem*. 2001;276:15445–15452.
63. Booden MA, Siderovski DP, Der CJ. Leukemia-associated Rho guanine nucleotide exchange factor promotes G alpha q-coupled activation of RhoA. *Mol Cell Biol*. 2002;22:4053–4061.

64. Soh UJ, Trejo J. Activated protein C promotes protease-activated receptor-1 cytoprotective signaling through beta-arrestin and dishevelled-2 scaffolds. *Proc Natl Acad Sci USA*. 2011;108:E1372–E1380.
65. Li Z, Dong X, Wang Z, et al. Regulation of PTEN by Rho small GTPases. *Nat Cell Biol*. 2005;7:399–404.
66. Papakonstanti EA, Ridley AJ, Vanhaesebroeck B. The p110delta isoform of PI 3-kinase negatively controls RhoA and PTEN. *EMBO J*. 2007;26:3050–3061.
67. Lima-Fernandes E, Enslen H, Camand E, et al. Distinct functional outputs of PTEN signalling are controlled by dynamic association with beta-arrestins. *EMBO J*. 2011;30:2557–2568.
68. Didsbury J, Weber RF, Bokoch GM, Evans T, Snyderman R. rac, a novel ras-related family of proteins that are botulinum toxin substrates. *J Biol Chem*. 1989;264: 16378–16382.
69. Haataja L, Groffen J, Heisterkamp N. Characterization of RAC3, a novel member of the Rho family. *J Biol Chem*. 1997;272:20384–20388.
70. Wennerberg K, Der CJ. Rho-family GTPases: it's not only Rac and Rho (and I like it). *J Cell Sci*. 2004;117:1301–1312.
71. Bosco EE, Mulloy JC, Zheng Y. Rac1 GTPase: a "Rac" of all trades. *Cell Mol Life Sci*. 2009;66:370–374.
72. Gavard J, Gutkind JS. VEGF controls endothelial-cell permeability by promoting the beta-arrestin-dependent endocytosis of VE-cadherin. *Nat Cell Biol*. 2006;8:1223–1234.
73. Evans T, Brown ML, Fraser ED, Northup JK. Purification of the major GTP-binding proteins from human placental membranes. *J Biol Chem*. 1986;261:7052–7059.
74. Shinjo K, Koland JG, Hart MJ, et al. Molecular cloning of the gene for the human placental GTP-binding protein Gp (G25K): identification of this GTP-binding protein as the human homolog of the yeast cell-division-cycle protein CDC42. *Proc Natl Acad Sci USA*. 1990;87:9853–9857.
75. Kvachnina E, Liu G, Dityatev A, et al. 5-HT7 receptor is coupled to G alpha subunits of heterotrimeric G12-protein to regulate gene transcription and neuronal morphology. *J Neurosci*. 2005;25:7821–7830.
76. Yamauchi J, Miyamoto Y, Kokubu H, et al. Endothelin suppresses cell migration via the JNK signaling pathway in a manner dependent upon Src kinase, Rac1, and Cdc42. *FEBS Lett*. 2002;527:284–288.
77. Slice LW, Chiu T, Rozengurt E. Angiotensin II and epidermal growth factor induce cyclooxygenase-2 expression in intestinal epithelial cells through small GTPases using distinct signaling pathways. *J Biol Chem*. 2005;280:1582–1593.
78. Kozma R, Ahmed S, Best A, Lim L. The Ras-related protein Cdc42Hs and bradykinin promote formation of peripheral actin microspikes and filopodia in Swiss 3 T3 fibroblasts. *Mol Cell Biol*. 1995;15:1942–1952.
79. Feoktistov I, Goldstein AE, Biaggioni I. Cyclic AMP and protein kinase A stimulate Cdc42: role of A(2) adenosine receptors in human mast cells. *Mol Pharmacol*. 2000;58:903–910.
80. Touchot N, Chardin P, Tavitian A. Four additional members of the ras gene superfamily isolated by an oligonucleotide strategy: molecular cloning of YPT-related cDNAs from a rat brain library. *Proc Natl Acad Sci USA*. 1987;84:8210–8214.
81. Gallwitz D, Donath C, Sander C. A yeast gene encoding a protein homologous to the human c-has/bas proto-oncogene product. *Nature*. 1983;306:704–707.
82. Seachrist JL, Ferguson SS. Regulation of G protein-coupled receptor endocytosis and trafficking by Rab GTPases. *Life Sci*. 2003;74:225–235.
83. Moore RH, Sadovnikoff N, Hoffenberg S, et al. Ligand-stimulated beta 2-adrenergic receptor internalization via the constitutive endocytic pathway into rab5-containing endosomes. *J Cell Sci*. 1995;108(Pt. 9):2983–2991.

84. Bremnes T, Paasche JD, Mehlum A, Sandberg C, Bremnes B, Attramadal H. Regulation and intracellular trafficking pathways of the endothelin receptors. *J Biol Chem.* 2000;275:17596–17604.
85. Esseltine JL, Dale LB, Ferguson SS. Rab GTPases bind at a common site within the angiotensin II type I receptor carboxyl-terminal tail: evidence that Rab4 regulates receptor phosphorylation, desensitization, and resensitization. *Mol Pharmacol.* 2011;79:175–184.
86. Hamelin E, Theriault C, Laroche G, Parent JL. The intracellular trafficking of the G protein-coupled receptor TPbeta depends on a direct interaction with Rab11. *J Biol Chem.* 2005;280:36195–36205.
87. Parent A, Hamelin E, Germain P, Parent JL. Rab11 regulates the recycling of the beta2-adrenergic receptor through a direct interaction. *Biochem J.* 2009;418:163–172.
88. Reid HM, Mulvaney EP, Turner EC, Kinsella BT. Interaction of the human prostacyclin receptor with Rab11: characterization of a novel Rab11 binding domain within alpha-helix 8 that is regulated by palmitoylation. *J Biol Chem.* 2010;285:18709–18726.
89. Oakley RH, Laporte SA, Holt JA, Barak LS, Caron MG. Association of beta-arrestin with G protein-coupled receptors during clathrin-mediated endocytosis dictates the profile of receptor resensitization. *J Biol Chem.* 1999;274:32248–32257.
90. Dale LB, Seachrist JL, Babwah AV, Ferguson SS. Regulation of angiotensin II type 1A receptor intracellular retention, degradation, and recycling by Rab5, Rab7, and Rab11 GTPases. *J Biol Chem.* 2004;279:13110–13118.
91. Kahn RA, Gilman AG. The protein cofactor necessary for ADP-ribosylation of Gs by cholera toxin is itself a GTP binding protein. *J Biol Chem.* 1986;261:7906–7911.
92. O'Neal CJ, Jobling MG, Holmes RK, Hol WG. Structural basis for the activation of cholera toxin by human ARF6-GTP. *Science.* 2005;309:1093–1096.
93. D'Souza-Schorey C, Chavrier P. ARF proteins: roles in membrane traffic and beyond. *Nat Rev Mol Cell Biol.* 2006;7:347–358.
94. Galas MC, Helms JB, Vitale N, Thierse D, Aunis D, Bader MF. Regulated exocytosis in chromaffin cells. A potential role for a secretory granule-associated ARF6 protein. *J Biol Chem.* 1997;272:2788–2793.
95. Cohen LA, Honda A, Varnai P, Brown FD, Balla T, Donaldson JG. Active Arf6 recruits ARNO/cytohesin GEFs to the PM by binding their PH domains. *Mol Biol Cell.* 2007;18:2244–2253.
96. Mitchell R, Robertson DN, Holland PJ, Collins D, Lutz EM, Johnson MS. ADP-ribosylation factor-dependent phospholipase D activation by the M3 muscarinic receptor. *J Biol Chem.* 2003;278:33818–33830.
97. Boulay PL, Cotton M, Melancon P, Claing A. ADP-ribosylation factor 1 controls the activation of the phosphatidylinositol 3-kinase pathway to regulate epidermal growth factor-dependent growth and migration of breast cancer cells. *J Biol Chem.* 2008;283:36425–36434.
98. Premont RT, Claing A, Vitale N, et al. Beta2-adrenergic receptor regulation by GIT1, a G protein-coupled receptor kinase-associated ADP ribosylation factor GTPase-activating protein. *Proc Natl Acad Sci USA.* 1998;95:14082–14087.
99. Houndolo T, Boulay PL, Claing A. G protein-coupled receptor endocytosis in ARF6-depleted cells. *J Biol Chem.* 2005;280:5598–5604.
100. Poupart ME, Fessart D, Cotton M, Laporte SA, Claing A. ARF6 regulates angiotensin II type 1 receptor endocytosis by controlling the recruitment of AP-2 and clathrin. *Cell Signal.* 2007;19:2370–2378.
101. Paleotti O, Macia E, Luton F, et al. The small G-protein Arf6GTP recruits the AP-2 adaptor complex to membranes. *J Biol Chem.* 2005;280:21661–21666.
102. Hoeppner CZ, Cheng N, Ye RD. Identification of a nuclear localization sequence in beta-arrestin-1 and its functional implications. *J Biol Chem.* 2012;287:8932–8943.
103. Xiao K, McClatchy DB, Shukla AK, et al. Functional specialization of beta-arrestin interactions revealed by proteomic analysis. *Proc Natl Acad Sci USA.* 2007;104:12011–12016.

CHAPTER SEVEN

Arrestins and Protein Ubiquitination

Reddy Peera Kommaddi*, **Sudha K. Shenoy***,†
*Department of Medicine, Duke University Medical Center, Durham, North Carolina, USA
†Department of Cell Biology, Duke University Medical Center, Durham, North Carolina, USA

Contents

1. Introduction 176
   1.1 Arrestins: Adaptors for receptors and effectors 176
   1.2 Ubiquitination: A pleiotropic posttranslational modification 178
   1.3 Deubiquitination 181
2. Ubiquitination of Arrestins 182
   2.1 Mechanisms: Modification by E3 ubiquitin ligase(s) 182
   2.2 Biology: Kinetics and stability 183
   2.3 Functional effects 186
3. Deubiquitination of Arrestins 187
4. Arrestins Act as Adaptors for Ubiquitination 189
   4.1 7TMRs 190
   4.2 Other receptors 191
   4.3 Ion channels 192
   4.4 Enzymes and other proteins 192
5. Arrestins and Seven-Transmembrane Receptor Deubiquitination 193
6. Arrestin-Like Proteins 194
   6.1 *Aspergillus nidulans* 194
   6.2 *Saccharomyces cerevisiae* 196
   6.3 Mammalian arrestin domain-containing proteins 196
7. Conclusions 198
Acknowledgment 199
References 199

Abstract

The adaptor proteins, β-arrestins 1 and 2, were originally identified as inhibitors of G protein signaling at the seven-transmembrane receptors (7TMRs, also called G protein-coupled receptors or GPCRs). Subsequent studies have established β-arrestins as critical multifunctional 7TMR adaptors that mediate receptor trafficking and activate G protein-independent signaling pathways. 7TMR activation leads not only to the recruitment of arrestin proteins upon phosphorylation by GPCR kinases but also to

Progress in Molecular Biology and Translational Science, Volume 118
ISSN 1877-1173
http://dx.doi.org/10.1016/B978-0-12-394440-5.00007-3

β-arrestin ubiquitination. This posttranslational modification of β-arrestin is appended by specific E3 ubiquitin ligases and reversed by deubiquitinases, which are also recruited in a receptor- and agonist-specific manner. β-Arrestin ubiquitination allows it to form protein complexes with activated 7TMRs, endocytic proteins such as clathrin, and phosphorylated ERK1/2. β-Arrestin ubiquitination is dependent on its activated conformation and likely regulates timing and subcellular localization of various protein interactions during receptor trafficking and signaling. β-Arrestins also serve as adaptors that escort E3 ubiquitin ligases to mediate ubiquitination of a wide list of substrate proteins including 7TMRs and provide an added layer of regulation for defining substrate specificity in the cellular ubiquitination pathway.

1. INTRODUCTION

1.1. Arrestins: Adaptors for receptors and effectors

Arrestins are multifunctional adaptor proteins that bind to diverse cell-surface receptors and regulate transmembrane signal transduction.[1–3] The arrestin gene family contains four members and is divided into two groups based on sequence identity, function, and tissue distribution: (1) visual arrestins that regulate the opsins: arrestin 1 (or S antigen) and arrestin 4 (X-arrestin, cone arrestin, or C-arrestin) and (2) nonvisual arrestins: β-arrestins 1 and 2 (arrestin 2 and arrestin 3).[4–7] Arrestin 1 is highly expressed in the retinal rods, whereas arrestin 4 is highly expressed in retinal cones. The two β-arrestins share ~78% amino acid identity and are expressed ubiquitously; at least one β-arrestin isoform is required for normal survival of mouse neonates because genetic deletion of both β-arrestin isoforms confers perinatal lethality.[8,9] The β-arrestins function as prototypical adaptors for the vast majority of heptahelical cell-surface receptors known as the seven-transmembrane receptors (G protein-coupled receptors or GPCRs).[10] Upon agonist stimulation, the 7TMRs activate heterotrimeric G proteins leading to changes in the cellular levels of second-messenger molecules (e.g., cAMP, inositol phosphate, Ca^{2+}), which bind and activate signaling kinases, ion channels, and other proteins leading to specific cellular responses.[11] However, concomitant processes, namely, phosphorylation within receptor intracellular domains by the serine–threonine kinases called G protein-coupled-receptor kinases (GRKs) and subsequent recruitment of cytoplasmic β-arrestins to plasma membrane-localized 7TMRs, cause desensitization of G protein signaling (Fig. 7.1; Ref. 12).

About a decade after the initial characterization, it became evident that β-arrestins are important adaptors for the internalization of activated

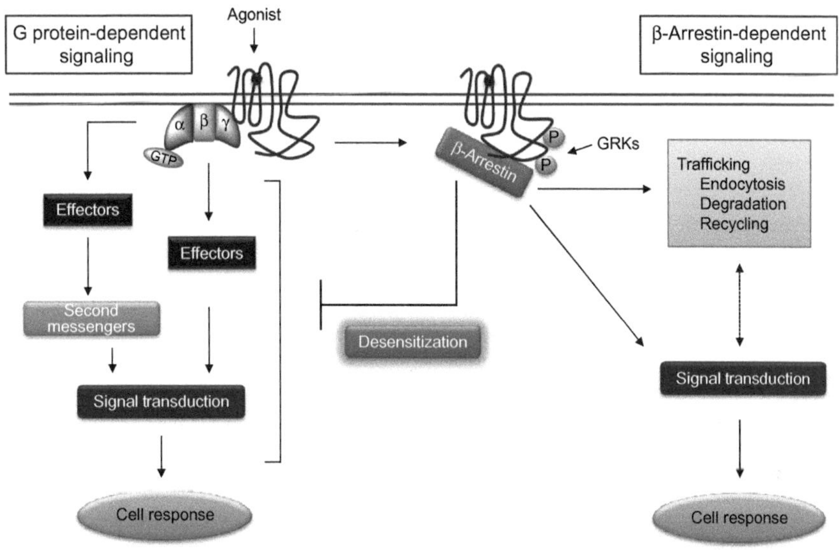

Figure 7.1 G protein and β-arrestin-dependent signaling elicited by 7TMRs. Agonist binding to extracellular and/or transmembrane regions of the receptors leads to the interaction and activation of heterotrimeric G proteins. The agonist-occupied 7TMR acts as a guanine nucleotide exchange factor (GEF), catalyzing the exchange of GDP for GTP on the G_α subunit and inducing dissociation of the G_α and $G_{\beta\gamma}$ subunits from the complex. The G protein subunits bind to enzymes or other effector molecules (e.g., ion channels) and modulate the levels of second messengers within the cell. This leads to the activation of kinase cascades culminating in a cell response. G protein-dependent signaling is terminated via 7TMR phosphorylation in the cytoplasmic domains by G protein-coupled receptor kinases (GRKs). Phosphorylated receptors bind β-arrestins with high affinity. β-Arrestin recruitment from the cytosol to the phosphorylated receptor at the plasma membrane leads to an immediate blockade of G protein coupling and G protein-dependent signaling (desensitization); subsequently, β-arrestin interacts with endocytic proteins such as clathrin and the clathrin adaptor protein 2 (AP2) and facilitates 7TMR internalization and endocytosis. Additionally, β-arrestin bound to the activated receptor can function as a signal transducer by initiating the β-arrestin-dependent signaling pathway(s) by interacting and activating signaling kinases. (For color version of this figure, the reader is referred to the online version of this chapter.)

7TMRs from the plasma membrane into intracellular vesicles or endosomes.[13,14] β-Arrestins were found to serve as adaptors for endocytic proteins, first shown for clathrin, which is a structural component of endocytic vesicles.[15–18] β-Arrestins also bind to adaptin protein 2, which functions together with clathrin for coat formation and transport of specific cargo through the endocytic pathway.[19,20] The endocytic adaptor function of β-arrestin is continually expanding to include new interacting proteins of the endocytic machinery and other families of cell-surface receptors, thus underscoring their importance in cellular trafficking.[14,21,22]

A significant turning point in our understanding about β-arrestin's functions was the discovery that they not only initiate receptor endocytosis but also "switch on" signal transduction as they "switch off" G protein coupling (Ref. 23; Fig. 7.1). Thus, through the second decade after their cloning, β-arrestins were shown to function as signaling adaptors for the nonreceptor tyrosine kinase c-Src, extracellular signal-regulated kinases 1 and 2 (ERK1/2), c-Jun-N-terminal kinase 3 (JNK3), p38 MAP kinase, and AKT.[23–27] By far the most critical landmark is the proposal and subsequent proving of the paradigm-changing hypotheses that β-arrestins are stand-alone signal transducers and that 7TMRs (and perhaps other receptors) could assume a unique set(s) of activated conformations to specifically signal through β-arrestins, independent of G protein activation.[28–32]

The ubiquitous protein posttranslational modification called ubiquitination, in which the protein ubiquitin becomes covalently appended to substrate proteins,[33] plays a critical role in 7TMR-β-arrestin endocytic and signaling pathways.[34] As will be described through this chapter, not only do β-arrestins become ubiquitinated, which remarkably impacts their adaptor and scaffolding roles, but they also serve as integral components of the ubiquitination machinery by functioning as indispensible adaptors that link substrate proteins including 7TMRs with the cognate E3 ubiquitin ligases that mediate ubiquitination.

1.2. Ubiquitination: A pleiotropic posttranslational modification

Ubiquitin is a small, ubiquitously expressed, 76-amino acid highly conserved protein found in all eukaryotic cells from yeast to mammals. It is the best characterized member of a class of small protein modifiers containing a characteristic ubiquitin fold in the tertiary structure.[35] The posttranslational covalent attachment of one or several ubiquitin molecules to a targeted protein is known as ubiquitination (also known as ubiquitylation or ubiquitinylation). It is a highly regulated process in which the carboxyl group of the C-terminal glycine within a di-glycine motif of the activated ubiquitin forms an amide bond with the ε-amino group of a lysine residue in the modified protein.[33] This process is carried out by the action of three distinct enzymes: an ubiquitin-activating enzyme (E1), an ubiquitin-conjugating enzyme (E2), and an ubiquitin-ligating enzyme (E3). E1 carries out the ATP-dependent activation of the carboxyl-terminal glycine residue of ubiquitin by the formation of ubiquitin adenylate intermediate. This is followed by the transfer of activated ubiquitin to the E1 active site cysteine to form a thioester linkage with concomitant release of AMP. Activated ubiquitin is then transferred to a thiol site of an E2 ubiquitin-conjugating enzyme via a transacylation reaction (Fig. 7.2). The last step in the ubiquitination cascade creates an isopeptide bond between a lysine residue

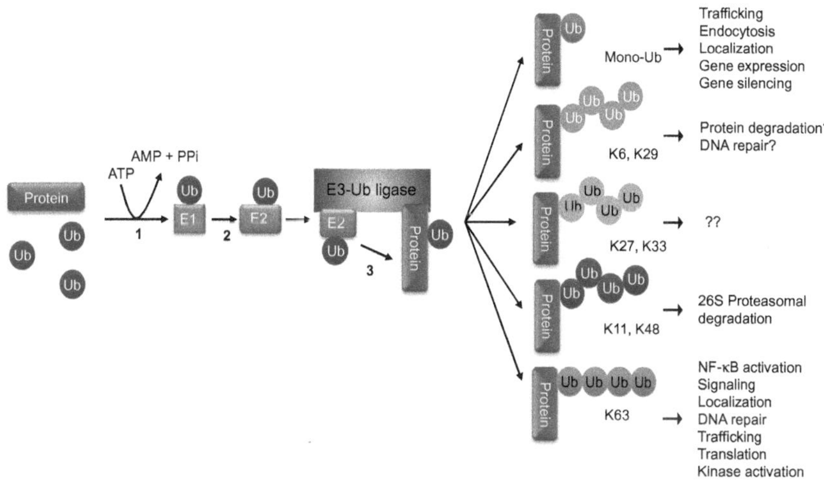

Figure 7.2 Schematic representation of ubiquitination. Ubiquitin (Ub), a highly conserved 76 amino acid polypeptide, is covalently attached to substrates by an isopeptide bond between a substrate lysine residue and the C-terminal glycine residue of ubiquitin. Ubiquitination is carried out by three distinct enzymes: an ubiquitin-activating enzyme (E1), an ubiquitin-conjugating enzyme (E2), and an ubiquitin-ligating enzyme (E3). As the first step, the C-terminal glycine residue of ubiquitin, G76, is activated by the E1 enzyme in an ATP-dependent reaction, resulting in an intermediary ubiquitin adenylate, release of pyrophosphate (PPi), and the attachment of ubiquitin to a cysteine residue in E1. The activated ubiquitin is subsequently transferred to an active cysteine in E2 (ubiquitin-conjugating enzyme). Next, a ubiquitin protein ligase E3 recruits the carboxyl terminus of charged ubiquitin on the E2 to an ε-amino group of a lysine residue in the substrate protein. Attachment of a single ubiquitin at a single site in the substrate is called monoubiquitination, whereas one ubiquitin attached at multiple sites is referred as multi-monoubiquitination. When successive additions of four or more ubiquitins are made to a lysine within the previously added ubiquitin, a polyubiquitin chain is formed. Substrates that are monoubiquitinated, multi-monoubiquitinated, or polyubiquitinated (excluding K48 poly ubiquitination chains) are not degraded, and such ubiquitination facilitates change in the activation state of the substrate, localization, or signaling. The consequences of different types of polyubiquitin chain linkages formed on substrates are listed. Conversely, substrates that are polyubiquitinated by K48-linked chains, with at least four or more ubiquitin molecules on the same lysine residue in the substrate, are generally rapidly degraded by 26S proteasome. (See Color Insert.)

in the substrate protein and the C-terminal glycine of activated ubiquitin. This ubiquitin transfer to the substrate protein requires the activity of a third enzyme called E3 ubiquitin ligase.

According to a recent genomic annotation, humans express two E1s, approximately 62 E2s, and nearly 600 E3s.[36] Ubiquitin-protein ligases (E3s) represent the largest and most diverse group of enzymes in the ubiquitination process.[37] E3 ubiquitin ligases are traditionally categorized into two main types: (1) HECT (*H*omologous to *E*6AP *C*-*t*erminus) and (2) RING (*R*eally *I*nteresting *N*ew *G*ene) E3 ligases. The RING E3 ligases are further classified into two subtypes: (1) enzymes represented by single polypeptides (examples of single-polypeptide RING E3s are the Cbl oncogene, seven in absentia homolog (Siah), and murine double minute 2 (Mdm2)); and (2) ligases made up of multiprotein complexes with separate units for substrate binding, E2 interaction, and ubiquitin transfer (examples of multimeric RING E3s include the anaphase-promoting complex (APC, also called the cyclosome), Skp1-Cullin-F-box (SCF), suppressor of cytokine signaling (SOCS), and the recently identified Broad complex, Tramtrack, and Bric-a-brac (BTB) families).[37–40] Other E3 ubiquitin ligases that contain variations within the RING domain have also been described (e.g., U-box, B-box, and PHD containing E3s).[41,42] The two E3 families differ in the general mechanism of ubiquitin transfer: HECT E3s accept ubiquitin from E2s and then ligate the ubiquitin onto substrates, while RING E3s facilitate ubiquitin transfer from E2 enzyme to substrate lysines.[37]

Some proteins become modified with a single ubiquitin resulting in monoubiquitination, which is considered as a tag that promotes endocytic trafficking of modified proteins (Ref. 43; Fig. 7.2). Multi-monoubiquitination is defined as the addition of single ubiquitin moieties to multiple lysine residues in a target protein. After ubiquitin is conjugated to a substrate, successive ubiquitin additions can occur on a lysine of the previously attached ubiquitin, resulting in polyubiquitination (Fig. 7.2). Ubiquitin has seven lysines at positions 6, 11, 27, 29, 33, 48, and 63 of its primary amino acid sequence and each lysine can be interlinked forming specific polyubiquitin chains, each with distinct roles (Fig. 7.2). Polyubiquitination consisting of successive ubiquitin moieties linked at lysine 48 is called "kiss of death" because it often targets the ubiquitinated proteins to multi-subunit, megadalton, barrel-shaped protein complexes called 26S proteasomes, which are the primary nonlysosomal enzymatic degradation machines in cells.[44] Ubiquitin itself is not degraded; rather, it is cleaved from the substrates by ubiquitin-specific proteases (USPs) or deubiquitinating enzymes.[45] In contrast to the lysine 48-linked modification, polyubiquitin chains consisting of lysine 63 linkages are considered as nondegradative signals for recruiting and activating kinases.[46] Ubiquitination was discovered and characterized in the context of nonlysosomal protein degradation by the 26S proteasomes. However, it has turned out to be a highly versatile protein modification because it can alter the protein three-dimensional structure, leading to changes in protein location and/or activity, thereby regulating a wide array of biological processes, including cell-cycle control, DNA repair, stress response, transcription, gene silencing, protein processing, apoptosis, endocytosis, and signal transduction.

1.3. Deubiquitination

Generally, biochemical pathways are reversible such that there are "on" and "off" cycles; this reversibility is critical for the regulation of biological and physiological outcomes. Substrate ubiquitination is an excellent example of such regulation because it is a dynamic and reversible modification. Ubiquitinated proteins are deubiquitinated by specific proteases known as deubiquitinases (DUBs) or USPs that cleave the iso-peptide bond between carboxyl terminus of ubiquitin and the amino group of the substrate. Bioinformatic analyses have identified that the human genome encodes ~100 putative DUBs, of which 79 are postulated to be functional. The DUBs have been divided into five distinct classes based on sequence and structural homology.[47] Four of the five classes identified to date are cysteine proteases with a classical papain active catalytic site structure encompassing the catalytic triad of cysteine, histidine, and a third residue consisting of aspartic acid, asparagine, or rarely, serine. These four classes are (1) ubiquitin carboxy-terminal hydrolases (4 members), (2) USPs (over 58 members), (3) Machado–Joseph Domain (5 members), and (4) the ovarian tumor-related family (14 members). The fifth class, called jab1/MPN domain-associated metalloisopeptidase motif DUBs, is composed of Zn^{2+}-containing metalloproteases (14 members). Studies have demonstrated that DUBs play an important role in several aspects of the ubiquitin–proteasome system, and mutations in DUBs have been implicated in a number of diseases, ranging from hereditary cancer to neurodegeneration. The abundance of DUB family members suggests diversity and specificity of DUB activity, although currently our knowledge about the physiological substrates of individual members of the DUB family is limited.

2. UBIQUITINATION OF ARRESTINS

2.1. Mechanisms: Modification by E3 ubiquitin ligase(s)

Ubiquitination of arrestin was first demonstrated for the mammalian β-arrestin2. Upon agonist-stimulation of the β_2AR, isolated β-arrestin2 immunoprecipitates showed a smeary laddering of ubiquitination when immunoblotted for ubiquitin.[48] This pattern is characteristic of protein polyubiquitination; however, unlike many polyubiquitinated proteins that are recognized by the 26S proteasomal proteases and degraded, β-arrestin2 ubiquitination does not lead to detectable degradation of the protein.[48] It was found that the RING domain-containing E3 ubiquitin ligase Mdm2 mediates this ubiquitination, and further that when the process is blocked with Mdm2 mutants that lack the catalytic domain, but could still bind β-arrestin2, there is a significant inhibition of β_2AR internalization.[48] Studies have also revealed that the restoration of Mdm2 expression in Mdm2/p53 double knock-out mouse embryonic fibroblasts resulted in the rescue of both isoproterenol-stimulated β-arrestin ubiquitination and rapid receptor internalization.[48] These effects are also observed in heterologous cells such as human embryonic kidney cells (HEK293) in which Mdm2 expression is depleted using Mdm2-specific siRNA or in Chinese hamster fibroblasts (CHW) in which Mdm2 activity is inhibited by exogenous expression of a truncated Mdm2 lacking the catalytic domain.[48–50] Isoproterenol-stimulated Mdm2-dependent ubiquitination of β-arrestin is therefore an essential molecular event required for rapid internalization of the β_2AR into clathrin-coated vesicles. The importance of β-arrestin2 ubiquitination in 7TMR regulation is corroborated by several alternative experimental approaches, such as mutant β-arrestins that are stabilized and impaired in ubiquitination[51] and regulation by deubiquitination.[49]

In addition to Mdm2, other E3 ubiquitin ligases, namely, Nedd4,[50] AIP4,[52] and Parkin,[53] have also been shown to bind β-arrestin2; however, these interactions do not seem to result in β-arrestin ubiquitination. β-Arrestin has also been shown to be SUMOylated upon β_2AR stimulation. SUMO or small ubiquitin-like modifier is a ubiquitin-like protein that is also appended to lysines in substrate proteins.[54] SUMOylation has analogous effects to ubiquitination in modulating protein interactions, signaling, and gene transcription, but unlike ubiquitin, it does not promote proteasomal degradation. Additionally, unlike ubiquitination, SUMOylation is generally targeted to the canonical motif ψ-K-x-E which represents the recognition domain for binding of the sole SUMO-E2 enzyme, UBC9. In bovine β-arrestin2, SUMOylation predominantly occurs at a specific lysine residue and is important for β-arrestin interaction with the endocytic adaptor protein β-adaptin2.[54] The SUMO-E3 ligase that mediates this modification remains to be identified.

2.2. Biology: Kinetics and stability

7TMR-stimulated β-arrestin2 ubiquitination can be categorized into two patterns: (1) transient ubiquitination which occurs rapidly (e.g., acute stimulation of the β_2AR by the agonist isoproterenol leads to transient β-arrestin ubiquitination signals which are rapidly reduced to baseline levels within a few minutes) and (2) sustained ubiquitination in which the signals evoked by agonist treatment do not deteriorate for up to 1 h or more (e.g., stimulation of $AT_{1a}R$ by angiotensin II leads to sustained β-arrestin ubiquitination; Fig. 7.3A and B; Refs. 55,56). Apart from routinely used Western blot techniques mentioned above, both transient and sustained ubiquitination of β-arrestin as induced by the β_2AR and the V_2 vasopressin receptor (V_2R) are detectable as dynamic modifications in live cells by monitoring BRET (bioluminescence resonance energy transfer) between β-arrestin2 fused with luciferase and GFP^2-tagged ubiquitin.[57] The two patterns of β-arrestin ubiquitination parallel previously shown trafficking profiles of green fluorescent protein (GFP)-tagged β-arrestins, in which transient recruitment to activated β_2ARs at the plasma membrane defines "class A" and stable recruitment to activated AT_{1a}Rs, first to the plasma membrane followed by cointernalization of receptor–β-arrestin complexes, defines "class B" receptor.[58] In general, class A receptors recycle and resensitize at the plasma membrane more rapidly than the class B receptors, and their affinities for the two β-arrestin isoforms are also different. Class A receptors preferentially recruit β-arrestin2, whereas class B receptors bind both β-arrestins with equal affinity. These internalization, recycling, β-arrestin binding, and ERK activation patterns are attributed to particular serine–threonine-rich phosphorylation motifs found only in the class B receptors.[58,59] Interestingly, when the C-terminal region of a class A receptor is replaced with that of a class B receptor, trafficking, ERK activation, and ubiquitination patterns are transformed to the class B type.[55,58,59] Translational fusion of Ub to β-arrestin2 promotes its sustained binding to even a class A receptor, whereas removal of all ubiquitination sites results in transient recruitment to even class B receptors (Fig. 7.3C and D; Refs. 51,55,56). These findings

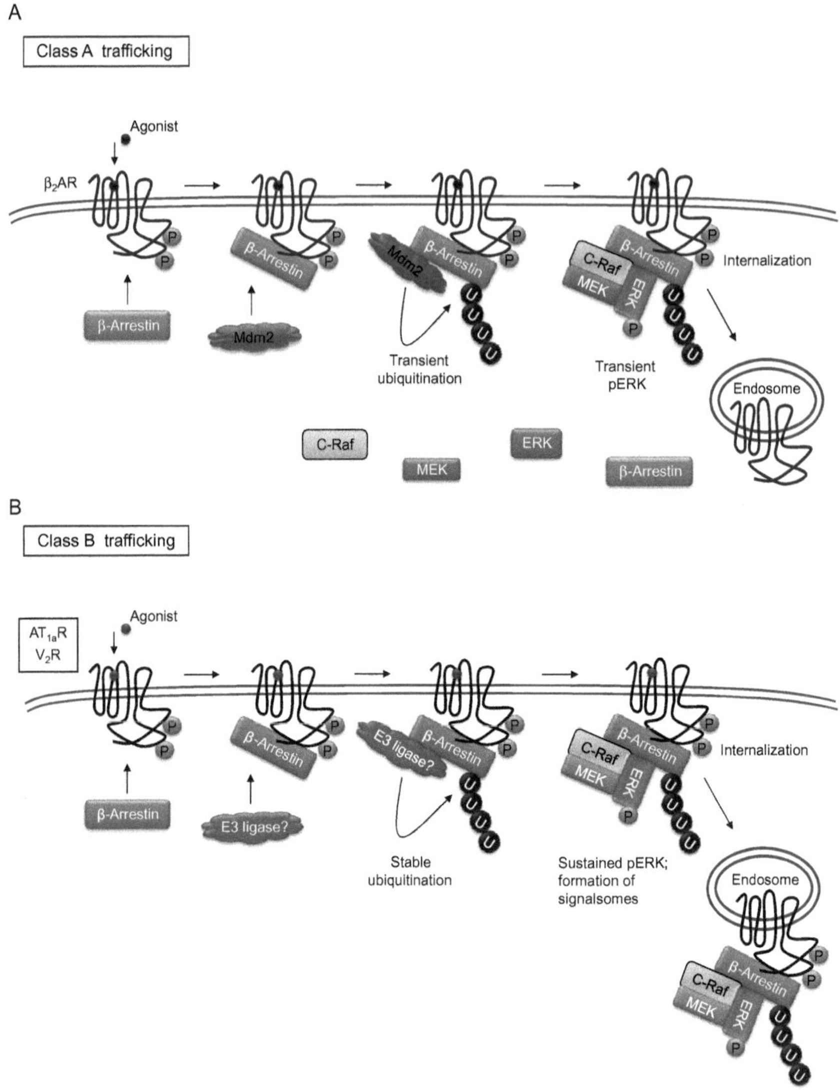

Figure 7.3 Ubiquitination and trafficking profiles of β-arrestin2. Agonist stimulation of 7TMRs leads to rapid phosphorylation of the receptor by GRKs. Phosphorylated receptors recruit the cytosolic adaptor protein β-arrestin. Most 7TMRs are capable of recruiting β-arrestin. However, the nature of β-arrestin/7TMR complexes can be transient (class A receptors) or stable (class B receptors). (A) Activation of the class A β_2 adrenergic receptor (β_2AR) induces rapid and transient ubiquitination of β-arrestin by an E3 ubiquitin ligase, Mdm2. The ubiquitinated β-arrestin–receptor complex engages an ERK-signaling scaffold at the plasma membrane and enhances transient phosphorylation of ERK. The ubiquitinated β-arrestins are rapidly deubiquitinated and dissociated from the internalizing receptor, while 7TMRs are internalized into endocytic vesicles. (B) Activation of class B receptors induces sustained β-arrestin ubiquitination by an unknown E3 ligase, leading to the formation of stable endocytic complexes. These stable complexes are localized to perinuclear endosomal compartments and ERK phosphorylation is sustained on signalsomes (signaling endosomes).

(Continued)

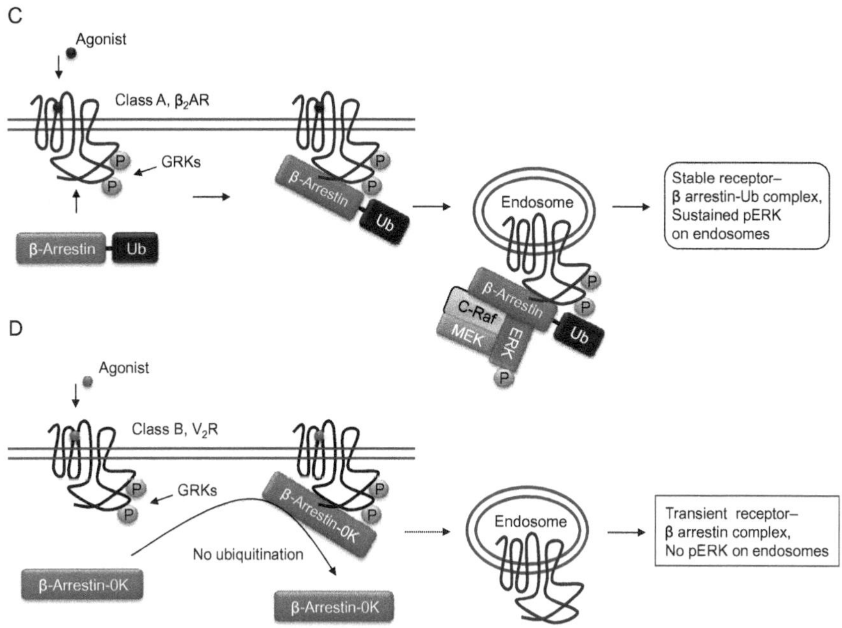

Figure 7.3 Cont'd (C) A β-arrestin2-Ub chimera, which cannot be deubiquitinated, forms a stable complex with the class A β_2AR and scaffolds active ERK on endosomes. (D) The mutant β-arrestin2-0K that lacks ubiquitination sites forms a highly unstable complex even with a class B receptor. (See Color Insert.)

support the idea that the β-arrestin-binding pattern, as well as the stability and signaling of receptor–β-arrestin complexes, is dictated by the molecular signatures on the receptor carboxyl terminus as well as on β-arrestins, namely, distinct phosphorylation motifs corresponding to either class A or class B receptors, and ubiquitin chains on β-arrestin. The 7TMR recycling and resensitization kinetics can be further dictated by ubiquitination of the receptor itself.

Ubiquitination occurs on the ε-amino group of lysine residues, and there is no consensus sequence for substrate ubiquitination. Both β-arrestin isoforms have many lysines (e.g., 35 in rat β-arrestin1 and 31 in rat β-arrestin2) dispersed along the length of the protein which presents a challenging task to identify specific lysines targeted for ubiquitination. By undertaking series of mutagenesis steps, the lysine residues in β-arrestin2 targeted for stable ubiquitination as induced by the class B type $AT_{1a}R$ have been mapped; here, the stable ubiquitination occurs primarily at the vicinal lysines 11 and 12 in rat β-arrestin2.[56] Mutation of these lysines to arginines leads to reversal of angiotensin II-stimulated β-arrestin ubiquitination from a sustained to a transient pattern, with a corresponding reversal of $AT_{1a}R$-β-arrestin binding from stable endosome-localized complexes to transiently associated complexes seen only at the plasma membrane. Quite unexpectedly, however, the same β-arrestin mutant is stably ubiquitinated and associated tightly with two other activated class B receptors, the V_2R and neurokinin 1 receptor. Furthermore, V_2R-stimulated β-arrestin ubiquitination is impaired in a quintuple lysine mutant β-arrestin2 in which five lysines at positions 18, 107, 108, 207, and 296 are altered to ariginines.[56] These provocative findings suggest that different 7TMRs are capable of inducing sustained β-arrestin ubiquitination at distinct sites on β-arrestin2. Whether the susceptibility of target lysines in β-arrestin is a result of distinct conformational changes or whether it reflects the activity of distinct E3 ubiquitin ligases that bind and modify different domains in β-arrestin remains an important topic for future studies.

Unlike stable ubiquitination that is targeted to specific lysines in β-arrestin2, transient ubiquitination appears to be a random modification because it is eliminated only when all the lysines in β-arrestin2 are mutated to arginines (β-arrestin2-0K).[51] Although β-arrestin2-0K is capable of equivalent protein–protein interactions with the β_2AR as the non-ubiquitinated wild-type protein *in vitro*, in a cellular context it shows impairment in binding because unlike the wild type, it cannot be ubiquitinated at proper site(s). In fact, the recruitment of β-arrestin2-0K-GFP to either β_2AR or V_2R is very transient and detected only when GRKs are overexpressed to augment receptor phosphorylation or when activated receptor complexes are forcibly retained at the plasma membrane by blocking internalization.[51] Accordingly, ubiquitination of β-arrestin2 is not required for its initial recruitment to the cell-surface 7TMR, but it is required for prolonging and stabilizing the β-arrestin–receptor interaction.

2.3. Functional effects

Apart from stabilizing receptor interaction as discussed above, ubiquitination of β-arrestin has been shown to promote both its endocytic and scaffolding functions. While earlier studies had indicated a dependence of β_2AR internalization on β-arrestin ubiquitination status, later studies carried out with two modified β-arrestins, namely, β-arrestin2-0K that lacks ubiquitin acceptor sites and β-arrestin2-Ub which remains stably ubiquitinated have provided important insights.[51] β-Arrestin2-0K does not promote internalization of either β_2AR or V_2R, confirming that ubiquitination of β-arrestin is crucial for its role in promoting receptor endocytosis (Fig. 7.3D). In contrast, a β-arrestin-Ub chimera that remains ubiquitinated can enhance β_2AR internalization and is stably recruited to endosomal compartments with the β_2AR (Fig. 7.3C; Refs. 51,56,60). Correlating with these effects on internalization, these two forms of β-arrestins show reciprocal binding affinities with clathrin, suggesting that ubiquitin moieties on β-arrestin2 are crucial "tags" for its efficient binding to clathrin and entry into the endocytic pathway.[51]

In cellular coimmunoprecipitation assays, β-arrestin2-0K displays similar binding with overexpressed c-Raf and GFP-ERK2 as the wild type, whereas β-arrestin2-Ub exhibits enhanced interaction.[51] However, pERK contained in isolated β_2AR complexes is significantly decreased when β-arrestin2-0K is coexpressed when compared with β-arrestin2 wild type. In contrast, β-arrestin2-Ub coexpression markedly increases pERK levels in receptor complexes and further stabilizes the kinase activity on endosomes (Fig. 7.3C). Thus, although both ubiquitinated and nonubiquitinated forms of β-arrestin can form complexes with pERK, only the ubiquitinated form is capable of this function in a receptor complex.[51] Subcellular fractionation experiments indicate that ubiquitination of β-arrestin2 favors its membrane association, suggesting that although ubiquitination is dispensable for β-arrestin's interactions with cytososlic partners, it may be necessary to facilitate the formation of functional 7TMR-β-arrestin endocytic and signaling complexes in a membrane environment.[51]

3. DEUBIQUITINATION OF ARRESTINS

The DUB, ubiquitin-specific protease 33 (USP33, also known as von-Hippel Lindau interacting deubiquitinase1 or VDU1; Ref. 61), was identified as a β-arrestin-interacting protein in a yeast two-hybrid screen and its binding promotes deubiquitination of β-arrestin2.[49] Upon USP33 overexpression, V_2R-stimulated effects, namely, stable β-arrestin binding, sustained β-arrestin2 ubiquitination, and persistent ERK activation, are all inhibited. On the other hand, targeted gene silencing of USP33 promotes stable interaction and cointernalization of β-arrestin2 with the β_2AR, sustained ubiquitination of β-arrestin2, and prolonged activation of ERK. Accordingly, the kinetics of β-arrestin ubiquitination and deubiquitination are tightly regulated by USP33 ensuring the appropriate duration and magnitude of β-arrestin-biased signaling.

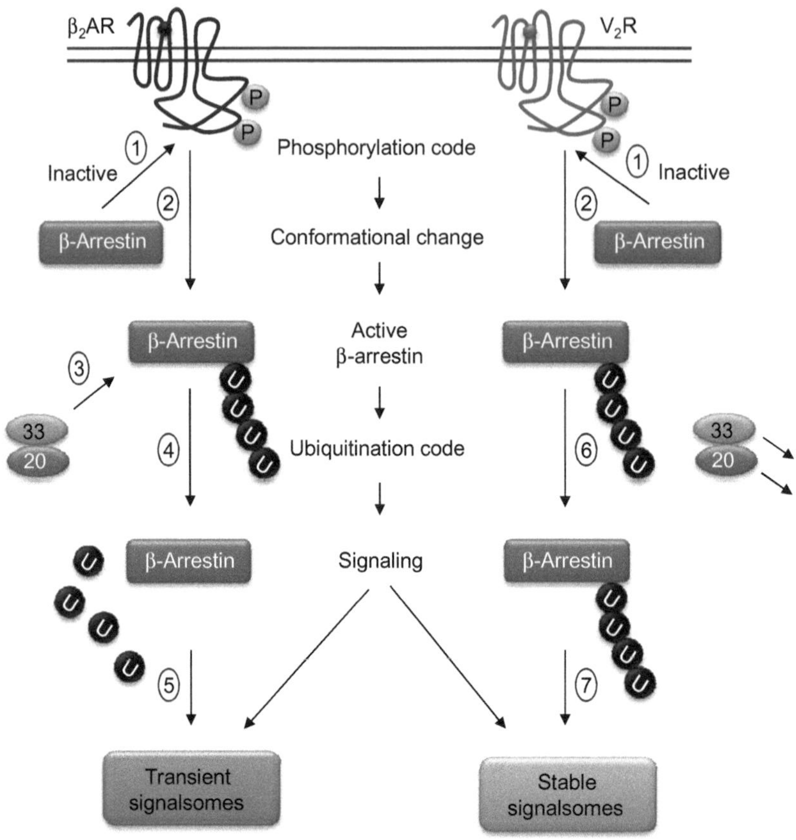

Figure 7.4 Effects of posttranslational modifications in 7TMR signaling. Step 1: β-arrestin2 resides in a basal state in the cytoplasm and is recruited to the plasma membrane and binds phosphorylated C-termini of 7TMRs. The sites of phosphorylation differ among the two representative receptors shown. Step 2: Upon binding to each receptor, β-arrestin2 undergoes a distinct conformational reorientation, allowing distinct regions to be modified by ubiquitination. Step 3: The β_2AR-induced conformation promotes β-arrestin2–USP33 interaction. Step 4: USP33 deubiquitinates β-arrestin leading to the dissociation of β-arrestin from the β_2AR. Step 5: β_2AR-β-arrestin2 signalsomes are short lived and promote transient ERK activity that is predominantly nonendosomal. Step 6: The β-arrestin2 conformation induced by the V_2R activation prevents USP33 binding, thus protecting β-arrestin ubiquitination, allowing tight binding to activated receptors. Step 7: V_2R-β-arrestin2 signalsomes are stable and result in robust ERK activity that is predominantly localized on endosomes. *This figure and legend are adapted from reference 49.* (For color version of this figure, the reader is referred to the online version of this chapter.)

At a molecular level, these effects are dependent upon conformation-specific interaction of USP33 and β-arrestin2. The β_2AR-induced conformation of "active" β-arrestin2 enhances its binding with USP33, whereas the V_2R-induced "active" β-arrestin2 conformation reduces it (Fig. 7.4). These conformation-specific interactions are supported by observations from two sets of assays conducted in vitro: (1) interaction of purified β-arrestin2 and USP33 is augmented by addition of a synthetic peptide mimicking the phosphorylated carboxyl tail of the β_2AR, but not that of the V_2R or nonphosphorylated forms of either tail peptide and (2) differential trypsin-digestion profiles of purified β-arrestin2 in the presence of phosphorylated peptides representing the carboxyl tails of β_2AR and the V_2R. Accordingly, β-arrestin–receptor interaction is regulated by the phosphorylation motifs on the carboxyl tail of 7TMRs, which is followed by ubiquitination of specific residues by E3 ubiquitin ligase(s), and recruitment and conformation-specific interaction with deubiquitinating enzymes. Therefore, two types of unique codes dictate β-arrestin-dependent cellular effects: a "phosphorylation code" at the receptors generated by specific GRK-mediated phosphorylation, followed by "ubiquitination code" on β-arrestin conferred by E3 ubiquitin ligase(s). Although receptor interaction causes a conformational change in β-arrestin, further addition of ubiquitin moieties would result in a second phase of conformational modification ensuing binding or dissociation of specific cellular proteins.

4. ARRESTINS ACT AS ADAPTORS FOR UBIQUITINATION

The E3 ubiquitin ligases or DUBs recognize their substrates via specific protein interaction domains or conformational cues appended to the substrate by another posttranslation modification (e.g., phosphorylation/dephosphorylation).[33] Nonetheless, a majority of substrates are predicted to bind these enzymes via adaptor molecules, which function as accessory proteins in escorting the E3 ubiquitin ligases or scaffolding both E2 and E3 enzymes.[22,33,62] These adaptor proteins provide an added layer of regulation, increase specificity as well as efficiency, and define timing and localization of substrate ubiquitination.

Both β-arrestin isoforms act as indispensable E3 ligase adaptors for 7TMRs and non-7TMRs to mediate ubiquitination, and they can play an equally important role in escorting DUBs.[22] As adaptors, β-arrestins can potentially determine the dynamics of ubiquitination/deubiquitination of endocytic proteins that are important for proper steering of internalizing cargo across various vesicular pathways. Substrates that have been shown to require a β-arrestin isoform as an E3 ubiquitin ligase adaptor are described below.

4.1. 7TMRs

Agonist activation of β_2ARs induces ubiquitination of the receptor protein itself in addition to β-arrestin ubiquitination described earlier.[48,50,63–65] Receptor ubiquitination is detected within minutes and remains stable for several hours, diminishing as subsequent lysosomal degradation of the protein ensues. Although the activated β_2ARs internalize into endosomes within minutes, their trafficking to lysosomes is detected only after several hours.[50,63] In the commonly employed HEK-293 or COS-7 cells, the receptor protein remains in these compartments for prolonged periods and is slowly degraded.[50,66,67] When receptor ubiquitination is prevented, as in a lysine-less mutant, internalization of the β_2AR proceeds normally, but lysosomal trafficking and receptor degradation are impaired.[48,50,64] Additional studies with four β_2AR mutants each with lysines restricted to a single intracellular domain showed that both the third intracellular loop and the carboxyl tails of the β_2AR are ubiquitinated.[64] The lysines targeted for ubiquitination have been mapped utilizing mass spectrometry approaches to residues in the third intracellular loop (lys-263, lys-270) and in the carboxyl tail (lys-348, lys-372, lys-375) of human β_2AR.[64]

β_2AR ubiquitination is not detectable in β-arrestin1/2 null mouse embryonic fibroblasts, but robust receptor ubiquitination is detected when β-arrestin2 is expressed or added back into these cells.[48] In contrast, reexpression of β-arrestin1 does not promote β_2AR ubiquitination. Subsequent studies showed that β-arrestin2 binds and recruits the HECT domain-containing E3 ubiquitin ligase Nedd4 to mediate β_2AR ubiquitination and lysosomal degradation.[50] Although Nedd4 enzymes and their yeast homolog Rsp5 are known to interact with substrate or adaptor proteins via WW domains (in Nedd4) and proline-rich motifs (PPXY in substrate or adaptors), elimination of all the WW domains in Nedd4 did not affect its interaction with β-arrestin2 suggesting other modes of adaptor binding.[50] In fact, for some substrates, WW domain-independent interactions of Nedd4 have been reported.[68–71]

Ubiquitination of CXCR4, a chemokine receptor, involves modification of lysine residues 327, 331, and 333 in the carboxyl tail and requires phosphorylation of nearby serine residues.[72] Unlike the β_2AR modification, CXCR4 is modified with only one or small number of ubiquitin moieties. However, similar to the β_2AR, a CXCR4 lysine mutant defective in ubiquitination is internalized at a normal rate but not degraded in lysosomes. CXCR4 ubiquitination is mediated by the HECT domain E3 ligase atrophin-interacting protein 4 (AIP4, also known as Itch), but here arrestins may not be functioning as E3 ligase adaptors; depletion of β-arrestin1 has no effect on CXCR4 ubiquitination.[52,73] Further, AIP4 can bind directly to CXCR4: this binding involves WW domains in AIP4 and phosphorylated serines in the CXCR4 carboxyl tail.[74] On the other hand, the N-terminus of β-arrestin1 can bind to the WW domains of AIP4 and β-arrestin1 depletion blocks CXCR4 lysosomal degradation, suggesting that arrestin–AIP4 interaction regulates postendocytic sorting of CXCR4.[52] In later studies, however, β-arrestin1 was reported to inhibit degradation of ubiquitinated CXCR4 by interacting with signal-transducing adaptor molecule-1, a component of the postendocytic sorting complexes collectively called as endosomal sorting complex required for transport 0 (ESCRT 0).[75,76] Although β-arrestins may not act as ubiquitin ligase adaptors for CXCR4, they may be required for this function toward an endocytic protein or cofactor at a downstream sorting step during CXCR4 lysosomal trafficking.

For other 7TMRs, β-arrestins have been shown to be required for receptor ubiquitination, but the E3 ligase(s) involved is not known. β-Arrestin2 is required for the ubiquitination of V2R, which occurs on lysine 268 in the third intracellular loop of the receptor.[77] V2R ubiquitination is dispensable for internalization as well as for slow agonist-independent protein turnover, but it is required for rapid degradation of the receptor upon agonist stimulation.[77] Ubiquitination of the μ-opioid receptor is induced by the agonist [D-Ala2,N-Me-Phe4,Gly5-ol]-enkephalin and requires β-arrestin1, but not β-arrestin2 expression.[78] Although a growing number of 7TMRs have been reported to be regulated by ubiquitination,[34,79] conclusive evidence regarding the role of β-arrestins in this process remains an open question.

4.2. Other receptors

As indicated above, the endocytic functions of β-arrestins are not limited to regulation of 7TMRs; rather, they play critical roles in mediating endocytosis of other cell-surface receptors including growth factor receptors.[80] In the case of the insulin-like growth factor 1 receptor (IGF-1R), β-arrestin1 functions as an essential adaptor to promote receptor internalization via clathrin-coated vesicles.[81] In addition, β-arrestin1 escorts the E3 ubiquitin ligase Mdm2 to mediate ubiquitination of the IGF-1R leading to receptor degradation.[82] This role is specific for β-arrestin1 because depletion of β-arrestin2 does not diminish IGF-1R ubiquitination. The β-arrestin homolog Kurtz expressed in Drosophila also acts as an E3 ligase adaptor.[83] Kurtz binds to both the Notch receptor and the HECT domain E3 ligase Deltex and mediates Notch ubiquitination and downregulation.[83] β-Arrestin2 forms a complex with the androgen receptor (AR) and acts as an AR corepressor in androgen-dependent prostate cancer cells by promoting Mdm2-dependent ubiquitination and degradation of AR.[84]

4.3. Ion channels

An adaptor role of β-arrestin1 has recently been discovered for transient receptor potential (TRP) ion channel family member TRPV4 and Na^+/H^+ exchanger 1 (NHE1) ubiquitination, in which the E3 ligases AIP4 and Nedd4-1 are involved, respectively.[85,86] TRPV4 is widely expressed in various tissues and initiates Ca^{2+} influx upon activation. Interestingly, TRPV4 and $AT_{1a}R$ are found to coexist as complexes at the plasma membrane.[85] Upon $AT_{1a}R$ activation, β-arrestin1 is recruited to this complex resulting in the internalization of TRPV4 and inhibition of Ca^{2+} influx into cells. This functional downregulation requires ubiquitination of TRPV4 mediated by AIP4 which is escorted specifically by β-arrestin1.[85] NHE1 is expressed ubiquitously and controls cell volume and pH; protein turnover of this ion transport protein requires ubiquitination of the intracellular domain by Nedd4 which is recruited in a β-arrestin1-dependent manner.[86] Because NHE1 shares homology with other 11 members of this protein family and internalization of NHE5 is mediated by β-arrestins, it is tempting to speculate that β-arrestin-dependent ubiquitination would be a common theme in the regulation of these ion-transport proteins.[87]

4.4. Enzymes and other proteins

Ubiquitination of GRK2 is triggered by β_2AR stimulation and is mediated by Mdm2.[88,89] Association between GRK2 and Mdm2 is augmented upon agonist activation of β_2ARs and is regulated by phosphorylation on GRK2 at Serine 670 by MAPKs. Under agonist-induced conditions, β-arrestin1 and/or β-arrestin2 scaffolded Mdm2 ubiquitinates GRK2 and promotes its degradation via 26S proteasomes. Interestingly, β-arrestins also facilitate an alternative mode of degradation that is augmented when c-Src tyrosine phosphorylates GRK2, an event that occurs independent of Mdm2 under basal or agonist-independent conditions.[88] Thus for GRK2, β-arrestins can scaffold multiple E3 ligases, suiting the signaling context prevalent in cells and fine-tuning GRK2 levels. Because GRK2 also phosphorylates multiple substrates in addition to 7TMRs, these effects would have a broader impact in cell signaling.[77,90,91]

cAMP in cells is rapidly degraded by phosphodiesterases (PDE), and upon β_2AR stimulation, the recruited β-arrestin2 was shown to scaffold PDE4 isoforms to prime cAMP degradation in membrane microdomains, increasing the efficiency of desensitization.[92] Intriguingly, PDE4D5 is rapidly and transiently ubiquitinated in response to agonist stimulation of β_2ARs, and this is mediated by Mdm2 scaffolded by β-arrestin2.[93] PDE4D5 ubiquitination serves to enhance its interaction with β-arrestin2, while simultaneously decreasing interaction with the alternate scaffold RACK1, thus increasing the fidelity of specific protein–protein interactions.[93]

β-Arrestins act as 7TMR-regulated scaffolds for MAPK activation as shown for both the JNK3 and ERK1/2 pathways.[94] For JNK3 activation, β-arrestin brings together the upstream MAPKKK (ASK1) and the MAPKK (MKK4), thus increasing proximity of the individual components of this kinase cascade.[26,27] During H_2O_2-induced apoptosis, β-arrestins promote ubiquitination and 26S proteasomal degradation of bound ASK1 by recruiting the U-box containing E3 ubiquitin ligase C-terminus of Hsc70-interacting protein (CHIP).[95] CHIP activity requires the additional molecular chaperones Hsp-90 and HSc-70. In the above stress-induced pathway, β-arrestins act to suppress apoptosis by decreasing ASK1 levels.

Recent studies have shown that chronic β_2AR activation by catecholamines triggers DNA damage and downregulation of p53.[96] The molecular mechanism involves β-arrestin1-dependent ubiquitination of p53 by the E3 ubiquitin ligase Mdm2 leading to p53 degradation, which results in impaired apoptosis of cells carrying DNA damage and accumulation of DNA damage. This study exemplifies the emerging role of β-arrestin1 as an E3-ligase adaptor in the nucleus and highlights the pathophysiological consequences of obliterating this β-arrestin function *in vivo*.

5. ARRESTINS AND SEVEN-TRANSMEMBRANE RECEPTOR DEUBIQUITINATION

Plasma membrane-localized β_2ARs can bind to two closely related DUBs, USP33 and USP20 (also known as VDU2). Agonist stimulation of the β_2AR leads to ubiquitination and lysosomal degradation of the receptor, but overexpression of USP33 and USP20 counteracts these effects and promotes receptor recycling and resensitization.[63] Additionally, knockdown of both USP33 and USP20 abolishes receptor recycling and resensitization, but enhances ubiquitination and lysosomal degradation. USP20 and USP33 thus act as novel regulators that dictate the postendocytic fate of internalized β_2ARs. These DUBs are constitutively bound to the cell-surface β_2ARs; however, β_2AR–USP association decreases upon agonist activation, while simultaneously agonist stimulation leads to the recruitment of β-arrestins to the β_2AR with a resultant increase in β-arrestin–USP33 binding. Thus, isoproterenol stimulation induces a reciprocal pattern of USP33 interaction with the β_2AR and β-arrestin2: dissociation of USP33 from the β_2AR and association of USP33 with β-arrestin2.[63] These data support the idea that while β-arrestin2 facilitates β_2AR ubiquitination by recruiting Nedd4, it might also serve to remove the DUBs from the activated β_2AR to facilitate receptor ubiquitination.

Reassociation of the DUBs during β_2AR trafficking leads to receptor deubiquitination, preventing lysosomal degradation of the receptors while concomitantly promoting receptor recycling and resensitization (Fig. 7.5). Thus, USP20 and 33 function as a tag team to separate internalizing β_2AR and β-arrestin, leading to a tight regulation and balancing of signaling and internalization processes. For the 7TMR protease-activated receptor 2, lysosomal trafficking has been reported to be regulated by the DUBs AMSH and UBPY, and the cell-surface trafficking of A2 adenosine receptor requires USP4-mediated deubiquitination; however, the role of β-arrestin in these processes is not known.[97,98]

6. ARRESTIN-LIKE PROTEINS

Although nonmetazoans do not express a homolog of mammalian arrestins, proteins that lack sequence homology and yet show structural conservation with the arrestin N and C domains have been recently described.[99,100] Similar arrestin domain-containing proteins also exist in mammals, and the following sections describe their known functions and analogy to β-arrestins.

6.1. *Aspergillus nidulans*

In the model fungus *A. nidulans*, the arrestin-like protein PalF forms a critical component of the fungal ambient pH-signaling pathway.[101] Increase in extracellular pH activates the 7TMR PalH leading to recruitment,

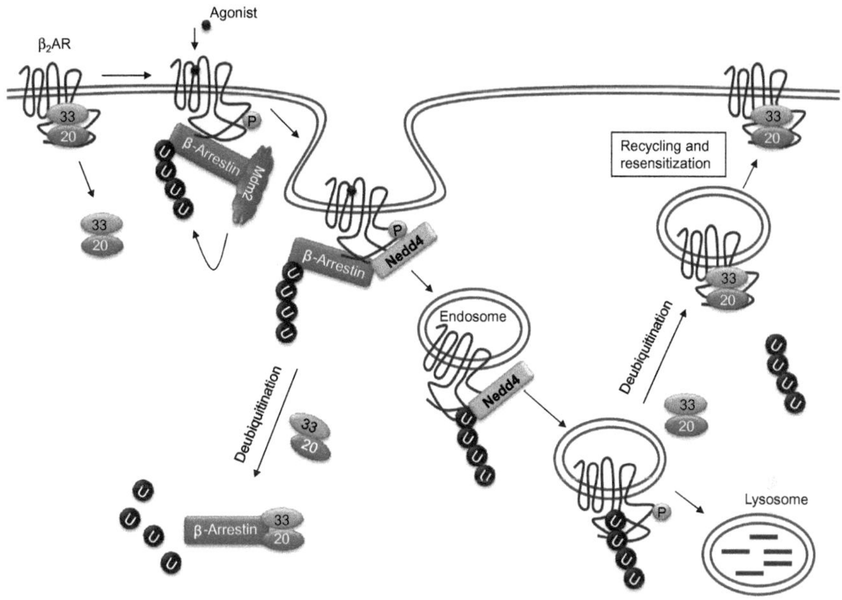

Figure 7.5 Receptor trafficking and DUBs. The deubiquitinating enzymes (DUBs), USP20 and USP33 are associated with the β_2AR in quiescent cells. Agonist stimulation of the β_2AR leads to recruitment and ubiquitination of β-arrestin2, which is mediated by the E3 ubiquitin ligase Mdm2. Subsequently, ubiquitinated β-arrestin2 interacts with the receptor and recruits Nedd4 to the activated β_2AR leading to β_2AR ubiquitination. The DUBs show reciprocal binding patterns with activated receptor versus β-arrestins; they dissociate from the activated β_2ARs and bind with increasing affinity with agonist-activated β-arrestins, catalyze reversal of β-arrestin ubiquitination, which decreases receptor interaction, retards internalization, and disassembles β arrestin complexes. The ubiquitinated β_2ARs are trafficked into early endosomes and subsequently to late endosomal/lysosomal compartments. The pools of USP20 and USP33 that reside on vesicles and cytoplasm reassociate with internalized β_2ARs, reverse receptor ubiquitination, and promote recycling and resensitization of the β_2ARs at the plasma membrane. (See Color Insert.)

ubiquitination, and phosphorylation of PalF, all of which are crucial in linking receptor internalization to the multivesicular body-sorting pathway. The downstream pH-signaling pathway elicited by the PalH is dependent on the arrestin-like protein and its posttranslational modifications but does not require any G protein activity. These features present a parallel with the mammalian β-arrestins where phosphorylation and ubiquitination play crucial roles in endocytic and signaling pathways. Interestingly, the PalF proteins do not have the conserved polyproline motifs that are signatures of the other arrestin-related adaptors discussed below.

6.2. *Saccharomyces cerevisiae*

In the yeast *S. cerevisiae*, ubiquitination of plasma membrane proteins serves as the major signal for their endocytosis.[102] In yeast, ubiquitination of plasma membrane substrates is mediated by the HECT domain E3 ubiquitin ligase, Rsp5 (a homolog of mammalian Nedd4) but most yeast membrane proteins do not contain a polyproline motif, which is a major recognition domain for binding with the WW domains in Rsp5. The conundrum of how Rsp5 can recognize and modify these substrates was recently resolved by studies that uncovered the existence of a family of 10 adaptor proteins called arrestin-related trafficking adaptors (ARTs 1–10) that carry polyproline motifs that bind to the WW domains in Rsp5.[103–106] Although ARTs do not share sequence homology with mammalian arrestins, structural modeling predicts that they fold into three-dimensional configurations carrying the characteristic amino and C-terminal arrestin domain architecture. ARTs function as E3 ubiquitin ligase adaptors for various yeast plasma membrane transporters: ART1 (also known as Cvs7 or Ldb19) for methionine transporter Mup1 and arginine transporter Can1; ART3 (also known as Aly2) for dicarboxylic acid transporter Dip5; ARTs 1 and 2 (ART2 is also termed Ecm21) for lysine transporter Lyp1; ARTs 2 and 8 (ART8 is also termed Csr2) for the manganese transporter Smf1; ART5 for the inositol transporter Itr1; and ART4 (ART4 is also termed Rod1); and ART8 for the high-affinity hexose transporter Hxt6. In addition to these ARTs, yeast express other proteins with partial or no conservation of arrestin domain sequences, but which contain polyproline motifs and are substrates or adaptors for Rsp5.[71] ART9 (also called Rim8 and related to PalF of *A. nidulans* discussed above) is a unique member in having arrestin domains but no polyproline motifs. ART9 deletion proves to have a harsher effect on yeast growth than a combined deletion of all the other ARTs.[105] Although more partners for these ART adaptors are emerging, their role in the regulation of yeast 7TMRs Ste2 and Ste3 remains to be documented.

6.3. Mammalian arrestin domain-containing proteins

The three-dimensional X-ray crystal structure of VPS26, which is a subunit of the mammalian retromer complex, shows an unprecedented resemblance to that of arrestin(s), although there is no sequence conservation between the two proteins.[107,108] Previous structure–function studies of mammalian arrestins support the idea that these molecules exist in a basal rigid conformation due to the restrictions imposed by the polar amino acid interactions within the core of the molecule, which are disrupted upon binding of phosphate moieties on the phosphorylated 7TMR intracellular domains leading to "activation" or conformational changes in arrestins that allow them to exhibit agonist-induced effects.[109–113] Although VPS26 has a polar core, the composition and surrounding structural elements are different and similar 7TMR-dependent conformational change and activation do not appear to be plausible for VPS26.[108] Nonetheless, these similar topologies predict some functional analogy between VPS26 and arrestins in terms of cargo trafficking, which awaits further characterization.

Arrestin domain structure is also predicted for Down syndrome critical region 3, a nuclear protein involved in partial or complete trisomy found in Down syndrome.[100] This protein has not been studied in any great detail so far, but there are indications of a role in transcription and/or MAPK signaling.

Six related proteins called arrestin domain-containing proteins 1–5 (ARRDC1–5) and thioredoxin-interacting protein or TXNIP (also known as vitamin D up regulated protein, VDUP1) also contain regions with the arrestin N- and C-terminal domain signatures but share very little sequence identity (11–15%) with arrestins.[100] With the exception of ARRDC5, these proteins have polyproline motifs and interact with WW domain-containing E3 ubiquitin ligases.[114–117] TXNIP, with predominant nuclear distribution, and ARRDC1, with plasma membrane localization, do not interact with the agonist-activated β_2AR.[118] However, three members that show endosomal distribution, ARRDC2, ARRDC3, and ARRDC4, can bind with internalized β_2ARs on endosomes.[118] These ARRDC proteins are localized in endocytic vesicles that are associated with a protein called hepatocyte growth factor-regulated substrate (HRS), and this binding also requires polyproline motif in the ARRDC proteins. Studies show that in the case of the β_2AR, β-arrestin2 acts as the primary adaptor for recruiting Nedd4 to ubiquitinate the β_2AR and that overexpressed ARRDC proteins (one or more of ARRDC2, -3, and -4) are subsequently recruited to the internalized β_2AR on endosomes.[118] This recruitment to the internalized β_2AR is abolished when the polyproline motifs in these ARRDC proteins are altered such that their binding to the Nedd4 WW domain is lost.[118] Thus, although domain–domain interactions predict recruitment of Nedd4 to the β_2AR via ARRDC proteins, experimental evidence points to an alternate mechanism: Nedd4 bound to the β_2AR complex recruits ARRDC positive endosomes to traffic-ubiquitinated β_2ARs into HRS-positive endosomes. Furthermore, gene silencing of ARRDC3 had no effect on β_2AR degradation or lysosomal trafficking, which could be attributed to the functional complementation by other two ARRDCs. Because of the redundant nature of ARRDC binding with the internalized β_2AR, deciphering their exact role in β_2AR trafficking poses the technical challenge of simultaneous and efficient silencing of all three genes in mammalian cells. Although yeast ARTs function as primary adaptors for cell-surface transporters, mammalian ARRDCs most likely function as secondary adaptors for 7TMRs and their functional interaction with the 7TMRs could require β-arrestin-mediated receptor internalization, as demonstrated for the β_2AR.[118] Although the ARRDCs share common features with the β-arrestins, each appears to have unique roles in mediating cellular homeostasis and embryonic development.[9,117]

7. CONCLUSIONS

Ubiquitination is a highly versatile, precisely timed, dynamic, and ubiquitous posttranslational modification, which plays a crucial regulatory role in not only determining levels but also activity and subcellular localization of modified proteins. The mammalian multifunctional adaptor protein, β-arrestin, which regulates various facets of 7TMR signal transduction and intracellular trafficking, is both a substrate and adaptor for ubiquitination. Ubiquitination of β-arrestin promotes its tight binding to activated 7TMRs as well as to nonreceptor binding partners such as clathrin, c-Raf, and pERK. The time course of β-arrestin ubiquitination is dictated by conformational changes induced by 7TMR activation and is regulated by binding or dissociation of DUBs. The E3 ubiquitin ligases that append the modification on β-arrestin and the site(s) of modification appear to be specific for receptor-β–arrestin pair. Differential ubiquitination on β-arrestin, consisting of distinct types of chain architecture occurring in a dynamic fashion, can be one mechanism that allows interactions with various protein partners. Because of their binding to DUBs, β-arrestins may be able to change their interaction surfaces to accommodate necessary protein recognition domains as demanded by the cellular context by ubiquitin editing (complete or partial removal of ubiquitin linkages). Thus, ubiquitination could be a critical molecular modification of β-arrestin needed for its multifaceted adaptor functions and perhaps for its activity in scaffolding and stabilizing various signaling kinases.

Remarkably, besides their own modification, β-arrestins also serve as E3 ubiquitin ligase adaptors to mediate ubiquitin-dependent regulation of a growing list of cellular proteins including 7TMRs, ion channels, kinases, and other regulatory proteins such as the tumor suppressor p53. This adaptor role appears to have been conserved through evolution as is inferred from the striking structural similarities and functional analogies shown by the yeast and fungal arrestin-related proteins with the mammalian arrestins.

ACKNOWLEDGMENT

This work was supported by grants from the NIH and the American Heart Association (to SKS).

REFERENCES

1. Lefkowitz RJ, Shenoy SK. Transduction of receptor signals by beta-arrestins. *Science*. 2005;308:512–517.
2. Luttrell LM, Lefkowitz RJ. The role of beta-arrestins in the termination and transduction of G-protein-coupled receptor signals. *J Cell Sci*. 2002;115:455–465.
3. Gurevich VV, Gurevich EV. The molecular acrobatics of arrestin activation. *Trends Pharmacol Sci*. 2004;25:105–111.
4. Pfister C, Chabre M, Plouet J, et al. Retinal S antigen identified as the 48 K protein regulating light-dependent phosphodiesterase in rods. *Science*. 1985;228:891–893.
5. Murakami A, Yajima T, Sakuma H, McLaren MJ, Inana G. X-arrestin: a new retinal arrestin mapping to the X chromosome. *FEBS Lett*. 1993;334:203–209.
6. Lohse MJ, Benovic JL, Codina J, Caron MG, Lefkowitz RJ. Beta-arrestin: a protein that regulates beta-adrenergic receptor function. *Science*. 1990;248:1547–1550.
7. Attramadal H, Arriza JL, Aoki C, et al. Beta-arrestin2, a novel member of the arrestin/beta-arrestin gene family. *J Biol Chem*. 1992;267:17882–17890.
8. Kohout TA, Lin FS, Perry SJ, Conner DA, Lefkowitz RJ. Beta-arrestin 1 and 2 differentially regulate heptahelical receptor signaling and trafficking. *Proc Natl Acad Sci USA*. 2001;98:1601–1606.
9. Premont RT, Gainetdinov RR. Physiological roles of G protein coupled receptor kinases and arrestins. *Annu Rev Physiol*. 2007;69:511–534.
10. Fredriksson R, Lagerstrom MC, Lundin LG, Schioth HB. The G-protein-coupled receptors in the human genome form five main families. Phylogenetic analysis, paralogon groups, and fingerprints. *Mol Pharmacol*. 2003;63:1256–1272.
11. Neves SR, Ram PT, Iyengar R. G protein pathways. *Science*. 2002;296:1636–1639.
12. Pitcher JA, Freedman NJ, Lefkowitz RJ. G protein-coupled receptor kinases. *Annu Rev Biochem*. 1998;67:653–692.
13. Ferguson SS, Downey 3rd WE, Colapietro AM, Barak LS, Menard L, Caron MG. Role of beta-arrestin in mediating agonist-promoted G protein-coupled receptor internalization. *Science*. 1996;271:363–366.
14. Moore CA, Milano SK, Benovic JL. Regulation of receptor trafficking by GRKs and arrestins. *Annu Rev Physiol*. 2007;69:451–482.
15. Goodman Jr OB, Krupnick JG, Gurevich VV, Benovic JL, Keen JH. Arrestin/clathrin interaction. Localization of the arrestin binding locus to the clathrin terminal domain. *J Biol Chem*. 1997;272:15017–15022.
16. Goodman Jr OB, Krupnick JG, Santini F, et al. Beta-arrestin acts as a clathrin adaptor in endocytosis of the beta2-adrenergic receptor. *Nature*. 1996;383:447–450.
17. Krupnick JG, Goodman Jr OB, Keen JH, Benovic JL. Arrestin/clathrin interaction. Localization of the clathrin binding domain of nonvisual arrestins to the carboxy terminus. *J Biol Chem*. 1997;272:15011–15016.

18. Pearse BM. Clathrin: a unique protein associated with intracellular transfer of membrane by coated vesicles. *Proc Natl Acad Sci USA*. 1976;73:1255–1259.
19. Laporte SA, Oakley RH, Holt JA, Barak LS, Caron MG. The interaction of beta-arrestin with the AP-2 adaptor is required for the clustering of beta 2-adrenergic receptor into clathrin-coated pits. *J Biol Chem*. 2000;275:23120–23126.
20. Kirchhausen T, Nathanson KL, Matsui W, et al. Structural and functional division into two domains of the large (100- to 115-kDa) chains of the clathrin-associated protein complex AP-2. *Proc Natl Acad Sci USA*. 1989;86:2612–2616.
21. Drake MT, Shenoy SK, Lefkowitz RJ. Trafficking of G protein-coupled receptors. *Circ Res*. 2006;99:570–582.
22. Shenoy SK, Lefkowitz RJ. Beta-arrestin-mediated receptor trafficking and signal transduction. *Trends Pharmacol Sci*. 2011;32:521–533.
23. Luttrell LM, Ferguson SS, Daaka Y, et al. Beta-arrestin-dependent formation of beta2 adrenergic receptor-Src protein kinase complexes. *Science*. 1999;283:655–661.
24. Beaulieu JM, Sotnikova TD, Marion S, Lefkowitz RJ, Gainetdinov RR, Caron MG. An Akt/beta-arrestin 2/PP2A signaling complex mediates dopaminergic neurotransmission and behavior. *Cell*. 2005;122:261–273.
25. DeFea KA, Zalevsky J, Thoma MS, Dery O, Mullins RD, Bunnett NW. Beta-arrestin-dependent endocytosis of proteinase-activated receptor 2 is required for intracellular targeting of activated ERK1/2. *J Cell Biol*. 2000;148:1267–1281.
26. McDonald PH, Chow CW, Miller WE, et al. Beta-arrestin 2: a receptor-regulated MAPK scaffold for the activation of JNK3. *Science*. 2000;290:1574–1577.
27. Sun Y, Cheng Z, Ma L, Pei G. Beta-arrestin2 is critically involved in CXCR4-mediated chemotaxis, and this is mediated by its enhancement of p38 MAPK activation. *J Biol Chem*. 2002;277:49212–49219.
28. DeWire SM, Ahn S, Lefkowitz RJ, Shenoy SK. Beta-arrestins and cell signaling. *Annu Rev Physiol*. 2007;69:483–510.
29. Gesty-Palmer D, Flannery P, Yuan L, et al. A beta-arrestin-biased agonist of the parathyroid hormone receptor (PTH1R) promotes bone formation independent of G protein activation. *Sci Transl Med*. 2009;1:1ra1.
30. Shenoy SK, Drake MT, Nelson CD, et al. Beta-arrestin-dependent, G protein-independent ERK1/2 activation by the beta2 adrenergic receptor. *J Biol Chem*. 2006;281:1261–1273.
31. Violin JD, Lefkowitz RJ. Beta-arrestin-biased ligands at seven-transmembrane receptors. *Trends Pharmacol Sci*. 2007;28:416–422.
32. Wei H, Ahn S, Shenoy SK, et al. Independent beta-arrestin 2 and G protein-mediated pathways for angiotensin II activation of extracellular signal-regulated kinases 1 and 2. *Proc Natl Acad Sci USA*. 2003;100:10782–10787.
33. Hershko A, Ciechanover A. The ubiquitin system. *Annu Rev Biochem*. 1998;67:425–479.
34. Shenoy SK. Seven-transmembrane receptors and ubiquitination. *Circ Res*. 2007;100: 1142–1154.
35. Hochstrasser M. Evolution and function of ubiquitin-like protein-conjugation systems. *Nat Cell Biol*. 2000;2:E153–157.
36. Deshaies RJ, Joazeiro CA. RING domain E3 ubiquitin ligases. *Annu Rev Biochem*. 2009;78:399–434.
37. Joazeiro CA, Weissman AM. RING finger proteins: mediators of ubiquitin ligase activity. *Cell*. 2000;102:549–552.
38. Willis MS, Zungu M, Patterson C. Cardiac muscle ring finger-1—friend or foe? *Trends Cardiovasc Med*. 2010;20:12–16.
39. Willems AR, Schwab M, Tyers M. A hitchhiker's guide to the cullin ubiquitin ligases: SCF and its kin. *Biochim Biophys Acta*. 2004;1695:133–170.

40. Skaar JR, Pagano M. Control of cell growth by the SCF and APC/C ubiquitin ligases. *Curr Opin Cell Biol.* 2009;21:816–824.
41. Patterson C. A new gun in town: the U box is a ubiquitin ligase domain. *Sci STKE.* 2002;2002:pe4.
42. Coscoy L, Ganem D. PHD domains and E3 ubiquitin ligases: viruses make the connection. *Trends Cell Biol.* 2003;13:7–12.
43. Schnell JD, Hicke L. Non-traditional functions of ubiquitin and ubiquitin-binding proteins. *J Biol Chem.* 2003;278:35857–35860.
44. Eytan E, Ganoth D, Armon T, Hershko A. ATP-dependent incorporation of 20S protease into the 26S complex that degrades proteins conjugated to ubiquitin. *Proc Natl Acad Sci USA.* 1989;86:7751–7755.
45. Wilkinson KD. Ubiquitination and deubiquitination: targeting of proteins for degradation by the proteasome. *Semin Cell Dev Biol.* 2000;11:141–148.
46. Ikeda F, Dikic I. Atypical ubiquitin chains: new molecular signals. 'Protein Modifications: Beyond the Usual Suspects' review series. *EMBO Rep.* 2008;9:536–542.
47. Nijman SM, Luna-Vargas MP, Velds A, et al. A genomic and functional inventory of deubiquitinating enzymes. *Cell.* 2005;123:773–786.
48. Shenoy SK, McDonald PH, Kohout TA, Lefkowitz RJ. Regulation of receptor fate by ubiquitination of activated beta 2-adrenergic receptor and beta-arrestin. *Science.* 2001;294:1307–1313.
49. Shenoy SK, Modi AS, Shukla AK, et al. Beta-arrestin-dependent signaling and trafficking of 7-transmembrane receptors is reciprocally regulated by the deubiquitinase USP33 and the E3 ligase Mdm2. *Proc Natl Acad Sci USA.* 2009;106:6650–6655.
50. Shenoy SK, Xiao K, Venkataramanan V, Snyder PM, Freedman NJ, Weissman AM. Nedd4 mediates agonist-dependent ubiquitination, lysosomal targeting, and degradation of the beta2-adrenergic receptor. *J Biol Chem.* 2008;283:22166–22176.
51. Shenoy SK, Barak LS, Xiao K, et al. Ubiquitination of beta-arrestin links seven-transmembrane receptor endocytosis and ERK activation. *J Biol Chem.* 2007;282: 29549–29562.
52. Bhandari D, Trejo J, Benovic JL, Marchese A. Arrestin-2 interacts with the ubiquitin-protein isopeptide ligase atrophin-interacting protein 4 and mediates endosomal sorting of the chemokine receptor CXCR4. *J Biol Chem.* 2007;282:36971–36979.
53. Ahmed MR, Zhan X, Song X, Kook S, Gurevich VV, Gurevich EV. Ubiquitin ligase parkin promotes Mdm2-arrestin interaction but inhibits arrestin ubiquitination. *Biochemistry.* 2011;50:3749–3763.
54. Wyatt D, Malik R, Vesecky AC, Marchese A. Small ubiquitin-like modifier modification of arrestin-3 regulates receptor trafficking. *J Biol Chem.* 2011;286:3884–3893.
55. Shenoy SK, Lefkowitz RJ. Trafficking patterns of beta-arrestin and G protein-coupled receptors determined by the kinetics of beta-arrestin deubiquitination. *J Biol Chem.* 2003;278:14498–14506.
56. Shenoy SK, Lefkowitz RJ. Receptor-specific ubiquitination of beta-arrestin directs assembly and targeting of seven-transmembrane receptor signalosomes. *J Biol Chem.* 2005;280:15315–15324.
57. Perroy J, Pontier S, Charest PG, Aubry M, Bouvier M. Real-time monitoring of ubiquitination in living cells by BRET. *Nat Methods.* 2004;1:203–208.
58. Oakley RH, Laporte SA, Holt JA, Caron MG, Barak LS. Differential affinities of visual arrestin, beta arrestin1, and beta arrestin2 for G protein-coupled receptors delineate two major classes of receptors. *J Biol Chem.* 2000;275:17201–17210.
59. Tohgo A, Choy EW, Gesty-Palmer D, et al. The stability of the G protein-coupled receptor-beta-arrestin interaction determines the mechanism and functional consequence of ERK activation. *J Biol Chem.* 2003;278:6258–6267.

60. Shenoy SK, Lefkowitz RJ. Multifaceted roles of beta-arrestins in the regulation of seven-membrane-spanning receptor trafficking and signalling. *Biochem J.* 2003;375: 503–515.
61. Li Z, Na X, Wang D, Schoen SR, Messing EM, Wu G. Ubiquitination of a novel deubiquitinating enzyme requires direct binding to von Hippel-Lindau tumor suppressor protein. *J Biol Chem.* 2002;277:4656–4662.
62. Leon S, Haguenauer-Tsapis R. Ubiquitin ligase adaptors: regulators of ubiquitylation and endocytosis of plasma membrane proteins. *Exp Cell Res.* 2009;315:1574–1583.
63. Berthouze M, Venkataramanan V, Li Y, Shenoy SK. The deubiquitinases USP33 and USP20 coordinate beta2 adrenergic receptor recycling and resensitization. *EMBO J.* 2009;28:1684–1696.
64. Xiao K, Shenoy SK. Beta2-adrenergic receptor lysosomal trafficking is regulated by ubiquitination of lysyl residues in two distinct receptor domains. *J Biol Chem.* 2011;286:12785–12795.
65. Liang W, Fishman PH. Resistance of the human beta1-adrenergic receptor to agonist-induced ubiquitination: a mechanism for impaired receptor degradation. *J Biol Chem.* 2004;279:46882–46889.
66. Gagnon AW, Kallal L, Benovic JL. Role of clathrin-mediated endocytosis in agonist-induced down-regulation of the beta2-adrenergic receptor. *J Biol Chem.* 1998;273:6976–6981.
67. Barak LS, Caron MG. Modeling of sequestration and down regulation in cells containing beta2-adrenergic receptors. *J Recept Signal Transduct Res.* 1995;15:677–690.
68. Huang Q, Szebenyi DM. Structural basis for the interaction between the growth factor-binding protein GRB10 and the E3 ubiquitin ligase NEDD4. *J Biol Chem.* 2010;285:42130–42139.
69. Ingham RJ, Colwill K, Howard C, et al. WW domains provide a platform for the assembly of multiprotein networks. *Mol Cell Biol.* 2005;25:7092–7106.
70. Malik R, Marchese A. Arrestin-2 interacts with the endosomal sorting complex required for transport machinery to modulate endosomal sorting of CXCR4. *Mol Biol Cell.* 2010;21:2529–2541.
71. Sullivan JA, Lewis MJ, Nikko E, Pelham HR. Multiple interactions drive adaptor-mediated recruitment of the ubiquitin ligase rsp5 to membrane proteins in vivo and in vitro. *Mol Biol Cell.* 2007;18:2429–2440.
72. Marchese A, Benovic JL. Agonist-promoted ubiquitination of the G protein-coupled receptor CXCR4 mediates lysosomal sorting. *J Biol Chem.* 2001;276:45509–45512.
73. Marchese A, Raiborg C, Santini F, Keen JH, Stenmark H, Benovic JL. The E3 ubiquitin ligase AIP4 mediates ubiquitination and sorting of the G protein-coupled receptor CXCR4. *Dev Cell.* 2003;5:709–722.
74. Bhandari D, Robia SL, Marchese A. The E3 ubiquitin ligase atrophin interacting protein 4 binds directly to the chemokine receptor CXCR4 via a novel WW domain-mediated interaction. *Mol Biol Cell.* 2009;20:1324–1339.
75. Henne WM, Buchkovich NJ, Emr SD. The ESCRT pathway. *Dev Cell.* 2011;21:77–91.
76. Malik R, Soh UJ, Trejo J, Marchese A. Novel roles for the E3 ubiquitin ligase atrophin-interacting protein 4 and signal transduction adaptor molecule 1 in G protein-coupled receptor signaling. *J Biol Chem.* 2012;287:9013–9027.
77. Martin NP, Lefkowitz RJ, Shenoy SK. Regulation of V2 vasopressin receptor degradation by agonist-promoted ubiquitination. *J Biol Chem.* 2003;278:45954–45959.
78. Groer CE, Schmid CL, Jaeger AM, Bohn LM. Agonist-directed interactions with specific beta-arrestins determine mu-opioid receptor trafficking, ubiquitination, and dephosphorylation. *J Biol Chem.* 2011;286:31731–31741.
79. Dores MR, Trejo J. Ubiquitination of GPCRs. Functional implications and drug discovery. *Mol Pharmacol.* 2012;82:563–570.

80. Lefkowitz RJ, Rajagopal K, Whalen EJ. New roles for beta-arrestins in cell signaling: not just for seven-transmembrane receptors. *Mol Cell.* 2006;24:643–652.
81. Lin FT, Daaka Y, Lefkowitz RJ. Beta-arrestins regulate mitogenic signaling and clathrin-mediated endocytosis of the insulin-like growth factor I receptor. *J Biol Chem.* 1998;273:31640–31643.
82. Girnita L, Shenoy SK, Sehat B, et al. {beta}-Arrestin is crucial for ubiquitination and down-regulation of the insulin-like growth factor-1 receptor by acting as adaptor for the MDM2 E3 ligase. *J Biol Chem.* 2005;280:24412–24419.
83. Mukherjee A, Veraksa A, Bauer A, Rosse C, Camonis J, Artavanis-Tsakonas S. Regulation of notch signalling by non-visual beta-arrestin. *Nat Cell Biol.* 2005;7:1191–1201.
84. Lakshmikanthan V, Zou L, Kim JI, et al. Identification of betaArrestin2 as a corepressor of androgen receptor signaling in prostate cancer. *Proc Natl Acad Sci USA.* 2009;106:9379–9384.
85. Shukla AK, Kim J, Ahn S, et al. Arresting a transient receptor potential (TRP) channel: beta-arrestin 1 mediates ubiquitination and functional down-regulation of TRPV4. *J Biol Chem.* 2010;285:30115–30125.
86. Simonin A, Fuster D. Nedd4-1 and beta-arrestin-1 are key regulators of Na+/H + exchanger 1 ubiquitylation, endocytosis, and function. *J Biol Chem.* 2010;285: 38293–38303.
87. Szabo EZ, Numata M, Lukashova V, Iannuzzi P, Orlowski J. Beta-arrestins bind and decrease cell-surface abundance of the Na+/H + exchanger NHE5 isoform. *Proc Natl Acad Sci USA.* 2005;102:2790–2795.
88. Nogues L, Salcedo A, Mayor Jr F, Penela P. Multiple scaffolding functions of {beta}-arrestins in the degradation of G protein-coupled receptor kinase 2. *J Biol Chem.* 2011;286:1165–1173.
89. Salcedo A, Mayor Jr F, Penela P. Mdm2 is involved in the ubiquitination and degradation of G-protein-coupled receptor kinase 2. *EMBO J.* 2006;25:4752–4762.
90. Evron T, Daigle TL, Caron MG. GRK2: multiple roles beyond G protein-coupled receptor desensitization. *Trends Pharmacol Sci.* 2012;33:154–164.
91. Penela P, Murga C, Ribas C, Lafarga V, Mayor Jr F. The complex G protein-coupled receptor kinase 2 (GRK2) interactome unveils new physiopathological targets. *Br J Pharmacol.* 2010;160:821–832.
92. Perry SJ, Baillie GS, Kohout TA, et al. Targeting of cyclic AMP degradation to beta 2-adrenergic receptors by beta-arrestins. *Science.* 2002;298:834–836.
93. Li X, Baillie GS, Houslay MD. Mdm2 directs the ubiquitination of beta-arrestin-sequestered cAMP phosphodiesterase-4D5. *J Biol Chem.* 2009;284:16170–16182.
94. Miller WE, Lefkowitz RJ. Expanding roles for beta-arrestins as scaffolds and adapters in GPCR signaling and trafficking. *Curr Opin Cell Biol.* 2001;13:139–145.
95. Zhang Z, Hao J, Zhao Z, et al. Beta-arrestins facilitate ubiquitin-dependent degradation of apoptosis signal-regulating kinase 1 (ASK1) and attenuate H2O2-induced apoptosis. *Cell Signal.* 2009;21:1195–1206.
96. Hara MR, Kovacs JJ, Whalen EJ, et al. A stress response pathway regulates DNA damage through beta2-adrenoreceptors and beta-arrestin-1. *Nature.* 2011;477: 349–353.
97. Milojevic T, Reiterer V, Stefan E, et al. The ubiquitin-specific protease Usp4 regulates the cell surface level of the A2A receptor. *Mol Pharmacol.* 2006;69:1083–1094.
98. Hasdemir B, Murphy JE, Cottrell GS, Bunnett NW. Endosomal deubiquitinating enzymes control ubiquitination and down-regulation of protease-activated receptor 2. *J Biol Chem.* 2009;284:28453–28466.
99. Alvarez CE. On the origins of arrestin and rhodopsin. *BMC Evol Biol.* 2008;8:222.
100. Aubry L, Guetta D, Klein G. The arrestin fold: variations on a theme. *Curr Genomics.* 2009;10:133–142.

101. Herranz S, Rodriguez JM, Bussink HJ, et al. Arrestin-related proteins mediate pH signaling in fungi. *Proc Natl Acad Sci USA*. 2005;102:12141–12146.
102. Hicke L. Protein regulation by monoubiquitin. *Nat Rev Mol Cell Biol*. 2001;2:195–201.
103. Lin CH, MacGurn JA, Chu T, Stefan CJ, Emr SD. Arrestin-related ubiquitin-ligase adaptors regulate endocytosis and protein turnover at the cell surface. *Cell*. 2008;135:714–725.
104. Macgurn JA, Hsu PC, Emr SD. Ubiquitin and membrane protein turnover: from cradle to grave. *Annu Rev Biochem*. 2012;81:231–259.
105. Nikko E, Pelham HR. Arrestin-mediated endocytosis of yeast plasma membrane transporters. *Traffic*. 2009;10:1856–1867.
106. Nikko E, Sullivan JA, Pelham HR. Arrestin-like proteins mediate ubiquitination and endocytosis of the yeast metal transporter Smf1. *EMBO Rep*. 2008;9:1216–1221.
107. Shi H, Rojas R, Bonifacino JS, Hurley JH. The retromer subunit Vps26 has an arrestin fold and binds Vps35 through its C-terminal domain. *Nat Struct Mol Biol*. 2006;13:540–548.
108. Collins BM, Norwood SJ, Kerr MC, et al. Structure of Vps26B and mapping of its interaction with the retromer protein complex. *Traffic*. 2008;9:366–379.
109. Charest PG, Terrillon S, Bouvier M. Monitoring agonist-promoted conformational changes of beta-arrestin in living cells by intramolecular BRET. *EMBO Rep*. 2005;6:334–340.
110. Gurevich VV, Benovic JL. Visual arrestin interaction with rhodopsin. Sequential multisite binding ensures strict selectivity toward light-activated phosphorylated rhodopsin. *J Biol Chem*. 1993;268:11628–11638.
111. Kisselev OG, McDowell JH, Hargrave PA. The arrestin-bound conformation and dynamics of the phosphorylated carboxy-terminal region of rhodopsin. *FEBS Lett*. 2004;564:307–311.
112. Xiao K, Shenoy SK, Nobles K, Lefkowitz RJ. Activation-dependent conformational changes in {beta}-arrestin 2. *J Biol Chem*. 2004;279:55744–55753.
113. Nobles KN, Guan Z, Xiao K, Oas TG, Lefkowitz RJ. The active conformation of beta-arrestin1: direct evidence for the phosphate sensor in the N-domain and conformational differences in the active states of beta-arrestins1 and -2. *J Biol Chem*. 2007;282:21370–21381.
114. Patwari P, Lee RT. An expanded family of arrestins regulate metabolism. *Trends Endocrinol Metab*. 2012;23:216–222.
115. Nabhan JF, Pan H, Lu Q. Arrestin domain-containing protein 3 recruits the NEDD4 E3 ligase to mediate ubiquitination of the beta2-adrenergic receptor. *EMBO Rep*. 2010;11:605–611.
116. Rauch S, Martin-Serrano J. Multiple interactions between the ESCRT machinery and arrestin-related proteins: implications for PPXY-dependent budding. *J Virol*. 2011;85:3546–3556.
117. Patwari P, Emilsson V, Schadt EE, et al. The arrestin domain-containing 3 protein regulates body mass and energy expenditure. *Cell Metab*. 2011;14:671–683.
118. Han SO, Kommaddi RP, Shenoy SK. Distinct roles for β-arrestin2 and ARRDC proteins in β2AR trafficking and signaling. *EMBO Rep*. 2013;14:164–171.

CHAPTER EIGHT

Arrestins in Actin Reorganization and Cell Migration

Kathryn A. DeFea
Biomedical Sciences Division, School of Medicine, University of California Riverside, Riverside, California, USA

Contents

1. Introduction 206
2. β-Arrestins as Regulators of Gradient Sensing for Chemokine Receptors 206
   2.1 Receptor turnover 208
   2.2 Differential signaling in response to ligand multimers 208
3. β-Arrestins as Regulators of Actin Assembly 209
   3.1 The cofilin pathway: Creating new actin seeds 210
   3.2 Filamin and other actin-binding proteins 212
   3.3 RhoA GTPases 212
4. Regulation of Kinase Activities by β-Arrestins 214
   4.1 β-Arrestin-dependent ERK1/2 activity 215
   4.2 β-Arrestin/Src complexes and chemotaxis 216
5. Additional Roles for β-Arrestins and Chemotaxis *In Vivo* 217
6. Role of β-Arrestin-Dependent Chemotaxis in Health and Disease 218
7. Concluding Remarks 219
References 220

Abstract

Arrestins have emerged as important regulators of actin reorganization and cell migration. Both in their classical roles as mediators of receptor desensitization and internalization, and in their newer role as signaling scaffolds, β-arrestins help orchestrate the cellular response to chemotactic signals. However, there is still a considerable amount to be learned about the precise molecular mechanisms underlying these processes. This review discusses how, by regulating receptor internalization and by scaffolding of signaling molecules in discrete cellular locations, arrestins facilitate gradient sensing and cytoskeletal reorganization, ultimately resulting in cell migration. In addition, putative new targets of β-arrestin regulation that may play important roles in cell migration are discussed, as continued research on these targets may provide important details to fill in the current gaps in our understanding of these processes.

Progress in Molecular Biology and Translational Science, Volume 118
ISSN 1877-1173
http://dx.doi.org/10.1016/B978-0-12-394440-5.00008-5

1. INTRODUCTION

Cell migration requires a number of spatially controlled events involving reorganization of the actin cytoskeleton, formation of a leading edge, assembly and disassembly of focal contacts, and contraction of the cell cortex. Actin cytoskeletal reorganization and cell migration, downstream of numerous receptors, require either or both β-arrestins, leading to a great deal of interest in the mechanism by which they regulate these processes.[1] The first step in cell migration involves the sensing of a gradient. While many cells can randomly migrate through a uniform concentration of agonist due to the activation of receptors that mediate morphological changes, the ability to sense a gradient is the first step for most cell migration pathways *in vivo*. The second step is the formation of a leading edge, which involves reorganization of the actin cytoskeleton and disassembly of focal contacts. Third, new focal contacts are formed and the cortex contracts, dragging the cell body forward. This process continues until the cell reaches the chemokine source. Once it is surrounded by a high concentration of the chemokine and the cell no longer senses a gradient, migration will cease.[2] β-Arrestins have been implicated at all stages of this process: β-arrestin scaffolds have been implicated in actin assembly events necessary for the formation of gradient-sensing filipodia and the lamellipodia of the leading edge. Additionally, β-arrestin-dependent desensitization of chemokine receptors in response to high concentrations of agonist has been implicated in the ability of migrating immune cells to sense a chemokine gradient (Fig. 8.1). Impaired desensitization results in random rather than directed migration. Finally, β-arrestins scaffold a number of proteins involved in endocytosis and many of the signaling proteins regulated by β-arrestins require endocytotic machinery for correct subcellular localization. Thus, while the differing roles of β-arrestin signaling and endocytosis in cell migration and cytoskeletal regulation can be separated *in vitro*, they are likely inseparable *in vivo*.

2. β-ARRESTINS AS REGULATORS OF GRADIENT SENSING FOR CHEMOKINE RECEPTORS

Cancer cells and immune cells both migrate toward a chemotactic agent and a primary receptor family that mediates this migration is the chemokine receptor family. Chemokine receptors are GPCRs and a number

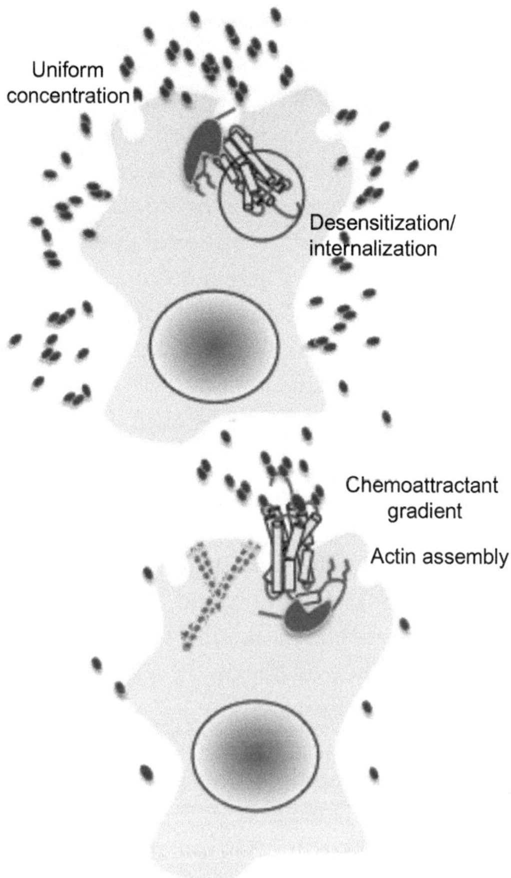

Figure 8.1 Model for β-arrestin regulation of receptor turnover and actin assembly. At the top of a chemoattractant gradient, the cell will be surrounded, thus experiencing a uniform concentration. At these high concentrations, β-arrestin-mediated desensitization and internalization aid in receptor turnover. As the cell moves toward a gradient, the concentration of agonist is lower and only receptors closest to the gradient will be activated. At these concentrations, β-arrestin-mediated actin assembly dominates, allowing the cell to move toward the source of the chemoattractant. (See Color Insert.)

have been shown to require β-arrestins, both for receptor internalization and for chemotaxis. Furthermore, chemokines oligomerize and the oligomers can signal in a fashion distinct from the monomers, which can affect the receptor response. Both of these features factor into β-arrestin-dependent gradient sensing.

2.1. Receptor turnover

The most obvious mechanism by which β-arrestins would affect receptor turnover is through their most well-known role as adaptors for clathrin and clathrin-binding proteins, facilitating receptor endocytosis. As discussed in Chapter 4, along with receptors, β-arrestins can bind both clathrin and clathrin adaptor proteins to facilitate their incorporation into clathrin-coated pits and removal from the cell surface. In some cases, this process also is an early step in the formation of "signaling endosomes" that scaffold signaling molecules and localize them to discrete cellular locations. β-Arrestins are required for the desensitization and internalization of several chemokine receptors, including CCR2, activated by MCP-1 (monocyte chemoattractant protein-1); CXCR1, activated by IL-8 (interleukin-8); CXCR2, activated by IL-8; CXCR4, activated by CXCL12; CXCR7, activated by CXCL12; and CCR5, activated by RANTES (regulated upon activation T cell expressed and secreted).[1,3] Early studies investigating the role of β-arrestins in chemotaxis revealed that many of the receptors that require β-arrestins for internalization show impaired chemotaxis in their absence.[1] Receptors that were allowed to signal constitutively, as they would in the absence of β-arrestins, also showed impaired chemotaxis. Recent proteomic screens for β-arrestin-binding proteins have led to the identification of more clathrin adaptor proteins, such as AGEP and ARF, with additional evidence suggesting that chemokine receptor turnover via β-arrestins through association with the endocytotic machinery is required for migrating cells to respond to an agonist gradient.[1,4,5] These studies strongly suggest a role for β-arrestin-mediated desensitization and internalization in chemotaxis. However, *in vivo*, these results are not always replicated.[1] These studies have taught us that it is dangerous to generalize the role of β-arrestins in chemokine receptor signaling, as each receptor appears to utilize β-arrestins differently in the process of chemotaxis.

2.2. Differential signaling in response to ligand multimers

CXCR4 is a crucial mediator of cell migration in both leukocytes and tumor cells and was one of the first to show impaired cell migration in the absence of β-arrestin-2.[6] CXCR4 recruits β-arrestins and receptor internalization is impaired in their absence. There is another aspect to this mechanism though. A migrating cell will climb "up" a gradient of chemokine, a feature that allows it to home to the site at which the agonist was released. At higher concentrations, CXCL12 starts to dimerize, and dimers elicit a different signal from the monomers. While CXCL12 monomers can promote β-arrestin recruitment and elicit β-arrestin-dependent signals such as actin assembly and chemotaxis, the dimers do not. Thus, in addition to simply promoting receptor turnover, β-arrestins may mediate the primary chemotactic signal at lower CXCL12 concentrations and their signaling may be dampened at higher concentrations, contributing to the cell's ability to sense a gradient. These studies are particularly interesting as they suggest that altered forms of the chemokines observed at high concentrations, rather than receptor desensitization, play a role in the inhibition of cell migration observed as cells reach the higher end of the concentration gradient. It is important to note that there is a reciprocal relationship between endocytosis and the scaffolding functions of β-arrestins discussed in the remainder of this chapter. β-Arrestins can form scaffolds containing proteins involved in actin assembly in response to activation of receptors that promote cell migration, and these scaffolds can facilitate the formation of protrusions that initiate movement toward the chemoattractant. Simultaneously, β-arrestins can facilitate internalization of these same receptors through association with endocytotic machinery.

3. β-ARRESTINS AS REGULATORS OF ACTIN ASSEMBLY

Other GPCRs, such as protease-activated receptor-2 (PAR2) and angiotensin receptor (AT1AR), have been demonstrated to promote the formation of β-arrestin scaffolds containing key actin assembly proteins, supporting a model wherein active β-arrestin signaling is involved in chemotaxis. β-Arrestins are required for both actin reorganization and chemotaxis by these and other GPCRs. Actin assembly within a cell is the primary driving force behind directed cell movement and can be regulated, both directly and indirectly, by various proteins. Actin polymerization from monomers is a spontaneous but slow process and the rate-limiting step is the formation of a stable nucleus, or actin seed, consisting of three or more actin monomers. Addition of actin monomers is always at the ATP-binding or barbed end of the actin molecule. Provided that the barbed end of a filament is free from capping proteins, addition of monomers onto preassembled filaments or actin seeds is very rapid. These seeds can be created in two major ways: (1) activation of proteins that break existing filaments into smaller fragments creating a free barbed end at each break or (2) activation of nucleators, that is, proteins that overcome the rate-limiting step in actin assembly by facilitating association of actin monomers into filament seeds. Actin nucleators can also bind to the sides of existing filaments to facilitate branching of actin filaments such as make up the broad lamellipodia of the leading edge. These processes are essential to provide directionality during migration, as well as for the assembly and maintenance of cortical actin filaments that interact with contractile proteins. Cortical actin filaments allow the cell to contract against the substrate over which it is migrating. These actin structures must be dynamically remodeled, requiring an intricate balance of input from various signaling pathways. We will now examine these different processes, the proteins that drive them and the reported role of β-arrestins.

3.1. The cofilin pathway: Creating new actin seeds

Proteins of the cofilin pathway (cofilin, chronophin, slingshot, and LIM kinase (LIMK)) have long been known to play crucial roles in establishing a leading edge downstream of multiple receptors. Cofilin is one of the primary actin filament severing proteins and its activation is often an early event in cell migration.[7,8] Cofilin is activated by dephosphorylation on serine 3, carried out by the phosphatases slingshot and chronophin and inactivated by LIMK. Dephosphorylated cofilin binds to the side of actin filaments, destabilizing them such that they break apart creating multiple actin seeds to which actin monomers can add. Evidence that all of these proteins bind β-arrestins has led to the hypothesis that β-arrestins are responsible for spatial control of their activities. The dendritic model for actin remodeling predicts that cofilin rapidly disassembles existing filaments, providing free barbed ends for elongation and, in coordination with activation of nucleating proteins, can lead to the formation of a leading edge that drives the direction of cell migration.[8] Spatial control over cofilin activity is essential to allow for treadmilling of filaments. Two GPCRs, PAR2 and AT1AR,[9–11] have been shown to promote β-arrestin-dependent cofilin activation through scaffolding of its upstream regulators. β-Arrestins can also bind LIMK and inhibit its activity downstream of PAR2, to facilitate cofilin dephosphorylation.[9] Association with β-arrestin also facilitates dephosphorylation of cofilin by chronophin and slingshot, and localizes cofilin activity to the leading edge of the cell to promote barbed-end formation and membrane protrusion[10,11] (Fig. 8.2). This β-arrestin/cofilin scaffold is important for leukocyte migration and may contribute to extravasation of leukocytes into tissue during inflammation.[10] When cofilin activity is not spatially controlled, protrusions can form randomly and cells lose their ability to move toward a chemotactic

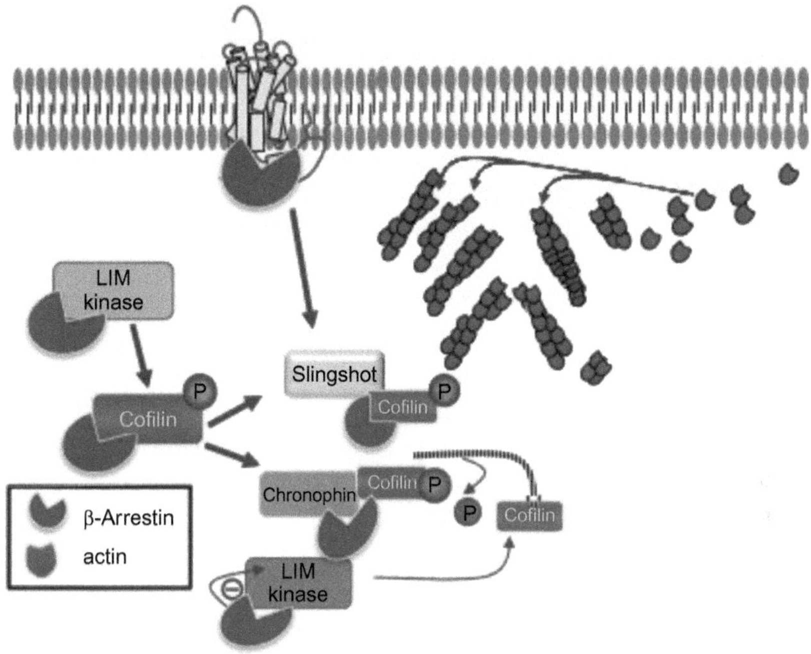

Figure 8.2 Model for regulation of cofilin activity by β-arrestins. Upon recruitment to certain GPCRs, β-arrestins can activate cofilin through various mechanisms. First, β-arrestin-2 can directly bind the two cofilin phosphatases, slingshot and chronophin, as well as cofilin. Association with β-arrestin-2 appears to scaffold cofilin to either slingshot or chronophin to facilitate dephosphorylation. β-Arrestin-1 can directly bind LIMK and interfere with its phosphorylation of cofilin. Together, both β-arrestins can facilitate GPCR-mediated cofilin dephosphorylation, which is essential for the creation of free barbed ends at the leading edge for actin polymerization. (For color version of this figure, the reader is referred to the online version of this chapter.)

gradient. Thus, it is likely that β-arrestins contribute to the ability of a cell to sense a gradient as well as form a leading edge through the regulation of cofilin activity.

Reorganization of actin filaments by cofilin is important for other cellular activities as well. In neurons, actin-rich dendritic spines form the postsynaptic sites for most excitatory neurons and are important for normal learning and memory processes. Dendritic spines are comprised primarily of filamentous actin and dynamic reorganization of these structures is functionally linked to synaptic plasticity. Many studies have demonstrated that spatially constrained cofilin activity is crucial to the dynamic nature of dendritic spines and, subsequently, to synaptic plasticity. For example, the NMDA receptor promotes remodeling of dendritic spines and translocation of active cofilin from the cell body to the spines. In hippocampal neurons from β-arrestin-2 knockout mice, translocation of cofilin in response to NMDA receptor activation is abolished and spines show impaired remodeling activity. Subsequently, β-arrestin-2 knockout mice have spatial learning deficits, in keeping with the reported role for cofilin activity in spine remodeling. Thus, spatial regulation of cofilin activity by β-arrestins is crucial not only for actin assembly events leading to the formation of a leading edge and cell migration but also for actin remodeling necessary for neuronal plasticity.[12]

3.2. Filamin and other actin-binding proteins

The actin-binding protein, filamin, has been identified in a complex with β-arrestins and ERK1/2. The role of filamin in cell migration is a compound one, and the manner in which it is controlled by β-arrestins is still not completely clear. It plays an important role in the turnover of adhesion proteins such as integrins during cell migration,[13] a process that is important for cell attachment and contraction. It has also been reported to regulate the internalization of GPCRs such as the dopamine receptor[14,15] and CXCR4.[13] Downstream of the AT1AR, filamin associates with β-arrestin-2 and ERK1/2 and this complex is thought to play a role in membrane ruffling.[16] Thus, β-arrestin-dependent regulation of filamin may contribute to both focal adhesion remodeling and receptor turnover, proposed to be important for gradient sensing. However, it also appears to be involved in the sequestration of ERK1/2 activity at the leading edge,[17] which likely facilitates localized phosphorylation of key substrates involved in chemotaxis.

3.3. RhoA GTPases

RhoA GTPases are often upstream regulators of both cofilin and other actin assembly activities. The three most commonly studied members of the Rho family are RhoA, Rac-1, and Cdc42, and each is associated with a different actin structure. RhoA typically causes stress-fiber formation, Cdc42 induces filipodia formation, and Rac-1 is important for membrane ruffling and lamellipodia formation.[18] β-Arrestins can regulate the activity of many monomeric GTPases including those of the Rho family.[19–21] For example, knockdown of β-arrestin-1, but not β-arrestin-2, with siRNA significantly reduces RhoA activation and stress-fiber formation by the angiotensin AT1AR.[19] β-Arrestin-1-dependent p38MAPK activation can elicit F-actin rearrangement via a Rac-1-dependent mechanism, downstream of β_2-adrenergic receptors,[22] raising the possibility that Rac-1 could be a β-arrestin-dependent p38MAPK target. β-Arrestins can also negatively regulate RhoA GTPases. The type III TGF-β receptor (TβRIII) forms a β-arrestin-2 scaffolding complex with Cdc42, leading to inhibition of lamellipodia formation in both cancer and epithelial cells.[23,24]

How is Rho GTPase activity affected by β-arrestins? To answer this question, one must consider how monomeric GTPases are regulated in general. Their activity is increased by binding to guanine nucleotide exchange factors (GEFs) and turned off by association with GTPase activating proteins (GAPs). β-Arrestin-1 directly binds and inhibits the RhoGAP (ARHGAP21) downstream of AT1AR activation. Disruption of this complex inhibits AT1AR-mediated stress-fiber formation and RhoA activation.[25] Phosphatidylinositol-3-phosphate (PIP3), which is generated by phosphatidylinositol-3 kinase (PI3K), can also activate a number of GEFs for RhoA, Cdc42, and Rac. β-Arrestins have been shown to regulate PI3K activity both positively and negatively depending on the activating receptor[26,27] (Fig. 8.3A). β-Arrestin-1-dependent p38MAPK activation was reported to elicit F-actin rearrangement via a Rac-1-dependent mechanism, downstream of β2-adrenergic receptors,[22] so regulation of RhoA GTPases may also lie downstream of the MAPK scaffolds. β-Arrestin-dependent regulation of RhoA GTPases has also been implicated in inhibition of cell migration. The type III TGF-β receptor (TβRIII) inhibits migration and alters actin cytoskeleton via forming a β-arrestin-2 scaffolding complex with Cdc42 in both cancer and epithelial cells.[23,24]

Another small GTPase, RalA, induces membrane blebbing in response to the fMLP receptor in neutrophils and LPA in cancer cells.[20,21,28] Ral-GDS is a GEF that activates RalA and can exist in an inactive complex with β-arrestin-1 in the cytosol of resting cells. Activation of the fMLP or LPA receptor recruits the β-arrestin-1/Ral-GDS complex to the membrane and upon receptor/β-arrestin-1 binding where Ral-GDS is released and activates RalA. fMLP receptor-induced membrane ruffling is blocked by a mutant Ral-GDS which cannot bind β-arrestin-1, suggesting that the ability of β-arrestin-1 to traffic it to the membrane is crucial for its activity (Fig. 8.3B). Subsequent studies have demonstrated that expression of RalA and β-arrestins-1 and 2 is increased in metastatic cancers and expression of a Ral-GDS mutant that is deficient in β-arrestin binding inhibits RalA activation in tumor cells, leading to decreased cell migration. RalA itself is important for targeting another actin-binding protein regulated by β-arrestins, filamin, to the membrane.[29]

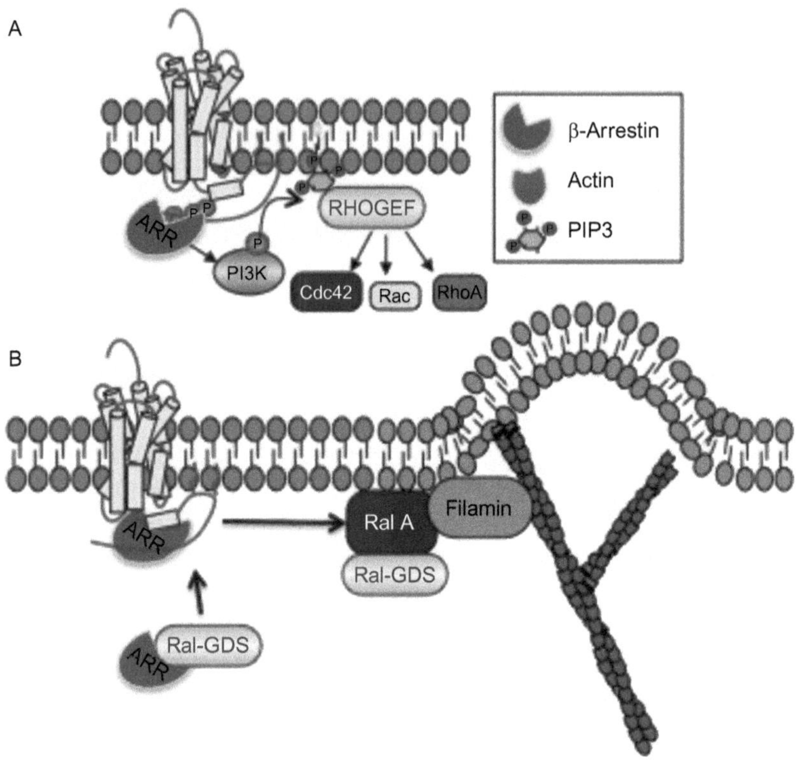

Figure 8.3 Model for regulation of RhoA GTPases by β-arrestins. (A) Upon binding to a variety of GPCRs, β-arrestins can indirectly facilitate the activity of PI3K or directly inhibit its activity. PIP3, the product of PI3K activity, can then activate a number of GEFs for the Rho GTPases, Cdc42, Rac and RhoA. In this fashion, β-arrestins can either activate or inhibit Rho GTPase activity in a manner dependent upon the activating receptor. (B) β-Arrestin-2 binds constitutively to the guanine exchange factor for the GTPase, Ral1 (Ral-GDS). Upon binding to the activated fMLP receptor, Ral-GDS is released and free to activate membrane-associated Ral, which associates with filamin to promote actin filament reorganization and membrane blebbing. (For color version of this figure, the reader is referred to the online version of this chapter.)

4. REGULATION OF KINASE ACTIVITIES BY β-ARRESTINS

β-Arrestins are required for ERK1/2 and JNK activation downstream of numerous GPCRs, and β-arrestins can sequester ERK1/2 activity within the leading edge of migrating cells. Furthermore, β-arrestins can facilitate activation of Src and other non-receptor-tyrosine kinases and can both activate and inhibit PI3Ks. Because both ERK1/2 and β-arrestins are required for cell migration downstream of many of these receptors, it is likely that

ERK1/2 might phosphorylate proteins important for cell migration, altering their activity. The search for candidate targets of β-arrestin-dependent ERK1/2 activity is ongoing, but phosphoproteomics screens have identified a number of putative candidates. Some of these ERK1/2 targets are actin nucleators and Wiscott–Aldrich syndrome proteins (WASps), focal adhesion proteins, and cofilin.

4.1. β-Arrestin-dependent ERK1/2 activity

Despite the mystery as to what ERK1/2 is doing at the leading edge to enhance cell motility, there is a significant body of evidence pointing to an essential role for β-arrestin-sequestration of ERK1/2 in chemotaxis. Many of the receptors first demonstrated to promote β-arrestin-dependent ERK1/2 activation also require both for chemotaxis. We previously discussed chemokine receptors as prime examples of utilizing β-arrestins for receptor turnover. However, not only can CXCR4 respond differently to monomers versus dimers of its ligand, CCL12, but it can also form heterodimers with the decoy receptor CXCR7, leading to dominance of β-arrestin-dependent ERK1/2 phosphorylation and chemotaxis over G-protein signaling.[30] Thus, CXCR4 stands out as an argument for the importance of both the ERK1/2 scaffolding and internalization functions of β-arrestins in chemotaxis. A number of proteins involved in actin assembly have been identified as β-arrestin-dependent ERK1/2 phosphorylation targets; however, for most of these, a definitive role in chemotaxis downstream of β-arrestins has not been proven.

4.1.1 Proposed role of ERK1/2 in actin nucleation

Regulation of actin nucleation by β-arrestins has not been directly shown, although several proteins involved in actin nucleation have been identified as β-arrestin-dependent ERK1/2 targets. Arp2/3 complex components and WASp-family proteins have been identified in a proteomics screen as potential β-arrestin interacting proteins[5] and phosphorylation targets of MAPK[31] and activation of WASp-family proteins is enhanced by ERK1/2 phosphorylation. The Arp2/3 complex, along with formins and p150spir, are the primary nucleation factors in mammalian cells.[32,33] Arp2/3 is crucial for the formation of branched actin filaments such as are observed in the leading edge of migrating cells. Activation by WASps, which bind Arp2/3 and induce a conformational change resulting in apposition of Arp2 and 3, is essential for the formation of branched filaments. Patients with Wiscott–Aldrich syndrome have defective lymphocyte trafficking and function, due in part to the disruption of actin nucleation. These studies have raised the possibility that β-arrestin-dependent phosphorylation of WASp-family proteins might play a role in chemotaxis.[34,35] The actin nucleating proteins, formins and p150spir, are primarily responsible for *de novo* assembly of unbranched actin filaments.[33] Recently, a formin-like protein was also identified as a putative β-arrestin-dependent phosphorylation target.[31] In the years to come, a more definitive role for regulation of actin nucleation by β-arrestins may emerge.

4.1.2 ERK1/2 phosphorylation of other actin assembly proteins

Another protein identified as a putative β-arrestin-dependent ERK1/2 substrate in two separate proteomics screens is, the barbed-end capping protein, adducin. Phosphorylation of adducin by ser/thr kinases, PKC and Rho-activated kinase(ROCK), diminishes its affinity for actin, increasing the pool of free barbed ends and facilitating actin polymerization. Thus, phosphorylation by ERK1/2 may have a similar effect. Cofilin was also identified as a β-arrestin-dependent ERK1/2 substrate. The known regulatory site on cofilin (S3) was not the site identified in the phosphoproteomics screen. Thus, if ERK1/2 phosphorylation contributes to β-arrestin-dependent coflin dephosphorylation, it would be through a distinct mechanism, possibly by stabilizing the active form.

4.2. β-Arrestin/Src complexes and chemotaxis

Src was identified as a β-arrestin-binding partner downstream of several GPCRs, linking them to ERK1/2 activation.[36] Downstream of prostaglandin E2 (PGE2) receptor, cell migration involves β-arrestin-dependent recruitment of Src into a signaling complex that then transactivates the epidermal growth factor receptor.[37] Similar β-arrestin-dependent transactivation pathways have been linked to proliferation as well. In another example, activation of CCR5 with MIP1 leads to recruitment of Pyk2, the p85 regulatory subunit of PI3K, and the tyrosine kinase, Lyn, into a complex with β-arrestin-2 that is thought to be essential for macrophage chemotaxis.[38] In these examples, the role of β-arrestins is relatively indirect, essentially linking a GPCR-elicited signal to a tyrosine kinase cascade. Whether these complexes are localized to the leading edge or lead to spatial regulation of actin assembly proteins has not been examined.

5. ADDITIONAL ROLES FOR β-ARRESTINS AND CHEMOTAXIS *IN VIVO*

Because chemotaxis *in vivo* is a complex process involving not just the ability of cells to migrate, but to migrate in the correct direction, extravasate into the tissue, and, in some cases, reattach at a distal site, *in vitro* models of chemotaxis may not always predict the *in vivo* response correctly. In many cases, the *in vitro* models for chemokine receptor chemotaxis are supported with *in vivo* studies. Multiple groups have shown that β-arrestins-2 is required for infiltration of white blood cells to sites of infection.[39,40] However, an *in vivo* study on the role of β-arrestin-2 in neutrophil chemotaxis found that, contrary to *in vitro* chemotaxis studies, CXCR2-mediated neutrophil recruitment to sites of inflammation was increased in β-arrestin-$2^{-/-}$ mice.[41]

One of the key features of chemotaxis *in vivo* is that the first group of cells to arrive at a target can release chemotactic factors attracting other cells. Thus, release of chemokines from migrating cells, such as macrophages, neutrophils, and lymphocytes, plays an important role on chemotaxis *in vivo*. Furthermore, during migration *in vivo*, such as is observed during inflammation, there are factors released that inhibit migration of certain cell types, so as to control the cellular makeup in a given locale. In addition to playing a role in regulating receptor turnover and actin assembly during cell migration, β-arrestins can promote release of chemokines and other factors that then stimulate the migration of additional cells. β-Arrestins were first implicated in neutrophil degranulation, which is important for leukocyte chemotaxis *in vivo*, as some of the factors released are chemotactic factors that attract additional immune cells to the site of inflammation.[42] The tyrosine kinases Hck and Fgr are part of a signaling cascade required for IL-8 and zymosan granule release from neutrophils.[42,43] IL-8 promotes the recruitment of Hck and Fgr into a complex with β-arrestin-1, and dominant-negative β-arrestin-1 blocks both neutrophil degranulation and tyrosine phosphorylation, suggesting a role for β-arrestin scaffolding in the secretion of chemotactic agents. In a cultured leukemia cell line, GRK3-mediated receptor phosphorylation of the chemokine receptor CC3A and association with β-arrestins was essential for expression of the chemokine CCL2, supporting the hypothesis that β-arrestins may contribute to chemotaxis *in vivo* on multiple levels.

6. ROLE OF β-ARRESTIN-DEPENDENT CHEMOTAXIS IN HEALTH AND DISEASE

The two main disease scenarios in which β-arrestin-dependent chemotaxis has been implicated are tumor cell metastasis and inflammation. Tumor metastasis requires migration of malignant cells from the original tumor to other sites within the body. This process requires a number of chemotactic signals that allow the cells to migrate to nearby vessels, enter the vasculature, and extravasate at distal sites. Similarly, inflammation involves recruitment of leukocytes and other inflammatory cells from the vasculature to injured tissue, and numerous changes within the tissue such as epithelial proliferation, extracellular matrix deposition, and functional alterations in both the invading and host cells. Several studies over the last decade in tumor cell lines have demonstrated that constitutive migration of many cancer cells is dependent upon β-arrestin. Likewise, studies have demonstrated that certain inflammatory processes are impaired in the absence of β-arrestin-2. Furthermore, β-arrestin-dependent regulation of actin cytoskeletal proteins and signaling pathways that affect cell migration has been demonstrated in human cancer cell lines and leukocytes.

Recent *in vivo* studies have shed more light on the multiple mechanisms by which β-arrestins can promote tumor cell migration and metastasis. Interestingly, these studies suggest that both tumor cell and host β-arrestin play important roles in this process, and that β-arrestins are pro-metastatic in some scenarios and antimetastatic in others. In many cases, β-arrestin complexes described in the previous sections play a major role in the metastatic progression of cancers *in vivo*. CXCR4, which we discussed earlier in this chapter, is upregulated in numerous malignant cancers, and CXCL12 is expressed in many of the tissues to which tumor cells commonly metastasize. While CXCL12 monomers promote metastasis, CXCL12 dimers have tumor suppressor functions, effectively inhibiting cell migration through a β-arrestin-independent pathway.[44] Thus, development of biased ligands that act like CXCL12 dimers may have therapeutic value for the treatment of some cancers.

Several reports have also indicated that β-arrestins are required for the recruitment of immune cells to the airways during asthma. Influx of leukocytes to the lungs was not only decreased in β-arrestin-2 knockout mice, but in wild-type mice receiving knockout bone marrow. Thus, the role of β-arrestins in asthma appears to be mediating the chemotaxis of invading inflammatory cells.[39,40,45] Matrix deposition, on the other hand, involves fibroblasts that invade the extracellular matrix and secrete matrix components; when uncontrolled, this process contributes to fibrosis. In a mouse model of pulmonary fibrosis, β-arrestin-1 or -2 knockout mice were protected from excessive matrix deposition, resulting in protected lung function and heightened survival. Knockdown of either β-arrestin in fibroblasts from patients with pulmonary fibrosis inhibited their migration and invasive behavior.[46] Thus, β-arrestin-dependent chemotaxis *in vivo* contributes to recruitment of inflammatory cells and matrix depositing fibroblasts, which can result in chronic inflammation and fibrosis.

In contrast to the studies on airway inflammation, sepsis-induced inflammation may utilize β-arrestins in the opposite capacity. Recent studies reveal that β-arrestin-2 appears to negatively regulate tissue damage and mortality resulting from sepsis.[47] These studies did not deal with β-arrestin-stimulated chemotaxis, but rather production of inflammatory mediators important for resolving bacterial infections. Release of these cytokines may be dependent upon the G-protein signaling arm of chemokine receptors and thus inhibited by β-arrestins. It is important to bear in mind that the role of β-arrestins in processes requiring chemotaxis *in vivo* is far more complicated than their role in chemotaxis *in vitro*. Depending on the physiological scenario and the receptor being activated, the classical role of β-arrestins as terminators of G-protein signaling or their role as facilitators of chemotaxis may dominate.

7. CONCLUDING REMARKS

Although it has been over a decade since the demonstration that β-arrestins are important for chemotaxis downstream of numerous GPCRs, much remains to be elucidated regarding the underlying molecular mechanisms. Clearly, receptor turnover at the hands of β-arrestin-dependent endocytosis is important, but evidence suggests that the story is far more complicated. β-Arrestins are capable of regulating a barrage of cellular activities essential for cell migration and, in so doing, controlling the localization of these activities. Since so many of the signals generated during cell migration must be tightly controlled both spatially and temporally, it stands to reason that the role of β-arrestins as signaling scaffolds is equally important in cell migration. Given the plethora of biological responses, from inflammation to neuronal development to cancer, that are controlled by β-arrestin-dependent regulation of actin assembly and chemotaxis, gaining a deeper understanding of these processes at the molecular level will enhance our understanding of many disease processes.

REFERENCES

1. DeFea KA. Stop that cell! Beta-arrestin-dependent chemotaxis: a tale of localized actin assembly and receptor desensitization. *Annu Rev Physiol.* 2007;69:535–560.
2. Iglesias PA, Devreotes PN. Navigating through models of chemotaxis. *Curr Opin Cell Biol.* 2008;20:35–40.
3. Luker KE, Steele JM, Mihalko LA, Ray P, Luker GD. Constitutive and chemokine-dependent internalization and recycling of CXCR7 in breast cancer cells to degrade chemokine ligands. *Oncogene.* 2010;29:4599–4610.
4. Bouschet T, Martin S. The calcium-sensing receptor changes cell shape via a beta-arrestin-1 ARNO ARF6 ELMO protein network. *J Cell Sci.* 2007;120:2489–2497.
5. Xiao K, McClatchy DB, Shukla AK, et al. Functional specialization of beta-arrestin interactions revealed by proteomic analysis. *Proc Natl Acad Sci USA.* 2007;104: 12011–12016.
6. Fong AM, Premont RT, Richardson RM, Yu YRA, Lefkowitz RJ, Patel DD. Defective lymphocyte chemotaxis in beta-arrestin2-and GRK6-deficient mice. *Proc Natl Acad Sci USA.* 2002;99:7478–7483.
7. Wang W, Eddy R, Condeelis J, Condeelis J. The cofilin pathway in breast cancer invasion and metastasis. Ins and outs of ADF/cofilin activity and regulation. *Nat Rev Cancer.* 2007;7:429–440.
8. Oser M, Condeelis J. The cofilin activity cycle in lamellipodia and invadopodia. *J Cell Biochem.* 2009;108:1252–1262.
9. Zoudilova M, Kumar P, Ge L, Wang P, Bokoch GM, DeFea KA. Beta-arrestin-dependent regulation of the cofilin pathway downstream of protease-activated receptor-2. *J Biol Chem.* 2007;282:20634–20646.
10. Zoudilova M, Min J, Richards HL, Carter D, Huang T, DeFea KA. Beta-arrestins scaffold cofilin with chronophin to direct localized actin filament severing and membrane protrusions downstream of protease-activated receptor-2. *J Biol Chem.* 2010;285: 14318–14329.
11. Xiao K, Sun J, Kim J, et al. Global phosphorylation analysis of beta-arrestin-mediated signaling downstream of a seven transmembrane receptor (7TMR). *Proc Natl Acad Sci USA.* 2010;107:15299–15304.
12. Pontrello CG, Sun MY, Lin A, Fiacco TA, DeFea KA, Ethell IM. Cofilin under control of beta-arrestin-2 in NMDA-dependent dendritic spine plasticity, long-term depression (LTD), and learning. *Proc Natl Acad Sci USA.* 2012;109:E442–E451.
13. Kim H, McCulloch CA. Filamin A mediates interactions between cytoskeletal proteins that control cell adhesion. *FEBS Lett.* 2011;585:18–22.
14. Cho EY, Cho DI, Park JH, Kurose H, Caron MG, Kim KM. Roles of protein kinase C and actin-binding protein 280 in the regulation of intracellular trafficking of dopamine D3 receptor. *Mol Endocrinol.* 2007;21:2242–2254.
15. Kim KM, Gainetdinov RR, Laporte SA, Caron MG, Barak LS. G protein-coupled receptor kinase regulates dopamine D3 receptor signaling by modulating the stability of a receptor-filamin-beta-arrestin complex. A case of autoreceptor regulation. *J Biol Chem.* 2005;280:12774–12780.
16. Scott MG, Pierotti V, Storez H, et al. Cooperative regulation of extracellular signal-regulated kinase activation and cell shape change by filamin A and beta-arrestins. *Mol Cell Biol.* 2006;26:3432–3445.

17. Ge L, Ly Y, Hollenberg M, DeFea K. A beta-arrestin-dependent scaffold is associated with prolonged MAPK activation in pseudopodia during protease-activated receptor-2-induced chemotaxis. *J Biol Chem.* 2003;278:34418–34426.
18. Hall A. Rho GTPases and the actin cytoskeleton. *Science.* 1998;279:509–514.
19. Barnes WG, Reiter E, Violin JD, Ren X-R, Milligan G, Lefkowitz RJ. β-Arrestin 1 and Gαq/11 coordinately activate RhoA and stress fiber formation following receptor stimulation. *J Biol Chem.* 2005;280:8041–8050.
20. Bhattacharya M, Wang J, Ribeiro FM, et al. Analysis of a missense variant of the human N-formyl peptide receptor that is associated with agonist-independent [beta]-arrestin association and indices of inflammation. *Pharmacogenomics J.* 2006;7:190–199.
21. Bhattacharya M, Anborgh PH, Babwah AV, et al. [beta]-Arrestins regulate a Ral-GDS-Ral effector pathway that mediates cytoskeletal reorganization. *Nat Cell Biol.* 2002;4:547–555.
22. Gong K, Li Z, Xu M, Du J, Lv Z, Zhang Y. A novel protein kinase A-independent, beta-arrestin-1-dependent signaling pathway for p38 mitogen-activated protein kinase activation by beta2-adrenergic receptors. *J Biol Chem.* 2008;283:29028–29036.
23. Mythreye K, Blobe GC. The type III TGFbeta receptor regulates directional migration: new tricks for an old dog. *Cell Cycle.* 2009;8:3069–3070.
24. Mythreye K, Blobe GC. The type III TGF-beta receptor regulates epithelial and cancer cell migration through beta-arrestin2-mediated activation of Cdc42. *Proc Natl Acad Sci USA.* 2009;106:8221–8226.
25. Anthony DF, Sin YY, Vadrevu S, et al. {beta}-Arrestin 1 inhibits the GTPase-activating protein function of ARHGAP21, promoting activation of RhoA following angiotensin II type 1A receptor stimulation. *Mol Cell Biol.* 2011;31:1066–1075.
26. Wang P, DeFea KA. Protease-activated receptor-2 simultaneously directs beta-arrestin-1-dependent inhibition and Galphaq-dependent activation of phosphatidylinositol 3-kinase. *Biochemistry.* 2006;45:9374–9385.
27. Povsic TJ, Kohout TA, Lefkowitz RJ. Beta-arrestin1 mediates insulin-like growth factor 1 (IGF-1) activation of phosphatidylinositol 3-kinase (PI3K) and anti-apoptosis. *J Biol Chem.* 2003;278:51334–51339.
28. Li TT, Alemayehu M, Aziziyeh AI, et al. Beta-arrestin/Ral signaling regulates lysophosphatidic acid-mediated migration and invasion of human breast tumor cells. *Mol Cancer Res.* 2009;7:1064–1077.
29. Ohta Y, Suzuki N, Nakamura S, Hartwig JH, Stossel TP. The small GTPase RalA targets filamin to induce filopodia. *Proc Natl Acad Sci USA.* 1999;96:2122–2128.
30. Decaillot FM, Kazmi MA, Lin Y, Ray-Saha S, Sakmar TP, Sachdev P. CXCR7/CXCR4 heterodimer constitutively recruits β-arrestin to enhance cell migration. *J Biol Chem.* 2011;286:32188–32197.
31. Christensen GL, Kelstrup CD, Lyngsø C, et al. Quantitative phosphoproteomics dissection of 7TM receptor signaling using full and biased agonists. *Mol Cell Proteomics.* 2010;9:1540–1553.
32. Campellone KG, Welch MD. A nucleator arms race: cellular control of actin assembly. *Nat Rev Mol Cell Biol.* 2010;11:237–251.
33. Firat-Karalar EN, Welch MD. New mechanisms and functions of actin nucleation. *Curr Opin Cell Biol.* 2011;23:4–13.
34. Nakanishi O, Suetsugu S, Yamazaki D, Takenawa T. Effect of WAVE2 phosphorylation on activation of the Arp2/3 complex. *J Biochem.* 2007;141:319–325.
35. Mendoza Michelle C, Er EE, Zhang W, et al. ERK-MAPK drives lamellipodia protrusion by activating the WAVE2 regulatory complex. *Mol Cell.* 2011;41:661–671.
36. Luttrell LM, Gesty-Palmer D. Beyond desensitization: physiological relevance of arrestin-dependent signaling. *Pharm Rev.* 2010;62:305–330.

37. Kim JI, Lakshmikanthan V, Frilot N, Daaka Y. Prostaglandin E2 promotes lung cancer cell migration via EP4-βarrestin1-c-Src signalsome. *Mol Cancer Res*. 2010;8:569–577.
38. Cheung R, Malik M, Ravyn V, Tomkowicz B, Ptasznik A, Collman RG. An arrestin-dependent multi-kinase signaling complex mediates MIP-1 beta/CCL4 signaling and chemotaxis of primary human macrophages. *J Leukocyte Biol*. 2009;86:833–845.
39. Nichols HL, Saffeddine M, Theriot BS, et al. β-Arrestin-2 mediates the proinflammatory effects of proteinase-activated receptor-2 in the airway. *Proc Natl Acad Sci USA*. 2012;109:16660–16665.
40. Walker JK, Fong AM, Lawson BL, et al. Beta-arrestin-2 regulates the development of allergic asthma. *J Clin Invest*. 2003;112:566–574.
41. Su YJ, Raghuwanshi SK, Yu YC, Nanney LB, Richardson RM, Richmond A. Altered CXCR2 signaling in beta-arrestin-2-deficient mouse models. *J Immunol*. 2005;175: 5396–5402.
42. Vines CM, Xue M, Maestas DC, Cimino DF, Prossnitz ER. Regulation of N-formyl peptide-mediated degranulation by receptor phosphorylation. *J Immunol*. 2002;169: 6760–6766.
43. Barlic J, Khandaker MH, Mahon E, et al. Beta-arrestins regulate interleukin-8-induced CXCR1 internalization. *J Biol Chem*. 1999;274:16287–16294.
44. Drury LJ, Ziarek JJ, Gravel S, et al. Monomeric and dimeric CXCL12 inhibit metastasis through distinct CXCR4 interactions and signaling pathways. *Proc Natl Acad Sci USA*. 2011;108:17655–17660.
45. Hollingsworth JW, Theriot BS, Li Z, et al. Both hematopoietic-derived and non-hematopoietic-derived {beta}-arrestin-2 regulates murine allergic airway disease. *Am J Respir Cell Mol Biol*. 2010;43:269–275.
46. Lovgren AK, Kovacs JJ, Xie T, et al. Deficiency protects against pulmonary fibrosis in mice and prevents fibroblast invasion of extracellular matrix. *Sci Trans Med*. 2011;3:74ra23.
47. Fan HK, Bitto A, Zingarelli B, et al. Beta-arrestin 2 negatively regulates sepsis-induced inflammation. *Immunology*. 2010;130:344–351.

PART III

The Physiological Roles of Arrestins

CHAPTER NINE

The Role of Arrestins in Development

Melanie Philipp*, **Tama Evron**[†], **Marc G. Caron**[†,‡,§]

*Institute of Biochemistry and Molecular Biology, Ulm University, Ulm, Germany
[†]Department of Cell Biology, Duke University Medical Center, Durham North Carolina, USA
[‡]Department of Medicine, Duke University Medical Center, Durham, North Carolina, USA
[§]Department of Neurobiology, Duke University Medical Center, Durham, North Carolina, USA

Contents

1. Introduction 226
2. β-Arrestins in Model Organisms 228
3. The Function of Arrestins in Invertebrates 229
4. β-Arrestins in Vertebrate Development 230
   4.1 βarr1 in hematopoiesis 230
   4.2 βarr2 in Wnt signaling 231
   4.3 βarrs in Hh signaling 234
5. Conclusions 238
Acknowledgments 238
References 238

Abstract

β-Arrestins are versatile scaffolding proteins that are involved in orchestrating a large number of signaling cascades. Because β-arrestin 1 and β-arrestin 2 are individually dispensable during development, it has long been assumed that β-arrestins do not play an important role during embryogenesis. Nonetheless, there is growing evidence from both invertebrate and vertebrate animal models that β-arrestins are integral regulators of developmental pathways. They are involved in diverse processes such as early hematopoiesis, establishment of axial development axes, cell convergence and extension movements leading to axial elongation, musculoskeletal and craniofacial development, cell proliferation, and neuronal development. Most of the developmental roles of arrestins involve interactions with or regulation of novel non-G protein-coupled receptor partners, such as atypical seven-transmembrane receptors, small G proteins, and nuclear transcription factors, revealing surprising diversity in arrestin function. In contrast to their largely overlapping roles in G protein-coupled receptor desensitization and internalization, study of the role of arrestins in development has demonstrated marked functional specialization of the β-arrestin 1 and 2 isoforms.

Progress in Molecular Biology and Translational Science, Volume 118
ISSN 1877-1173
http://dx.doi.org/10.1016/B978-0-12-394440-5.00009-7

1. INTRODUCTION

β-Arrestins (βarrs) are scaffolding proteins, which assemble signaling complexes and thereby orchestrate a large number of signaling cascades (reviewed in Ref. 1). By doing so, they are fundamental regulators of multiple physiological as well as pathological processes in our bodies. Due to their rather ubiquitous expression[2–5] and the normal fertility of βarr1 or βarr2 knockout (KO) mice,[4,5] it has long been assumed that βarrs play no significant role during embryogenesis. However, there is growing evidence in animal models other than mice demonstrating that βarrs implement their role in cellular signaling already during embryonic development. This chapter gives an overview of the different functions of βarrs in development that have been uncovered up to date.

In the adult, βarrs facilitate a large variety of signaling cascades. Besides their traditional role in G protein-coupled receptor (GPCR) desensitization and internalization, βarrs enable G protein-independent signal transduction by the assembly of signaling complexes.[6] Furthermore, they have been found to be central regulators of the so-called atypical GPCRs (i.e. Smoothened (Smo) or Frizzled proteins) as well as molecules with structures completely different from seven-transmembrane receptors.[7] Considering their impact upon such a large repertoire of pathways, it is surprising that mice lacking one or the other βarr would develop normally. In fact, βarr1 KO mice are viable and fertile. They show normal cardiac parameters but are hypersensitive to adrenergic stimulation.[5] Continuous treatment of βarr1 KO mice with the β-adrenergic receptor agonist isoproterenol results in DNA damage through suppression of p53.[8] Similarly, mice lacking βarr2 appear to be normal. However, they display prolonged analgesic effects when treated with the opioid agonist morphine.[4]

βarr1 and 2 are highly conserved proteins, which can be found throughout vertebrate as well as invertebrate species (Fig. 9.1). They share great homology with each other, which is likely to be the reason why single KOs for either βarr are viable. The combined loss of both βarrs, however, is incompatible with life. Newborn βarr1/2 double KO mice are smaller than their wild-type littermates and die soon after birth from respiratory failure as they do not develop fully mature lungs.[9] When looking more thoroughly, it appears that their lungs arrest in differentiation, which manifests in reduced proliferation, abnormal accumulation of glycogen, and disturbed production of surfactant proteins. In addition, this phenotype is associated

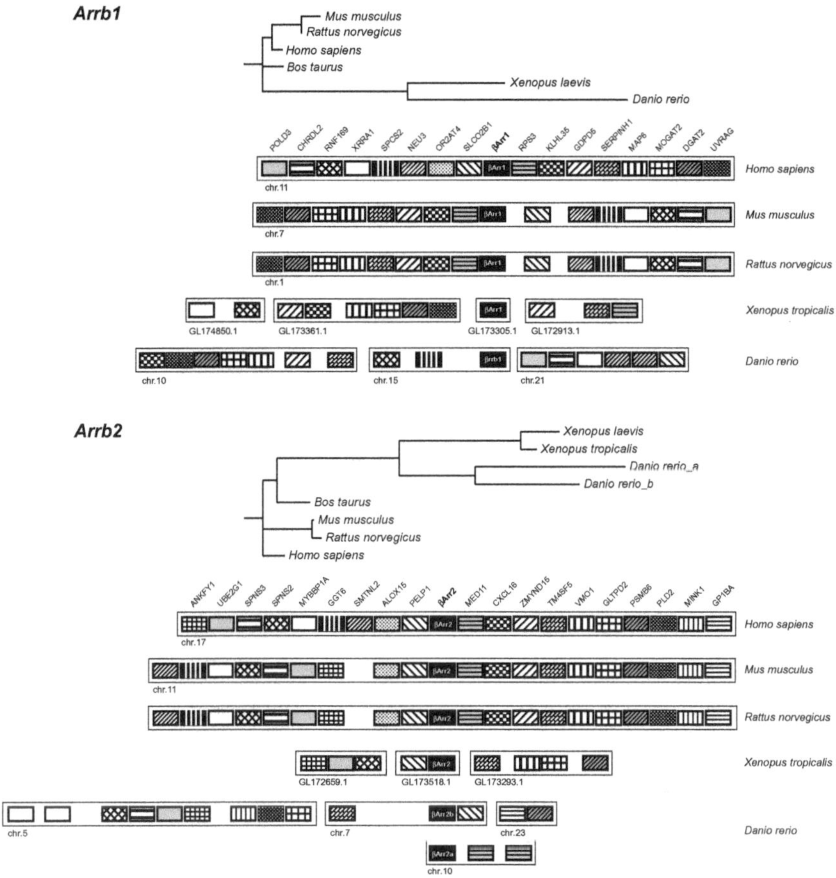

Figure 9.1 Conservation of vertebrate β-arrestins (βarrs). Phylogenetic and synteny analyses show the significant conservation of vertebrate arrestins. However, the phylogenetic analyses clearly indicate that βarrs from aquatic species are somewhat distinct from mammalian βarrs. Upper panel: Comparison of βarr1 (*Arrb1*) in different species. On the chromosomal level, evolutionary conservation of *βarr1* can be seen. The fragmented appearance of the *Xenopus* and zebrafish locus is due to the partially assembled genomes in these species. Lower panels: comparison of βarr2 (*Arrb2*). Note that in zebrafish, the *βarr2* gene is duplicated.

with reduced expression of the glucocorticoid receptor.[9,10] Consistent with these findings, mice deficient in glucocorticoid receptors die perinatally of respiratory failure,[11] similar to the βarr1/2 double KO mice. Furthermore, glucocorticoid receptors in the lung epithelium have been implicated in the production of surfactant.[12] At present, however, little is known about the signaling defects involved in the perinatal lethality of βarr1/2 double KO mice. Most studies to date regarding the function of βarrs during embryogenesis have been performed in lower model organisms.

2. β-ARRESTINS IN MODEL ORGANISMS

Most studies aiming to understand the function of βarrs during embryogenesis have been undertaken in *Xenopus*, zebrafish, and flies. All these models feature at least one *βarr* gene (Fig. 9.1). While *Drosophila* exhibits a single *βarr* gene, which is named *kurtz*,[13] two *βarrs* have been reported in *Xenopus*.[14,15] The zebrafish genome harbors three βarrs, namely *βarr1*, *βarr2a*, and *βarr2b* due to *βarr2* gene duplication. So far, however, only two of the three zebrafish βarrs have been analyzed, and it is not clear if the two *βarr2* genes differ functionally. Despite the duplication, βarrs in lower vertebrates are closely related to their mammalian homologs, as can be seen by phylogenetic comparison (Fig. 9.1). In addition, the genomic arrangement of the *βarr* locus is rather conserved between species. Thus, aquatic models, such as frogs and zebrafish, can be considered valid models for the analysis of βarr function in embryonic development.

This comes in handy, particularly as the ease of manipulation and rapid generation of loss-of-function animals make *Xenopus* and zebrafish very attractive model organisms for developmental studies. The embryos of both frogs and zebrafish develop extrauterine, with accelerated development during early embryogenesis. Both species produce embryos in great numbers, which by simple injection of nucleic acids can be used for knockdown as well as overexpression studies. Functional loss- or gain-of-function studies can thus be realized in a time frame similar to cell culture studies but with the advantage of an *in vivo* context. Interestingly, loss of single βarrs in aquatic models does result in developmental phenotypes (see below; Refs. 14–19). This sets *Xenopus* and zebrafish apart from mice, where deletion of both βarrs is required to interfere with organogenesis.[9,10] One explanation for this discrepancy might be that all studies in zebrafish and *Xenopus* so far have been done using transient knockdown approaches rather than complete loss-of-function technologies. Knockdowns do not result in 100% depletion of protein. Hence, residual protein in knockdown animals may be sufficient to prevent the activation of compensating mechanisms by another βarr isoform. In KO mice, on the other hand, the targeted gene is not expressed at all and the developing embryo will try to activate molecular bypasses such as compensation by other genes of the same protein family. Furthermore, both *Xenopus* and zebrafish carry maternal protein from their yolk. In most cases, knockdown embryos are therefore indistinguishable from controls up to gastrulation. Knockdown may be observed only at the point where embryonic protein synthesis outpaces maternally supplied protein content. To date, no loss-of-function mutant for either βarr in aquatic models has been reported.

3. THE FUNCTION OF ARRESTINS IN INVERTEBRATES

A growing number of studies in *Drosophila* suggest that the fly ortholog of vertebrate βarr is only partially functionally conserved from invertebrates to vertebrates. *kurtz*, the only nonvisual βarr in *Drosophila*, is ubiquitously expressed during early embryogenesis. At later stages, it becomes more localized to the larval central nervous system and fat bodies, a tissue fulfilling a similar function as the vertebrate liver. Similar to βarr2b knockdown in zebrafish[18] and the βarr1/2 double KO in mice,[9] loss of *kurtz* is incompatible with normal embryonic development.[13] Fly embryos deficient in *kurtz* are immobile and do not survive the larval–pupal transition. Prior to their death, the *kurtz* mutants display melanotic tumors, which are formed on the surface of their abnormal fat bodies. Interestingly, reconstitution of *kurtz* exclusively in the central nervous system rescues the mutant lethality and phenotypes, although not fully.[13] Furthermore, maternal loss of *kurtz* was found to interfere with receptor tyrosine kinase signaling toward MAP kinases as well as Toll/NFκB signaling.[20] By binding to inactive ERK, *kurtz* sequesters ERK and prevents its activation. This is in contrast to vertebrates, where the βarr-mediated assembly of signaling complexes propagates ERK activation and signaling.[21]

On the other hand, very akin to mammalian βarr1 and 2,[22] *kurtz* limits NFκB signaling by association with cactus, the *Drosophila* homolog of IκB.[20] Another striking difference with the vertebrate actions of βarr on signal transduction can be found in Smo-mediated Hedgehog (Hh) signal transduction. *kurtz* appears to limit Hh signaling, rather than promoting it as seen in different vertebrate systems.[17,18,23] Loss of *kurtz* has no effect on Hh signaling, whereas ectopic expression of *kurtz* phenocopies Hh loss of function.[24] Flies overexpressing *kurtz* develop smaller wings and display downregulation of *patched*, the negative feedback loop target of the Hh pathway. This is because *kurtz* promotes the internalization of Smo by binding to its C tail,[25] subsequently leading to its proteasomal degradation.[24] Interestingly, this action is independent of GPCR kinases (GRKs), which were shown to facilitate Hh signaling in vertebrates.[26,27] In fact, combining loss of Gprk2, the *Drosophila* homolog of vertebrate GRKs 4–6, with ectopic expression of *kurtz* exacerbates the Hh phenotype.[24]

In keeping with its inhibitory roles in signal transduction during fly development, *kurtz* has also been reported to attenuate Notch signaling.[28] It has been suggested that *kurtz* forms a complex with Notch and Deltex, which induces ubiquitination and degradation of Notch.[28] Although these examples substantiate the complexity of βarr function, they also underline the importance of further studies to fully define the functional differences as well as overlaps of βarrs in flies and vertebrates.

4. β-ARRESTINS IN VERTEBRATE DEVELOPMENT

4.1. βarr1 in hematopoiesis

Despite the common characteristic of βarrs to act as scaffolding proteins in receptor signaling, βarr1 and 2 also have distinct functions. This becomes very apparent during embryonic development. Experiments in zebrafish using a morpholino (MO)-based knockdown strategy revealed an unanticipated role for βarr1, but not for βarr2.[19] Depletion of βarr1 protein in zebrafish embryos selectively interferes with early hematopoiesis. This is especially interesting, as βarr1 appears to be uniformly expressed in the early embryo and not just in hematopoietic lineages.[19] Knockdown of βarr1, but not βarr2, markedly decreases the expression of genes that are essential for blood development, such as *gata1* and *scl* (*stem cell leukemia hematopoietic transcription factor*).[29,30] Hh target genes on the other hand, for which expression depends on βarr2,[18] are unchanged when βarr1 is lost. Microarray analysis revealed that βarr1 depletion is reflected in the reduction of *cdx4* and certain *hox* genes that are associated with blood development.[31] Coinjection of βarr1 MO and mRNA encoding *cdx4*, *hox4a*, or *hox9a* could in turn rescue the βarr1 knockdown phenotype.

Subsequent yeast-2-hybrid experiments uncovered that βarr1 interacts with Yin Yang 1 (YY1), a protein that was originally identified as a human GLI–Krüppel-related zinc finger protein and thought to act as a transcriptional repressor.[32] However, as it turns out, YY1 is able to modulate transcription in different ways. Depending on the respective context, YY1 can function as a repressor,[32] as an activator,[33] or as a recruitment factor for other proteins such as Polycomb repressors, modulating transcriptional activity.[34] In addition, binding of YY1 to certain DNA sequences may result in disruption of binding sites for other DNA-interacting molecules and can even change the conformation of DNA.[35] Interestingly, *hox* gene expression is regulated by Polycomb protein binding to *hox* promoters.[36] Zebrafish βarr1, like mammalian βarr1,[37] shows both cytoplasmic and nuclear localization. In the above study, it was shown that βarr1 acts as a positive replacement factor of YY1. When βarr1 interaction is lost, YY1 is preferentially localized to the nucleus of transfected cells and zebrafish embryos where it binds to promoter regions of *cdx4* and *hox* factors and causes the recruitment of Polycomb repressors. As a result, the transcription of *hox* genes in halted and blood development is hindered. Further studies investigating if βarr1 has a general impact on Polycomb-mediated gene repression would be of high interest. It would also be worthwhile to test whether the reported actions are restricted to embryos of certain embryonic developmental stages or if the described actions of βarr1 have an impact on embryonic or even adult stem cells as well.

4.2. βarr2 in Wnt signaling

The Wnt family of secreted lipoglycoproteins represents an important class of signaling molecules in embryonic development as well as in cancer and aging. This family of proteins modulates cell fate by changing transcriptional activity, cell differentiation, proliferation, migration, and polarization. Vertebrate genomes encode for more than 10 different Wnt proteins, which depend on temporal or spatial contexts trigger three different signaling scenarios (reviewed in Ref. 38). Canonical Wnt signaling functions through the stabilization of cytosolic β-catenin that translocates to the nucleus. In the nucleus, β-catenin activates transcription factors of the Tcf/Lef family.[39] Noncanonical Wnt pathways, on the other hand, do not rely on β-catenin. These can be further subdivided into two cascades. Noncanonical Wnt stimulation through the calcium pathway leads to the release of intracellular calcium ions. This in turn activates calcium–calmodulin-dependent kinase II, protein kinase C, or calcineurin.[40] The second noncanonical cascade acts through the activation of the small G proteins RhoA and Rac. Subsequent activation of RhoA kinase and JNK results in cytoskeletal rearrangement.[41] The latter pathway is also referred to the planar cell polarity (PCP) pathway, as it regulates the polarization of cells with respect to the anterior–posterior embryonic axis and results in the so-called convergent extension movements that are indispensable for axis elongation.[42] All three Wnt pathways are transduced by Frizzled proteins, which were first identified in *Drosophila* more than 20 years ago.[43] Frizzleds are seven transmembrane spanning proteins that have been considered as GPCR-like molecules.[44–47] Studies using G protein inhibitors in zebrafish embryos as well as experiments in cultured cells using a chimeric protein consisting of the N-terminal portion of the β1-adrenergic receptor (βAR1) and the C terminal domain of Frizzled1 demonstrated that Frizzled proteins signal through $G_{i/o}$ proteins.[47–49] Similar to classical GPCRs, they also associate with βarrs,[50] which can already be observed during embryogenesis.

4.2.1 *βarr2 and canonical Wnt signaling*

In the absence of Wnt ligands, β-catenin is caught in a multimeric destruction complex. This complex consists of the scaffolding protein axin, APC, and glycogen synthase kinase 3 (GSK3),[51,52] which phosphorylates β-catenin and thereby primes it for proteasomal degradation.[53] However, upon binding of Wnt to Frizzled and its coreceptor proteins LRP5/6[54], the cytoplasmic protein disheveled (Dsh) is recruited to the cell membrane.[55] This in turn facilitates the disassembly of the destruction complex. β-Catenin is then stabilized and can translocate to the nucleus.[56–58] Several groups have reported that βarr2 can associate with Dsh proteins.[14,50,59] Thus, the notion that βarr2 also functions in Wnt/beta-catenin signaling is not surprising.

A classical readout for canonical Wnt signaling is the axis duplication assay in *Xenopus* embryos.[60] Injection of *Wnt8a* mRNA into two ventral blastomeres at the four-cell stage induces a second organizer and consequently two body axes. Simultaneous knockdown of βarr2, however, is sufficient to block secondary axis formation.[14] It also inhibits axis duplication induced by a Dsh construct lacking the DEP domain, which has been shown to activate β-catenin signaling.[61] In addition, Wnt target genes are less transcribed when embryos are depleted for βarr2. Complementary experiments in βarr2 KO mouse embryonic fibroblast (MEFs) further helped to build a hierarchy of Wnt signaling and βarr2.[16] When KO MEFs are stimulated with Wnt3a, β-catenin accumulation is inhibited and transcriptional activity as measured using the TOPFlash assay[62] is reduced. Thus, βarr2 must act upstream of the destruction complex. At the membrane, Wnt signal transduction is initiated by the phosphorylation of LRP6 at PPPSP motifs by GSK3 and GRKs 5 and 6.[63,64] Wnt3a-treated βarr2 KO MEFs displayed similar levels of PPPSP phosphorylation as wild-type cells, implying βarr2 acts downstream of LRP6. Thus, βarr2 is a positive modulator of canonical Wnt signaling during development (Fig. 9.2, left panel).

4.2.2 *βarr2 and noncanonical Wnt signaling*

Interestingly, βarr2 actions are not limited to canonical Wnt/β-catenin signaling. It is also involved in a noncanonical Wnt pathway (Fig. 9.2, right

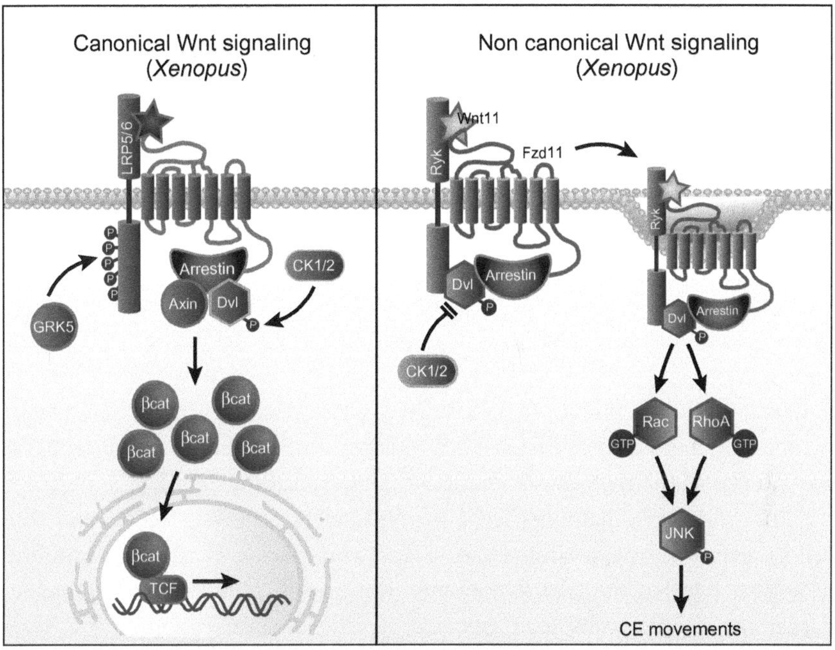

Figure 9.2 β-Arrestin 2 and vertebrate Wnt signaling. βarr2 is a regulator of both canonical (left panel) as well as noncanonical (right panel) Wnt signaling. The studies summarized in this model were mostly conducted in *Xenopus*. Other evidence originates from experiments using βarr KO MEFs. (See Color Insert.)

panel). During gastrulation, βarr2 is expressed in cells of the dorsal marginal zone (DMZ) of *Xenopus* embryos.[15] This is a group of cells that undergoes extensive convergence and extension (CE) movements, during which cells from lateral parts of the embryo move toward the midline.[65] At the midline, they intercalate, by which a change in cell shape is induced and the anterior–posterior axis becomes longer through this process. This elongation persists during the following neurulation and organogenesis stages, when βarr2 is expressed in presumptive neural tissues.[15] At later stages, it becomes localized to the brain, retina, the branchial arches, and the otic vesicle.[15]

The molecular basis for CE movements was first investigated in invertebrates. It depends on the PCP pathway,[66,67] which in vertebrates is analogous to noncanonical Wnt/JNK signaling. MO-mediated knockdown of βarr2 in *Xenopus* interferes with this pathway, as embryos with reduced βarr2 expression are shorter than wild-type embryos and display a bent body curvature.[15] Strikingly, ectopic expression of βarr2 induces the same phenotype, implying that βarr2 levels must be very delicately regulated or the pathway is deranged. Closer analysis revealed that βarr2 is localized to cell membranes of DMZ cells while they undergo CE movements. In DMZ explants, βarr2 was shown to activate RhoA and JNK via interaction with Dsh. This requires βarr2 and clathrin-coated pit-mediated internalization of Frizzled proteins and Dsh.[68] Interestingly, Wnt11 stimulation of Frizzled7 in the presence of Dsh and βarr2 is not sufficient to induce internalization. However, if Frizzled forms a ternary complex with Wnt11 and the transmembrane protein Ryk, internalization and signaling are restored.[68] When βarr2 is then depleted, Dsh stays at the cell membrane. In addition, knockdown of Ryk causes CE defects that can be rescued by overexpression of βarr2, while CE phenotypes induced by ectopic expression of Ryk can be reversed by depletion of βarr2. This underscores that certain components of the Wnt/JNK pathway must be fine-tuned, including βarr2.

The Wnt/JNK pathway leads to cytoskeletal rearrangement, and thus cell movement, by activating RhoA, Rac, and Cdc42. Further experiments in MEFs revealed that βarr2-mediated noncanonical Wnt signal transduction is also mediated by Rac, but not by Cdc42. Consequently, the βarr2 extension phenotype could be rescued by constitutively active RhoA or Rac1, but not by active Cdc42. Neither was it possible to reverse the effect by overexpression of canonical Wnt target genes, indicating that the observed effects of βarr2 on CE movements are not due to its role in canonical Wnt signaling, which can counteract the Wnt/JNK pathway.

4.3. βarrs in Hh signaling

The Hh signaling pathway is one of the key regulators of embryonic development, crucial for many aspects of organogenesis and growth control. Hh signaling is transduced by the atypical GPCR Smo (reviewed in Ref. 69). Binding of Hh ligand to its transmembrane receptor Patched 1 (Ptch1) relieves Ptch1 inhibition of Smo and allows Smo to signal to a complex of proteins that includes the transcription factors Ci (in *Drosophila*) and Gli (in vertebrates).[70] Data from mammalian cell culture, mice, and zebrafish have indicated that Smo trafficking into the primary cilium by intraflagellar transport (IFT) particles is an essential step in Hh downstream signaling (reviewed in Refs. 71,72). Smo translocation results in the accumulation of transcriptionally active Gli at the tip of the cilium, from which is then transported out of the cilium to promote Hh target gene expression[71] (Fig. 9.3).

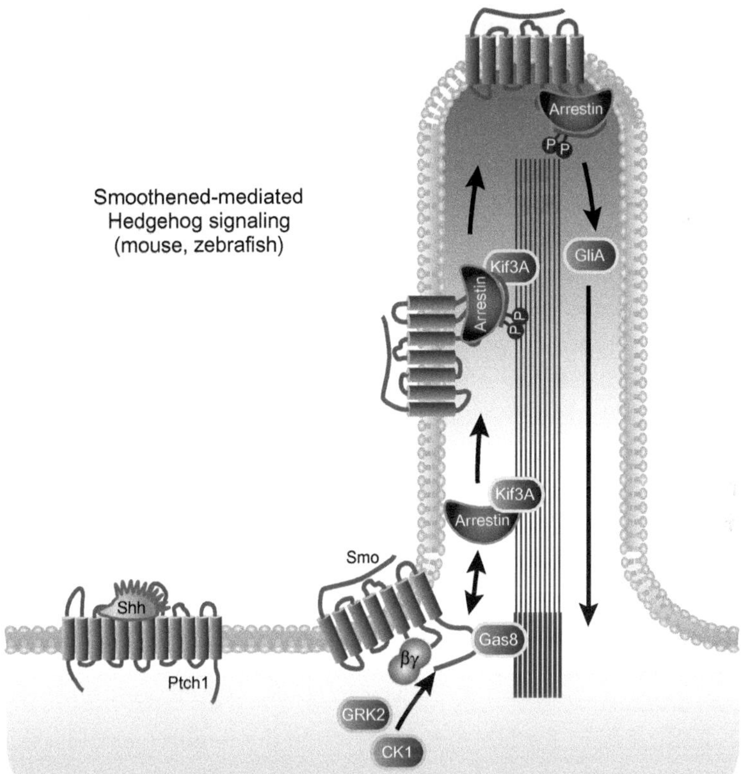

Figure 9.3 β-Arrestin 2 and vertebrate Hedgehog signaling. Vertebrate Hedgehog signaling depends greatly on the translocation of required signaling molecules to primary cilia. Importantly, βarr2 physically links Smoothened to a kinesin motor and thus facilitates the ciliary shuttling of Smoothened. (See Color Insert.)

4.3.1 βarr2 in Hh signaling

Similar to classic GPCRs, phosphorylation of active Smo by GRK2 and recruitment of βarr2 to the plasma membrane results in Smo internalization through clathrin-mediated endocytosis in nonciliated HEK293 cells.[73] Interestingly, both GRK2 and βarr2 actively promote Hh signaling, as GRK2 and βarr2 knockdown in zebrafish embryos recapitulates Hh loss-of-function phenotypes.[17,18,27] Similar to the typical phenotypes of *smu* mutant fish, βarr2 and GRK2 morphants display a ventrally curved body and misshaped somites at 24 hours post fertilization (hpf).[18,27] βarr2 and GRK2 morphants also show impaired muscle development, manifest by reduced numbers of muscle pioneer cells and slow muscle fibers,[17,18,27] whose formation requires Shh secretion from the notochord.[74] These phenotypes are accompanied by reduced expression of Hh downstream target genes as measured by *in situ* hybridization experiments. βarr2 MO effects can be rescued by expression of *βarr2* mRNA, by constitutive activation of the Hh pathway using a dominant negative form of *PKA* mRNA, or by injecting a MO against su(fu), a Hh function inhibitor. Additional phenotypes of βarr2 morphants become apparent later in development, including partial cylopia and reduced head size as well as failure of the optic nerve to form the optic chiasm at 48 hpf. At 72 hpf, βarr2 morphants lack craniofacial muscle and floor plate development and at 120 hpf they fail to develop normal cartilage and pectoral fin.[18]

Consistent with their positive roles in this cascade, casein kinase CK1α and GRK2-mediated phosphorylation of Smo facilitates βarr2 interaction with Smo, preferentially near or at the basal body of the primary cilium.[75] Upon their interaction, βarr scaffolds the kinesin motor protein Kif3A,[59] which has been reported to be essential for Hh signaling,[76–78] and all three proteins translocate into the primary cilium.[23] Inconsistent with their divergent roles during zebrafish development, Kovacs *et al.*[23] reported that in mammalian cell cultures, both βarr1 and βarr2 are required for the interaction between Smo with Kif3A, Smo translocation into the primary cilium, and the subsequent activation of Gli. The reason for this difference between zebrafish and mammalian cells might be explained by their distinct dependence on the cilium for Hh signaling. While the requirement for cilia and IFT protein for proper Hh signaling in mice is well documented,[76,77,79,80] such dependence in zebrafish remains uncertain or not fully understood. A study in zebrafish lacking cilia has suggested a dampening of Hh signaling,[81] while another reported no signaling effect in IFT mutant and morphant fish.[82] Thus, whether and to what extent both βarr1 and βarr2 would act downstream of the mammalian Smo *in vivo* is still not fully understood.

Moreover, *kurtz* in *Drosophila* and βarr2 in vertebrates apparently function differently regarding Hh signaling. As discussed earlier, *kurtz* has a limiting effect on Hh pathway, whereas vertebrate βarr2 is a positive regulator. This phenomenon probably represents more fundamental differences in the mechanisms underlying Hh signaling between these species, as Hh signaling depends on primary cilia only in vertebrates. Interestingly, many invertebrates such as flies possess cilia only on sensory neurons. In line with this notion, it would be fascinating to uncover new roles of βarr2 downstream of the vertebrate Smo. Perhaps Smo can signal through yet unidentified signaling molecules based on the ability of βarr2 to recruit secondary scaffolds, similar to the case for many other GPCRs,[21] even independent of the actions of Gli. For example, βarr2 is localized to the primary cilium of MEFs derived from wild-type mice and its absence in MEFs derived from βarr2 KO mice has been reported to cause uncontrolled cell proliferation and defects in ciliogenesis.[83] Whether this effect is regulated by Smo has not been examined in detail, and thus, the mechanism of this putative role of βarr2 is not clear yet. It could potentially involve the scaffolding of molecules that are necessary for cell cycle regulation.

4.3.2 βarr1 in Hh signaling

At the time the role of βarr2 in the Hh pathway was revealed in zebrafish, there was no information regarding the zebrafish homolog of βarr1. Therefore, a potential role for βarr1 in Hh signaling was not studied at the time.[18] Five years later, however, Yue *et al.*[19] published a study on zebrafish βarr1 and its developmental phenotypes. In this study, βarr1 depletion attenuated hematopoiesis but did not affect the expression of Hh target genes as measured by *in situ* hybridization. These findings suggest that βarr1 and βarr2 might play distinct roles during embryonic development. However, an influence of βarr1 depletion on Hh signaling cannot be fully ruled out because *in situ* hybridization may not be sensitive enough to detect mild changes in gene expression and other measures were not explored.

Despite the results reported by Yue *et al.* an interesting but negative role in Shh-mediated cell proliferation has recently been uncovered for βarr1 in mammalian neuronal development.[84] As previously shown, βarr1 translocates to the nucleus of transfected fibroblasts in response to opioid receptor activation.[37] In the nucleus, βarr1 is enriched at specific promoters, such as that of the cyclin-dependent kinase (CDK) inhibitor *p27* and *c-fos*. Consequently, βarr1 facilitates the recruitment of histone acetyltransferase p300, resulting in local chromatin reorganization and transcription of these genes.[37] To date, no apparent phenotype that could be explained by defects in cell cycle control has been reported for βarr1 KO mice, suggesting that the role of βarr1 on gene expression *in vivo* is either limited or compensated. However, a potential physiological role for βarr1 may still exist in cell cycle regulation as reported by Parathath *et al.*[84] In this study, *ex vivo* Shh stimulation of cerebellar granule neuron precursors (CGNPs) isolated from postnatal day 5 (P5) mice upregulated βarr1, but not βarr2, levels and facilitated βarr1 translocation to the nucleus of these developing cells.[84] There, βarr1 interacts with P300 and the transcription factor CREB selectively at the promoter of the *p27* gene, causing enhanced *p27* transcription. This Shh-mediated transcription is accompanied by reduced cell proliferation and drives CGNPs cell cycle exit. Accordingly, blocking βarr1 with shRNA leads to the opposite effect and increased cell proliferation.[84] Thus, in addition to the well-known negative feedback of Shh signaling through increased transcription of the Smo inhibitor Ptch1, Shh also induces βarr1 translocation to the nucleus, where it facilitates transcription of a CDK inhibitor, thereby negatively regulating mitogenic Shh signaling.

5. CONCLUSIONS

When βarrs were first identified more than 20 years ago, they were believed to simply put a halt on G protein-coupled receptor signaling. Since then βarrs have been studied extensively. Approaches with purified proteins, heterologous cell systems, and finally genetically engineered mice illustrated that βarrs are in fact multifunctional integrators of many different signaling cascades. Apart from their function in the regulation of canonical GPCR signaling, βarrs modulate the signaling function of atypical 7TM proteins such as Smo and Frizzled proteins. This function has gained intense interest in the arrestin community. Experiments using frogs, flies, and zebrafish have further emphasized the importance of βarrs in the Hh and Wnt pathways. Even more, these reports have emphasized that arrestins are important signaling molecules during embryonic development. So far, we know that βarrs steer hematopoiesis and regulate cell cycle progression, in part because they are regulated themselves. In addition, βarrs drive lung morphogenesis and muscle and cartilage development. Not all findings obtained in lower model organisms can be related to mammals. But nevertheless, it has become apparent that βarrs have multiple functions in cellular signaling and thus govern many physiological as well as pathological processes in the embryo.

ACKNOWLEDGMENTS

Work from the authors' laboratories reviewed in this chapter was supported in part by grants from the National Institutes of Health: NS-019576 and MH-073853 (M. G. C.). T. E. was supported by a fellowship from the Machiah Foundation. Melanie Philipp is the recipient of a Marie Curie Reintegration Grant (268333) and a grant by the Deutsche Stiftung für Herzforschung (F/09/11).

REFERENCES

1. Shenoy SK, Lefkowitz RJ. Beta-Arrestin-mediated receptor trafficking and signal transduction. *Trends Pharmacol Sci.* 2011;32:521–533.
2. Lohse MJ, Benovic JL, Codina J, Caron MG, Lefkowitz RJ. Beta-Arrestin: a protein that regulates beta-adrenergic receptor function. *Science.* 1990;248:1547–1550.

3. Parruti G, Peracchia F, Sallese M, et al. Molecular analysis of human beta-arrestin-1: cloning, tissue distribution, and regulation of expression. Identification of two isoforms generated by alternative splicing. *J Biol Chem*. 1993;268:9753–9761.
4. Bohn LM, Lefkowitz RJ, Gainetdinov RR, Peppel K, Caron MG, Lin FT. Enhanced morphine analgesia in mice lacking beta-arrestin 2. *Science*. 1999;286:2495–2498.
5. Conner DA, Mathier MA, Mortensen RM, et al. Beta-Arrestin1 knockout mice appear normal but demonstrate altered cardiac responses to beta-adrenergic stimulation. *Circ Res*. 1997;81:1021–1026.
6. Lefkowitz RJ, Shenoy SK. Transduction of receptor signals by beta-arrestins. *Science*. 2005;308:512–517.
7. Lefkowitz RJ, Rajagopal K, Whalen EJ. New roles for beta-arrestins in cell signaling: not just for seven-transmembrane receptors. *Mol Cell*. 2006;24:643–652.
8. Hara MR, Kovacs JJ, Whalen EJ, et al. A stress response pathway regulates DNA damage through beta2-adrenoreceptors and beta-arrestin-1. *Nature*. 2011;477:349–353.
9. Zhang M, Liu X, Zhang Y, Zhao J. Loss of beta arrestin1 and beta arrestin2 contributes to pulmonary hypoplasia and neonatal lethality in mice. *Dev Biol*. 2010;339:407–417.
10. Zhang M, Teng H, Shi J, Zhang Y. Disruption of beta-arrestins blocks glucocorticoid receptor and severely retards lung and liver development in mice. *Mech Dev*. 2011;128:368–375.
11. Schmid W, Cole TJ, Blendy JA, Schutz G. Molecular genetic analysis of glucocorticoid signalling in development. *J Steroid Biochem Mol Biol*. 1995;53:33–35.
12. Manwani N, Gagnon S, Post M, et al. Reduced viability of mice with lung epithelial-specific knockout of glucocorticoid receptor. *Am J Respir Cell Mol Biol*. 2010;43:599–606.
13. Roman G, He J, Davis RL. Kurtz, a novel nonvisual arrestin, is an essential neural gene in Drosophila. *Genetics*. 2000;155:1281–1295.
14. Bryja V, Gradl D, Schambony A, Arenas E, Schulte G. Beta-arrestin is a necessary component of Wnt/beta-catenin signaling in vitro and in vivo. *Proc Natl Acad Sci USA*. 2007;104:6690–6695.
15. Kim GH, Han JK. Essential role for beta-arrestin 2 in the regulation of Xenopus convergent extension movements. *EMBO J*. 2007;26:2513–2526.
16. Bryja V, Schambony A, Cajanek L, Dominguez I, Arenas E, Schulte G. Beta-arrestin and casein kinase 1/2 define distinct branches of non-canonical WNT signalling pathways. *EMBO Rep*. 2008;9:1244–1250.
17. Evron T, Philipp M, Lu J, et al. Growth Arrest Specific 8 (Gas8) and G protein-coupled receptor kinase 2 (GRK2) cooperate in the control of Smoothened signaling. *J Biol Chem*. 2011;286:27676–27686.
18. Wilbanks AM, Fralish GB, Kirby ML, Barak LS, Li YX, Caron MG. Beta-arrestin 2 regulates zebrafish development through the hedgehog signaling pathway. *Science*. 2004;306:2264–2267.
19. Yue R, Kang J, Zhao C, et al. Beta-arrestin1 regulates zebrafish hematopoiesis through binding to YY1 and relieving polycomb group repression. *Cell*. 2009;139:535–546.
20. Tipping M, Kim Y, Kyriakakis P, Tong M, Shvartsman SY, Veraksa A. Beta-Arrestin Kurtz inhibits MAPK and Toll signalling in Drosophila development. *EMBO J*. 2010;29:3222–3235.
21. DeWire SM, Ahn S, Lefkowitz RJ, Shenoy SK. Beta-arrestins and cell signaling. *Annu Rev Physiol*. 2007;69:483–510.
22. Witherow DS, Garrison TR, Miller WE, Lefkowitz RJ. Beta-Arrestin inhibits NF-kappaB activity by means of its interaction with the NF-kappaB inhibitor IkappaBalpha. *Proc Natl Acad Sci USA*. 2004;101:8603–8607.
23. Kovacs JJ, Whalen EJ, Liu R, et al. Beta-arrestin-mediated localization of smoothened to the primary cilium. *Science*. 2008;320:1777–1781.

24. Molnar C, Ruiz-Gomez A, Martin M, Rojo-Berciano S, Mayor F, de Celis JF. Role of the Drosophila non-visual β-arrestin kurtz in hedgehog signalling. *PLoS Genet.* 2011;7: e1001335.
25. Li S, Chen Y, Shi Q, Yue T, Wang B, Jiang J. Hedgehog-regulated ubiquitination controls smoothened trafficking and cell surface expression in Drosophila. *PLoS Biol.* 2012;10:e1001239.
26. Meloni AR, Fralish GB, Kelly P, et al. Smoothened signal transduction is promoted by G protein-coupled receptor kinase 2. *Mol Cell Biol.* 2006;26:7550–7560.
27. Philipp M, Fralish GB, Meloni AR, et al. Smoothened signaling in vertebrates is facilitated by a G protein-coupled receptor kinase. *Mol Biol Cell.* 2008;19:5478–5489.
28. Mukherjee A, Veraksa A, Bauer A, Rosse C, Camonis J, Artavanis-Tsakonas S. Regulation of Notch signalling by non-visual beta-arrestin. *Nat Cell Biol.* 2005;7:1191–1201.
29. Detrich 3rd HW, Kieran MW, Chan FY, et al. Intraembryonic hematopoietic cell migration during vertebrate development. *Proc Natl Acad Sci USA.* 1995;92:10713–10717.
30. Liao EC, Paw BH, Oates AC, Pratt SJ, Postlethwait JH, Zon LI. SCL/Tal-1 transcription factor acts downstream of cloche to specify hematopoietic and vascular progenitors in zebrafish. *Genes Dev.* 1998;12:621–626.
31. Davidson AJ, Ernst P, Wang Y, et al. cdx4 mutants fail to specify blood progenitors and can be rescued by multiple hox genes. *Nature.* 2003;425:300–306.
32. Shi Y, Seto E, Chang LS, Shenk T. Transcriptional repression by YY1, a human GLI-Kruppel-related protein, and relief of repression by adenovirus E1A protein. *Cell.* 1991;67:377–388.
33. Seto E, Shi Y, Shenk T. YY1 is an initiator sequence-binding protein that directs and activates transcription in vitro. *Nature.* 1991;354:241–245.
34. Srinivasan L, Atchison ML. YY1 DNA binding and PcG recruitment requires CtBP. *Genes Dev.* 2004;18:2596–2601.
35. Natesan S, Gilman MZ. DNA bending and orientation-dependent function of YY1 in the c-fos promoter. *Genes Dev.* 1993;7:2497–2509.
36. Brown JL, Mucci D, Whiteley M, Dirksen ML, Kassis JA. The Drosophila Polycomb group gene pleiohomeotic encodes a DNA binding protein with homology to the transcription factor YY1. *Mol Cell.* 1998;1:1057–1064.
37. Kang J, Shi Y, Xiang B, et al. A nuclear function of beta-arrestin1 in GPCR signaling: regulation of histone acetylation and gene transcription. *Cell.* 2005;123:833–847.
38. Rao TP, Kuhl M. An updated overview on Wnt signaling pathways: a prelude for more. *Circ Res.* 2010;106:1798–1806.
39. Korinek V, Barker N, Willert K, et al. Two members of the Tcf family implicated in Wnt/beta-catenin signaling during embryogenesis in the mouse. *Mol Cell Biol.* 1998;18:1248–1256.
40. Saneyoshi T, Kume S, Amasaki Y, Mikoshiba K. The Wnt/calcium pathway activates NF-AT and promotes ventral cell fate in Xenopus embryos. *Nature.* 2002;417:295–299.
41. Habas R, Dawid IB, He X. Coactivation of Rac and Rho by Wnt/Frizzled signaling is required for vertebrate gastrulation. *Genes Dev.* 2003;17:295–309.
42. Yamanaka H, Moriguchi T, Masuyama N, et al. JNK functions in the non-canonical Wnt pathway to regulate convergent extension movements in vertebrates. *EMBO Rep.* 2002;3:69–75.
43. Vinson CR, Conover S, Adler PN. A Drosophila tissue polarity locus encodes a protein containing seven potential transmembrane domains. *Nature.* 1989;338:263–264.
44. Adler PN, Vinson C, Park WJ, Conover S, Klein L. Molecular structure of frizzled, a Drosophila tissue polarity gene. *Genetics.* 1990;126:401–416.
45. Josefsson LG. Evidence for kinship between diverse G-protein coupled receptors. *Gene.* 1999;239:333–340.

46. Sheldahl LC, Park M, Malbon CC, Moon RT. Protein kinase C is differentially stimulated by Wnt and Frizzled homologs in a G-protein-dependent manner. *Curr Biol.* 1999;9:695–698.
47. Slusarski DC, Corces VG, Moon RT. Interaction of Wnt and a Frizzled homologue triggers G-protein-linked phosphatidylinositol signalling. *Nature.* 1997;390:410–413.
48. DeCostanzo AJ, Huang XP, Wang HY, Malbon CC. The Frizzled-1/beta(2)-adrenergic receptor chimera: pharmacological properties of a unique G protein-linked receptor. *Naunyn Schmiedebergs Arch Pharmacol.* 2002;365:341–348.
49. Koval A, Purvanov V, Egger-Adam D, Katanaev VL. Yellow submarine of the Wnt/Frizzled signaling: submerging from the G protein harbor to the targets. *Biochem Pharmacol.* 2011;82:1311–1319.
50. Chen W, ten Berge D, Brown J, et al. Dishevelled 2 recruits beta-arrestin 2 to mediate Wnt5A-stimulated endocytosis of Frizzled 4. *Science.* 2003;301:1391–1394.
51. Fagotto F, Jho E, Zeng L, et al. Domains of axin involved in protein-protein interactions, Wnt pathway inhibition, and intracellular localization. *J Cell Biol.* 1999;145:741–756.
52. Hart MJ, de los Santos R, Albert IN, Rubinfeld B, Polakis P. Downregulation of beta-catenin by human Axin and its association with the APC tumor suppressor, beta-catenin and GSK3 beta. *Curr Biol.* 1998;8:573–581.
53. van Noort M, Meeldijk J, van der Zee R, Destree O, Clevers H. Wnt signaling controls the phosphorylation status of beta-catenin. *J Biol Chem.* 2002;277:17901–17905.
54. Tamai K, Semenov M, Kato Y, et al. LDL-receptor-related proteins in Wnt signal transduction. *Nature.* 2000;407:530–535.
55. Axelrod JD, Miller JR, Shulman JM, Moon RT, Perrimon N. Differential recruitment of Dishevelled provides signaling specificity in the planar cell polarity and Wingless signaling pathways. *Genes Dev.* 1998;12:2610–2622.
56. Aberle H, Bauer A, Stappert J, Kispert A, Kemler R. Beta-Catenin is a target for the ubiquitin-proteasome pathway. *EMBO J.* 1997;16:3797–3804.
57. Huber O, Korn R, McLaughlin J, Ohsugi M, Herrmann BG, Kemler R. Nuclear localization of beta-catenin by interaction with transcription factor LEF-1. *Mech Dev.* 1996;59:3–10.
58. Papkoff J, Rubinfeld B, Schryver B, Polakis P. Wnt-1 regulates free pools of catenins and stabilizes APC-catenin complexes. *Mol Cell Biol.* 1996;16:2128–2134.
59. Chen W, Hu LA, Semenov MV, et al. Beta-Arrestin1 modulates lymphoid enhancer factor transcriptional activity through interaction with phosphorylated dishevelled proteins. *Proc Natl Acad Sci USA.* 2001;98:14889–14894.
60. Kuhl M, Pandur P. Dorsal axis duplication as a functional readout for Wnt activity. *Methods Mol Biol.* 2008;469:467–476.
61. Moriguchi T, Kawachi K, Kamakura S, et al. Distinct domains of mouse dishevelled are responsible for the c-Jun N-terminal kinase/stress-activated protein kinase activation and the axis formation in vertebrates. *J Biol Chem.* 1999;274:30957–30962.
62. Molenaar M, van de Wetering M, Oosterwegel M, et al. XTcf-3 transcription factor mediates beta-catenin-induced axis formation in Xenopus embryos. *Cell.* 1996;86:391–399.
63. Chen M, Philipp M, Wang J, et al. G Protein-coupled receptor kinases phosphorylate LRP6 in the Wnt pathway. *J Biol Chem.* 2009;284:35040–35048.
64. Zeng X, Tamai K, Doble B, et al. A dual-kinase mechanism for Wnt co-receptor phosphorylation and activation. *Nature.* 2005;438:873–877.
65. Keller RE, Danilchik M, Gimlich R, Shih J. The function and mechanism of convergent extension during gastrulation of Xenopus laevis. *J Embryol Exp Morphol.* 1985;89 (Suppl.):185–209.
66. Strutt DI, Weber U, Mlodzik M. The role of RhoA in tissue polarity and Frizzled signalling. *Nature.* 1997;387:292–295.

67. Theisen H, Purcell J, Bennett M, Kansagara D, Syed A, Marsh JL. Dishevelled is required during wingless signaling to establish both cell polarity and cell identity. *Development.* 1994;120:347–360.
68. Kim GH, Her JH, Han JK. Ryk cooperates with Frizzled 7 to promote Wnt11-mediated endocytosis and is essential for Xenopus laevis convergent extension movements. *J Cell Biol.* 2008;182:1073–1082.
69. Philipp M, Caron MG. Hedgehog signaling: is Smo a G protein-coupled receptor? *Curr Biol.* 2009;19:R125–R127.
70. Huangfu D, Anderson KV. Signaling from Smo to Ci/Gli: conservation and divergence of Hedgehog pathways from Drosophila to vertebrates. *Development.* 2006;133:3–14.
71. Goetz SC, Anderson KV. The primary cilium: a signalling centre during vertebrate development. *Nat Rev Genet.* 2010;11:331–344.
72. Wilson CW, Stainier DY. Vertebrate Hedgehog signaling: cilia rule. *BMC Biol.* 2010;8:102.
73. Chen W, Ren XR, Nelson CD, et al. Activity-dependent internalization of smoothened mediated by beta-arrestin 2 and GRK2. *Science.* 2004;306:2257–2260.
74. Barresi MJ, Stickney HL, Devoto SH. The zebrafish slow-muscle-omitted gene product is required for Hedgehog signal transduction and the development of slow muscle identity. *Development.* 2000;127:2189–2199.
75. Chen Y, Yue S, Xie L, Pu XH, Jin T, Cheng SY. Dual phosphorylation of suppressor of fused (Sufu) by PKA and GSK3beta regulates its stability and localization in the primary cilium. *J Biol Chem.* 2011;286:13502–13511.
76. Huangfu D, Anderson KV. Cilia and Hedgehog responsiveness in the mouse. *Proc Natl Acad Sci USA.* 2005;102:11325–11330.
77. Huangfu D, Liu A, Rakeman AS, Murcia NS, Niswander L, Anderson KV. Hedgehog signalling in the mouse requires intraflagellar transport proteins. *Nature.* 2003;426:83–87.
78. Liu A, Wang B, Niswander LA. Mouse intraflagellar transport proteins regulate both the activator and repressor functions of Gli transcription factors. *Development.* 2005;132:3103–3111.
79. Haycraft CJ, Banizs B, Aydin-Son Y, Zhang Q, Michaud EJ, Yoder BK. Gli2 and Gli3 localize to cilia and require the intraflagellar transport protein polaris for processing and function. *PLoS Genet.* 2005;1:e53.
80. Qin J, Lin Y, Norman RX, Ko HW, Eggenschwiler JT. Intraflagellar transport protein 122 antagonizes Sonic Hedgehog signaling and controls ciliary localization of pathway components. *Proc Natl Acad Sci USA.* 2011;108:1456–1461.
81. Huang P, Schier AF. Dampened Hedgehog signaling but normal Wnt signaling in zebrafish without cilia. *Development.* 2009;136:3089–3098.
82. Lunt SC, Haynes T, Perkins BD. Zebrafish ift57, ift88, and ift172 intraflagellar transport mutants disrupt cilia but do not affect hedgehog signaling. *Dev Dyn.* 2009;238:1744–1759.
83. Molla-Herman A, Boularan C, Ghossoub R, et al. Targeting of beta-arrestin2 to the centrosome and primary cilium: role in cell proliferation control. *PLoS One.* 2008;3:e3728.
84. Parathath SR, Mainwaring LA, Fernandez LA, Guldal CG, Nahle Z, Kenney AM. Beta-Arrestin-1 links mitogenic sonic hedgehog signaling to the cell cycle exit machinery in neural precursors. *Cell Cycle.* 2010;9:4013–4024.

CHAPTER TEN

The Role of Arrestins in Visual and Disease Processes of the Eye

W. Clay Smith
Department of Ophthalmology, University of Florida, Gainesville, Florida, USA

Contents

1. Visual Arrestin Nomenclature 244
2. The Function of Visual Arrestins in Quenching Phototransduction 244
3. Splice Variants of Arrestin1 245
4. Arrestins in Cone Photoreceptors 246
5. Translocation of Visual Arrestins 247
6. New Roles for Arrestin1 in the Retina 252
   6.1 Microtubules 253
   6.2 Kinases 254
   6.3 Ubiquitin ligases 254
   6.4 Calmodulin 255
   6.5 *N*-ethylmaleimide-sensitive factor 255
   6.6 Enolase1 256
7. Arrestins in Disease Processes in the Eye 257
   7.1 Oguchi disease 257
   7.2 Retinitis pigmentosa 258
   7.3 Autoimmune diseases 258
8. The Future for Visual Arrestins 259
References 259

Abstract

Visual arrestins are well known for their function in quenching the phototransduction process in rods and cones. Perhaps not as well known is their participation in multiple other processes in the normal and disease states of the eye. This chapter covers the range of the known functions of the visual arrestins, beginning with their classical role in quenching light-activated visual pigments. The role of visual arrestins is also reviewed from the perspective of their dynamic mobility whereby they redistribute significantly between the compartments of highly polarized photoreceptor cells. Additional roles of the visual arrestins are also reviewed based on new interacting partners that have been discovered over the past decade. Finally, the contribution of the visual arrestins to diseases of the visual system is explored.

Progress in Molecular Biology and Translational Science, Volume 118
ISSN 1877-1173
http://dx.doi.org/10.1016/B978-0-12-394440-5.00010-3

1. VISUAL ARRESTIN NOMENCLATURE

It is always a good idea to begin any discussion with a clear understanding of the vocabulary being used. Because the visual arrestins were discovered independently by laboratories that were studying different processes, there has been a proliferation of arrestin synonyms. For example, immunologists focusing on the immunogenic properties of arrestin referred to arrestin as "soluble antigen," or "S-antigen," owing to arrestin's ability to induce autoimmune uveoretinitis.[1,2] Physiologists studying the function of an abundant retinal protein called it the "48-kDa protein", based on its apparent molecular mass measured by gel electrophoresis.[3] In both cases, the same protein was being studied and was proved to be the first member of what has become the arrestin family.[4,5] This protein is now designated as arrestin1, the abundant retinal arrestin principally localized in rod photoreceptors, and thus has been variously indicated in the literature as rod arrestin, visual arrestin, S-antigen, and 48-kDa protein. Similarly, arrestin4 was originally identified as being expressed in pinealocytes and cone photoreceptors and is variously known as cone arrestin[6,7] or X-arrestin, because of its genetic localization to the X chromosome.[8,9] To maintain clarity in this chapter, we use the more widely accepted designation of these members of the arrestin family as arrestin1 and arrestin4, occasionally referring to the aggregate of the two proteins as the visual arrestins, owing to their nearly exclusive expression in visual tissues.

2. THE FUNCTION OF VISUAL ARRESTINS IN QUENCHING PHOTOTRANSDUCTION

Much of what is known about arrestin function in visual signaling was learned in retinal photoreceptors, owing to the natural concentration of visual pigments and associated components. Phototransduction, the process of converting the energy of a photon into a change in membrane potential that can be synaptically transmitted to the brain, begins with the absorption of light by the 11-*cis* retinoid chromophore of rhodopsin. In this canonical G-protein-coupled receptor, changes in the configuration of the retinoid to all-*trans* induces conformational changes in rhodopsin that allow the activation of the guanine nucleotide-binding protein, transducin, and the activation of the subsequent phototransduction cascade components that ultimately signal the quantum event (for a recent review of the phototransduction cascade, see Ref. 10).

In addition to permitting interaction with transducin, light activation of rhodopsin also leads to interaction with rhodopsin kinase (also known as G-protein coupled receptor kinase 1), thus initiating the process of rhodopsin quenching and limiting its effective lifetime. Rhodopsin kinase phosphorylates the carboxy-terminal tail of rhodopsin at multiple serine and threonine residues.[11–13] Numerous studies have shown that this phosphorylation of rhodopsin is one of the key regulatory steps that leads to the binding of arrestin1 with high affinity, thus sterically occluding the interaction of rhodopsin with transducin and "arresting" the light signal.[14–17] Although rhodopsin can be phosphorylated at six to eight sites, it is clear that in most species just a few (two or three) are preferred phosphorylation sites.[18–20] The binding affinity of arrestin1 for rhodopsin increases with each phosphorylation,[21] plateauing at three phosphates.[21,22] The determination that arrestin1 binding requires phosphorylation of light-activated rhodopsin clarified one of the important roles of ATP in phototransduction.[23,24]

Once bound by arrestin1, rhodopsin remains in the inactivated state until the all-*trans* retinal chromophore is released from the phospho-opsin and reduced by retinol dehydrogenase.[25] Recent studies have shown that the release of the all-*trans* retinal chromophore from rhodopsin is slowed by arrestin1, suggesting that arrestin1 may also function to regulate the amount of free all-*trans* retinal that forms under bright lighting conditions, thus preventing the accumulation of cytotoxic all-*trans* retinal.[26] Arrestin1 binding of phosphorylated rhodopsin also protects rhodopsin from dephosphorylation by protein phosphatase 2A until the arrestin is released.[27]

3. SPLICE VARIANTS OF ARRESTIN1

In addition to the approximately 400 amino acid version of arrestin1, many species appear to have a shortened version of arrestin1 generated by alternative splicing. The first variant was identified in bovine arrestin1 and termed p44 for its apparent molecular mass of 44 kDa.[28] This arrestin1 variant was demonstrated to be a truncated version of full-length arrestin, formed by replacement of the final exon that coded for 35 amino acids of the arrestin1 C-terminus with an alternative exon that coded for a single alanine residue.[29] Investigations in other species showed that splice variants of arrestin1 could also be identified in mouse and human retinas,[30,31] although the variation occurred at different splicing sites, resulting in different polypeptides. In all cases, however, the splice variants are present in the retina at approximately 5–10% of the amount of the full-length arrestin1. Interestingly, splice variations also occur in arrestin4,[32] and in the nonvisual arrestin2 and arrestin3,[33,34] and are similarly confined to the C-terminal 100 amino acids.

In vitro binding studies showed that like full-length arrestin1, this truncated form was selective for activated rhodopsin. But unlike the full-length version, the splice variant form also binds unphosphorylated, activated rhodopsin with relatively high affinity.[28,35] The p44 variant of arrestin1 has different binding affinities for various forms of differentially phosphorylated rhodopsin.[36] A recent crystal structure of bovine p44 shows that p44 is very similar to the structure of arrestin1, with some differences in the loop structures that likely influence the different selectivities of the truncated splice variant for unphosphorylated, activated rhodopsin.[37] All these studies suggested that the splice variants might serve an auxiliary role in regulating rhodopsin activity. However, a study of transgenic mice engineered to selectively express either the full-length version of arrestin1 or the mouse analogue of the bovine p44 splice form showed that either variant of arrestin1 could quench the photoresponse equally well.[38] Surprisingly, only the full-length version of arrestin1 could quench unphosphorylated, photoactivated rhodopsin in contrast to what was predicted from *in vitro* binding assays. The only other difference observed between the arrestin1 variants was under very bright flash conditions, at which point the flash response in rods expressing the splice variant of arrestin1 returned to baseline much more slowly than in those expressing the full-length arrestin1. Perhaps this difference in quenching relates to the slower rate of translocation seen in the splice variants of arrestin1 (see section 5 on *Arrestin Translocation* below) with the amount of activated rhodopsin exceeding that of the available p44 and exceeding the rate at which it can translocate into the outer segments to quench rhodopsin.

4. ARRESTINS IN CONE PHOTORECEPTORS

Similar to rod photoreceptors, cone photoreceptors also have a phototransduction cascade that is regulated by arrestin family members. The initial discovery of an arrestin family member, arrestin4, in pinealocytes that was also highly expressed in cone photoreceptors led to the hypothesis that rod phototransduction was regulated by arrestin1 and cone phototransduction by arrestin4.[6,7,39] Recent studies, however, have shown that some cone photoreceptors contain both arrestin1 and arrestin4. Evidence of this coexpression was first published in immunohistochemical analyses of primate retinas in which arrestin1 immunoreactivity was clearly detected in S-cones, but not in LM-cones.[40] This finding was more systematically examined and confirmed in a detailed study showing that arrestin1 is present in S-cones, and, surprisingly, is about 50-fold more concentrated in these cones than arrestin4.[41] The cellular distributions of arrestin1 and arrestin4 in these cones are relatively similar with the exception that arrestin4 is more highly concentrated in the synaptic pedicle than arrestin1.

The functional differences between arrestin1 and arrestin4 in cones are not completely deciphered, but several facts are clear. First, cones require at least one arrestin for normal recovery from strong flashes of light as cones without either arrestin1 or arrestin4 saturate at lower light levels and have prolonged recovery phases consistent with extended phosphodiesterase activity.[41] Further, it is clear that either arrestin1 or arrestin4 is sufficient to provide relatively normal response recovery dynamics to cones.[41] This observation actually provides the first definitive evidence that arrestin4 functions in the inactivation of cone visual pigments. What is not clear is whether the two arrestins are fully redundant or have complementary roles. Perhaps the function of the two arrestins in cones is related to the relative stability of the arrestins in complex with the cone opsins, as arrestin4 has a much higher rate of dissociation from activated cone opsin compared with arrestin1 that forms more stable complexes.[42,43] Although there is not yet any definitive evidence to support this conjecture, perhaps these different stabilities provide a portion of the biochemical mechanism underlying the enormous range of light intensities to which cones can respond, with arrestin1 removing available visual pigment from potential activation and arrestin4 remaining available to quench newly activated cone opsins.

5. TRANSLOCATION OF VISUAL ARRESTINS

In both rods and cones, the outer segment is a highly elaborated sensory cilium specialized for transducing a photon into a change in membrane potential that can be propagated to the visual cortex. Considering this adaptation, it is not surprising that the biochemical components of the phototransduction cascade are abundant in the outer segment. What is surprising, however, is that several of these components, most notably the visual arrestins and transducin, undergo massive redistributions between the inner and outer segments in response to lighting conditions. In dark-adapted rods, arrestin1 principally localizes to the inner segments and transducin to the outer segments. In response to light adaptation, arrestin1 moves to the outer segments while transducin moves to the inner segments. This translocation has been documented in every vertebrate retina that has been examined to date, including rodents, primates, amphibians, and fish. Although arrestin1 translocation is most dramatic in rods, where the movement involves more than 90% of the protein, arrestin4 translocation has also been demonstrated in cones under appropriately bright illumination conditions.[44–48]

Although first documented nearly three decades ago,[49] the functional consequences of arrestin1 translocation for photoreceptors remain unresolved. Over the years, several hypotheses have been proposed. One possibility is that the translocation of visual arrestins provides a mechanism to regulate the dynamic sensitivity of photoreceptors, allowing a maximal quantum response in dark-adapted photoreceptors when arrestin1 concentration is lower in the outer segments, and either reducing response amplitude or accelerating response recovery in light-adapted photoreceptors when arrestin1 concentration is higher in outer segments (see Ref. 50 for a discussion of photoreceptor adaptation). In mouse rods, arrestin1 translocation is initiated at light levels causing a 3% bleach of the visual pigment, which is nearly at the end of the dynamic range for rods at which point vision shifts to cones.[51] Thus, there is some small potential for adaptive extension of the rod response, although the time scale of arrestin1 translocation is likely too slow to have a significant impact on rod adaptation.[52,53]

Another potential function of arrestin1 translocation is preservation of the rod photoreceptors during lighting conditions that are beyond the dynamic response range of rods.[53] According to this hypothesis, the increased concentration of arrestin1 in the light-adapted rod outer segment would prevent phototransduction signaling from activated rhodopsin or signaling by excess all-*trans* retinal that can also complex to opsin and activate transducin.[54] This hypothesis is supported by studies in humans showing that there is very slow recovery of rod function after a bright stimulation (reviewed in Ref. 55). Such signaling inactivation would prevent the long-term depletion of calcium in photoreceptors from extended cyclic nucleotide-gated (CNG) channel closure that is known to initiate apoptosis in photoreceptors.[56] Although this hypothesis is logical for rod photoreceptors, it is unclear what advantage is conferred on cones, which do not saturate their light response and also support arrestin4 translocation. Clearly, much work yet remains to decipher the functional consequences of arrestin1 and arrestin4 translocation in rod and cone photoreceptors.

Although the function of translocation of arrestins in photoreceptors is not yet understood, the mechanism underlying arrestin1 translocation has been partially elucidated. Several elegant studies have shown that the transition zone of the cilium does not present a significant barrier to diffusion of soluble proteins, including fusions of arrestin1 with GFP, and that diffusional rates alone are sufficient to account for the rate at which arrestin1 translocates during both light and dark adaptations.[57–59] The diffusion model suggests that different binding partners primarily drive the diffusion-mediated translocation of arrestin1. During light exposure, activated rhodopsin serves as the principal light-driven impetus to draw arrestin1 by diffusion through the transition zone, and then be retained in the outer segment by binding to light-activated rhodopsin. Direct evidence supporting this model comes from studies using transgenic mice expressing various levels of rhodopsin. In these mice, the fraction of arrestin1 that translocates to the outer segments during light adaptation is directly proportional to the level of rhodopsin expressed.[60] One puzzling observation that does not fit with this model is that in *Xenopus* that are continuously exposed to light, arrestin1 is not retained in the outer segments as occurs in mammalian rods, but returns to the inner segments after illuminations lasting longer than 60 min.[52] How this observation reconciles with the idea of arrestin1 binding to activated rhodopsin as the principal driving force for arrestin1 translocation to the outer segment is not clear.

For selective partitioning of arrestin1 in the inner segment of dark-adapted rods, two potential explanations have put forward—steric exclusion of arrestin1 tetramers[59] and inner segment binding partners.[61] In a recent analysis of the cytoplasmic space in the disc-filled outer segment, Calvert and his collaborators have shown that steric exclusion can result in as much as 90% of selective partitioning of proteins to the inner segments for cytosolic proteins greater than 80 kDa.[59] Although arrestin1 is only 45 kDa in size, it forms multimers, forming dimers and tetramers at ~40 μM and ~10 μM, respectively,[62–64] well within the physiological millimolar concentration of arrestin1 measured in photoreceptors.[51,65] This tetrameric size (180 kDa) would be excluded from most of the cytosolic volume of the outer segment and could thus account for much of the dark-adapted distribution of arrestin1 that is observed. Note that arrestin1 binds activated rhodopsin in its monomeric form[64] and would thus not be excluded from the space between discs in the outer segments. Direct interaction of arrestin1 with inner segment proteins could also account for selective partitioning of arrestin1 to the inner segment in the absence of binding to rhodopsin (i.e., in the dark). To this end, arrestin1 has been shown to interact directly with microtubules[61,64,66,67] and with enolase1,[68] both of which are predominantly inner segment proteins. Interestingly, the tetrameric form of arrestin1 has been shown to bind microtubules without dissociation of the tetramer.[64] The potential contribution of these binding partners to selective partitioning of arrestin1 to the inner segments during dark adaptation has not yet been tested. In the end, the dark partitioning of arrestin1 to the inner segments is likely complex, probably involving a combination of both cytosolic exclusion and selective binding partners.

The structural elements of the arrestin1 protein that drive translocation are primarily localized in the C-terminus of arrestin1. The first suggestion of this structural feature was noted in the splice variants of arrestin1 which showed a preferential localization to the outer segments in dark-adapted bovine retinas compared to the full-length arrestin1.[29] Subsequent analysis of the C-terminus showed that the C-terminal 30 amino acids of arrestin1 are essential for its proper localization, with the truncated version of arrestin1 preferentially partitioning to the outer segments of rod outer segments regardless of lighting conditions.[69] Studies in transgenic mice expressing a truncated form of arrestin1 missing the C-terminal 26 amino acids also demonstrated that this truncated version of arrestin1 was significantly slower in moving to the outer segments compared with the full-length arrestin1.[61] In combination with the investigations of the diffusion-mediated movement of arrestin1, these studies suggest that the C-terminus of arrestin1 is important for interacting with whatever binding partner(s) is utilized for the partitioning of arrestin1 to the inner segments during dark adaptation.

Although diffusion is sufficient to explain much of the observed arrestin1 translocation, it appears that the complete process of arrestin1 translocation is more complex and likely involves contributions from additional mechanisms. Several studies have shown that disruption of cytoskeletal components can affect certain aspects of arrestin1 translocation. In retinas treated with microtubule depolymerizing agents, translocation of arrestin1 to the outer segments during light adaptation is restricted to the proximal portion of the outer segments in both mouse and frogs.[69,70] It is not clear if these limits on translocation are a consequence of losing architectural elements that maintain appropriate spacing of the distal discs for diffusional access or are a result of disrupting a microtubule motor process that facilitates arrestin1 translocation through the distal region of the outer segments. It is intriguing that the distal outer segments use a different kinesin motor, Kif17, in the same region where arrestin1 translocation is limited.[71]

Additionally, depolymerization of microtubules in photoreceptors also blocks arrestin1 translocation to the inner segments during dark adaptation.[69,70] This disruption is apparently not the result of nonspecifically clogging the transition zone of the cilium since the same duration of treatment with the microtubule depolymerizing agent still permits translocation of both transducin α and arrestin1 through the transition zone during light adaptation. Clearly, there is something unique about microtubules as it relates to translocation of arrestin1 through the transition zone to the inner segments. Perhaps the answer relates to arrestin's direct binding to the polymerized microtubules,[61,64,66,67] although this seems unlikely as arrestin1 binds the αβ microtubule subunits of unpolymerized microtubules with nearly the same affinity as polymerized microtubules.[66] Exactly how these contributions of the cytoskeletal elements mesh with the diffusion of arrestin1 in light-driven translocation have not been resolved and await further investigation.

In addition to these nuances of translocation, it is also apparent that translocation in photoreceptors is controlled by a signaling cascade that regulates the diffusion of arrestin1. Evidence for this signaling cascade was first noted by Strissel *et al.* where they showed that at low levels of illumination, a threshold for arrestin1 translocation exists, corresponding to approximately 1.5% bleaching of rhodopsin.[51] Below this level of illumination, no arrestin1 translocation was observed. Above this level, the amount of arrestin1 translocation increased approximately linearly with increased bleaching of rhodopsin. Interestingly, at threshold illumination levels, nearly 30 times more arrestin1 molecules translocate to the outer segments than the number of rhodopsin molecules that are activated. This superstoichiometric translocation at a threshold level is a hallmark of a signaling cascade.

Investigations into this signaling cascade have implicated phospholipase C (PLC) and protein kinase C (PKC) as key elements in signaling arrestin1 translocation in both amphibian and mammalian rods.[72] In this study, both PLC and PKC agonists were effective at initiating arrestin1 translocation in the absence of light. Additionally, PLC and PKC antagonists also slowed arrestin1 translocation when eyes were exposed to light. Findings from this study are also consistent with the diffusion hypothesis for arrestin1 translocation. When PLC and PKC agonists were applied to photoreceptors in the dark, arrestin1 translocation to the outer segments was initiated, but arrestin1 was not retained in the outer segments. These results suggest that in the absence of activated rhodopsin, there is no high affinity binding partner in the outer segments to drive diffusion to the outer segments, and thus arrestin1 returns to the inner segments.

There remain two obvious unknowns associated with this signaling pathway—how is PLC activated by light and how does PKC activation initiate translocation? The most parsimonious explanation for PLC activation is by rhodopsin signaling either through transducin or an alternative G-protein. Evidence supporting this idea comes from animal models lacking R9AP, a regulator of G-protein signaling, which have prolonged G-protein activation.[51] In these animals, the light threshold for arrestin1 translocation is significantly reduced, implicating the involvement of a G-protein. The identity of this G-protein, however, is not precisely known, although there are suggestive data. For example, animals lacking transducin α subunits show an absence of threshold signaling of arrestin1 translocation, implicating transducin as the signaling protein.[51] Further, pharmacological studies indicate that arrestin1 translocation is sensitive to cholera toxin,[72] as is transducin.[73] Additional studies will help define if transducin activation or activation of another G-protein triggers arrestin1 translocation.

With regard to how PKC activation regulates the mobility of arrestin1, there is currently no indication as to how this might occur although presumably it is through PKC-mediated phosphorylation of a substrate that likely resides in the transition zone and along the axoneme of the ciliary structure. Although an earlier report indicated that arrestin1 translocation did not require ATP,[61] subsequent reinvestigation in which ATP was more completely depleted determined that ATP is essential for arrestin1 translocation, consistent with PKC signaling through phosphorylation of an unidentified target substrate.[72]

6. NEW ROLES FOR ARRESTIN1 IN THE RETINA

Although many researchers thought the function of arrestin1 and arrestin4 were well described when their role in visual pigment quenching was identified, some puzzling features about the cell biology of visual arrestins suggested that their function might be more complex. For example, as described earlier, the localization of arrestin1 and arrestin4 is not static in photoreceptors but changes between inner and outer segments in response to lighting conditions. Further, when arrestin1 translocates to the outer segments during light adaptation, a significant fraction of arrestin1 remains localized to the synaptic region in the inner plexiform layer.[45,52,74–76]

As Nature is parsimonious and typically does not put proteins in remote locations or change a protein's location without a purpose, these types of observations have led various laboratories to expand their search for potential interactions for arrestin1 beyond activated rhodopsin. The list of these binding partners for arrestin1 is growing annually.

6.1. Microtubules

Visual arrestin binding to microtubules was first demonstrated for arrestin1 and its shorter splice variant.[77] Subsequently, binding to microtubules has been demonstrated for all members of the arrestin family.[67] The association of arrestin1 with microtubules occurs through the same surface of arrestin1 that interacts with activated rhodopsin, and is thus mutually exclusive.[66,77] The affinity of arrestin1 is comparatively lower for microtubules (40 μM[66]) than for activated phosphorhodopsin (50 nM[78]). The binding occurs nearly as well for the unpolymerized αβ dimers of tubulin as for the polymerized microtubules.[66] Unlike the binding of arrestin1 to rhodopsin, which occurs only through the monomeric form of arrestin1, the binding to microtubules can occur through either the monomeric or the tetrameric forms of arrestin1.[64]

The functional consequences of arrestin's interaction with microtubules is unclear, but several hypotheses have been proposed. First, as described earlier, the binding of arrestin1 to microtubules has been proposed as a mechanism for contributing to the concentration of arrestin1 to the inner segments during dark adaptation.[61] Perhaps this serves as a reservoir that can meet the rod's quenching requirements during high levels of photon flux to preserve the cell from metabolic rundown since the relatively low affinity of arrestin1 for microtubules allows it to quickly diffuse and bind activated rhodopsin in the outer segments. Alternatively, the association of arrestin1 with microtubules in photoreceptors could also function as a scaffolding agent, a well-known function for arrestin2 and arrestin3 (see Ref. 79, for a recent review). For example, as previously discussed, arrestin1 associates with enolase1, perhaps forming a scaffolding structure that helps maintain the glycolytic complex and retain its demonstrated association with microtubules.[80,81] In addition, other binding partners have been identified for arrestin1, such as parkin and JNK3 (discussed below), which apparently associate with microtubules via their interaction with arrestin1 without modification of their catalytic activity[82,83] consistent with a scaffolding function. Experimental evidence supports both of these proposed functions, but no animal model in which the binding of arrestin1 to microtubules is specifically disrupted currently exists that permits direct testing of these functions. It is worth noting, however, that there is no reason to consider these potential functions as mutually exclusive.

6.2. Kinases

Scaffolding interactions for arrestin1 and arrestin4 with kinases in the mitogen-activated protein kinase (MAPK) pathway have been demonstrated similar to those previously demonstrated for arrestin2 and arrestin3.[84–87] Using a fluorescence relocalization assay and coimmunoprecipitation, Song *et al.* have convincingly demonstrated that both arrestin1 and arrestin4 bind all three kinases in the c-Jun N-terminal kinase stress response arm of the MAPK cascade—apoptosis signal-regulating kinase 1, MAPK4, and c-Jun N-terminal kinase 3(JNK3).[82,88] Interestingly, although the kinases are bound in a complex, they do not lead to the activation of JNK3, in contrast to arrestin3 which both binds and activates JNK3.[82]

Arrestin1 also binds extracellular signal-related kinases 1 and 2 (ERK1/2) of the classical arm of the MAPK pathway. In this case, interaction with arrestin1 leads to the sequestration of ERK1/2 to microtubules and, unlike the association with JNK3, appears to functionally reduce the catalytic activity of the kinases.[67] This association with ERK1/2 appears to be specific for arrestin1, as ERK1/2 does not bind to arrestin4. Although these associations between the visual arrestins and the MAP kinases have been clearly demonstrated in cell culture in COS-7 and HEK-293 cells, there is currently no indication if scaffolding of these kinases by arrestin1 and 4 has a role in rod and cone photoreceptors.

6.3. Ubiquitin ligases

In addition to the interaction with above-described kinases, arrestin1 and 4 also have the potential to associate with the ubiquitin ligases parkin and murine double minute 2 (Mdm2).[67,83,88,89] In the case of parkin, an E3 ubiquitin ligase classically implicated in familial recessive Parkinson's disease, both arrestin1 and arrestin4 can be coimmunoprecipitated with parkin.[83] Importantly, using purified arrestin1 and parkin proteins, a direct interaction between these proteins was demonstrated, indicating that no additional elements are required to facilitate the association between the visual arrestins and this ubiquitin ligase.[83] Although arrestin2 and arrestin3 are ubiquitinylated by parkin,[83,90] this has not yet been demonstrated for either arrestin1 or arrestin4.

The visual arrestins have also been shown to bind to Mdm2, another ubiquitin ligase classically involved in ubiquitination of the p53 tumor suppressor.[83,88,89] Although both arrestin1 and arrestin4 associate with Mdm2 in tissue culture,[88,89] only arrestin1 appears to act as an adaptor molecule, associating Mdm2 with microtubules.[67] Accordingly, only association of Mdm2 with arrestin1, and not arrestin4, results in elevated ubiquitination levels of proteins in the cytoskeletal fractions of transfected HEK-293 cells.[67] Although the interactions between visual arrestins and ubiquitin ligases are clearly supported by strong empirical evidence, like the MAP kinases, there has not yet been any clear indication if these interactions occur within photoreceptors, and if so, what the functional consequences might be.

6.4. Calmodulin

Both arrestin1 and arrestin4 have been shown to bind calcium-occupied calmodulin.[91] Despite the relatively low affinity of visual arrestins for calmodulin (~7 μM), the high concentration of arrestin1 and arrestin4 in photoreceptors (millimolar range) makes this association physiologically relevant. Although calmodulin binds on the same surface of arrestin1 as does activated rhodopsin, calmodulin does not appear to have any effect on receptor binding. Consequently, the role of the arrestin–calmodulin interaction remains undefined, but perhaps it functions to either differentially localize calmodulin in photoreceptors or perhaps regulate the availability of calmodulin for other interactions, including regulating/modulating CNG channels.

6.5. *N*-ethylmaleimide-sensitive factor

Recently, the Craft laboratory identified a novel interaction for arrestin1 that likely explains the localization of arrestin1 in the synapses of photoreceptors. Using anti-arrestin1 antibodies to pull-down arrestin1-interacting partners, Huang *et al.* identified an interaction between arrestin1 and *N*-ethylmaleimide-sensitive factor (NSF).[92] The interaction with NSF was validated by reversed immunoprecipitation assays as well as pull-down assays with heterologously expressed and purified components. NSF is an ATPase involved in membrane fusion and thus plays a critical role in the transfer of membrane vesicles.[93] The interaction of arrestin1 with NSF increases the ATPase activity of NSF, thus increasing the rate of disassembly of the SNARE complexes involved in synaptic exocytosis. Because NSF and arrestin1 interactions are at their peak in dark-adapted rods, the hypothesized role of arrestin1 interaction with NSF is to sustain the release of synaptic vesicles required by rod photoreceptors which are depolarized in the dark and constantly releasing L-glutamate until a light response leads to a graded diminution of transmitter release. Consistent with this hypothesis, in mice deficient for arrestin1, there is no increase in photopic b-wave amplitude of the electroretinogram in response to increasing light intensities, indicating an absence of change in the synaptic transmission from rod and cone cells to the bipolar cells.[92,94] Interestingly, NSF also localizes to the photoreceptor ciliary transition zone, raising the potential that its multiple interactions with proteins may have a role in protein trafficking through the cilium.[95]

Despite the overall structural homology between arrestin1 and arrestin4, only arrestin1 interacts with NSF as demonstrated *in vitro* and *in vivo*.[92] Interestingly, an interaction with NSF has been previously noted in the arrestin family for arrestin2 (β-arrestin1), although here the functional mechanism seems different as arrestin2 appears to modulate clathrin-mediated receptor endocytosis[96] and there is no evidence for interaction of arrestin1 with clathrin.

6.6. Enolase1

Using an alternative approach to identifying potential interaction partners, chemical cross-linking agents were utilized to tether arrestin1 to interacting elements. Using dithiobis(succinimidylpropionate) as the cross-linker, Smith *et al.* showed an association of arrestin1 with enolase1 in dark-adapted rods.[68] This association was validated using a variety of techniques, including coimmunoprecipitation and surface plasmon resonance. Importantly, the surface plasmon resonance study used heterologously expressed and highly purified arrestin1 and enolase1, demonstrating that the interaction between these proteins is direct and does not require any additional scaffolding elements. An interaction between arrestin4 and enolase1 was also shown using coimmunoprecipitation. There were two interesting observations made about this interaction. First, arrestin1 modulates the glycolytic activity of enolase1, decreasing its catalytic rate by as much as 25%; this is the first known intersection of the phototransduction cascade and the glycolysis pathway. The second observation is that while arrestin1 translocates in response to light, enolase1 does not. Consequently, arrestin1 interacts with enolase1 in the dark, and upon translocating to the outer segments during light adaptation, releases its suppression of enolase1 activity. The potential function of this interaction is not clear as photoreceptors use 75% more energy in the dark than in the light to maintain the ionic polarization while the CNG sodium channels are open.[97,98] The current hypothesis is that increasing glycolysis in the light might have more to do with the production of NADH than with ATP production as there is clearly an increased need for reducing agents in light-exposed photoreceptors for reduction of all-*trans* retinal produced during rhodopsin activation and reduction of oxidation by-products from photon flux.

7. ARRESTINS IN DISEASE PROCESSES IN THE EYE

Because of the central role of arrestin1 in the phototransduction process, defects arising from mutations in the arrestin1 gene are the basis for a limited number of visual deficits. Perhaps not surprisingly, based on the coexpression of arrestin1 and arrestin4 in cones and their ability to substitute for each other, there have not yet been any diseases identified that associate with defects in arrestin4.

7.1. Oguchi disease

The most common visual defect associated with mutations in the arrestin1 gene is Oguchi disease, a relatively rare form of stationary night blindness.[99–101] Instead of the normal 30–40 min required for dark adaptation, patients with Oguchi disease typically require more than 2 h (and up to 5–7 h) to adapt to darkness.[102] Optical coherence tomography measurements indicate that at least for some Oguchi patients, there is a shortening of the outer segments associated with this arrestin1 deficit which correlates with delayed recovery in multifocal electroretinogram measurements.[103,104] This observation is recapitulated in a mouse arrestin1 knockout model that shows initial shortening of the outer segments prior to a slow retinal degeneration.[105] The underlying mechanism for delayed dark adaptation in these patients is not clearly defined but is likely related to the removal of all activated rhodopsin and subsequent regeneration with 11-*cis* retinal. In the presence of arrestin1, activated rhodopsin (metarhodopsin) is bound and stabilized in its metarhodopsin II conformation,[78] which is accessible to the retinol dehydrogenase for reduction of its retinoid chromophore.[25] In the absence of arrestin1, activated rhodopsin accumulates in the metarhodopsin III form, in which the retinoid chromophore is more resistant to reduction and thus continues to signal for a longer period.[55,106]

7.2. Retinitis pigmentosa

In addition to Oguchi disease, some patients with defects in the arrestin1 gene present with a more progressive retinal degeneration known as retinitis pigmentosa.[100,107–109] Curiously, these patients can have the same mutation (adenine deletion at position 1147) common to many Oguchi patients who simply develop stationary night blindness. The nature of this variable penetrance is unclear but is a common feature of many eye diseases. Arrestin1 can also lead to retinitis pigmentosa if it forms stable complexes with rhodopsin. This observation was first made in *Drosophila*, where stable complexes of arrestin and rhodopsin led to severe forms of retinal degeneration.[110] The source of this degeneration appears to be twofold, resulting from overwhelming of the endocytic machinery as well as signaling of an apoptotic cascade through a guanine nucleotide-binding protein.[111] In vertebrates, a similar retinal degeneration has been observed with mutants of rhodopsin that are constitutively active and thus perpetually bound by arrestin1.[112] In K296E rhodopsin mutants, the arrestin1–rhodopsin complex mislocalizes to the inner segments and could trigger cell death via one of several conceivable mechanisms, including excessive endocytosis, excess proteasomal degradation, or activation of G-protein signaling pathway. Experiments to distinguish between the potential mechanisms leading to retinal degeneration have not been conducted.

7.3. Autoimmune diseases

In addition to retinal diseases, arrestin1 has also been implicated in several autoimmune diseases. Circulating antibodies directed against arrestin1 have been identified in both multiple sclerosis patients[113–115] and in patients suffering from cancer-associated retinopathy. In patients with multiple sclerosis, it is not clear if there is a direct pathology of the myelin by these autoantibodies. There is clear evidence, however, that arrestin1 leads to T-cell activation[115] which has a defined role in the development of lesions in multiple sclerosis.[116] Arrestin1 has also been implicated in cancer-associated retinopathy.[117] In cancer-associated retinopathy, antibodies typically develop against tumor antigens that either cross-react with retinal proteins or are present in both the tumor tissue and retina, leading to retinal degeneration. Arrestin1 is one of the common retinal proteins to which antibodies cross-react in patients with cancer-associated retinopathy.[118,119] Finally, arrestin1 is a causative agent in uveitis, a T-cell-mediated autoimmune disease that results in inflammation of the retina.[120] In fact, it is in this role that arrestin1 was first discovered and described as "retinal S-antigen," the *soluble antigen* causing uveitis.[1,2]

8. THE FUTURE FOR VISUAL ARRESTINS

Although much has been learned about the role of the visual arrestins in the eye over the past four decades since the arrestins were first discovered, it is clear from many of the recent discoveries that arrestin1 and arrestin4 have complex roles in rod and cone photoreceptors. Clearly, the visual arrestins are not only central to quenching phototransduction in rods and cones but also play diverse roles in regulating metabolism and scaffolding elements in other signaling cascades. Understanding how all of these functions of arrestin1 and arrestin4 are coordinated will be essential for not only understanding how photoreceptors function but also for elucidating the role of visual arrestins in visual system pathologies and how they might be used therapeutically in the future.

REFERENCES

1. Wacker WB, Donoso LA, Kalsow CM, Yankeelov JAJ, Organisciak DT. Experimental allergic uveitis. Isolation, characterization, and localization of a soluble uveitopathogenic antigen from bovine retina. *J Immunol.* 1977;119:1949–1958.
2. de Kozak Y, Sakai J, Thillaye B, Faure JP. S antigen-induced experimental autoimmune uveo-retinitis in rats. *Curr Eye Res.* 1981;1:327–337.
3. Kühn H. Light-regulated binding of rhodopsin kinase and other proteins to cattle photoreceptor membranes. *Biochemistry.* 1978;17:4389–4395.
4. Pfister C, Chabre M, Plouet J, et al. Retinal S antigen identified as the 48K protein regulation light-dependent phosphodiesterase in rods. *Science.* 1985;228:891–893.
5. Zuckerman R, Cheasty JE. A 48 kDa protein arrests cGMP phosphodiesterase activation in retinal rod disk membranes. *FEBS Lett.* 1986;207:35–41.
6. Craft CM, Whitmore DH, Wiechmann AF. Cone arrestin identified by targeting expression of a functional family. *J Biol Chem.* 1994;269:4613–4619.
7. Craft CM, Whitmore DH. The arrestin superfamily: cone arrestins are a fourth family. *FEBS Lett.* 1995;362:247–255.
8. Murakami A, Yajima T, Sakuma H, McLaren MJ, Inana G. X-arrestin: a new retinal arrestin mapping to the X chromosome. *FEBS Lett.* 1993;334:203–209.
9. Sakuma H, Inana G, Murakami A, Higashide T, McLaren MJ. Immunolocalization of X-arrestin in human cone photoreceptors. *FEBS Lett.* 1996;382:105–110.
10. Ripps H. Light to sight: milestones in phototransduction. *FASEB J.* 2010;24:970–975.
11. Kühn H, Dreyer WJ. Light dependent phosphorylation of rhodopsin by ATP. *FEBS Lett.* 1972;20:1–6.

12. McDowell JH, Kühn H. Light-induced phosphorylation of rhodopsin in cattle photoreceptor membranes: substrate activation and inactivation. *Biochemistry*. 1977;16: 4054–4060.
13. Wilden U, Kühn H. Light-dependent phosphorylation of rhodopsin: number of phosphorylation sites. *Biochemistry*. 1982;21:3014–3022.
14. Kühn H, Hall SW, Wilden U. Light-induced binding of 48-kDa protein to photoreceptor membranes is highly enhanced by phosphorylation of rhodopsin. *FEBS Lett*. 1984;176:473–478.
15. Miller JL, Fox DA, Litman BJ. Amplification of phosphodiesterase activation is greatly reduced by rhodopsin phosphorylation. *Biochemistry*. 1986;25:4983–4988.
16. Wilden U, Hall SW, Kühn H. Phosphodiesterase activation by photoexcited rhodopsin is quenched when rhodopsin is phosphorylated and binds the intrinsic 48-kDa protein of rod outer segments. *Proc Natl Acad Sci USA*. 1986;83:1174–1178.
17. Palczewski K, Buczylko J, Imami NR, McDowell JH, Hargrave PA. Role of the carboxyl-terminal region of arrestin in binding to phosphorylated rhodopsin. *J Biol Chem*. 1991;266:15334–15339.
18. McDowell JH, Nawrocki JP, Hargrave PA. Phosphorylation sites in bovine rhodopsin. *Biochemistry*. 1993;32:4968–4974.
19. Ohguro H, Palczewski K, Ericsson LH, Walsh KA, Johnson RS. Sequential phosphorylation of rhodopsin at multiple sites. *Biochemistry*. 1993;32:5718–5724.
20. Papac DI, Oatis JE, Crouch RK, Knapp DR. Mass spectrometric identification of phosphorylation sites in bleached bovine rhodopsin. *Biochemistry*. 1993;32:5930–5934.
21. Gibson SK, Parkes JH, Liebman PA. Phosphorylation modulates the affinity of light-activated rhodopsin for G protein and arrestin. *Biochemistry*. 2000;39:5738–5749.
22. Vishnivetskiy SA, Raman D, Wei J, Kennedy MJ, Hurley JB, Gurevich VV. Regulation of arrestin binding by rhodopsin phosphorylation level. *J Biol Chem*. 2007;282: 32075–32083.
23. Liebman PA, Pugh EN. The control of phosphodiesterase in rod disk membranes: kinetics, possible mechanisms and significance for vision. *Vision Res*. 1979;19:375–380.
24. Liebman PA, Pugh EN. ATP mediates rapid reversal of cyclic GMP phosphodiesterase activation in visual receptor membranes. *Nature*. 1980;287:734–736.
25. Hofmann KP, Pulvermüller A, Buczylko J, Van Hooser P, Palczewski K. The role of arrestin and retinoids in the regeneration pathway of rhodopsin. *J Biol Chem*. 1992;267:15701–15706.
26. Sommer ME, Smith WC, Farrens DL. Dynamics of arrestin-rhodopsin interactions: arrestin and retinal release are directly linked events. *J Biol Chem*. 2005;280:6861–6871.
27. Brown BM, Carlson BL, Zhu XM, Lolley RN, Craft CM. Light-driven translocation of the protein phosphatase 2A complex regulates light/dark dephosphorylation of phosducin and rhodopsin. *Biochemistry*. 2002;41:13526–13538.
28. Palczewski K, Buczylko J, Ohguro H, et al. Characterization of a truncated form of arrestin isolated from bovine rod outer segments. *Protein Sci*. 1994;3:314–324.
29. Smith WC, Milam AH, Dugger DR, Arendt A, Hargrave PA, Palczewski K. A splice variant of arrestin: molecular cloning and localization in bovine retina. *J Biol Chem*. 1994;269:15407–15410.
30. Palczewski K, Smith WC. Splice variants of arrestin. *Exp Eye Res*. 1996;63:599–602.
31. Smith WC. A splice variant of arrestin from human retina. *Exp Eye Res*. 1996;62:585–592.
32. Zhu X, Ma B, Babu S, Murage J, Knox BE, Craft CM. Mouse cone arrestin gene characterization: promoter targets expression to cone photoreceptors. *FEBS Lett*. 2002;524:116–122.
33. Parruti G, Ambrosini G, Sallese M, DeBlasi A. Comparative analysis of beta-adrenergic receptor kinase and beta-arrestin mRNA expression in human cells. *J Receptor Res*. 1993;13:609–618.

34. Sterne-Marr R, Gurevich VV, Goldsmith P, et al. Polypeptide variants of beta-arrestin and arrestin3. *J Biol Chem.* 1993;268:15640–15648.
35. Schroder K, Pulvermuller A, Hofmann KP. Arrestin and its splice variant Arr(1-370A) (P-44)—mechanism and biological role of their interaction with rhodopsin. *J Biol Chem.* 2002;277:43987–43996.
36. Ascano M, Robinson PR. Differential phosphorylation of the rhodopsin cytoplasmic tail mediates the binding of arrestin and its splice variant, p44. *Biochemistry.* 2006;45:2398–2407.
37. Granzin J, Cousin A, Weirauch M, Schlesinger R, Büldt G, Batra-Safferling R. Crystal structure of p44, a constitutively active splice variant of visual arrestin. *J Mol Biol.* 2012;416:611–618.
38. Burns ME, Mendez A, Chen C-K, et al. Deactivation of phosphorylated and nonphosphorylated rhodopsin by arrestin splice variants. *J Neurosci.* 2006;26: 1036–1044.
39. Sakuma H, Murakami A, Fujimaki T, Inana G. Isolation and characterization of the human X-arrestin gene. *Gene.* 1998;224:87–95.
40. Nir I, Ransom N. S-antigen in rods and cones of the primate retina: different labeling patterns are revealed with antibodies directed against specific domains in the molecule. *J Histochem Cytochem.* 1992;40:343–352.
41. Nikonov SS, Brown BM, Davis JA, et al. Mouse cones require an arrestin for normal inactivation of phototransduction. *Neuron.* 2008;59:462–474.
42. Smith WC, Gurevich EV, Dugger DR, et al. Cloning and functional characterization of salamander rod and cone arrestins. *Invest Ophthalmol Vis Sci.* 2000;41:2445–2455.
43. Sutton RB, Vishnivetskiy SA, Robert J, et al. Crystal structure of cone arrestin at 2.3A: evolution of receptor specificity. *J Mol Biol.* 2005;354:1069–1080.
44. Zhu X, Li A, Brown B, Weiss ER, Osawa S, Craft CM. Mouse cone arrestin expression pattern: light induced translocation in cone photoreceptors. *Mol Vis.* 2002;8:462–471.
45. Zhang HB, Cuenca N, Ivanova T, et al. Identification and light-dependent translocation of a cone-specific antigen, cone arrestin, recognized by monoclonal antibody 7G6. *Invest Ophthalmol Vis Sci.* 2003;44:2858–2867.
46. Zhang HB, Huang W, Zhang HK, et al. Light-dependent redistribution of visual arrestins and transducin subunits in mice with defective phototransduction. *Mol Vis.* 2003;9:231–237.
47. Coleman JE, Semple-Rowland SL. GC1 deletion prevents light-dependent arrestin translocation in mouse cone photoreceptor cells. *Invest Ophthalmol Vis Sci.* 2005;46:12–16.
48. Haire SE, Pang J, Boye SL, et al. Light-driven cone arrestin translocation in cones of postnatal guanylate cyclase-1 knockout mouse retina treated with AAV-GC1. *Invest Ophthalmol Vis Sci.* 2006;47:3745–3753.
49. Broekhuyse RM, Tolhuizen EFJ, Janssen APM, Winkens HJ. Light induced shift and binding of S-antigen in retinal rods. *Curr Eye Res.* 1985;4:613–618.
50. Pugh EN, Nikonov S, Lamb TD. Molecular mechanisms of vertebrate photoreceptor light adaptation. *Curr Opin Neurobiol.* 1999;9:410–418.
51. Strissel KJ, Sokolov M, Trieu LH, Arshavsky VY. Arrestin translocation Is induced at a critical threshold of visual signaling and is superstoichiometric to bleached rhodopsin. *J Neurosci.* 2006;26:1146–1153.
52. Peterson JJ, Tam BM, Moritz OL, et al. Arrestin migrates in photoreceptors in response to light: a study of arrestin localization using an arrestin-GFP fusion protein in transgenic frogs. *Exp Eye Res.* 2003;76:553–563.
53. Elias R, Sezate S, Cao W, McGinnis J. Temporal kinetics of the light/dark translocation and compartmentation of arrestin and alpha-transducin in mouse photoreceptor cells. *Mol Vis.* 2004;10:672–681.

54. Jager S, Palczewski K, Hofmann KP. Opsin/all-trans-retinal complex activates transducin by different mechanisms than photolyzed rhodopsin. *Biochemistry*. 1996; 35:2901–2908.
55. Lamb TD, Pugh Jr EN. Dark adaptation and the retinoid cycle of vision. *Prog Retin Eye Res*. 2004;23:307–380.
56. Maeda A, Maeda T, Golczak M, et al. Involvement of all-trans-retinal in acute light-induced retinopathy of mice. *J Biol Chem*. 2009;284:15173–15183.
57. Peet JA, Bragin A, Calvert PD, et al. Quantification of the cytoplasmic spaces of living cells with EGFP reveals arrestin-EGFP to be in disequlibrium in dark adapted rod photoreceptors. *J Cell Sci*. 2004;117:3049–3059.
58. Calvert PD, Schiesser WE, Pugh EN. Diffusion of a soluble protein, photoactivatable GFP, through a sensory cilium. *J Gen Physiol*. 2010;135:173–196.
59. Najafi M, Maza NA, Calvert PD. Steric volume exclusion sets soluble protein concentrations in photoreceptor sensory cilia. *Proc Natl Acad Sci USA*. 2012;109:203–208.
60. Hanson SM, Gurevich EV, Vishnivetskiy SA, Ahmed MR, Song X, Gurevich VV. Each rhodopsin molecule binds its own arrestin. *Proc Natl Acad Sci USA*. 2007;104:3125–3128.
61. Nair KS, Hanson SM, Mendez A, et al. Light-dependent redistribution of arrestin in vertebrate rods is an energy-independent process governed by protein–protein interactions. *Neuron*. 2005;46:555–567.
62. Shilton BH, McDowell JH, Smith WC, Hargrave PA. The solution structure and activation of visual arrestin studied by small-angle X-ray scattering. *Eur J Biochem*. 2002;269:3801–3809.
63. Imamoto Y, Tamura C, Kamikubo H, Kataoka M. Concentration-dependent tetramerization of bovine visual arrestin. *Biophys J*. 2003;85:1186–1195.
64. Hanson SM, Van Eps N, Francis DJ, et al. Structure and function of the visual arrestin oligomer. *EMBO J*. 2007;26:1726–1736.
65. Lyubarsky AL, Daniele LL, Pugh EN. From candelas to photoisomerizations in the mouse eye by rhodopsin bleaching in situ and the light-rearing dependence of the major components of the mouse ERG. *Vision Res*. 2004;44:3235–3251.
66. Hanson SM, Francis DJ, Vishnivetskiy SA, Klug CS, Gurevich VV. Visual arrestin binding to microtubules involves a distinct conformational change. *J Biol Chem*. 2006;281:9765–9772.
67. Hanson SM, Cleghorn WM, Francis DJ, et al. Arrestin mobilizes signaling proteins to the cytoskeleton and redirects their activity. *J Mol Biol*. 2007;368:375–387.
68. Smith WC, Bolch S, Dugger DR, et al. Interaction of arrestin with enolase1 in photoreceptors. *Invest Ophthalmol Vis Sci*. 2011;52:1832–1840.
69. Peterson JJ, Orisme W, Fellows J, et al. A role for cytoskeletal elements in the light-driven translocation of proteins in rod photoreceptors. *Invest Ophthalmol Vis Sci*. 2005;46:3988–3998.
70. Reidel B, Goldmann T, Giessl A, Wolfrum U. The translocation of signaling molecules in dark adapting mammalian rod photoreceptor cells is dependent on the cytoskeleton. *Cell Motil Cytoskel*. 2008;65:785–800.
71. Insinna C, Pathak N, Perkins B, Drummond I, Besharse JC. The homodimeric kinesin, Kif17, is essential for vertebrate photoreceptor sensory outer segment development. *Dev Biol*. 2008;316:160–170.
72. Orisme W, Li J, Goldmann T, Bolch S, Wolfrum U, Smith WC. Light-dependent translocation of arrestin in rod photoreceptors is signaled through a phospholipase C cascade and requires ATP. *Cell Signal*. 2010;22:447–456.
73. Bornancin F, Franco M, Bigay J, Chabre M. Functional modifications of transducin induced by cholera or pertussis-toxin-catalyzed ADP-ribosylation. *Eur J Biochem*. 1992;210:33–44.

74. Whelan JP, McGinnis JF. Light-dependent subcellular movement of photoreceptor proteins. *J Neurosci Res*. 1988;20:263–270.
75. Mirshahi M, DeKozak Y, Faure JP, Brisson JP, Falcon J, Collin JP. Influence of light on S-antigen localization in rods and cones and immunopathogenicity in rats. In: Secchi AG, Fregona IA, eds. *Modern Trends in Immunology and Immunopathology of the Eye*. Milano: Masson; 1989:45–49.
76. McGinnis JF, Matsumoto B, Whelan JP, Cao W. Cytoskeleton participation in subcellular trafficking of signal transduction proteins in rod photoroeceptor cells. *J Neurosci Res*. 2002;67:290–297.
77. Nair KS, Hanson SM, Kennedy MJ, Hurley JB, Gurevich VV, Slepak VZ. Direct binding of visual arrestin to microtubules determines the differential subcellular localization of its splice variants in rod photoreceptors. *J Biol Chem*. 2004;279:41240–41248.
78. Schleicher A, Kuhn H, Hofmann KP. Kinetics, binding constant, and activation energy of the 48-kDa protein-rhodopsin complex by extra-metarhodopsin II. *Biochemistry*. 1989;28:1770–1775.
79. DeFea KA. Beta-arrestins as regulators of signal termination and transduction: How do they determine what to scaffold? *Cell Signal*. 2011;23:621–629.
80. Volker KW, Reinitz CA, Knull HR. Glycolytic enzymes and assembly of microtubule networks. *Comp Biochem Physiol B Biochem Mol Biol*. 1995;112:503–514.
81. Srere PA, Knull HR. Location-location-location. *Trends Biochem Sci*. 1998;23: 319–320.
82. Song X, Coffa S, Fu H, Gurevich VV. How does arrestin assemble MAPKs into a signaling complex? *J Biol Chem*. 2009;284:685–695.
83. Ahmed MR, Zhan X, Song X, Kook S, Gurevich VV, Gurevich EV. Ubiquitin ligase parkin promotes Mdm2-arrestin interaction but inhibits arrestin ubiquitination. *Biochemistry*. 2011;50:3749–3763.
84. McDonald PH, Chow CW, Miller WE, et al. Beta-arrestin 2: a receptor-regulated MAPK scaffold for the activation of JNK3. *Science*. 2000;290:1574–1577.
85. Luttrell LM, Roudabush FL, Choy EW, et al. Activation and targeting of extracellular signal-regulated kinases by beta-arrestin scaffolds. *Proc Natl Acad Sci USA*. 2001;98:2449–2454.
86. Bruchas MR, Macey TA, Lowe JD, Chavkin C. Kappa opioid receptor activation of p38 mapk is GRK3- and arrestin-dependent in neurons and astrocytes. *J Biol Chem*. 2006;281:18081–18089.
87. Zhan X, Kaoud TS, Dalby KN, Gurevich VV. Nonvisual arrestins function as simple scaffolds assembling the MKK4-JNK3alpha2 signaling complex. *Biochemistry*. 2011; 50:10520–10529.
88. Song X, Gurevich EV, Gurevich VV. Cone arrestin binding to JNK3 and Mdm2: conformational preference and localization of interaction sites. *J Neurochem*. 2007; 103:1053–1062.
89. Song X, Raman D, Gurevich EV, Vishnivetskiy SA, Gurevich VV. Visual and both non-visual arrestins in their "inactive" conformation bind JNK3 and Mdm2 and relocalize them from the nucleus to the cytoplasm. *J Biol Chem*. 2006;281: 21491–21499.
90. Shenoy SK, McDonald PH, Kohout TA, Lefkowitz RJ. Regulation of receptor fate by ubiquitination of activated B2 adrenergic receptor and B-arrestin. *Science*. 2001;294: 1307–1313.
91. Wu N, Hanson SM, Francis DJ, et al. Arrestin binding to calmodulin: a direct interaction between two ubiquitous signaling proteins. *J Mol Biol*. 2006;364:955–963.
92. Huang S-P, Brown BM, Craft CM. Visual arrestin 1 acts as a modulator for N-ethylmaleimide-sensitive factor in the photoreceptor synapse. *J Neurosci*. 2010; 30:9381–9391.

93. Whiteheart SW, Rossnagel K, Buhrow SA, Brunner M, Jaenicke R, Rothman JE. N-ethylmaleimide-sensitive fusion protein: a trimeric ATPase whose hydrolysis of ATP is required for membrane fusion. *J Cell Biol.* 1994;126:945–954.
94. Brown BM, Ramirez T, Rife L, Craft CM. Visual arrestin 1 contributes to cone photoreceptor survival and light-adaptation. *Invest Ophthalmol Vis Sci.* 2010;51:2372–2380.
95. Huang SP, Craft CM. Potential cellular functions of N-ethylmaleimide sensitive factor in the photoreceptor. *Adv Exp Med Biol.* 2012;723:791–797.
96. McDonald PH, Cote NL, Lin FT, Premont RT, Pitcher JA, Lefkowitz RJ. Identification of NSF as a beta-arrestin1-binding protein—implications for beta(2)-adrenergic receptor regulation. *J Biol Chem.* 1999;274:10677–10680.
97. Okawa H, Sampath AP, Laughlin SB, Fain GL. ATP consumption by mammalian rod photoreceptors in darkness and in light. *Curr Biol.* 2008;18:1917–1921.
98. Winkler BS, Starnes CA, Twardy BS, Brault D, Taylor RC. Nuclear magnetic resonance and biochemical measurements of glucose utilization in the cone-dominant ground squirrel retina. *Invest Ophthalmol Vis Sci.* 2008;49:4613–4619.
99. Fuchs S, Nakazawa M, Maw M, Tamai M, Oguchi Y, Gal A. A homozygous 1-base pair deletion in the arrestin gene is a frequent cause of Oguchi disease in Japanese. *Nat Genet.* 1995;10:360–362.
100. Yoshii M, Murakami A, Akeo K, et al. Visual function and gene analysis in a family with Oguchi's disease. *Ophthalmic Res.* 1998;30:394–401.
101. Fujinami K, Tsunoda K, Nakamura M, Oguchi Y, Miyake Y. Oguchi disease with unusual findings associated with a heterozygous mutation in the SAG gene. *Arch Ophthalmol.* 2011;129:1375–1376.
102. Dryja TP. Molecular genetics of Oguchi disease, fundus albipunctatus, and other forms of stationary night blindness: LVII Edward Jackson Memorial Lecture. *Am J Ophthalmol.* 2000;130:547–563.
103. Hashimoto H, Kishi S. Shortening of the rod outer segment in Oguchi disease. *Graefes Arch Clin Exp Ophthalmol.* 2009;247:1561–1563.
104. Hayashi T, Tsuzuranuki S, Kozaki K, Urashima M, Tsuneoka H. Macular dysfunction in Oguchi disease with the frequent mutation 1147delA in the *SAG* gene. *Ophthalmic Res.* 2011;46:175–180.
105. Chen J, Simon MI, Matthes MT, Yasumura D, LaVail MM. Increased susceptibility to light damage in an arrestin knockout mouse model of Oguchi disease (stationary night blindness). *Invest Ophthalmol Vis Sci.* 1999;40:2978–2982.
106. Sommer ME, Farrens DL. Arrestin can act as a regulator of rhodopsin photochemistry. *Vision Res.* 2006;46:4532–4546.
107. Nakamachi Y, Nakamura M, Fujii S, Yamamoto M, Okubo K. Oguchi disease with sectoral retinitis pigmentosa harboring adenine deletion at position 1147 in the arrestin gene. *Am J Ophthalmol.* 1998;125:249–251.
108. Nakazawa M, Wada Y, Tamai M. Arrestin gene mutations in autosomal recessive retinitis pigmentosa. *Arch Ophthalmol.* 1998;116:498–501.
109. Isashiki Y, Ohba N, Kimura K, Sonoda S, Kakiuchi T, Ozawa T. Retinitis pigmentosa with visual fluctuations and arrestin gene mutation. *Br J Ophthalmol.* 1999;83:1197–1198.
110. Alloway PG, Howard L, Dolph PJ. The formation of stable rhodopsin-arrestin complexes induces apoptosis and photoreceptor cell degeneration. *Neuron.* 2000;28:129–138.
111. Iakhine R, Chorna-Ornan I, Zars T, et al. Novel dominant rhodopsin mutation triggers two mechanisms of retinal degeneration and photoreceptor desensitization. *J Neurosci.* 2004;24:2516–2526.

112. Chen J, Shi G, Concepcion FA, Xie G, Oprian D, Chen J. Stable rhodopsin/arrestin complex leads to retinal degeneration in a transgenic mouse model of autosomal dominant retinitis pigmentosa. *J Neurosci.* 2006;26:11929–11937.
113. Ohguro H, Chiba S, Igarashi Y, Matsumoto H, Akino T, Palczewski K. Beta-arrestin and arrestin are recognized by autoantibodies in sera from multiple sclerosis patients. *Proc Natl Acad Sci USA.* 1993;90:3241–3245.
114. Gorczyca WA, Ejma M, Witkowska D, et al. Retinal antigens are recognized by antibodies present in sera of patients with multiple sclerosis. *Ophthalmic Res.* 2004;36:120–123.
115. Forooghian F, Cheung RK, Smith WC, O'connor P, Dosch HM. Enolase and arrestin are novel nonmyelin autoantigens in multiple sclerosis. *J Clin Immunol.* 2007; 27:388–396.
116. Raine CS. Multiple sclerosis: a pivotal role for the T cell in lesion development. *Neuropathol Appl Neurobiol.* 1991;17:265–274.
117. Sawyer RA, Selhorst JB, Zimmerman LE, Hoyt WF. Blindness caused by photoreceptor degeneration as a remote effect of cancer. *Am J Ophthalmol.* 1976;81:606–613.
118. Adamus G, Chan CC. Experimental autoimmune uveitides: multiple antigens, diverse diseases. *Int Rev Immunol.* 2002;21:209–229.
119. Misiuk-Hojło M, Ejma M, Gorczyca W, et al. Cancer-associated retinopathy in patients with breast carcinoma. *Arch Immunol Ther Exp.* 2007;55:261–265.
120. Gery I, Mochizuki M, Nussenblatt RB. Retinal specific antigens and immunopathogenic processes they provoke. *Prog Retin Res.* 1986;5:75–109.

CHAPTER ELEVEN

β-Arrestins in the Central Nervous System

Camille Latapy[*,†], **Jean Martin Beaulieu**[*,†]
[*]Département de Psychiatrie et de Neurosciences, Faculté de Médecine, Université Laval, Pavillon Ferdinand-Vandry, Québec City, Quebec, Canada
[†]Institut Universitaire en Santé Mentale de Québec (IUSMQ), 2601, chemin de la Canardière, Québec City, Quebec, Canada

Contents

1. Introduction: Arrestins and GRK in the Central Nervous System 268
2. Arrestins and GRK in Dopaminergic Neurotransmission 270
   2.1 Generalities 270
   2.2 DA receptors and GRK 270
   2.3 Arrestins in DA receptor desensitization and internalization 272
   2.4 Alternative pathways 272
3. Arrestins in Serotonergic Neurotransmission 275
   3.1 Generalities 275
   3.2 5HT receptors and GRK 276
   3.3 Arrestins and 5HT receptor internalization 276
   3.4 Alternative pathways 278
4. Arrestins in Noradrenergic Neurotransmission 278
   4.1 Generalities 278
   4.2 Noradrenergic receptors and GRK 279
   4.3 Arrestins and noradrenergic receptor internalization 279
   4.4 Alternative pathways 280
5. Arrestins in Opioid Receptor Signaling 281
   5.1 Generalities 281
   5.2 Arrestins and MOR internalization 281
   5.3 Alternative pathways 283
6. Arrestins and Corticotropin Receptors 283
   6.1 Generalities 283
   6.2 CRF receptors and GRK 284
   6.3 Arrestins and CRF receptor internalization 285
7. Other β-arrestin-Dependent Behavior 285
   7.1 Cannabinoid receptors 285
   7.2 Ethanol preference 286
8. Conclusions 286
Acknowledgment 287
References 287

Progress in Molecular Biology and Translational Science, Volume 118
ISSN 1877-1173
http://dx.doi.org/10.1016/B978-0-12-394440-5.00011-5

Abstract

β-Arrestins 1 and 2 are ubiquitously expressed proteins that play a dual role in G protein-coupled receptor (GPCR) signaling. On the one hand, arrestins are central to the termination of G protein-mediated receptor signaling and subsequent clathrin-dependent internalization. On the other hand, these proteins act as molecular scaffolds for G protein-independent GPCR signaling. This review provides an overview of how these dual functions of arrestins contribute to the biological outcomes associated to the activation of different brain GPCR. It also explores recent evidence suggesting how the dual function of arrestins can lead to the development of more selective pharmacological approaches for the treatment of central nervous system disorders such as chronic pain, bipolar disorder, major depression, and schizophrenia. Development of such approaches may lead to new drugs having better clinical efficacy and lesser side effects.

1. INTRODUCTION: ARRESTINS AND GRK IN THE CENTRAL NERVOUS SYSTEM

Arrestins and G protein receptor kinases (GRK) are major players in cell signaling by G protein-coupled receptors (GPCR), the most abundant membrane receptor class.[1–4] Arrestins serve as scaffold proteins. Of the four arrestin proteins, isoforms 1 and 4 are predominant in the visual system, whereas arrestins 2 and 3 are ubiquitously expressed. In the adult brain, arrestin 2, classically named β-arrestin 1, is 10–20 times more abundant at the protein level despite only a two-fold difference in mRNA expression over arrestin 3, also known as β-arrestin 2.[5] Although expression of both isoforms is distributed throughout the brain, β-arrestin 2 is the predominant isoform in limbic areas such as the hypothalamus and extended amygdala. Therefore, the hypothalamus is the only major brain area where the β-arrestin 2 mRNA level is comparable to that β-arrestin 1.[5]

As hinted by their name, arrestins are major components of a protein complex shutting down G protein-mediated signaling by GPCR (Fig. 11.1). Following agonist binding, GPCR activation results in activation of G proteins, which regulate different downstream effectors through various combinations of Gα and Gβγ subunits. β-Arrestins are recruited to the receptor following its phosphorylation by GRK to trigger clathrin-mediated receptor internalization.[6–10] Arrestin recruitment to phosphorylated GPCR inhibits signal transduction by preventing further receptor coupling to G proteins. Following, or in parallel to, their action on G protein-mediated signaling, there is evidence that, at least for some

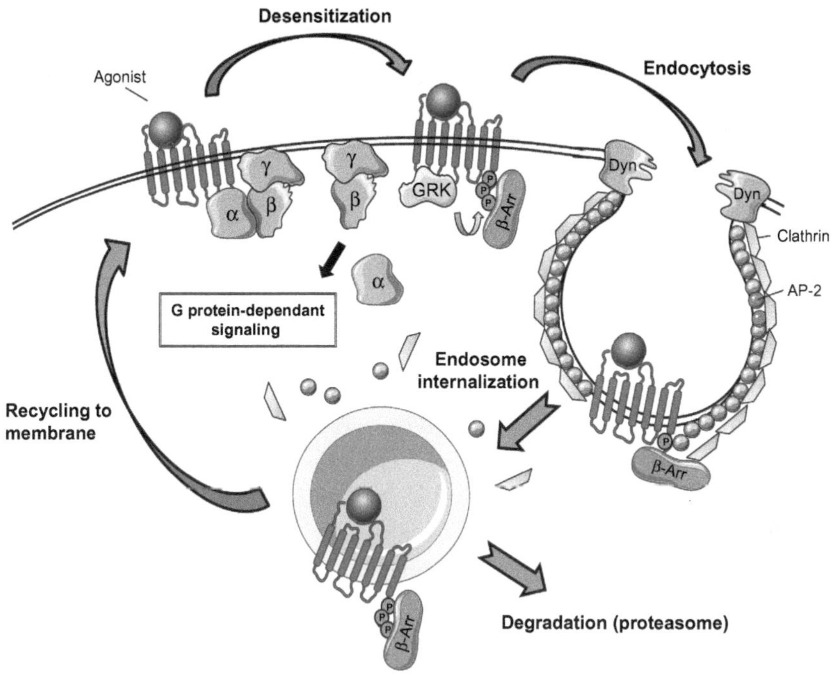

Figure 11.1 Schematic representation of β-arrestin-dependent regulation of GPCR signaling. In response to agonist stimulation, and considering the case of a monomeric receptor, the GPCR conformational change leads to G protein activation and G protein-dependent signaling, for example, mediated by adenylyl cyclase, cAMP, and PKA. Subsequently, the GRK induced receptor phosphorylation and recruitment of the multifunctional scaffolding protein, β-arrestin. The receptor/β-arrestin binding prevents further G protein activation, producing desensitization. The receptor can then be internalized via the formation of the receptor–protein complex: GPCR/β-arrestin/AP2/Clathrin. Once receptor endocytosis is completed via the action of dynamin (Dyn), the endosome is internalized and the complex disassembled. It can now be degradated or recycled to membrane.

GPCR, arrestins act as direct mediators of cell signaling by scaffolding various signaling molecules such as kinases and phosphatases. Therefore, a deficiency of one β-arrestin isoform can at the same time enhance the G protein-mediated signaling while abolishing the β-arrestin-mediated component of the signaling responses elicited by a given receptor.

Several neurotransmitters mediate their effects in the brain by activation of GPCR. Furthermore, more than 90% of all non-olfactory GPCR are expressed in the brain,[11] where they regulate multiple vital functions and control mood. Here, we will provide an overview of recent research examining the impact of the dual function of arrestins in relation to specific neurotransmitter systems. In doing so, we will concentrate mostly on work conducted directly in the brain tissue using various animal models of altered neurotransmission or arrestin expression.

2. ARRESTINS AND GRK IN DOPAMINERGIC NEUROTRANSMISSION

2.1. Generalities

Dopamine (DA) neurotransmission is implicated in a wide range of functions such as locomotion, cognition, emotions, neuroendocrine control, and reward.[12] It can thus be expected that abnormal DA neurotransmission is associated with several neurological and psychiatric disorders including, most notably, Parkinson's disease, Huntington's disease, and schizophrenia. DA signaling is mediated by five DA receptors separated in two families on the basis of their effect on cAMP production. Class 1 DA receptors consist of the D1 and D5 DA receptors (D1R and D5R) that stimulate the activity of adenylate cyclase (AC) and the production of cAMP[13,14] by coupling to Gαs subunit containing G proteins. Class 2 DA receptors are the D2R, D3R, and D4R receptors, which are coupled to Gαi and inhibit AC. In the striatum, these two classes of receptors are present in distinctive neuronal populations with a high level of segregation and their distinct projections distinguish between the direct and indirect striatonigral pathways.[15,16] Nevertheless, a small population of neurons expresses both D1 and D2 receptors, which may change the signaling properties of these receptors following the formation of receptor heterodimers.[17,18]

2.2. DA receptors and GRK

Various studies have suggested the importance of GRK, notably GRK2, for DA receptor regulation *in vitro*.[19–23] Mice lacking one allele of the GRK2 gene display a slight enhancement of responsiveness to cocaine, which increases DA tone by preventing its reuptake. This is suggestive of a role for GRK2 in the regulation of DA receptor desensitization *in vivo*. However, direct evidence for this is not currently available. A major reason is the embryonic lethality of GRK2 knockout mice, which has precluded the definitive examination of the role of this kinase in the regulation of DA receptor functions *in vivo*. While conditional GRK2 knockouts have been developed and used successfully for the study of the functions of GRK2 in other neurotransmitter systems,[24] this approach remains to be applied to the study of the role of GRK2 in DA receptor functions.

That being said, indirect evidence supports a role of GRK2 in DA receptor regulation. For instance, alterations in GRK2 expression levels have been reported in DA-related mood disorders and in response to pharmacologic treatments in different *in vivo* systems.[25,26] Furthermore, lesions of dopaminergic neurons in primates are followed by an increased expression of GRK2 that can be reversed by chronic levodopa treatment.[27] However, these data were not fully replicated in rodents[28] despite a modulation of GRK2 level by chronic or acute cocaine treatments.[29] There is also evidence that the neuronal calcium sensor 1, a regulator of GRK2 activity, is involved in D2R desensitization.[25] Finally, evidence obtained from mice lacking the fragile X mental retardation protein (FMRP) suggests a role of GRK2 in the regulation of forebrain D1R. In this model, the lack of an interaction between FMRP and GRK2 would result in hyperphosphorylation and desensitization of D1R that can be normalized following GRK2 inhibition.[30]

Despite its lower expression levels, GRK3 is also expressed in CNS and with a similar expression pattern as GRK2.[31,32] GRK3 expression can be enhanced by dopaminergic stimulation and reduced by unpredictable stress.[33,34] At the functional level, absence of GRK3 has been shown to decrease locomotor responses to cocaine and to the nonselective DA receptor agonist apomorphine. However, the molecular mechanism underlying these effects has not yet been elucidated.

GRK6 is also ubiquitously expressed in the brain but is principally found in the striatum and in the cell bodies from substantia nigra DA neurons.[31] In striatum, GRK6 is highly expressed in DA sensitive medium spiny neurons and in cholinergic interneurons.[35] Impaired desensitization of DA receptors was shown in GRK6-deficient mice. These mice are also hyperresponsive in tests of locomotor sensitivity to cocaine, amphetamine, and morphine which all result in an exacerbation of dopaminergic neurotransmission.[35,36] Studies performed with this model also revealed that the absence of GRK6 results in higher G protein coupling of striatal D2R without affecting D1R. This results in a greater locomotor response for direct D2R agonists and a rise in the affinity for D2R agonists in the absence of GRK6.[35,37] This suggests a GRK6-dependent regulation of postsynaptic striatal D2R desensitization.

2.3. Arrestins in DA receptor desensitization and internalization

Several *in vitro* studies support a role for β-arrestin 1[38,39] and β-arrestin 2 in D1R, D2R, and D3R desensitization.[4,38–40] Both β-arrestin isoforms are expressed along with DA receptors in striatal neurons.[41] Furthermore, there is evidence for a role of β-arrestin 1 in D2R internalization in striatal neurons.[42,43] However, behavioral studies of the role of β-arrestin isoforms have provided little support for this *in vivo*, since mice lacking either β-arrestin isoform are not hypersensitive to DA receptor agonists or to drugs such as amphetamine, cocaine, and morphine that exert part of their action by increasing DA tone. More specifically, mice deficient in β-arrestin 1 do not show any impairment beyond a reduction of cocaine and apomorphine-induced locomotor responses,[4] while β-arrestin 2-deficient mice show attenuated behavioral responses following amphetamine, morphine and apomorphine administration, and normal or slightly attenuated responses to cocaine.[44,45] A simplistic interpretation of these data would suggest that β-arrestin 1 and 2 are not directly involved in DA receptor desensitization. However, the apparent lack of effect β-arrestin also points toward more complex roles of β-arrestins in mediating both DA receptor desensitization and signaling.

2.4. Alternative pathways

Substantial evidence supports a role for β-arrestin in the DA-dependent regulation of AKT/glycogen synthase kinase 3 (GSK3) signaling. AKT, also termed protein kinase B, is involved in cellular processes such as transcription, glucose metabolism, proliferation, migration, and insulin and neurotrophin response via receptor tyrosine kinase (RTK) stimulation.[46–48] Following activation of RTK and some GPCR, phosphatidylinositol-3-kinase converts phosphatidylinositol-2-phosphate into phosphatidylinositol-3-phosphate (PIP3). PIP3 interacts with AKT and initiates its recruitment to the plasma membrane, where AKT is activated by phosphorylation of residues Thr308 and Ser473 by proline-directed kinase (PDK)1 and PDK2/rictor-mTOR, respectively.[48,49] Activated AKT phosphorylates in turn several proteins including GSK3[50] (Fig. 11.2). Originally identified as a component of the insulin signaling pathway, GSK3 is also implicated in a wide range of cellular processes, such as embryonic development, cell differentiation and survival, and Wnt signaling, as well as serotonergic and dopaminergic transmission.[51–54] Two GSK3 isoforms (GSK3α and GSK3β) are expressed in mammalian cells.

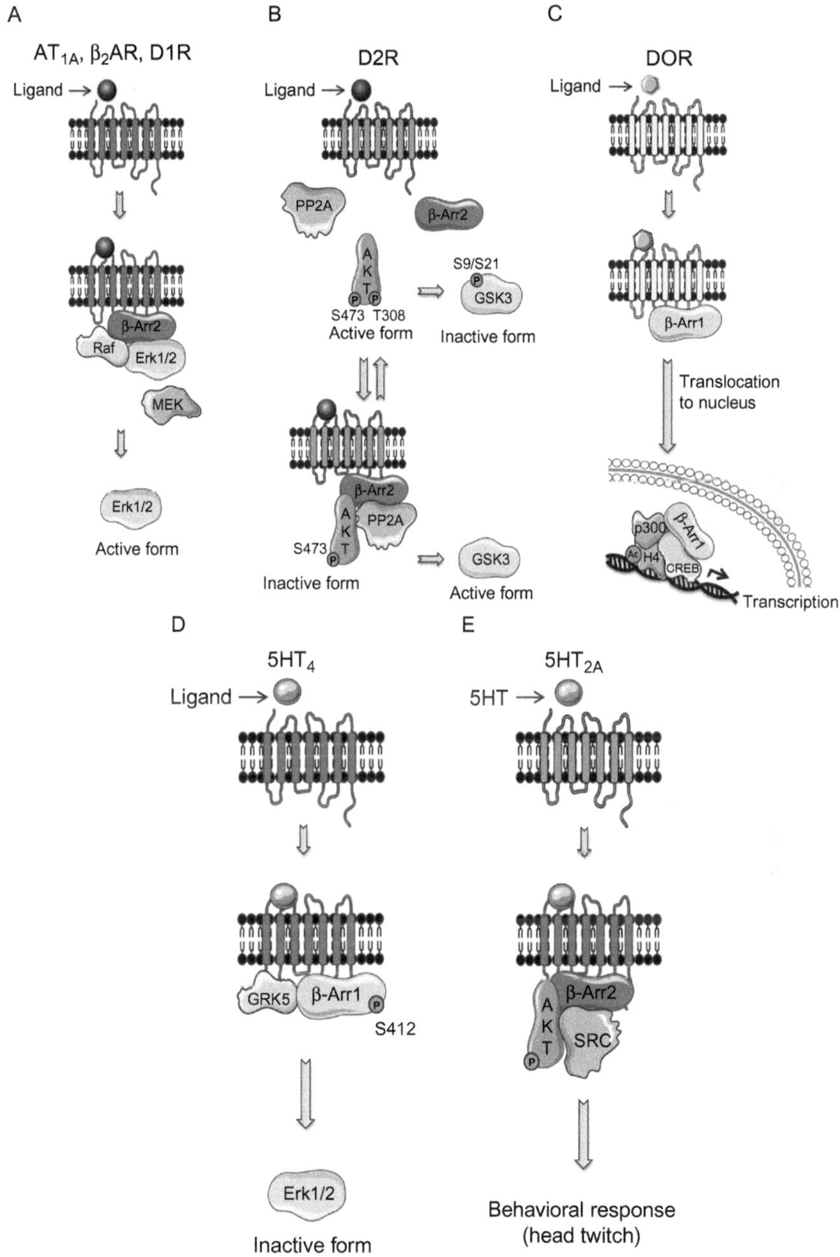

Figure 11.2 Illustration of some arrestin-dependent signaling complexes in the central nervous system. (A) The type 1 of dopamine receptor (D1R), AT1A, β2AR, as well as potentially 5HT receptors, regulates formation of a protein complex composed of β-arrestin 2/ERK/Raf, and putatively MEK, leading to ERK1/2 activation and regulation

(Continued)

They are constitutively active and negatively regulated by phosphorylation on Ser21 for GSK3α and Ser9 for GSK3β.[55–59]

The involvement of β-arrestin 2 in D2R-regulated AKT inhibition has been demonstrated.[45,52] The hypothesis that dopaminergic signaling is regulated by a β-arrestin 2-dependent D2R signaling pathway is supported by both behavioral and direct *in vivo* biochemical observations in pharmacological and genetic models. Thus, β-arrestin 2-deficient mice exhibit a lower locomotor response to DA-dependent drugs, and the increased locomotor activity observed in mice lacking the dopamine transporter (DAT) is abolished by the absence of β-arrestin 2 in the DAT/β-arrestin 2 double knockout mice.[4,45] While both amphetamine and apomorphine treatments inhibit AKT phosphorylation in the striatum of WT animals, this biochemical effect of dopaminergic drugs is lost in the absence of β-arrestin 2. Furthermore, while the genetically increased DA tone produced by DAT depletion affects AKT activity, it fails to act in the absence of β-arrestin 2, suggesting an important role of this scaffolding protein in AKT regulation by DA. This has been confirmed by the discovery that a protein complex formed between AKT, protein phosphatase 2A (PP2A), and β-arrestin 2 is involved in D2R signaling.[45] In this complex, PP2A dephosphorylates and deactivates AKT, which in turn fails to inhibit GSK3 and leaves the kinase activated.[45,52] These results showed for the first time an *in vivo* role of β-arrestin 2 scaffolds in GPCR regulation beyond in its well-known function in desensitization of receptors. Thus, β-arrestin 2 signaling has a functional role in the expression of DA-associated behaviors.

Among the mood stabilizers used clinically, lithium is known for its therapeutic effects in the treatment of bipolar affective disorder.[60–63] These

Figure 11.2 Cont'd—of downstream MAPK pathway effectors. (B) D2R activation modulates a protein complex composed of β-arrestin 2/PP2A/AKT. This complex induces inhibition of AKT via PP2A and the subsequent activation of GSK3 and related pathways. (C) Besides the effect of β-arrestin 2 on ERK1/2, β-arrestin 1 can translocate to the nucleus in response to activation of DOR, and potentially other GPCR. There, a protein complex is formed containing the histone acetyltransferase p300 and the transcription factor CREB. This complex promotes chromatin reorganization and β-arrestin-dependent gene transcription. (D) GRK5 is a direct interaction partner of 5HT4 receptors and a negative regulator of the ERK1/2 signaling pathway. GRK5 inhibits G protein-independent signaling pathways, including activation of ERK1/2, by phosphorylating β-arrestin 1 on Ser412, inhibiting both β-arrestin 1 binding to clathrin and receptor internalization. (E) Serotonin 5HT2A receptors induce β-arrestin 2-dependent ERK1/2 activation and also regulate the AKT pathway by forming a complex containing c-Src and PI3 kinase. This complex has been implicated in serotonergic behavioral responses like head twitch.

therapeutic effects are thought to be mediated, at least in part, by a direct or indirect inhibition of GSK3 activity.[64,65] In mouse models, the indirect inhibition of GSK3 through AKT and β-arrestin 2 in response to lithium has been confirmed. The indirect inhibition of GSK3 by lithium is mediated by the disruption of the β-arrestin 2/AKT/PP2A protein complex, at least in the striatum.[66] In β-arrestin 2-deficient mice, lithium fails to activate AKT and inhibit GSK3 in the striatum, and does not produce the antidepressant and antimanic effects observed in control mice. The hypothesis that lithium acts through β-arrestin 2/AKT/GSK3 modulation of dopaminergic signaling is further supported by recent findings that the AKT1 isoform is essential for behavioral regulation by lithium in mice.[67] More and more evidence implicates AKT/GSK3 pathway modulation in the effects of mood stabilizers such as valproic acid and lamotrigine.[68,69] All these mood stabilizers decrease phosphorylation of GSK3 substrates.[70] Thus, β-arrestin 2-mediated regulation of the AKT/GSK3 pathway may be a general mechanism of action for a variety of chemically distinct mood stabilizers, and targeting this pathway could be a pharmacologic approach for the treatment of neurologic or psychiatric dopaminergic diseases.[45,52]

3. ARRESTINS IN SEROTONERGIC NEUROTRANSMISSION

3.1. Generalities

Serotonin (5HT) is a neurotransmitter widely implicated in biological functions such as circadian rhythm, food intake, pain, learning, mood, sexual behaviors, aggressiveness, and anxiety. Except the third type of serotonergic receptors (5-HT3R), all serotonergic receptors are GPCR and are widely expressed throughout the central nervous system. Like other GPCR, serotonergic receptors are modulated by arrestins, notably receptors 5HT2A, 5HT2C, and 5HT4.[71] Like the 5HT2A receptor, 5HT2C is expressed in neurons[72] and is thought to be implicated in mental diseases such as schizophrenia, anxiety, and depression (reviewed in Ref. 73).

In humans, 5HT2A receptor activation is associated with hallucinations, and all serotonergic hallucinogens have affinity for this receptor.[74–76] Deregulation of 5HT2A receptors has been shown in various mental health diseases such as depression, suicide, and schizophrenia.[77] Both β-arrestin isoforms and 5HT2A receptors are coexpressed in a population of cortical pyramidal neurons, notably in intracellular vesicles where colocalization of β-arrestin 1 and the receptor has been demonstrated.[78] *In vivo* and *in vitro* studies demonstrate an interaction between arrestins and the 5HT2A receptor. Thus, the third intracellular domain of murine 5HT2A receptors interacts with both β-arrestin 1 and 2.[78] These studies indicate a potential *in vivo* role of β-arrestin 1 in 5HT2A receptor regulation and modulation in the frontal cortex. In mental diseases, the level of β-arrestin 1 is reduced in lymphocytes of depressed patients, and the level of diminution is correlated with the severity of symptoms.[79] In a reciprocal way, chronic treatment with antidepressant compounds such as imipramine, desipramine (DMI), and fluvoxamine is associated with an augmentation of β-arrestin 1 expression in the cortex and hippocampus in rat brains.[79]

Elevation of 5HT in rodent brains via administration of 5-hydroxytryptophan (5HTP), or 5HT2A receptor activation in response to hallucinogenic drugs, induces a characteristic behavior in control animals, head twitch.[80,81] This serotonin-specific behavior can be observed following an elevation of 5HT levels in the brain and is induced by administration of the serotonin precursors, tyrptophan or 5HTP, inhibition of serotonin reuptake by selective serotonin reuptake inhibitors, or inhibitors of serotonin degradation like inhibitors of monoamine oxidase A inhibitors.[80] Considering this, it is not surprising that blockade of 5HT2A receptors is a critical action underlying the antipsychotic potential of clozapine, risperidone, and olanzapine compounds.[82]

3.2. 5HT receptors and GRK

Serotonin receptor internalization is likely to be β-arrestin and GRK dependent, as it was described for 5HT4 receptors associated with the stress response.[83] Overexpression of GRK2 significantly diminished cAMP production, suggesting a negative regulation of the receptor by GRK2. This mechanism has been observed for different receptor variants expressed in COS-7 cells, including the 5HT4A, 5HT4B, 5HT4E, and 5HT4F receptors.[84]

3.3. Arrestins and 5HT receptor internalization

In HEK293 cells expressing 5HT4R, serotonin stimulation initiates the recruitment of β-arrestin 1 and 2 to the plasma membrane and induces receptor internalization.[85] After the serotonin stimulation, the recruitment of β-arrestin 2 is dependent on phosphorylation by GRK2 of Ser/Thr clusters on the receptor C-terminal tail.[84] In humans, 9 5HT4 receptor variants are produced by alternative splicing. These variants differ in their C-terminal regions. The pharmacologic properties seem to be shared among the variants, but the downstream signaling and regulation of these receptors can be different.[84,86] 5HT4 receptors can be internalized and desensitized in a conventional way via GRK and β-arrestin. Nevertheless, 5HT4A and 5HT4B signaling may depend on different internalization processes according to their different subcellular distribution.[87]

5HT2A receptors can also be internalized via a β-arrestin-dependent pathway. Studies using HEK293 cell lines demonstrate colocalization of 5HT2A receptors and β-arrestin, and membrane recruitment of both β-arrestin isoforms in response to agonists or the serotonin precursor, 5HTP.[88–90] In cortical neurons of control animals treated with 5HT, 5HT2A receptors are almost entirely intracellular, whereas in β-arrestin 1 and 2-deficient mice, or β-arrestin 2 knockout neurons, the receptors are present at the soma membrane, suggesting a role for arrestins in 5HT2A receptor trafficking. However, the internalization of receptors following treatment with the selective 5HT2A receptor agonist, 2,5-dimethoxy-4-iodoamphetamine (DOI), even in absence of β-arrestin isoforms.

From a phenotypic point of view, this interaction between 5HT2A receptors and β-arrestin results in signaling and trafficking processes that lead to behavior modulation such as head twitching.[71] The head twitch response induced by 5HTP administration appears to be β-arrestin 2 dose dependent. It fails to reach the wild-type level in β-arrestin 2 knockout mice and heterozygous mice display an intermediate phenotype.[91] Nevertheless, some physiological responses to a sudden rise in 5HT level associated with serotonin receptor activation, such as hypothermia and diarrhea, do not differ between control and β-arrestin 2-deficient mice.[91]

Other pharmacologic evidence indicates a β-arrestin 2-independent response to hallucinogenic compounds. While 5HTP has impaired efficacy to produce a head twitch response in β-arrestin 2 knockout mice, the selective hallucinogenic 5HT2A receptor agonist DOI does not show genotype-dependent efficacy for this behavior.[91] This suggests that while the head twitch response induced by 5HTP is a β-arrestin 2-dependent behavior, the one induced by DOI is not. This suggests β-arrestin 2-dependent signal transduction downstream of 5HT2A receptors exists but does not constitute an essential component of this 5HT2A receptor-modulated behavior.

3.4. Alternative pathways

The activation of 5HT2C receptors also leads to ERK1/2 activation in different cell lines[92,93] via both β-arrestin-dependent and -independent processes.[93] Activation of the ERK1/2 pathway by 5HT is dependent on 5HT2A receptor and β-arrestin in a MEF cellular model as well as *in vivo*.[91] In the mouse frontal cortex, serotonin and the 5HT2A receptor selective agonist DOI both induce ERK1/2 activation, but in the absence of β-arrestin 2 only DOI activates the pathway. Furthermore, the serotonin receptor-mediated activation of the AKT pathway via SRC and PI3 kinase is also β-arrestin dependent in cultured cortical neurons and *in vivo*[90] (Fig. 11.2).

It has also been shown that GRK5 is a direct interaction partner of 5HT4 receptors and a negative regulator of the ERK1/2 signaling pathway in response to serotonergic stimulation[85] (Fig. 11.2). In HEK cells and neurons, overexpression of GRK5 inhibits G protein-independent signaling pathways, including activation of ERK1/2. GRK5 phosphorylates the Ser412 residue of β-arrestin 1, leading to simultaneous inhibition of receptor internalization and β-arrestin 1 binding to clathrin, which is dependent upon Ser412 dephosphorylation.[94]

4. ARRESTINS IN NORADRENERGIC NEUROTRANSMISSION

4.1. Generalities

Epinephrine and norepinephrine (NE) neurotransmission is mediated predominantly by adrenergic receptors (AR). These endogenous catecholamines activate two classes of AR: α- and β-AR. All AR are GPCR, and β-AR were the first metabotropic neurotransmitter receptors identified. There are two subclasses of α-AR. While α_1 AR activation induces a slow depolarization mediated by inhibition of K^+ channels, activation of the α_2AR subclass induces a slow hyperpolarization. There are three subclasses of β-AR: β_1-, β_2-, and β_3AR, of which the β_1- and β_2AR are the major species expressed in the brain.[95]

Among α_2AR, another subdivision has been made and we can distinguish three α_2AR subtypes: α_{2A}, α_{2B}, and α_{2C}. All of these receptors are coupled to Gαi, and their activation leads to inhibition of AC and inhibition of cAMP production. Conversely, β-AR are coupled to Gαs proteins. The activation of Gαs leads to the activation of AC, generation of cAMP, and activation of the cAMP-dependent protein kinase A (PKA).[96] In the case of the β_2AR, PKA can phosphorylate the intracellular loops of the receptor, leading to a switch from Gαs to Gαi protein coupling which inhibits cAMP production and affects the signaling outcome of receptor activation.[97]

Of the different types of α-AR, α_{2A}ARs are the predominant subtype in the CNS.[98–100] Presynaptic α_{2A}AR located in the locus coeruleus function as key regulators of the noradrenergic system, controlling the synthesis and release of NE and also the release of other neurotransmitters. β-AR are expressed in both gray and white matter.[101–103] In the hippocampus, β-AR play an important role in regulating synaptic plasticity and memory consolidation.[104,105] Almost all nuclei are positive for β_1AR and β_2AR.[106] Of the three types of β-AR expressed in astrocytes, the β_2AR are the predominant one.[103,107–111] In astrocytes, β_2AR promote glucose uptake though GLUT1 and accelerate glycogen degradation via coupling to a Gαs and cAMP-dependent pathway.[112] From a clinical point of view, studies involving patients with depression have shown an upregulation of α_{2A}AR in the brain.[113–117]

4.2. Noradrenergic receptors and GRK

As observed with other GPCR in transfected cells, after agonist stimulation, the α_{2A}AR undergo rapid desensitization mediated by receptor phosphorylation by GRK.[118,119] While α_{2A}AR phosphorylation appears to be dependent on GRK2 and GRK3, the situation for the α_{2C}AR is less clear.[119,120] For this receptor, it seems that there is no agonist-induced promoted desensitization nor phosphorylation by GRK2, 3, 5, or 6. The formation of α_{2A}AR–α_{2C}AR heterodimers has been observed in cells and the α_{2C}AR component of the heterodimer seems to decrease the phosphorylation by GRK and subsequent recruitment of β-arrestin 2. From a mechanistic point of view, it appears that the α_{2A}AR conformation is less favorable for GRK2 phosphorylation when in an α_{2A}AR–α_{2C}AR heterodimer compared to a α_{2A}AR homodimer.[121]

4.3. Arrestins and noradrenergic receptor internalization

As has been previously described for other GPCR, the rapid agonist-induced desensitization of AR is driven by receptor phosphorylation and the subsequent binding of β-arrestins. The relationship between receptor phosphorylation by GRK and the ability to recruit β-arrestin 2 has been clearly demonstrated in a α_{2A}AR–α_{2C}AR dimerization model.[121]

In α_{2A}AR expressing cells, the β-arrestin is rapidly redistributed from the cytosol to the plasma membrane after agonist stimulation. This mechanism is absent in α_{2C}AR-expressing cells. When the receptors are coexpressed, there is a dose-dependent reduction in agonist-induced redistribution of β-arrestin 2 as the ratio of α_{2C}AR–α_{2A}AR expression is increased. This confirms that the loss of β-arrestin 2 recruitment observed with the heterodimer is due to the loss of GRK phosphorylation.[121]

In a pharmacological context, DMI, an antidepressant classically characterized as an inhibitor of NE reuptake, induces a robust internalisation of α_{2A}AR in transfected cells and primary neuron cultures that is indistinguishable from the one induced by NE.[122] Because this effect is lost in β-arrestin 1 and 2 knockout cells, it suggests that DMI is able to drive arrestin recruitment to the receptor and drive α_{2A}AR internalization via the canonical arrestin-dependent pathway. DMI induces a dose-dependent interaction between α_{2A}AR and β-arrestin 2 over its entire therapeutic range. It also promotes an interaction with β-arrestin 1 but with a weaker efficacy.[122]

4.4. Alternative pathways

In parallel to receptor internalization and contrary to what is observed with DMI stimulation, stimulation with NE in a α_{2A}AR-expressing cell model induces a robust activation ofboth ERK1/2 and AKT (Fig. 11.2). However, this effect is lost if the NE stimulation follows a long-term DMI treatment in control, but not in β-arrestin 1/2 knockout, murine embryo fibroblasts. This indicates that the DMI-induced α_{2A}AR downregulation of ERK1/2 signaling also requires arrestins.[122] The same treatment reduces ERK1/2 signaling in primary cultures of neurons. *In vivo* studies have shown that chronic DMI treatment leads to a clear reduction in synaptosomal α_{2A}AR expression in wild type, but not in β-arrestin 2 knockout mice. This indicates that long-term DMI exposure can downregulate synaptic α_{2A}AR expression in a β-arrestin 2-dependent fashion.[122]

β_2AR expression is also found in the nucleus, suggesting a specific role for this receptor in nuclear functions.[106] Thus, a chronic stimulation of β_2AR leads to an activation of Mdm2 and the subsequent degradation of p53 in a β-arrestin 1-dependent manner.[123] In this case, β-arrestin appears to play a dual role. While in the cytosol, it mediates catecholamine-induced activation of AKT and Mdm2, while in the nucleus as adaptor for Mdm2–p53 interaction, leading to p53 ubiquitination and degradation.

Sedation can be induced in mice by stimulation of α_2AR with compounds such as UK14304 [5-bromo-6-(2-imidazolin-2-ylamino) quinoxaline]. This sedation effect can be assessed in mice with the rotarod test, which determines performance in motor coordination. While α_2AR stimulation using UK14304 decreases performance on this test in wild-type animals, β-arrestin 2 knockout animals are more resistant. To assess if this sedation resistance in β-arrestin 2 knockout animals is specific to α_2AR, the test has been performed using R(−)N6-(2-phenylisopropyl)adenosine (R-PIA), an adenosine 1 receptor agonist. R-PIA-induced sedation is preserved in β-arrestin 2 knockout mice, demonstrating that these mice are not resistant to adenosine 1 receptor-mediated sedation, and that unlike the adenosine receptor, α_2AR responsiveness is β-arrestin 2-dependent.[124]

5. ARRESTINS IN OPIOID RECEPTOR SIGNALING

5.1. Generalities

Mu-opioid receptors (MOR) are coupled to G proteins, and their activity in periaqueductal gray matter and brainstem is associated with analgesic effects. MOR signaling in the CNS is complex, and its effects on such diverse processes as analgesia, constipation, respiratory depression, thermal regulation, locomotor activity, and narcotic addiction, dependence, and tolerance indicate that MOR signaling is highly context dependent. Among the physiological effects induced by morphine treatment, DA release in the striatum, drug reinforcement, and hypothermia, but not physical dependence, is affected in β-arrestin 2 knockout mice.[44,125,126] Furthermore, opiate-induced constipation, respiratory suppression, and hyperlocomotor activity are diminished in β-arrestin 2 knockout animals.[44,127] These data suggest that β-arrestin 2 can act as both a negative and a positive MOR regulator.

5.2. Arrestins and MOR internalization

Among the behavioral responses induced by β-arrestin 2 deletion, the first one reported is related to morphine-induced analgesia. When wild-type and β-arrestin 2 knockout mice are challenged in a hot plate test after morphine administration, the mutant mice exhibit prolonged latency.[125] This reveals defects in nociceptive perception of a thermal stimulus mediated by MOR.[128,129] MOR-G protein coupling is enhanced in β-arrestin 2 knockout mice as compared to wild-type animals, indicating a negative regulation of MOR by β-arrestin 2.[125,126] Thus, *in vivo* studies demonstrate that the absence of β-arrestin 2 delays the onset of MOR desensitization and amplifies its G protein coupling efficacy.[130] Furthermore, β-arrestin 2 knockout mice are resistant to opiate tolerance following repeated administration. After chronic exposure, MOR in wild-type animals are desensitized to the antinociceptive effect of morphine, but this phenomenon fails to occur in β-arrestin 2 knockout mice. The knockout mice do not develop antinociceptive tolerance to morphine in the hot plate test, as the lack of β-arrestin 2 allows MOR to engage in persistent G protein coupling.[126]

The tail flick test can be used to assess the role of β-arrestin 2 in nociception at the spinal cord level. In β-arrestin 2 knockout mice, results in this test are consistent with those obtained with the hot plate studies. Mutant mice exhibit enhanced and prolonged morphine-induced antinociception.[131] These results, and the role of β-arrestin 2 as a negative MOR regulator, have been confirmed by intrathecal injection of β-arrestin 2 antisense mRNA in rats. This selective β-arrestin 2 knockdown also enhances the response to morphine in the tail flick test.[132] After 7 days of morphine treatment, β-arrestin 2 knockout mice retain two-fold higher sensitivity to morphine as compared to wild-type mice but do develop some tolerance to morphine-induced analgesia in the tail flick test.[131] The tolerance develops by β-arrestin 2 knockout mice is less dramatic and delayed in onset when compared with wild-type animals and can be reversed by injection of chelerythrine, a PKC inhibitor. Thus, β-arrestin 2 also plays a significant role in morphine tolerance in at the spinal cord level, but in the absence of β-arrestin 2-mediated desensitization, other MOR effectors, such as PKC, are able to mediate a degree of tolerance.[133] Thus, regulation of MOR by β-arrestin 2 differs between the brain and the spinal cord indicating that the capacity of β-arrestins to modulate different aspects of MOR signaling varies depending on the cellular context.

MOR stimulation by different agonists such as morphine, etorphine, methadone, and fentanyl induces G protein coupling and analgesia in mice. Consistent with the concept that structurally distinct agonists can effect different downstream pathways with different efficacy, MOR regulation differs from one agonist to another. Morphine is a relatively weak activator of MOR desensitization. In contrast, etorphine, methadone, and fentanyl are very efficient at inducing MOR phosphorylation, β-arrestin 2 recruitment, and MOR internalization.[134–136] Thus, while knocking out β-arrestin 2 dramatically enhances the antinociceptive response to morphine, the responses of β-arrestin 2 knockout mice after etorphine, methadone, and fentanyl administration are similar to wild-type animals. This discrepancy between drugs is likely due to the efficiency with which the ligands stimulate receptor phosphorylation. While etorphine, methadone, and fentanyl induce a robust phosphorylation, morphine stimulation induces a weaker response. In the latter case, recruitment of β-arrestin 1 may permit near normal MOR desensitization and provide compensatory regulation.[137] Nonetheless, these data suggest that "biased" MOR ligands with reduced ability to stimulate β-arrestin recruitment might exhibit the desirable clinical properties of prolonged analgesia and diminished opiate tolerance.

The reinforcing and psychomotor effects of morphine involve MOR stimulation of the dopaminergic system. β-Arrestin 2-mediated MOR desensitization also appears to negatively regulate these effects *in vivo*.[57] Whereas the ability of cocaine to affect locomotor activity, behavioral sensitization, conditioned place preference, and striatal DA release is very similar between wild-type and β-arrestin 2 knockout mice, all of these responses to morphine are significantly reduced in the knockouts, indicating the β-arrestin 2 plays a more important role in modulating the dopaminergic effects of morphine than those produced by cocaine.

5.3. Alternative pathways

Some evidence supports a role for β-arrestin 1 in nuclear signaling by δ-opioid receptors (DOR). In transfected HEK cells and neuronal SK-N-SH cells, DOR activation causes β-arrestin 1 to translocate to the nucleus, where it is selectively enriched at specific promoters, such as p27 and c-fos [147A]. Once there, it facilitates recruitment of the histone acetyltransferase, p300, leading to increased local histone H4 acetylation and gene transcription (Fig. 11.2). This novel form of GPCR signaling to the nucleus via signal-dependent histone modification may underlie some of the transcriptional effects of opioid receptors in the brain.

6. ARRESTINS AND CORTICOTROPIN RECEPTORS

6.1. Generalities

Stress events are characterized by a release of corticotropin-releasing factor (CRF) from the hypothalamic paraventricular nucleus, which activates pituitary CRF receptors to promote adrenocorticotropic hormone (ACTH) secretion. In turn, ACTH receptor activation in the adrenal cortex modulates the synthesis and release of the steroid hormone, cortisol, an important mediator of the stress response.[138,139]

CRF_1 and CRF_2 receptors mostly differ in their N-terminal domain, while except for the four last amino acids, their intracellular domains are highly homologous. This variability at the N-terminus appears to affect the affinity for and efficacy of CRF.

CRF affinity for CRF_1 receptors is higher than for CRF_2 receptors. Thus, at a low CRF secretion level, CRF_1-dependent signaling pathways are more active than CRF_2-dependent ones, but both receptors are activated at high CRF concentrations. The activation of CRF_1 receptors induces anxiety-like responses which have been widely implicated in affective disorders and major depression,[140] suggesting that CRF_1 receptor antagonists may prove beneficial in anxiety and depression treatment.[141,142]

Surprisingly, the importance of CRF_1 receptors in depressive-like behaviors seems inconsistent with the results obtained in behavioral despair tests, such as the Porsolt swim test and tail suspension test. In these tests, treatment with antidepressant drugs that act on monoamine neurotransmission reduce immobility, an effect that is interpreted as enhanced motivation to confront. Contrary to expectations, CRF_1 receptor stimulation via intracerebroventricular injection of a CRF_1 receptor agonist reduces immobility.[143,144] These differences may be in part explained by the choice of animal model and the interpretation of results. While in some cases prolonged immobility can be representative of a depression-like behavior, the same response could also be interpreted as a diminished stress response.

As with CRF_1 receptors, the *in vivo* role of CRF_2 receptors in stress and anxiety is controversial. The role in CRF_2 receptors in anxiety and depressive physiopathology is even less clear. CRF_2 receptor knockout mice have an increased or normal anxiety responses, while blocking receptor pharmacologically can both increase or decrease anxiety-like response.[138,140,145–147] While some evidence supports the idea that CRF_2 receptors reestablish homeostasis by counteracting CRF_1 responses, other theories implicate these receptors in the passive anxiety and depression response.[148]

6.2. CRF receptors and GRK

CRF receptors are widely regulated by arrestins and GRK. The receptor contains seven putative sites of GRK phosphorylation on the C-terminal domain that may be involved in β-arrestin recruitment. Among the different

GRK, GRK3 and GRK6 translocate to the membrane after CRF stimulation in a HEK cell model and contribute to CRF_1 receptor desensitization.[149,150] Reducing GRK3 expression reduces CRF_1 receptor desensitization,[150] while GRK3 overexpression in HEK cells promotes a more rapid desensitization.[148] Contrary to CRF_2, the extreme C-terminus of the CRF_1 receptor contains a type 1 PDZ domain. Considering that PDZ domains can be phosphorylated by GRK or bind to PSD-95 and other modulatory proteins, its presence introduces another potential element of response modulation.[151]

6.3. Arrestins and CRF receptor internalization

CRF_1 receptors are rapidly activated and phosphorylated in the presence of agonist. In HEK cells and mouse cortical neurons, the recruitment of β-arrestin 2 is more efficient than β-arrestin 1.[152,153] Truncating the CRF_1 receptor C-terminus at Ser^{386} inhibits β-arrestin 2 translocation to the membrane, but truncating it at Ser^{412} does not.[153] Because reducing GRK3 expression impairs CRF_1 receptor desensitization, the phosphorylation site that most efficiently induces β-arrestin recruitment may be preferentially phosphorylated by this GRK.[149,154]

Urocortin neuropeptide 2 (UCN2) strongly activates CRF_{2A} receptors in human retinoblastoma Y79 cells.[155] The efficiency with which CRF_2 receptors (CRF_{2A} and CRF_{2B}) desensitize varies with different agonists. While UCN2 induces rapid and strong CRF_2 receptor desensitization, UCN3 and CRF are less efficient.[154] As with the CRF_1 receptor, CRF_{2B} receptors efficiently recruit β-arrestins. But unlike CRF_1 receptors, CRF_2 receptor regulation is more dependent on the β-arrestin 1 isoform, since decreasing cellular levels of β-arrestin 1 by siRNA reduces UNC2-induced desensitization and internalization to a greater degree than silencing β-arrestin 2.[154]

7. OTHER β-ARRESTIN-DEPENDENT BEHAVIOR

7.1. Cannabinoid receptors

Studies indicate that β-arrestin negatively regulates cannabinoid CB_1 receptor functions.[156] Δ^9-Tetrahydrocannabinol (Δ^9-THC) and CP55,940, two CB_1 receptor agonists, induce antinociception and decrease in body temperature in mice. While the response to Δ^9-THC is enhanced in β-arrestin 2 knockout mice for both of these phenotypes, CP55,940 induces these responses to the same extent in wild-type and β-arrestin 2 knockout animals.

7.2. Ethanol preference

Among other behaviors related to mood, emotional state, and psychiatric diseases, β-arrestins also seem to modulate ethanol consumption. Evidence points toward β-arrestin 2 as a positive regulator for ethanol abuse. In the absence of β-arrestin 2, a decreased behavioral response to ethanol is observed in rodents and, according to microarray experiments performed on rats, genetic variations in the *ARRB2* gene can influence alcohol preference.[157,158] Despite no changes in sucrose taste preference and quinine taste aversion in β-arrestin 2 knockout mice compared to wild-type animals, β-arrestin 2 knockout mice, as well as heterozygotes, exhibit reduced ethanol preference and voluntary intake.[159] β-Arrestin 2-depleted animals exhibit decreased ethanol-induced locomotion[159] consistent with the lack of characteristic locomotor responses induced by dopamine in β-arrestin 2 knockout mice.[45] This suggests that β-arrestins can modulate the acute response to ethanol in the same way that they modulate response to other psychoactive compounds, such as amphetamine and morphine.

8. CONCLUSIONS

GPCR represents the single largest group of pharmaceutical targets in the CNS. However, their broad involvement in overall physiological processes makes drug development free from side effects extremely difficult. Maintaining homeostasis across the wide range of functions that depend on GPCR-regulated pathways, GRK and arrestins depends on a fragile equilibrium. Relatively subtle changes in the levels of protein expression or coexpression, protein localization, or receptor function can disrupt this balance. Despite a potential for compensatory overlap among protein isoforms and signaling pathways, it appears that once disequilibrium occurs, the intracellular molecular consequences as well as the physiological ones can be dramatic. Psychiatric drugs used for clinical treatment modulate various aspects of this delicate equilibrium in an attempt to restore normal function in terms of receptor activity and protein complex formation[66] or even transcriptionally induced changes.[160] At present, biased agonists targeting the dopaminergic[161] and serotonergic systems appear to be the most promising field for the development of new therapeutics. Their potential to selectively modulate signaling pathways, whether G protein- or arrestin-dependent, downstream of the targeted GPCR constitutes a new field of

CNS research. Along with a better understanding of the roles of GPCR-heterodimerization-dependent signaling[162] and transactivation processes,[163,164] recent discoveries in the field of β-arrestin functions in the CNS offer the best hope of discovering novel therapeutics with fewer side effects.

ACKNOWLEDGMENT

J. M. B. is supported by a Canada Research Chair in Molecular Psychiatry.

REFERENCES

1. Premont RT. Once and future signaling: G protein-coupled receptor kinase control of neuronal sensitivity. *Neuromol Med.* 2005;7(1–2):129–147.
2. Penn RB, Pronin AN, Benovic JL. Regulation of G protein-coupled receptor kinases. *Trends Cardiovasc Med.* 2000;10(2):81–89.
3. Penela P, Ribas C, Mayor Jr F. Mechanisms of regulation of the expression and function of G protein-coupled receptor kinases. *Cell Signal.* 2003;15(11):973–981.
4. Gainetdinov RR, Premont RT, Bohn LM, Lefkowitz RJ, Caron MG. Desensitization of G protein-coupled receptors and neuronal functions. *Annu Rev Neurosci.* 2004;27:107–144.
5. Gurevich EV, Benovic JL, Gurevich VV. Arrestin2 and arrestin3 are differentially expressed in the rat brain during postnatal development. *Neuroscience.* 2002;109 (3):421–436.
6. Krupnick JG, Goodman Jr OB, Keen JH, Benovic JL. Arrestin/clathrin interaction. Localization of the clathrin binding domain of nonvisual arrestins to the carboxy terminus. *J Biol Chem.* 1997;272(23):15011–15016.
7. Laporte SA, Miller WE, Kim KM, Caron MG. beta-Arrestin/AP-2 interaction in G - protein-coupled receptor internalization: identification of a beta-arrestin binging site in beta 2-adaptin. *J Biol Chem.* 2002;277(11):9247–9254.
8. Kern RC, Kang DS, Benovic JL. Arrestin2/clathrin interaction is regulated by key N- and C-terminal regions in arrestin2. *Biochemistry.* 2009;48(30):7190–7200.
9. Kang DS, Kern RC, Puthenveedu MA, von Zastrow M, Williams JC, Benovic JL. Structure of an arrestin2-clathrin complex reveals a novel clathrin binding domain that modulates receptor trafficking. *J Biol Chem.* 2009;284(43):29860–29872.
10. Moore CA, Milano SK, Benovic JL. Regulation of receptor trafficking by GRKs and arrestins. *Annu Rev Physiol.* 2007;69:451–482.
11. Vassilatis DK, Hohmann JG, Zeng H,, et al. The G protein-coupled receptor repertoires of human and mouse. *Proc Natl Acad Sci USA.* 2003;100(8):4903–4908.
12. Carlsson A. A paradigm shift in brain research. *Science.* 2001;294(5544):1021–1024.
13. Spano PF, Govoni S, Trabucchi M. Studies on the pharmacological properties of dopamine receptors in various areas of the central nervous system. *Adv Biochem Psychopharmacol.* 1978;19:155–165.
14. Kebabian JW, Calne DB. Multiple receptors for dopamine. *Nature.* 1979;277 (5692):93–96.
15. Shuen JA, Chen M, Gloss B, Calakos N. Drd1a-tdTomato BAC transgenic mice for simultaneous visualization of medium spiny neurons in the direct and indirect pathways of the basal ganglia. *J Neurosci.* 2008;28(11):2681–2685.
16. Gerfen CR, Surmeier DJ. Modulation of striatal projection systems by dopamine. *Annu Rev Neurosci.* 2011;34:441–466.
17. Rashid AJ, So CH, Kong MM, et al. D1-D2 dopamine receptor heterooligomers with unique pharmacology are coupled to rapid activation of Gq/11 in the striatum. *Proc Natl Acad Sci USA.* 2007;104(2):654–659.

18. Valjent E, Bertran-Gonzalez J, Herve D, Fisone G, Girault JA. Looking BAC at striatal signaling: cell-specific analysis in new transgenic mice. *Trends Neurosci.* 2009;32 (10):538–547.
19. Ito K, Haga T, Lameh J, Sadee W. Sequestration of dopamine D2 receptors depends on coexpression of G-protein-coupled receptor kinases 2 or 5. *Eur J Biochem.* 1999;260 (1):112–119.
20. Iwata K, Ito K, Fukuzaki A, Inaki K, Haga T. Dynamin and rab5 regulate GRK2-dependent internalization of dopamine D2 receptors. *Eur J Biochem.* 1999;263 (2):596–602.
21. Tiberi M, Nash SR, Bertrand L, Lefkowitz RJ, Caron MG. Differential regulation of dopamine D1A receptor responsiveness by various G protein-coupled receptor kinases. *J Biol Chem.* 1996;271(7):3771–3778.
22. Sedaghat K, Tiberi M. Cytoplasmic tail of D1 dopaminergic receptor differentially regulates desensitization and phosphorylation by G protein-coupled receptor kinase 2 and 3. *Cell Signal.* 2011;23(1):180–192.
23. Evron T, Daigle TL, Caron MG. GRK2: multiple roles beyond G protein-coupled receptor desensitization. *Trends Pharmacol Sci.* 2012;33(3):154–164.
24. Daigle TL, Caron MG. Elimination of GRK2 from cholinergic neurons reduces behavioral sensitivity to muscarinic receptor activation. *J Neurosci.* 2012;32 (33):11461–11466.
25. Kabbani N, Negyessy L, Lin R, Goldman-Rakic P, Levenson R. Interaction with neuronal calcium sensor NCS-1 mediates desensitization of the D2 dopamine receptor. *J Neurosci.* 2002;22(19):8476–8486.
26. Koh PO, Undie AS, Kabbani N, Levenson R, Goldman-Rakic PS, Lidow MS. Up-regulation of neuronal calcium sensor-1 (NCS-1) in the prefrontal cortex of schizophrenic and bipolar patients. *Proc Natl Acad Sci USA.* 2003;100(1):313–317.
27. Bezard E, Gross CE, Qin L, Gurevich VV, Benovic JL, Gurevich EV. L-DOPA reverses the MPTP-induced elevation of the arrestin2 and GRK6 expression and enhanced ERK activation in monkey brain. *Neurobiol Dis.* 2005;18(2):323–335.
28. Ahmed MR, Bychkov E, Gurevich VV, Benovic JL, Gurevich EV. Altered expression and subcellular distribution of GRK subtypes in the dopamine-depleted rat basal ganglia is not normalized by l-DOPA treatment. *J Neurochem.* 2008;104(6):1622–1636.
29. Schroeder JA, McCafferty MR, Unterwald EM. Regulation of dynamin 2 and G protein-coupled receptor kinase 2 in rat nucleus accumbens during acute and repeated cocaine administration. *Synapse.* 2009;63(10):863–870.
30. Wang H, Wu U, Kim SS, et al. FMRP acts as a key messenger for dopamine modulation in the forebrain. *Neuron.* 2008;59(4):634–647.
31. Erdtmann-Vourliotis M, Mayer P, Ammon S, Riechert U, Hollt V. Distribution of G-protein-coupled receptor kinase (GRK) isoforms 2, 3, 5 and 6 mRNA in the rat brain. *Brain Res Mol Brain Res.* 2001;95(1–2):129–137.
32. Arriza JL, Dawson TM, Simerly RB, et al. The G-protein-coupled receptor kinases beta ARK1 and beta ARK2 are widely distributed at synapses in rat brain. *J Neurosci.* 1992;12(10):4045–4055.
33. Niculescu 3rd AB, Segal DS, Kuczenski R, Barrett T, Hauger RL, Kelsoe JR. Identifying a series of candidate genes for mania and psychosis: a convergent functional genomics approach. *Physiol Genomics.* 2000;4(1):83–91.
34. Taneja M, Salim S, Saha K, et al. Differential effects of inescapable stress on locus coeruleus GRK3, alpha2-adrenoceptor and CRF1 receptor levels in learned helpless and non-helpless rats: a potential link to stress resilience. *Behav Brain Res.* 2011;221(1):25–33.
35. Gainetdinov RR, Bohn LM, Sotnikova TD, et al. Dopaminergic supersensitivity in G protein-coupled receptor kinase 6-deficient mice. *Neuron.* 2003;38(2):291–303.

36. Sotnikova TD, Budygin EA, Jones SR, Dykstra LA, Caron MG, Gainetdinov RR. Dopamine transporter-dependent and -independent actions of trace amine beta-phenylethylamine. *J Neurochem.* 2004;91(2):362–373.
37. Seeman P, Weinshenker D, Quirion R, et al. Dopamine supersensitivity correlates with D2High states, implying many paths to psychosis. *Proc Natl Acad Sci USA.* 2005;102(9):3513–3518.
38. Kim KM, Valenzano KJ, Robinson SR, Yao WD, Barak LS, Caron MG. Differential regulation of the dopamine D2 and D3 receptors by G protein-coupled receptor kinases and beta-arrestins. *J Biol Chem.* 2001;276(40):37409–37414.
39. Oakley RH, Laporte SA, Holt JA, Barak LS, Caron MG. Molecular determinants underlying the formation of stable intracellular G protein-coupled receptor-beta-arrestin complexes after receptor endocytosis. *J Biol Chem.* 2001;276 (22):19452–19460.
40. Lan H, Teeter MM, Gurevich VV, Neve KA. An intracellular loop 2 amino acid residue determines differential binding of arrestin to the dopamine D2 and D3 receptors. *Mol Pharmacol.* 2009;75(1):19–26.
41. Bychkov E, Zurkovsky L, Garret MB, Ahmed MR, Gurevich EV. Distinct cellular and subcellular distributions of G protein-coupled receptor kinase and arrestin isoforms in the striatum. *PLoS One.* 2012;7(11):e48912.
42. Macey TA, Gurevich VV, Neve KA. Preferential interaction between the dopamine D2 receptor and arrestin2 in neostriatal neurons. *Mol Pharmacol.* 2004;66(6):1635–1642.
43. Skinbjerg M, Ariano MA, Thorsell A, et al. Arrestin3 mediates D(2) dopamine receptor internalization. *Synapse.* 2009;63(7):621–624.
44. Bohn LM, Gainetdinov RR, Sotnikova TD, et al. Enhanced rewarding properties of morphine, but not cocaine, in beta(arrestin)-2 knock-out mice. *J Neurosci.* 2003;23 (32):10265–10273.
45. Beaulieu JM, Sotnikova TD, Marion S, Lefkowitz RJ, Gainetdinov RR, Caron MG. An Akt/beta-arrestin 2/PP2A signaling complex mediates dopaminergic neurotransmission and behavior. *Cell.* 2005;122(2):261–273.
46. Cross DA, Alessi DR, Cohen P, Andjelkovich M, Hemmings BA. Inhibition of glycogen synthase kinase-3 by insulin mediated by protein kinase B. *Nature.* 1995;378 (6559):785–789.
47. Alessi DR, Andjelkovic M, Caudwell B, et al. Mechanism of activation of protein kinase B by insulin and IGF-1. *EMBO J.* 1996;15(23):6541–6551.
48. Scheid MP, Woodgett JR. PKB/AKT: functional insights from genetic models. *Nat Rev Mol Cell Biol.* 2001;2(10):760–768.
49. Jacinto E, Facchinetti V, Liu D, et al. SIN1/MIP1 maintains rictor-mTOR complex integrity and regulates Akt phosphorylation and substrate specificity. *Cell.* 2006;127 (1):125–137.
50. Rossig L, Badorff C, Holzmann Y, Zeiher AM, Dimmeler S. Glycogen synthase kinase-3 couples AKT-dependent signaling to the regulation of p21Cip1 degradation. *J Biol Chem.* 2002;277(12):9684–9689.
51. Woodgett JR. Physiological roles of glycogen synthase kinase-3: potential as a therapeutic target for diabetes and other disorders. *Curr Drug Targets Immune Endocr Metabol Disord.* 2003;3(4):281–290.
52. Beaulieu JM, Sotnikova TD, Yao WD, et al. Lithium antagonizes dopamine-dependent behaviors mediated by an AKT/glycogen synthase kinase 3 signaling cascade. *Proc Natl Acad Sci USA.* 2004;101(14):5099–5104.
53. Beaulieu JM, Zhang X, Rodriguiz RM, et al. Role of GSK3beta in behavioral abnormalities induced by serotonin deficiency. *Proc Natl Acad Sci USA.* 2008;105 (4):1333–1338.

54. Li X, Zhu W, Roh MS, Friedman AB, Rosborough K, Jope RS. In vivo regulation of glycogen synthase kinase-3beta (GSK3beta) by serotonergic activity in mouse brain. *Neuropsychopharmacology*. 2004;29(8):1426–1431.
55. Woodgett JR. Molecular cloning and expression of glycogen synthase kinase-3/factor A. *EMBO J*. 1990;9(8):2431–2438.
56. Frame S, Cohen P. GSK3 takes centre stage more than 20 years after its discovery. *Biochem J*. 2001;359(Pt. 1):1–16.
57. Yamada T, Katagiri H, Asano T, et al. Role of PDK1 in insulin-signaling pathway for glucose metabolism in 3T3-L1 adipocytes. *Am J Physiol Endocrinol Metab*. 2002;282(6): E1385–E1394.
58. Altar CA, Hunt RA, Jurata LW, et al. Insulin, IGF-1, and muscarinic agonists modulate schizophrenia-associated genes in human neuroblastoma cells. *Biol Psychiatry*. 2008;64 (12):1077–1087.
59. Woodgett JR. Regulation and functions of the glycogen synthase kinase-3 subfamily. *Semin Cancer Biol*. 1994;5(4):269–275.
60. Cade JF. Lithium salts in the treatment of psychotic excitement. *Med J Aust*. 1949;2 (10):349–352.
61. Schou M, Juel-Nielsen N, Stromgren E, Voldby H. The treatment of manic psychoses by the administration of lithium salts. *J Neurol Neurosurg Psychiatry*. 1954;17(4):250–260.
62. Phiel CJ, Klein PS. Molecular targets of lithium action. *Annu Rev Pharmacol Toxicol*. 2001;41:789–813.
63. Blanco C, Laje G, Olfson M, Marcus SC, Pincus HA. Trends in the treatment of bipolar disorder by outpatient psychiatrists. *Am J Psychiatry*. 2002;159(6):1005–1010.
64. Stambolic V, Ruel L, Woodgett JR. Lithium inhibits glycogen synthase kinase-3 activity and mimics wingless signalling in intact cells. *Curr Biol*. 1996;6(12):1664–1668.
65. O'Brien WT, Klein PS. Validating GSK3 as an in vivo target of lithium action. *Biochem Soc Trans*. 2009;37(Pt. 5):1133–1138.
66. Beaulieu JM, Marion S, Rodriguiz RM, et al. A beta-arrestin 2 signaling complex mediates lithium action on behavior. *Cell*. 2008;132(1):125–136.
67. Pan JQ, Lewis MC, Ketterman JK, et al. AKT kinase activity is required for lithium to modulate mood-related behaviors in mice. *Neuropsychopharmacology*. 2011;36(7):1397–1411.
68. Chen G, Huang LD, Jiang YM, Manji HK. The mood-stabilizing agent valproate inhibits the activity of glycogen synthase kinase-3. *J Neurochem*. 1999;72(3):1327–1330.
69. Li X, Bijur GN, Jope RS. Glycogen synthase kinase-3beta, mood stabilizers, and neuroprotection. *Bipolar Disord*. 2002;4(2):137–144.
70. Gould TD, Manji HK. Glycogen synthase kinase-3: a putative molecular target for lithium mimetic drugs. *Neuropsychopharmacology*. 2005;30(7):1223–1237.
71. Bohn LM, Schmid CL. Serotonin receptor signaling and regulation via beta-arrestins. *Crit Rev Biochem Mol Biol*. 2010;45(6):555–566.
72. Millan MJ. Serotonin 5-HT2C receptors as a target for the treatment of depressive and anxious states: focus on novel therapeutic strategies. *Therapie*. 2005;60(5):441–460.
73. Iwamoto K, Bundo M, Kato T. Serotonin receptor 2C and mental disorders: genetic, expression and RNA editing studies. *RNA Biol*. 2009;6(3):248–253.
74. Jakab RL, Goldman-Rakic PS. 5-Hydroxytryptamine2A serotonin receptors in the primate cerebral cortex: possible site of action of hallucinogenic and antipsychotic drugs in pyramidal cell apical dendrites. *Proc Natl Acad Sci USA*. 1998;95(2):735–740.
75. Aghajanian GK, Marek GJ. Serotonin and hallucinogens. *Neuropsychopharmacology*. 1999;21(2 Suppl.):16S–23S.
76. Nichols DE. Hallucinogens. *Pharmacol Ther*. 2004;101(2):131–181.
77. Meltzer HY. The role of serotonin in schizophrenia and the place of serotonin-dopamine antagonist antipsychotics. *J Clin Psychopharmacol*. 1995;15(1 Suppl. 1):2S–3S.

78. Gelber EI, Kroeze WK, Willins DL, et al. Structure and function of the third intracellular loop of the 5-hydroxytryptamine2A receptor: the third intracellular loop is alpha-helical and binds purified arrestins. *J Neurochem*. 1999;72(5):2206–2214.
79. Avissar S, Matuzany-Ruban A, Tzukert K, Schreiber G. Beta-arrestin-1 levels: reduced in leukocytes of patients with depression and elevated by antidepressants in rat brain. *Am J Psychiatry*. 2004;161(11):2066–2072.
80. Corne SJ, Pickering RW. A possible correlation between drug-induced hallucinations in man and a behavioural response in mice. *Psychopharmacologia*. 1967;11(1):65–78.
81. Corne SJ, Pickering RW, Warner BT. A method for assessing the effects of drugs on the central actions of 5-hydroxytryptamine. *Br J Pharmacol Chemother*. 1963;20: 106–120.
82. Meltzer HY. Action of atypical antipsychotics. *Am J Psychiatry*. 2002;159(1):153–154 author reply 154–5.
83. Bockaert J, Claeysen S, Compan V, Dumuis A. 5-HT4 receptors. *Curr Drug Targets CNS Neurol Disord*. 2004;3(1):39–51.
84. Barthet G, Gaven F, Framery B, et al. Uncoupling and endocytosis of 5-hydroxytryptamine 4 receptors. Distinct molecular events with different GRK2 requirements. *J Biol Chem*. 2005;280(30):27924–27934.
85. Barthet G, Carrat G, Cassier E, et al. Beta-arrestin1 phosphorylation by GRK5 regulates G protein-independent 5-HT4 receptor signalling. *EMBO J*. 2009;28 (18):2706–2718.
86. Claeysen S, Sebben M, Becamel C, Bockaert J, Dumuis A. Novel brain-specific 5-HT4 receptor splice variants show marked constitutive activity: role of the C-terminal intracellular domain. *Mol Pharmacol*. 1999;55(5):910–920.
87. Mnie-Filali O, Amraei MG, Benmbarek S, et al. Serotonin 4 receptor (5-HT4R) internalization is isoform-specific: effects of 5-HT and RS67333 on isoforms A and B. *Cell Signal*. 2010;22(3):501–509.
88. Bhatnagar A, Willins DL, Gray JA, Woods J, Benovic JL, Roth BL. The dynamin-dependent, arrestin-independent internalization of 5-hydroxytryptamine 2A (5-HT2A) serotonin receptors reveals differential sorting of arrestins and 5-HT2A receptors during endocytosis. *J Biol Chem*. 2001;276(11):8269–8277.
89. Bhattacharya A, Sankar S, Panicker MM. Differences in the C-terminus contribute to variations in trafficking between rat and human 5-HT(2A) receptor isoforms: identification of a primate-specific tripeptide ASK motif that confers GRK-2 and beta arrestin-2 interactions. *J Neurochem*. 2010;112(3):723–732.
90. Schmid CL, Bohn LM. Serotonin, but not N-methyltryptamines, activates the serotonin 2A receptor via a ss-arrestin2/Src/Akt signaling complex in vivo. *J Neurosci*. 2010;30(40):13513–13524.
91. Schmid CL, Raehal KM, Bohn LM. Agonist-directed signaling of the serotonin 2A receptor depends on beta-arrestin-2 interactions in vivo. *Proc Natl Acad Sci USA*. 2008;105(3):1079–1084.
92. Werry TD, Gregory KJ, Sexton PM, Christopoulos A. Characterization of serotonin 5-HT2C receptor signaling to extracellular signal-regulated kinases 1 and 2. *J Neurochem*. 2005;93(6):1603–1615.
93. Labasque M, Reiter E, Becamel C, Bockaert J, Marin P. Physical interaction of calmodulin with the 5-hydroxytryptamine2C receptor C-terminus is essential for G protein-independent, arrestin-dependent receptor signaling. *Mol Biol Cell*. 2008;19 (11):4640–4650.
94. Lin FT, Krueger KM, Kendall HE, et al. Clathrin-mediated endocytosis of the beta-adrenergic receptor is regulated by phosphorylation/dephosphorylation of beta-arrestin1. *J Biol Chem*. 1997;272(49):31051–31057.

95. MacDonald E, Scheinin M. Distribution and pharmacology of alpha 2-adrenoceptors in the central nervous system. *J Physiol Pharmacol.* 1995;46(3):241–258.
96. Pierce KL, Premont RT, Lefkowitz RJ. Seven-transmembrane receptors. *Nat Rev Mol Cell Biol.* 2002;3(9):639–650.
97. Daaka Y, Luttrell LM, Lefkowitz RJ. Switching of the coupling of the beta2-adrenergic receptor to different G proteins by protein kinase A. *Nature.* 1997;390(6655):88–91.
98. De Vos H, Vauquelin G, De Keyser J, De Backer JP, Van Liefde I. Regional distribution of alpha 2A- and alpha 2B-adrenoceptor subtypes in postmortem human brain. *J Neurochem.* 1992;58(4):1555–1560.
99. Sastre M, Garcia-Sevilla JA. Alpha 2-adrenoceptor subtypes identified by [3H]RX821002 binding in the human brain: the agonist guanoxabenz does not discriminate different forms of the predominant alpha 2A subtype. *J Neurochem.* 1994;63(3):1077–1085.
100. Wang R, Macmillan LB, Fremeau Jr RT, Magnuson MA, Lindner J, Limbird LE. Expression of alpha 2-adrenergic receptor subtypes in the mouse brain: evaluation of spatial and temporal information imparted by 3 kb of 5′ regulatory sequence for the alpha 2A AR-receptor gene in transgenic animals. *Neuroscience.* 1996;74 (1):199–218.
101. Aoki C. Beta-adrenergic receptors: astrocytic localization in the adult visual cortex and their relation to catecholamine axon terminals as revealed by electron microscopic immunocytochemistry. *J Neurosci.* 1992;12(3):781–792.
102. Sutin J, Shao Y. Resting and reactive astrocytes express adrenergic receptors in the adult rat brain. *Brain Res Bull.* 1992;29(3–4):277–284.
103. Zeinstra E, Wilczak N, De Keyser J. [3H]dihydroalprenolol binding to beta adrenergic receptors in multiple sclerosis brain. *Neurosci Lett.* 2000;289(1):75–77.
104. Gibbs ME, Hutchinson DS, Summers RJ. Noradrenaline release in the locus coeruleus modulates memory formation and consolidation; roles for alpha- and beta-adrenergic receptors. *Neuroscience.* 2010;170(4):1209–1222.
105. Schutsky K, Ouyang M, Castelino CB, Zhang L, Thomas SA. Stress and glucocorticoids impair memory retrieval via beta2-adrenergic, Gi/o-coupled suppression of cAMP signaling. *J Neurosci.* 2011;31(40):14172–14181.
106. Guo NN, Li BM. Cellular and subcellular distributions of beta1- and beta2-adrenoceptors in the CA1 and CA3 regions of the rat hippocampus. *Neuroscience.* 2007;146(1):298–305.
107. Liu Y, Jia WG, Strosberg AD, Cynader M. Morphology and distribution of neurons and glial cells expressing beta-adrenergic receptors in developing kitten visual cortex. *Brain Res Dev Brain Res.* 1992;65(2):269–273.
108. Salm AK, McCarthy KD. The evidence for astrocytes as a target for central noradrenergic activity: expression of adrenergic receptors. *Brain Res Bull.* 1992;29(3–4): 265–275.
109. Shao Y, Sutin J. Expression of adrenergic receptors in individual astrocytes and motor neurons isolated from the adult rat brain. *Glia.* 1992;6(2):108–117.
110. Mantyh PW, Rogers SD, Allen CJ, et al. Beta 2-adrenergic receptors are expressed by glia in vivo in the normal and injured central nervous system in the rat, rabbit, and human. *J Neurosci.* 1995;15(1 Pt. 1):152–164.
111. Catus SL, Gibbs ME, Sato M, Summers RJ, Hutchinson DS. Role of beta-adrenoceptors in glucose uptake in astrocytes using beta-adrenoceptor knockout mice. *Br J Pharmacol.* 2011;162(8):1700–1715.
112. Dong JH, Chen X, Cui M, Yu X, Pang Q, Sun JP. Beta2-adrenergic receptor and astrocyte glucose metabolism. *J Mol Neurosci.* 2012;48(2):456–463.
113. Meana JJ, Barturen F, Garcia-Sevilla JA. Alpha 2-adrenoceptors in the brain of suicide victims: increased receptor density associated with major depression. *Biol Psychiatry.* 1992;31(5):471–490.

114. Callado LF, Meana JJ, Grijalba B, Pazos A, Sastre M, Garcia-Sevilla JA. Selective increase of alpha2A-adrenoceptor agonist binding sites in brains of depressed suicide victims. *J Neurochem*. 1998;70(3):1114–1123.
115. Garcia-Sevilla JA, Escriba PV, Ozaita A, et al. Up-regulation of immunolabeled alpha2A-adrenoceptors, Gi coupling proteins, and regulatory receptor kinases in the prefrontal cortex of depressed suicides. *J Neurochem*. 1999;72(1):282–291.
116. Ordway GA, Schenk J, Stockmeier CA, May W, Klimek V. Elevated agonist binding to alpha2-adrenoceptors in the locus coeruleus in major depression. *Biol Psychiatry*. 2003;53(4):315–323.
117. Escriba PV, Ozaita A, Garcia-Sevilla JA. Increased mRNA expression of alpha2A-adrenoceptors, serotonin receptors and mu-opioid receptors in the brains of suicide victims. *Neuropsychopharmacology*. 2004;29(8):1512–1521.
118. Eason MG, Liggett SB. Subtype-selective desensitization of alpha 2-adrenergic receptors. Different mechanisms control short and long term agonist-promoted desensitization of alpha 2C10, alpha 2C4, and alpha 2C2. *J Biol Chem*. 1992;267 (35):25473–25479.
119. Kurose H, Lefkowitz RJ. Differential desensitization and phosphorylation of three cloned and transfected alpha 2-adrenergic receptor subtypes. *J Biol Chem*. 1994;269 (13):10093–10099.
120. Jewell-Motz EA, Liggett SB. G protein-coupled receptor kinase specificity for phosphorylation and desensitization of alpha2-adrenergic receptor subtypes. *J Biol Chem*. 1996;271(30):18082–18087.
121. Small KM, Schwarb MR, Glinka C, et al. Alpha2A- and alpha2C-adrenergic receptors form homo- and heterodimers: the heterodimeric state impairs agonist-promoted GRK phosphorylation and beta-arrestin recruitment. *Biochemistry*. 2006;45 (15):4760–4767.
122. Cottingham C, Chen Y, Jiao K, Wang Q, et al. The antidepressant desipramine is an arrestin-biased ligand at the alpha(2A)-adrenergic receptor driving receptor down-regulation in vitro and in vivo. *J Biol Chem*. 2011;286(41):36063–36075.
123. Hara MR, Kovacs JJ, Whalen EJ, et al. A stress response pathway regulates DNA damage through beta2-adrenoreceptors and beta-arrestin-1. *Nature*. 2011;477(7364):349–353.
124. Wang Q, Zhao J, Brady AE, et al. Spinophilin blocks arrestin actions in vitro and in vivo at G protein-coupled receptors. *Science*. 2004;304(5679):1940–1944.
125. Bohn LM, Lefkowitz RJ, Gainetdinov RR, Peppel K, Caron MG, Lin FT. Enhanced morphine analgesia in mice lacking beta-arrestin 2. *Science*. 1999;286 (5449):2495–2498.
126. Bohn LM, Gainetdinov RR, Lin FT, Lefkowitz RJ, Caron MG. Mu-opioid receptor desensitization by beta-arrestin-2 determines morphine tolerance but not dependence. *Nature*. 2000;408(6813):720–723.
127. Raehal KM, Walker JK, Bohn LM. Morphine side effects in beta-arrestin 2 knockout mice. *J Pharmacol Exp Ther*. 2005;314(3):1195–1201.
128. Heinricher MM, McGaraughty S, Farr DA. The role of excitatory amino acid transmission within the rostral ventromedial medulla in the antinociceptive actions of systemically administered morphine. *Pain*. 1999;81(1–2):57–65.
129. Kieffer BL, Gaveriaux-Ruff C. Exploring the opioid system by gene knockout. *Prog Neurobiol*. 2002;66(5):285–306.
130. Kohout TA, Lin FS, Perry SJ, Conner DA, Lefkowitz RJ. beta-Arrestin 1 and 2 differentially regulate heptahelical receptor signaling and trafficking. *Proc Natl Acad Sci USA*. 2001;98(4):1601–1606.
131. Bohn LM, Lefkowitz RJ, Caron MG. Differential mechanisms of morphine antinociceptive tolerance revealed in (beta)arrestin-2 knock-out mice. *J Neurosci*. 2002;22(23):10494–10500.

132. Przewlocka B, Sieja A, Starowicz K, Maj M, Bilecki W, Przewlocki R. Knockdown of spinal opioid receptors by antisense targeting beta-arrestin reduces morphine tolerance and allodynia in rat. *Neurosci Lett.* 2002;325(2):107–110.
133. Schmid CL, Bohn LM. Physiological and pharmacological implications of beta-arrestin regulation. *Pharmacol Ther.* 2009;121(3):285–293.
134. Zhang J, Ferguson SS, Barak LS, et al. Role for G protein-coupled receptor kinase in agonist-specific regulation of mu-opioid receptor responsiveness. *Proc Natl Acad Sci USA.* 1998;95(12):7157–7162.
135. Whistler JL, von Zastrow M. Morphine-activated opioid receptors elude desensitization by beta-arrestin. *Proc Natl Acad Sci USA.* 1998;95(17):9914–9919.
136. Bohn LM, Gainetdinov RR, Caron MG. G protein-coupled receptor kinase/beta-arrestin systems and drugs of abuse: psychostimulant and opiate studies in knockout mice. *Neuromolecular Med.* 2004;5(1):41–50.
137. Bohn LM, Dykstra LA, Lefkowitz RJ, Caron MG, Barak LS, et al. Relative opioid efficacy is determined by the complements of the G protein-coupled receptor desensitization machinery. *Mol Pharmacol.* 2004;66(1):106–112.
138. Hauger RL, Risbrough V, Brauns O, Dautzenberg FM. Corticotropin releasing factor (CRF) receptor signaling in the central nervous system: new molecular targets. *CNS Neurol Disord Drug Targets.* 2006;5(4):453–479.
139. Hauger RL, Olivares-Reyes JA, Dautzenberg FM, Lohr JB, Braun S, Oakley RH. Molecular and cell signaling targets for PTSD pathophysiology and pharmacotherapy. *Neuropharmacology.* 2012;62(2):705–714.
140. Bale TL, Vale WW. CRF and CRF receptors: role in stress responsivity and other behaviors. *Annu Rev Pharmacol Toxicol.* 2004;44:525–557.
141. Holsboer F, Ising M. Central CRH system in depression and anxiety—evidence from clinical studies with CRH1 receptor antagonists. *Eur J Pharmacol.* 2008;583(2–3):350–357.
142. Grigoriadis DE. The corticotropin-releasing factor receptor: a novel target for the treatment of depression and anxiety-related disorders. *Expert Opin Ther Targets.* 2005;9(4):651–684.
143. Cryan JF, Holmes A. The ascent of mouse: advances in modelling human depression and anxiety. *Nat Rev Drug Discov.* 2005;4(9):775–790.
144. Tezval H, Jahn O, Todorovic C, Sasse A, Eckart K, Spiess J. Cortagine, a specific agonist of corticotropin-releasing factor receptor subtype 1, is anxiogenic and antidepressive in the mouse model. *Proc Natl Acad Sci USA.* 2004;101(25):9468–9473.
145. Heinrichs SC, Koob GF. Corticotropin-releasing factor in brain: a role in activation, arousal, and affect regulation. *J Pharmacol Exp Ther.* 2004;311(2):427–440.
146. Coste SC, Murray SE, Stenzel-Poore MP. Animal models of CRH excess and CRH receptor deficiency display altered adaptations to stress. *Peptides.* 2001;22(5):733–741.
147. Muller MB, Holsboer F. Mice with mutations in the HPA-system as models for symptoms of depression. *Biol Psychiatry.* 2006;59(12):1104–1115.
148. Hauger RL, Risbrough V, Oakley RH, Olivares-Reyes JA, Dautzenberg FM. Role of CRF receptor signaling in stress vulnerability, anxiety, and depression. *Ann N Y Acad Sci.* 2009;1179:120–143.
149. Teli T, Markovic D, Levine MA, Hillhouse EW, Grammatopoulos DK. Regulation of corticotropin-releasing hormone receptor type 1alpha signaling: structural determinants for G protein-coupled receptor kinase-mediated phosphorylation and agonist-mediated desensitization. *Mol Endocrinol.* 2005;19(2):474–490.
150. Dautzenberg FM, Braun S, Hauger RL. GRK3 mediates desensitization of CRF1 receptors: a potential mechanism regulating stress adaptation. *Am J Physiol Regul Integr Comp Physiol.* 2001;280(4):R935–R946.

151. Drake MT, Shenoy SK, Lefkowitz RJ. Trafficking of G protein-coupled receptors. *Circ Res*. 2006;99(6):570–582.
152. Holmes KD, Babwah AV, Dale LB, Poulter MO, Ferguson SS. Differential regulation of corticotropin releasing factor 1alpha receptor endocytosis and trafficking by beta-arrestins and Rab GTPases. *J Neurochem*. 2006;96(4):934–949.
153. Oakley RH, Olivares-Reyes JA, Hudson CC, Flores-Vega F, Dautzenberg FM, Hauger RL. Carboxyl-terminal and intracellular loop sites for CRF1 receptor phosphorylation and beta-arrestin-2 recruitment: a mechanism regulating stress and anxiety responses. *Am J Physiol Regul Integr Comp Physiol*. 2007;293(1):R209–R222.
154. Markovic D, Punn A, Lehnert H, Grammatopoulos DK. Intracellular mechanisms regulating corticotropin-releasing hormone receptor-2beta endocytosis and interaction with extracellularly regulated kinase 1/2 and p38 mitogen-activated protein kinase signaling cascades. *Mol Endocrinol*. 2008;22(3):689–706.
155. Gutknecht E, Hauger RL, Van der Linden I, Vauquelin G, Dautzenberg FM. Expression, binding, and signaling properties of CRF2(a) receptors endogenously expressed in human retinoblastoma Y79 cells: passage-dependent regulation of functional receptors. *J Neurochem*. 2008;104(4):926–936.
156. Breivogel CS, Lambert JM, Gerfin S, Huffman JW, Razdan RK. Sensitivity to delta9-tetrahydrocannabinol is selectively enhanced in beta-arrestin2 −/− mice. *Behav Pharmacol*. 2008;19(4):298–307.
157. Arlinde C, Sommer W, Bjork K, et al. A cluster of differentially expressed signal transduction genes identified by microarray analysis in a rat genetic model of alcoholism. *Pharmacogenomics J*. 2004;4(3):208–218.
158. Sommer W, Hyytia P, Kiianmaa K. The alcohol-preferring AA and alcohol-avoiding ANA rats: neurobiology of the regulation of alcohol drinking. *Addict Biol*. 2006;11(3–4):289–309.
159. Bjork K, Rimondini R, Hansson AC, et al. Modulation of voluntary ethanol consumption by beta-arrestin 2. *FASEB J*. 2008;22(7):2552–2560.
160. Schreiber G, Golan M, Avissar S. Beta-arrestin signaling complex as a target for antidepressants and as a depression marker. *Drug News Perspect*. 2009;22(8):467–480.
161. Allen JA, Yost JM, Setola V, et al. Discovery of beta-arrestin-biased dopamine D2 ligands for probing signal transduction pathways essential for antipsychotic efficacy. *Proc Natl Acad Sci USA*. 2011;108(45):18488–18493.
162. Rozenfeld R, Devi LA. Exploring a role for heteromerization in GPCR signalling specificity. *Biochem J*. 2011;433(1):11–18.
163. Swift JL, Godin AG, Dore K, et al. Quantification of receptor tyrosine kinase transactivation through direct dimerization and surface density measurements in single cells. *Proc Natl Acad Sci USA*. 2011;108(17):7016–7021.
164. Oligny-Longpre G, Corbani M, Zhou J, Hogue M, Guillon G, Bouvier M. Engagement of beta-arrestin by transactivated insulin-like growth factor receptor is needed for V2 vasopressin receptor-stimulated ERK1/2 activation. *Proc Natl Acad Sci USA*. 2012;109(17):E1028–E1037.

CHAPTER TWELVE

Arrestins in the Cardiovascular System

Anastasios Lymperopoulos, Ashley Bathgate
Department of Pharmaceutical Sciences, Nova Southeastern University College of Pharmacy, Fort Lauderdale, Florida, USA

Contents

1. Introduction 298
2. Cardiovascular Adrenergic Receptors and βarrs 299
   2.1 α_1-Adrenergic receptors 299
   2.2 α_2-Adrenergic receptors 300
   2.3 β-Adrenergic receptors 305
3. Cardiovascular Angiotensin II Receptors and βarrs 309
   3.1 Angiotensin II type 1 receptors 309
   3.2 Angiotensin II type 2 receptors 314
4. Other Cardiovascular GPCRs and βarrs 315
   4.1 Endothelin receptors 315
   4.2 Vasopressin receptors 316
   4.3 Niacin receptor (GPR109A) 317
   4.4 P2Y Receptors 318
   4.5 Protease-activated receptors 320
5. Cardiovascular Non-GPCRs and βarrs 320
6. Biased Agonism/Antagonism of Cardiovascular GPCR Ligands 322
7. Perspectives and Future Directions 326
8. Conclusions 327
References 328

Abstract

Of the four mammalian arrestins, only the β-arrestins (βarrs; Arrestin2 and -3) are expressed throughout the cardiovascular system, where they regulate, as either desensitizers/internalizers or signal transducers, several G-protein-coupled receptors (GPCRs) critical for cardiovascular homeostasis. The cardiovascular roles of βarrs have been delineated at an accelerated pace via a variety of techniques and tools, such as knockout mice, siRNA knockdown, artificial or naturally occurring polymorphic GPCRs, and availability of new βarr "biased" GPCR ligands. This chapter summarizes the current knowledge of cardiovascular arrestin physiology and pharmacology, addressing the individual cardiovascular receptors affected by βarrs *in vivo*, as well as the individual cell types,

ISSN 1877-1173
http://dx.doi.org/10.1016/B978-0-12-394440-5.00012-7

tissues, and organs of the cardiovascular system in which βarr effects are exerted; for example, cardiac myocyte or fibroblast, vascular smooth muscle, adrenal gland and platelet. In the broader scope of cardiovascular βarr pharmacology, a discussion of the βarr "bias" of certain cardiovascular GPCR ligands is also included.

1. INTRODUCTION

Of the four mammalian arrestin isoforms known, only the two ubiquitous (outside the retinal system) βarrestins [i.e., βarrestin1 and -2 (βarr1 and -2) or Arrestin2 and -3 (Arr2 and -3)] are expressed in the cardiovascular system. Several of the seven transmembrane-spanning receptors (7TMRs) or G-protein-coupled receptors (GPCRs) that these two βarrs regulate play enormously important roles in cardiovascular physiology and homeostasis. For instance, cardiac function (contractility) is tightly controlled by the activity of β-adrenergic receptors (β_1- and β_2ARs) located in the membranes of cardiac myocytes. Cardiac structure and morphology are regulated by angiotensin II (AngII) type 1 receptors (AT_1Rs) present (mainly) in cardiac fibroblast and endothelial cell membranes, and also, to a lesser extent, in cardiomyocyte membranes. Heart rate is modulated by the balance between the activities of β-adrenergic and muscarinic cholinergic receptors located in various anatomical segments of the cardiac electrical conduction system. Vascular tone is regulated by α- and β-adrenergic, endothelin (ET), AngII, and bradykinin receptor activation; renal function is fine-tuned by AngII, adrenergic, vasopressin, and dopaminergic receptors; and systemic blood pressure is regulated by changes in all these parameters of cardiovascular function (i.e., heart rate, vascular resistance, and fluid/electrolyte balance). Each of these processes is kept in balance, to a large degree, by GPCRs. Furthermore, even the neurohormonal control of the circulatory system, whether it be catecholamine and corticosteroid release by the adrenal glands, activation of the renin–angiotensin–aldosterone system by the juxtaglomerular apparatus of the kidneys, or release of neurotransmitters by central and peripheral neurons innervating cardiovascular organs, is under tight regulation by various GPCRs.

Given that signaling from cardiovascular 7TMRs constitutes such an integral part of cardiovascular function, and that βarrs, both of which are expressed throughout the cardiovascular system, interact with and tightly regulate this signaling, it should come as no surprise that the roles βarrs play in cardiovascular biology, physiology, and pathology are both extremely significant and versatile. A more complete appreciation of the cardiovascular roles of βarrs has recently become possible due to a combination of techniques and tools, such as utilization of the (global) βarr1 and βarr2 knockout (KO) mice, usage of isoform-specific siRNA knockdown in *in vitro* systems, employment of *in vitro* and *in vivo* systems of artificially constructed 7TMR mutants incapable of signaling through G proteins, and, finally, synthesis and development of signaling pathway-selective (biased) 7TMR ligands that preferentially elicit βarr signaling over G protein signaling (and vice versa) from a given GPCR. This chapter gives a comprehensive account of the current knowledge in the field of the roles and functions of arrestins in the cardiovascular system, addressing each individual cardiovascular receptor type and cardiovascular organ/tissue type the βarrs act upon, and, whenever possible, discussing whether their effects are due to their ability to act as receptor desensitizers/internalizers or signal transducers.

2. CARDIOVASCULAR ADRENERGIC RECEPTORS AND βARRS

2.1. α_1-Adrenergic receptors

Cardiovascular α_1-adrenergic receptors (α_1ARs), consisting of three pharmacological subtypes (α_{1A}, α_{1B}, α_{1D}), function as stimulatory receptors and are the classical adrenoceptors mediating vascular smooth muscle (VSM) contraction, elevating systemic blood pressure via coupling to the $G_{q/11}$ protein–phospholipase C–Ca^{2+} mobilization pathway.[1] A fall in blood pressure due to causes such as hemorrhage will activate the baroreceptor reflex, increasing sympathetic outflow to cause vasoconstriction, via α_1ARs, of less vital vascular beds, especially splanchnic and skin, an effect that plays an important part in the "fight or flight" response produced by the sympathetic nervous system to situations of acute stress.[2] α_1AR antagonists lower blood pressure in hypertension and α_1AR agonist-mediated vasoconstriction can be used to treat hypotension. This vasoconstrictive property is also taken advantage of in alleviation of nasal congestion, a condition for which several preparations containing α_1AR agonists are used. On the other hand, α_1ARs present in coronary artery smooth muscle can cause vasoconstriction and precipitate angina pectoris.[2]

In addition to smooth muscle cells of various vascular beds, α_1ARs are also expressed in cardiac myocytes, albeit to a considerably lower extent than their βAR counterparts. Their cardiac effects are potentially harmful, since cardiac α_1ARs mediate pathological cardiac hypertrophy leading to heart failure (HF) by increasing expression of β-myosin heavy chain and other genes belonging to the so-called fetal gene program of cardiac adverse remodeling.[3] Carvedilol, a non-selective βAR blocker currently used with great success in HF (see Section 6), is also an α_1AR antagonist, and antagonism of the maladaptive cardiotoxic effects of cardiac α_1ARs might contribute to its beneficial actions in HF treatment.[4] However, not all actions of cardiac α_1ARs are bad; α_1ARs can also stimulate "adaptive" or "physiological" cardiac hypertrophy with no fibrosis or other manifestations of adverse remodeling, but rather with preservation or even augmentation of cardiac function and stimulation of myocyte proliferation and ventricular muscle wall thickening.[5,6]

Unfortunately, very little data exist with respect to interactions of the α_1ARs with βarrs, regardless of whether these interactions mediate α_1AR desensitization (i.e., termination of G protein signaling) or βarr-dependent protein scaffolding/signaling. All of the available data are from *in vitro* studies in heterologous cell systems (i.e., cell lines transfected with α_1AR subtypes) with virtually nothing on α_1ARs and βarrs specifically in the cardiovascular system.[7] Results from *in vitro* co-immunoprecipitation experiments and βarr translocation assays have indicated that the agonist-induced interaction of the α_{1A}AR with βarrs is much weaker than that of the α_{1B}AR.[8] The interaction of βarrs with the α_{1D}AR has not been directly explored, not even in *in vitro* systems, let alone in cardiovascular tissues.[7] Given that α_1ARs are βarr substrates and play such important roles in cardiovascular regulation, including stimulation of downstream effectors that can be activated by βarr-dependent signaling cascades (e.g., mitogen-activated protein kinases, MAPKs), it is evident that cardiovascular βarr biology with relation to α_1ARs has the potential for providing invaluable new information in the future.

2.2. α_2-Adrenergic receptors

2.2.1 Cardiovascular α_2ARs and βarrs as desensitizers

Of the three pharmacological α_2AR subtypes (α_{2A}, α_{2B}, α_{2C}), the α_{2B}AR is the predominant one expressed in smooth muscle cells of various vascular beds, wherein, like the α_1AR subtypes, it mediates vasoconstriction and can raise blood pressure. This distinguishes it from the other α_2AR subtype counterparts, which are present in the central sympathetic nervous system, that lower blood pressure by decreasing central sympathetic outflow.[9] Therefore, prolongation of centrally acting "sympatholytic" α_2AR signaling/function, for example, inhibition of βarr desensitizing actions on these receptors in the central nervous system (CNS), and/or antagonism of VSM-constrictive α_{2B}AR signaling/function, for example, enhancement of βarr desensitizing action on vascular α_{2B}ARs, is desired in treatment of hypertension.

On the latter point, a polymorphic variant of the human α_{2B}AR gene (Del 301–303 α_{2B}AR) containing a nine-nucleotide deletion that eliminates three Glu residues from the third intracellular loop of the encoded receptor, displays severely impaired GPCR kinase (GRK)-dependent phosphorylation and is incapable of undergoing desensitization.[10,11] This variant has been genetically associated with acute coronary vasospasm and hypertension in certain populations, as well as under specific conditions, such as stressful work environment.[12–14] Although the effect of this polymorphism on βarr binding to the α_{2B}AR has never been directly tested *in vitro* or *in vivo*, it is safe to surmise that βarrs fail to interact with and desensitize this α_{2B}AR deletion variant, given its complete failure to desensitize. Thus, this constitutes an example of a beneficial cardiovascular role of βarrs as receptor desensitizers, that is, blockade of vascular α_{2B}AR-dependent vasoconstriction, augmentation of which could confer a significant advantage in the treatment of cardiovascular disease (i.e., hypertension, in this particular case; Table 12.1).

The opposite is true for βarrs acting as α_2AR desensitizers in the adrenal gland. Here, as in the CNS, α_2ARs function as presynaptic inhibitory autoreceptors, blocking catecholamine release from the chromaffin cells of the adrenal medulla, a process normally dependent on tonic activation of nicotinic cholinergic receptors by acetylcholine.[9] βarrs, acting in concert with their cofactor in receptor desensitization, GRK2, which is significantly upregulated in the adrenal medulla during HF, promote desensitization and downregulation of adrenal α_2ARs. This increases adrenal catecholamine secretion, leading to the enhanced circulating levels of norepinephrine and epinephrine that accompany and aggravate HF.[15–18] Thus, in the case of adrenal α_2ARs, desensitization by βarrs plays a detrimental role in HF pathophysiology and blockade of adrenal βarr action on α_2ARs, either directly or indirectly, for example, by blocking adrenal GRK2, could be pursued for the treatment of HF (Table 12.1).

2.2.2 Cardiovascular α_2ARs and βarrs as signal transducers

The beneficial hypotensive response to activation of centrally located α_2ARs comes with an important caveat: the unwanted effect of sedation.[48] Sedation tests in βarr2KO mice treated with the α_2AR agonist 5-bromo-*N*-(4,5-dihydro-1*H*-imidazol-2-yl)-6-quinoxalinamine (brimonidine or UK14304)

Table 12.1 *In vivo* cardiovascular effects of βarrs, underlying molecular mechanisms, and therapeutic implications for cardiovascular diseases

Tissue and receptor	βarr isoform	Action on receptor	Signaling pathway(s) affected	*In vivo* effect	Desirable intervention (disease)	Refs.
VSM α_{2B}AR	βarr1/2	Desensitizer	N/A	Vasodilation; ↓ vasoconstriction	Stimulation (HTN)	10,11
Adrenal α_2AR	βarr1	Desensitizer	N/A	↑ Catecholamine secretion	Inhibition (HF)	15–18
Central SNS α_2AR	βarr2	Signal transducer	Unknown, spinophilin?	Sedation of antihypertensive centrally acting α_2AR agonists	Inhibition (HTN)	19,20
Cardiac β_1- and β_2ARs	βarr1/2 (mainly βarr1)	Desensitizer	N/A	↓ Cardiac contractility	Inhibition (HF)	21–24
Cardiac β_1AR	βarr2	Signal transducer	EGFR transactivation, ERK	↑ Cardiomyocyte survival–proliferation	Stimulation (HF)	25–27
Cardiac β_2AR	βarr1/2	Desensitizer	N/A	↓ Cardiomyocyte survival–proliferation	Inhibition (HF?)	21,24,27
Cardiac β_2AR	βarr1/2	Signal transducer	ERK, NF-κB inhibition	↑ Cardiomyocyte survival–proliferation	Stimulation (HF?)	28–30
Cardiac AT_1R	βarr2	Signal transducer	EGFR transactivation, ERK, Akt	↑ Cardiac contractility–function–survival–proliferation	Stimulation or inhibition of βarr1 (HF)	31–34

VSM AT_1R	βarr2	Signal transducer	EGFR transactivation–ERK, BAD inhibition, ↑ DNA synthesis	↑ VSM hyperplasia–hypertrophy	Inhibition or stimulation of βarr1 (atherosclerosis)	35–37
Adrenal AT_1R	βarr1	Signal transducer	DGK–PA–ERK–StAR upregulation	↑ Aldosterone synthesis and secretion	Inhibition (HF, HTN)	38,39
VSM ET_AR	βarr2	Signal transducer	ERK, p38 MAPK	↑ VSM hyperplasia–hypertrophy–cell migration	Inhibition (atherosclerosis)	40,41
Renal distal tubule V_2R	βarr1/2	Desensitizer	N/A	↓ Renal H_2O reabsorption, NDI, NSIAD	Inhibition (NDI-NSIAD, polymorphic V_2Rs)	42,43
VSM GPR109A	βarr1	Signal transducer	$_cPLA_2$–AA–PGD_2	Niacin-induced vasodilation–"flushing," ↓ niacin efficacy via GPR109A desensitization	Inhibition (hyperlipidemia)	44
Platelet P2YR	βarr1	Signal transducer	Src–Akt–integrin $\alpha_{IIb}\beta_3$	↑ Thrombosis	Inhibition (TE syndromes)	45
VSM P2YR	βarr1	Desensitizer	N/A	↓ Vasoconstriction-VSM cell migration	Stimulation (HTN, arteriosclerosis)	41,46

Continued

Table 12.1 *In vivo* cardiovascular effects of βarrs, underlying molecular mechanisms, and therapeutic implications for cardiovascular diseases—cont'd

Tissue and receptor	βarr isoform	Action on receptor	Signaling pathway(s) affected	*In vivo* effect	Desirable intervention (disease)	Refs.
Platelet PAR4	βarr1	Signal transducer	PI3K/Akt–fibrinogen binding	↑ Thrombosis	Inhibition (TE syndromes)	45,47
Cardiac EGFR	βarr2	Signal transducer	AT_1R-dependent transactivation	↑ Cardiomyocyte survival–proliferation	Stimulation (HF)	34
VSM EGFR	βarr2	Signal transducer	AT_1R-dependent transactivation	↑ VSM hyperplasia–hypertrophy	Inhibition (atherosclerosis)	36

Abbreviations/acronyms: AA, arachidonic acid; Akt, Akt kinase or protein kinase B (PKB); AR, adrenergic receptor; βarr, βarrestin; AT_1R, angiotensin II type 1 receptor; BAD, Bcl-2-associated death promoter; CA, catecholamine; DGK, diacylglycerol kinase; EGFR, epidermal growth factor receptor; ERK, extracellular signal-regulated kinase; ET_AR, endothelin-1 type A receptor; GPR109A, G-protein-coupled receptor 109A (nicotinic acid receptor); HF, heart failure; HTN, hypertension; MAPK, mitogen-activated protein kinase; N/A, not applicable; NDI, nephrogenic diabetes insipidus; NF-κB, nuclear factor-kappaB; NSIAD, nephrogenic syndrome of inappropriate antidiuresis; P2YR, metabotropic purinergic receptor; PA, phosphatidic acid; PAR4, protease-activated receptor type 4; PGD_2, prostaglandin D_2; PI3K, phosphoinositide 3-kinase; $_cPLA_2$, cytosolic phospholipase A_2; Refs, references; SNS, sympathetic nervous system; StAR, steroidogenic acute regulatory protein; TE, thromboembolic; V_2R, vasopressin type 2 receptor; VSM, vascular smooth muscle.

Notes: "βarr1/2" indicates that it is currently not known which βarr isoform exerts the effect in question; "?" indicates unknown, not investigated, or not completely clarified.

suggest that this sedation may be related to βarr2-mediated signaling of α_2ARs.[19] Thus, an α_2AR-biased agonist lacking βarr2 stimulation properties but retaining the G protein-stimulating ones might be of therapeutic value in hypertension[20] (Table 12.1).

2.3. β-Adrenergic receptors

2.3.1 Cardiac βARs and βarrs—General considerations

The principal role of βARs in the heart is regulation of cardiac rate and contractility in response to catecholamines. Of the three known mammalian βAR subtypes (β_1AR, β_2AR, β_3AR), the β_1AR is the predominant one in cardiac myocytes, representing 75–80% of total βAR density, followed by the β_2AR, which comprises about 15–18% of total cardiomyocyte βARs, and the β_3AR that makes up the remaining 2–3% (under normal conditions).[49] β_1AR stimulation by catecholamines results in the dissociation of the stimulatory G protein alpha subunit ($G\alpha_s$) from $G_{\beta\gamma}$. $G\alpha_s$ stimulates adenylyl cyclase (AC) to produce 3′-5′-adenosine monophosphate (cAMP), which in turn, by activating protein kinase A (PKA), regulates different intracellular, sarcolemmal, and myofibrillar substrates, exerting the cellular effects of β_1AR activation on cardiac chronotropy, inotropy, and lusitropy. In addition, $G_{\beta\gamma}$ can activate downstream effectors that participate in cardiac signaling regulation.[49] β_2AR also mediates the effects of catecholamines on the heart, but in a qualitatively different manner from β_1AR, as it can also couple to the AC inhibitory G protein (G_i). It is now generally accepted that in the heart, β_2ARs signal and function in a substantially different way than β_1ARs.[50–52] Importantly, whereas β_1AR activation enhances cardiomyocyte apoptosis, β_2ARs exert beneficial antiapoptotic effects in the heart,[50–53] purportedly through this G_i-mediated signaling.[51] Several studies using transgenic mice, β_2AR-selective stimulation, and adenoviral-mediated β_2AR overexpression have established a consensus that β_2AR signaling is predominantly cardioprotective, improving cardiac function and decreasing apoptosis, whereas β_1AR-elicited signaling has detrimental effects.[53,54]

Of note, the differences between the signaling properties of these two predominant cardiac βARs might take a quite different shape in the setting of HF. For instance, the β_1AR is selectively downregulated in HF, that is, the total cellular receptor number is reduced, thus shifting the stoichiometry of β_1AR:β_2AR from approximately 75:25 in the normal healthy heart toward 50:50 in the failing heart.[55,56] However, β_2AR is also nonfunctional and does not signal properly in the failing heart.[56,57] In addition, β_2AR signaling in the failing heart might differ substantially from the beneficial signaling pattern in the normal heart; in HF, it seems to be more "diffuse," that is, less "compartmentalized," and resembles more the proapoptotic cAMP signaling pattern of its β_1AR counterpart.[58] Therefore, the aforementioned stoichiometric shift in favor of the "good" β_2AR in HF appears unable to help the heart to adapt its structure and function.

2.3.2 Cardiac βARs and βarrs as desensitizers

βarrs are abundantly expressed in cardiac muscle.[59] As cofactors of GRKs in cardiomyocyte βAR desensitization and/or downregulation, they contribute to the diminished inotropic and adrenergic reserve of the failing heart and their inhibition should theoretically be beneficial in acute HF, as it would enhance the $G\alpha_s$–AC–PKA-mediated pro-contractile signaling of cardiac βARs, which increases cardiac contractility.[21–24] Indeed, indirect blockade of βarr desensitizing action on cardiac βARs, that is, by genetic blockade of GRK2 and GRK5, the main cardiac GRKs stimulating βarr-dependent cardiac βAR desensitization, results in several desirable therapeutic effects in HF, such as dramatic improvement in positive inotropy and cardiac β-adrenergic reserve, amelioration of adverse remodeling, and increased survival (reviewed in Ref. 21–24; Table 12.1). Studies in global βarr1KO mice have confirmed that cardiac βarr1 diminishes inotropic and adrenergic reserves by means of desensitizing cardiomyocyte β_1- and β_2ARs (Table 12.1). The contractile response of βarr1KO mice to isoproterenol, a βAR agonist cardio-stimulant (positive inotrope), is significantly augmented compared to that of wild-type mice, while basal contractility is unaffected.[60] Of note, cardiac βarr2 does not appear to undergo compensatory upregulation in βarr1KO mice or to compensate for the loss of βarr1 in the myocardium,[60] indicating that the two βarr isoforms are rarely (if at all) physiologically interchangeable. Data from our lab in the same mice under conditions of myocardial infarction (MI)-induced HF have confirmed this predominant βAR desensitizing/downregulating, and hence contractility-limiting, role of cardiac βarr1 (Table 12.1). However, the majority of studies on cardiac βAR desensitization and its inhibition for therapeutic purposes in HF have focused on blockade of GRK action and not directly on βarr action on cardiac βARs *per se*. Therefore, any conclusions on the potential therapeutic benefits of cardiac βarr blockade for reversing βAR desensitization and increasing cardiac function must be drawn with caution, especially since cardiac βarr interaction with βARs can have additional effects, that is, signal transduction, beyond merely conferring βAR desensitization/downregulation.

2.3.3 Cardiac βARs and βarrs as signal transducers

In contrast to their action on cardiac βARs as desensitizers, βarr actions on cardiac βARs as signal transducers have been studied directly and at an accelerating pace, and these studies have yielded several interesting, if not eye-opening, findings. A consensus for cardiac β_1AR signaling is beginning to unfold, according to which cardiac β_1AR signaling is beneficial for the heart when it is mediated by βarrs, but cardiotoxic when it is G-protein-dependent. In HF, chronic catecholaminergic stimulation of the β_1AR promotes cardiac hypertrophy, decreases contractility, and increases myocyte apoptosis.[61] As a result, administration of β-blockers is currently part of standard care in the clinical management of congestive HF.[62,63] βarrs have been shown *in vitro* to mediate the mitogenic signaling of EGF (epidermal growth factor) receptor (EGFR) transactivation by the β_1AR. As inhibition of EGFR contributes to dilated cardiomyopathy, β_1AR signaling via βarr2 appears to have protective rather than deleterious effects on the heart[25] and βarr2-dependent EGFR transactivation might exert a cardioprotective effect[26,27] (Table 12.1). The physiologic relevance of EGFR transactivation by β_1AR-bound βarr2 has been demonstrated in transgenic mice overexpressing wild-type β_1ARs or mutant β_1ARs lacking GRK phosphorylation sites that are unable to undergo GRK-mediated phosphorylation and bind βarrs. Under conditions of chronic catecholamine stress, transgenic mice overexpressing the mutant β_1AR are incapable of transactivating cardiac EGFRs and show marked myocyte apoptosis and left ventricular dilatation compared to mice expressing wild-type β_1ARs.[25] Further substantiating a role for EGFR transactivation in the mechanism underlying these findings, pretreatment of mice overexpressing the wild-type β_1AR in their hearts with the selective EGFR inhibitor, erlotinib, prevents any improvement in cardiac function upon chronic stimulation with catecholamines.[25] Finally, data from our lab in post-MI βarr1KO mice confirm that βarr2 (and not βarr1) is the βarr isoform that mediates EGFR transactivation by the cardiac β_1AR *in vivo*, and that cardiac βarr1 might physiologically antagonize this effect of βarr2. We found that compared to wild-type control mouse hearts, post-MI βarr1KO mouse hearts display significantly elevated EGFR activity levels *in vivo* in response to treatment with isoproterenol in the presence of the β_2AR-selective antagonist ICI 118,551, that is, a treatment that activates only cardiac β_1ARs.

With regard to physiologic relevance of β_2AR-elicited βarr signaling in the heart, very little (if anything) is known. It appears that βarr binding to this receptor subtype in the heart is predominantly deleterious, since, by means of classical desensitization, it blocks the G_s protein-mediated procontractile signaling and the G_i protein-mediated antiapoptotic signaling of the cardiac β_2AR (Table 12.1). On the other hand, βarr signaling stimulated by the β_2AR can exert some beneficial effects on the cardiac myocyte, as it can be antiapoptotic and also anti-inflammatory in its own right by promoting extracellular signal-regulated kinase (ERK) activation, which increases cardiomyocyte survival and proliferation, and by blocking nuclear factor-kappaB (NF-κB) activation, which leads to proinflammatory cytokine production[28–30] (Table 12.1).

Finally, as far as the cardiac β_3AR is concerned, this βAR subtype is pharmacologically unique. Its effects on the myocardium appear to be opposite to the effects of the two more abundant cardiac βAR subtypes, that is, it relaxes the myocardium and decreases its contractility, presumably via activation of nitric oxide muscle-relaxing pathways.[64] Again contrary to its βAR subtype counterparts, the β_3AR is not a βarr or GRK substrate, as it does not undergo agonist-dependent desensitization or internalization.[64] However, β_3AR can heterodimerize (or hetero-oligomerize) with the β_2AR and dramatically alter the latter's desensitization and internalization properties.[65] Thus, β_3AR might indirectly affect βarr actions on β_2ARs in the heart.

To sum up, cardiac βarrs (primarily βarr2) can convert the normally cardiotoxic G-protein-dependent signaling of the β_1AR into cardioprotective signaling via βarr2-dependent EGFR transactivation, an effect of potential therapeutic benefit in HF (Table 12.1). With regard to the β_2AR, cardiac βarrs can have damaging effects by opposing (desensitizing) the beneficial G-protein-dependent procontractile and antiapoptotic β_2AR signaling, but they might also help preserve it in situations where this signaling goes awry, such as in HF (Table 12.1). Finally, the cardiac β_3AR does not engage directly in βarr interaction/binding.

2.3.4 Other cardiovascular βARs and βarrs

It is widely known that βARs (mainly of the β_2 subtype) are highly expressed in VSM, where they exert vasodilation by increasing cAMP levels.[64] VSM βarrs are bound to limit this vasodilation via classical desensitization of these receptors, thereby increasing vascular tone and promoting hypertension. The fact that GRK2 activity is elevated in human hypertension,[66,67] although it is not known whether it is elevated in VSM *per se*, argues in favor of vascular βarrs opposing βAR-dependent VSM relaxation, since enhanced GRK activity generally translates into enhanced GPCR desensitization by βarrs. However, βarr mRNA levels are unaltered in circulating lymphocytes of hypertensive subjects[66] and studies directly testing the effects of VSM βarrs on βAR-mediated vasodilation are lacking and likely to be inconclusive, given the large number of vasoconstrictive and vasodilatory GPCRs present in VSM that are regulated by βarrs.

Finally, βARs (of both the β_1 and β_2 subtypes) are also expressed in the adrenal medulla of mice and rats, where, contrary to their α_2AR counterparts, they enhance catecholamine release, acting as presynaptic facilitatory autoreceptors in an autocrine positive feedback manner.[18,68,69] Therefore, adrenal βarrs would be expected to limit adrenal catecholamine release by desensitizing these βARs. However, adrenal α_2ARs appear to play the dominant role in regulation of catecholamine secretion, given that they are the only GPCRs known to date that suppress catecholamine release from this organ. Thus, adrenal βarr actions on α_2ARs (rather than on βARs) are the determining factor in the regulation of catecholamine secretion from the adrenal medulla.

3. CARDIOVASCULAR ANGIOTENSIN II RECEPTORS AND βARRS

3.1. Angiotensin II type 1 receptors

3.1.1 Cardiac AT_1Rs and βarrs as desensitizers

The AT_1R is a very important 7TMR in cardiovascular physiology and pathology. In the heart, it is mainly expressed in cardiac fibroblasts, where it stimulates cellular proliferation, thus promoting fibrosis, and in cardiac myocytes, where it stimulates growth, thus promoting cardiac hypertrophy.[70] Whether it can also promote cardiomycyte contractility is still a matter of debate.[70] Combined with other cellular actions that promote inflammation and oxidative stress, cardiac AT_1R effects are clearly maladaptive and damaging for both the structure and function of the cardiac muscle, and AT_1Rs play a pivotal role in the so-called adverse remodeling of the post-MI heart progressing to HF.[70] Therefore, targeting cardiac AT_1Rs, either directly with angiotensin receptor/AT_1R-specific blockers (ARBs) or indirectly with angiotensin-converting enzyme inhibitors, is a major therapeutic strategy for cardiac disease treatment.

The AT_1R is a classic $G_{q/11}$-coupled receptor that can also couple to $G_{i/o}$ proteins. With regard to their classical role as G-protein-dependent signaling terminators (desensitizers), very little is known about βarrs and AT_1Rs, and even less about cardiac βarrs and AT_1Rs. The AT_1R is a known GRK substrate; thus, cardiac βarrs are bound to promote its desensitization in the heart secondary to its phosphorylation by GRKs, which has been demonstrated *in vivo*.[71] However, βarr-mediated AT_1R desensitization *per se* has never been directly investigated *in vivo*. Intriguingly, the AT_1R displays a somewhat peculiar behavior in terms of its desensitization. Not only is it subject to phosphorylation by other kinases, for example, PKA and protein kinase C (PKC), in addition to GRKs,[72] sometimes its phosphorylation is not required for desensitization.[73] Thus, it apparently can desensitize through multiple mechanisms and interactions with various other proteins.[74–77] What is more, the signaling pathways it elicits display variable desensitization kinetics, for example, Ca^{2+} transients induced by AT_1Rs readily and rapidly desensitize, whereas ERK activation and Janus kinase/signal transducer and activator of transcription signaling emanating from this receptor persist for long periods of time.[77,78] Given the prominent role of cardiac AT_1Rs in cardiac disease, such as post-MI HF, enhancement of βarr-dependent desensitization would theoretically be beneficial/therapeutic. But since classical βarr-mediated desensitization of AT_1Rs has never been directly demonstrated, especially *in vivo*, there is no real evidence for its occurrence in the heart, let alone evidence that βarr-mediated AT_1R desensitization in the heart is physiologically or pathophysiologically relevant.

3.1.2 Cardiac AT_1Rs and βarrs as signal transducers

Much more has come to light over the past several years regarding the physiological roles of cardiac βarrs as mediators of G-protein-independent signal transduction by AT_1Rs in the heart. The first such study, by Zhai *et al.* in 2005, showed, remarkably, that myocardial overexpression of an artificially constructed $AT_{1A}R$ mutant (AT1-i2m), which was incapable of activating G proteins but retained the ability to interact with βarrs, produced significantly less myocardial apoptosis and fibrosis, and enhanced cardiomyocyte hypertrophy, bradycardia, and fetal cardiac gene expression, compared to wild-type AT_1Rs expressed at similar levels.[31] In primary cardiomyocytes, the βarr-biased AT_1R agonist peptide Sarcosine[1]–Isoleucine[4]–Isoleucine[8]–AngII (SII; see Section 6) stimulates cardiomyocyte proliferation independently of G protein activation, but not hypertrophy, which requires $G_{q/11}$ protein signaling.[32] In addition, SII produces positive inotropic and lusitropic effects in isolated murine cardiomyocytes through GRK6-mediated phosphorylation of the cardiomyocyte AT_1R and subsequent βarr2 activation.[33] Interestingly, GRK2-mediated phosphorylation of the AT_1R in cardiac myocytes leads to activation of the other βarr isoform (βarr1), and cardiac βarr1 seems to oppose the positive effects of βarr2 on AT_1R-elicited contractility and relaxation, that is, βarr1 promotes negative inotropy and lusitropy upon AT_1R activation in cardiac myocytes.[33] These findings are consistent with the specialized roles of the various GRK isoforms described in transfected systems,[48] and also with the concept of GRK-induced receptor "barcoding," that is, the hypothesis that different GRK isoforms acting on the same GPCR induce recruitment of different βarr isoforms, resulting in different downstream signaling events and cellular responses, by phosphorylating the receptor at different sites/residues.[79]

In contrast to isolated murine cardiac myocytes, SII-activated AT_1Rs (i.e., AT_1R-bound βarrs) do not seem to produce inotropic or chronotropic effects in isolated Langendorff-perfused cardiac preparations, despite the fact that ERK1/2, which presumably mediates the positive inotropic effects of βarr2 in isolated cardiac myocytes,[33] is also activated by AT_1R-bound βarrs in Langendorff preparations.[32] Thus, it seems that the positive inotropic effects of cardiac βarr2 are strongly cell type- and experimental condition-dependent. Nevertheless, a consensus has emerged, according to which cardiomyocyte AT_1Rs promote hypertrophy and cardiomyocyte proliferation via βarrs, as well as contractility via (at least) βarr2, whereas cardiac fibroblast AT_1Rs promote fibrosis and cardiac adverse remodeling via the classical $G_{q/11}$ protein–PKC–Ca^{2+} signaling pathway. Since βarr2 also terminates the G-protein-mediated signaling of the AT_1R, stimulation of cardiac βarr2 activity and/or blockade of cardiac βarr1 activity at the AT_1Rs of the heart might be beneficial for the treatment of post-MI HF and the cardiac hypertrophy and adverse remodeling that accompany this devastating disease (Table 12.1). Indeed, TRV120027, a compound analogous to SII, that is, a βarr-biased AT_1R agonist that selectively activates βarrs while blocking G protein signaling, has shown promising results in animal models, blocking the undesirable G-protein-mediated AT_1R-induced vasoconstriction, thereby preserving renal function while, at the same time, enhancing the desirable (in acute HF) βarr-dependent contractility of cardiac myocytes. TRV120027 is currently under development for the clinical treatment of HF.[80]

Another interesting example of cardiac AT_1R βarr-mediated signaling is mechanical stretch-activation of the AT_1R.[34] Rakesh *et al.* showed that simple mechanical stretch (in the absence of any ligand) can activate the $AT_{1A}R$, leading to selective βarr recruitment and signaling without concomitant G protein activation.[34] What is more, the authors showed in an *ex vivo* murine heart model that this stretch-activated $AT_{1A}R$-elicited βarr signaling resulted in enhanced ERK1/2 and Akt (protein kinase B; PKB) activation, as well as EGFR transactivation, effects believed to mediate enhanced cardiomyocyte survival and protection.[34] In mouse hearts lacking βarrs or $AT_{1A}Rs$, mechanical stretch failed to produce these responses and led, instead, to enhanced myocyte apoptosis.[34] Thus, it appears that the heart is capable of responding to acute increases in mechanical stress by activating cardiac βarr-mediated cell survival signals, which again argues in favor of a beneficial and therapeutically desirable physiological role for cardiac AT_1R βarr-dependent signaling (at least for the cardiac βarr2 isoform; Table 12.1).

3.1.3 Vascular AT_1Rs and βarrs

In VSM cells, the AT_1R is known to promote vasoconstriction and raise blood pressure via the classic $G_{q/11}$ protein–diacylglycerol–phosphoinositide signaling that increases intracellular Ca^{2+}.[81] In addition, AngII acting through the AT_1R is one of the most powerful growth signals for VSM cells, stimulating cell proliferation that leads to VSM hyperplasia, a major contributing factor in vascular restenosis as occurs, for instance, after percutaneous coronary intervention followed by placement of an intravascular stent. This hyperplasia, combined with vascular inflammation that is also promoted by AT_1Rs residing in VSM cells, also contributes to the development of atherosclerosis.[82]

βArr-dependent AT_1R signaling in VSM cells has recently been documented as contributing to the proatherosclerotic effects of AngII.[35] Neointimal hyperplasia after carotid endothelial injury is enhanced in βarr1KO mice but, notably, diminished in βarr2KO mice.[35] Loss of βarr2 appears to decrease GPCR-stimulated VSM cell proliferation and ERK1/2-dependent activation/migration, consistent with a role for βarr2 signaling in the injury response.[35] When the low-density lipoprotein receptor KO mouse, a genetic model of accelerated atherogenesis, is crossed onto a βarr2KO background, atheroma formation is significantly reduced.[35] *In vitro*, both G-protein- and βarr-dependent pathways elicited by the $AT_{1A}R$ converge on EGFR transactivation to stimulate proliferation of primary VSM cells.[36] Specifically, βarr2 seems to enhance EGFR transactivation by the $AT_{1A}R$ by engaging the nonreceptor tyrosine kinase, c-Src.[36] In addition, βarr2-dependent ERK1/2 activation downregulates the proapoptotic phospho-Bcl-2-associated death promoter protein, thus inducing cytoprotective antiapoptotic effects in rat VSM cells.[37]

Thus, it appears that βarr2 can promote $AT_{1A}R$-dependent VSM proliferation and hypertrophy, thus contributing to the development and progression of atheromas and vascular restenosis, whereas βarr1, as in cardiomyocyte AT_1R signaling, opposes the actions of βarr2 in VSM cells (Table 12.1). The only difference appears to be that in VSM, βarr2 appears to be the "damaging" βarr isoform and βarr1 the "beneficial" one, whereas in cardiac myocytes, it is the other way around, that is, βarr2 seems beneficial by promoting AT_1R-dependent contractility and βarr1 harmful by opposing it. In conclusion, βarr2 inhibition and/or βarr1 stimulation might be desirable for the treatment of VSM hyperplasia and atherosclerotic/arteriosclerotic disease, contrary to HF, for which βarr2 stimulation appears to be a valid therapeutic strategy (Table 12.1).

3.1.4 Adrenal AT_1Rs and βarrs

Aldosterone is one of a number of hormones whose levels are elevated in HF. It produces a multitude of negative effects on the failing heart, including promotion of salt retention, adverse post-MI cardiac remodeling, and HF progression.[83–85] Its importance in cardiovascular pathophysiology is highlighted by the huge success of mineralocorticoid (aldosterone) receptor blockers, such as spironolactone and eplerenone, in the treatment of HF and other heart diseases.[86] Aldosterone is produced and secreted by the adrenocortical zona glomerulosa cells in response to AT_1R activation by AngII.[87] Until recently, the general consensus as regards AT_1R signaling to aldosterone production in the adrenal cortex was that it also proceeded through the classical $G_{q/11}$ protein–phosphoinositide–intracellular Ca^{2+} pathway.[81] However, a crucial role for adrenal βarr1-dependent signaling in AT_1R-induced aldosterone synthesis and secretion has now been demonstrated.[38] In adrenocortical zona glomerulosa cells *in vitro*, βarr1 was found to stimulate sustained ERK1/2 activation and subsequent upregulation of the steroidogenic acute regulatory (StAR) protein, a steroid transport protein that catalyzes the rate-limiting step in adrenal steroid biosynthesis, that is, mitochondrial cholesterol uptake.[38] These effects could be produced by AT_1R activation by SII, indicating that they were G-protein-independent, and led to a marked increase in aldosterone synthesis and secretion from these cells.[38] Of note, since StAR is the rate-limiting enzyme in the synthesis of all adrenal steroids, adrenocortical βarr1-dependent signaling might also affect the biosynthesis of other adrenal steroids (e.g., glucocorticoids).[38] With regard to adrenal βarr2, this βarr isoform appears to be expressed at very low levels in the adrenal cortex and is barely detectable in human adrenocortical zona glomerulosa cells.[38] Whether βarr2-dependent signaling from the adrenal AT_1R also affects aldosterone production/secretion and whether it affects it in a similar manner as βarr1 or opposes this effect of βarr1 remain open questions.

Importantly, adrenal βarr1 appears to be a critical regulator of circulating aldosterone levels *in vivo*. Under normal conditions, its adrenal-specific upregulation causes hyperaldosteronism.[38] In experimental animals progressing to HF after experimental MI, adrenal overexpression of βarr1 elevates blood aldosterone levels, whereas its adrenal-specific blockade, achieved via gene therapy with a βarr1 carboxyl terminal-derived protein fragment that abolishes the trafficking and signaling activities of βarr1, markedly reduces circulating aldosterone levels toward normal values.[39] This tremendous sensitivity of circulating aldosterone levels to adrenal βarr1 signaling *in vivo* has dramatic consequences for cardiac function and structural remodeling post-MI. Adrenal βarr1-promoted hyperaldosteronism results in accelerated adverse cardiac remodeling and deterioration of left ventricular function, whereas normalization of aldosterone levels by adrenal βarr1-targeted gene therapy stops or even reverses the negative structural and functional effects of aldosterone on post-MI HF progression.[39]

Thus, it appears that in the adrenal gland, βarr1 is a major driving force behind the elevation of cardiotoxic aldosterone levels that accompanies and aggravates HF, and its inhibition might be of therapeutic value in combating the significant cardiovascular toxicity of this hormone, for instance, in post-MI HF, hypertension, and other cardiovascular diseases (Table 12.1). Taking into account the fact that adrenal βarr1 also causes elevation of catecholamine secretion from the adrenal medulla by mediating adrenal α_2AR desensitization and downregulation, adrenal βarr1 emerges as a pivotal regulator of the neurohormonal, that is, catecholaminergic and mineralocorticoid-derived, burden on the failing heart. Blockade of adrenal βarr1 signaling, via gene therapy or pharmacologically, might achieve a pharmacological "killing two birds with one stone" in the treatment of HF, that is, suppression of both catecholamine and aldosterone levels,[88] which would relieve the failing heart of a great deal of toxic hormonal stress (Table 12.1).

3.2. Angiotensin II type 2 receptors

In contrast to their AT_1R counterparts, AT_2Rs do not appear to interact with βarrs or even to internalize in response to AngII activation.[89] However, few studies have examined the actions of βarrs (if any) on AT_2Rs, and *in vivo* in particular, none. Consequently, whether the striking difference in the affinity of the two major types of AngII receptors for βarrs has any role in the differences in the effects they elicit in cells[81] is completely unknown but potentially worth exploring. In any case, the majority of AT_2R effects are diametrically opposite of those of the AT_1R. For example, the AT_2R is generally proapoptotic and anti-inflammatory, and specifically in VSM, it reduces hyperplasia and cell proliferation, beneficial effects in atherosclerosis and vascular stenosis.[81] Thus, βarrs, by selectively interacting with and desensitizing/internalizing AT_1Rs, may indirectly promote the effects of its "opponent" receptor, the AT_2R, which normally counterbalances AT_1R effects in cardiovascular cells and tissues. Such effects should always be considered when studying the effects of βarrs on cardiovascular AT_1Rs. For instance, part of the beneficial antiatherosclerotic actions of βarr1 in VSM cells might simply be due to a "reflex" elevation of the beneficial, proapoptotic activity of the AT_2R, triggered by the βarr1-exerted suppression of AT_1R signaling in these cells. Therefore, whenever the therapeutic potential of stimulating the activity of a given βarr isoform at the AT_1R is considered, especially in the cardiovascular system, the indirect impact such a strategy might have on the function of the AT_1R's "natural ballast," the AT_2R, must always be considered.

4. OTHER CARDIOVASCULAR GPCRS AND βARRS

4.1. Endothelin receptors

Like AngII, ETs are potent regulators of vascular tone and are also known to induce VSM hyperplasia and hypertrophy, leading to hypertension and vascular remodeling and occlusion (stenosis), which can severely complicate atherosclerosis.[90] Like AngII acting through the AT_1R, ET-1 acting through its type A (ET_A) receptor exerts all these effects (vasoconstriction and VSM hyperplasia) through the classic $G_{q/11}$ protein–phospholipase C–inositol 1,4,5-trisphosphate–diacylglycerol pathway leading to elevated intracellular Ca^{2+} concentration and activation of PKC. In mesenteric arterial smooth muscle cells, βarrs have been found to antagonize this prohypertrophic and procontractile G-protein-dependent signaling of the ET_A receptor, through their classical function as desensitizers of GRK2-phosphorylated receptors.[40]

Very recently, however, a novel role of βarrs as mediators of ET_A receptor signaling in rat aortic smooth muscle cells was uncovered. Similar to the VSM cell AT_1R, βarr2 was found to mediate not only the G protein uncoupling of the ET_A receptor, but also its signaling to sustained ERK and p38 MAPK activation.[41] Notably, βarr1 does not seem to affect these properties of the ET_A receptor.[41] However, both βarr isoforms appear to be required for ET_A receptor-dependent arterial smooth muscle cell migration in response to ET-1.[41] Given that this process underlies vascular remodeling and occlusion, this finding might have important implications for treatment of atherosclerosis and vascular (re)stenosis. Thus, it appears that VSM cell βarr2 is instrumental not only in AngII/AT_1R-dependent VSM remodeling and vascular stenosis, but also in analogous ET-1/ET_A receptor-dependent effects, which suggests that inhibiting vascular βarr2 signaling might be beneficial in treatment of vascular diseases, such as atherosclerosis and post-PCI vessel restenosis (Table 12.1). Finally, ET receptors play important roles in pulmonary vasculature, where they mediate vasoconstriction and can contribute to pulmonary arterial hypertension.[91] Therefore, βarrs, as either ET receptor desensitizers or signal transducers, might be involved in the pathophysiology of this disease as well.

4.2. Vasopressin receptors

Vasopressin is another important hormone for the cardiovascular system that exerts its effects through two different GPCR types, V_1 and V_2.[92] V_1 vasopressin receptor (V_1R) is a $G_{q/11}$-coupled receptor located in VSM cells, where it promotes vasoconstriction and thus raises blood pressure, whereas V_2R is a G_s protein-coupled receptor, located mainly in the renal tubular epithelial cells of the distal nephron, where it regulates renal water reabsorption.[92] Both vasopressin receptor types interact with βarrs *in vitro*. The V_2R is a classic example of a Class B GPCR in terms of its βarr-dependent endocytosic properties, that is, it binds and internalizes with both βarrs equally, and its interaction with βarrs is strong and long-lasting.[93]

So far, the effects of βarrs on vasopressin receptor signaling relevant to the cardiovascular system have been studied only in the context of naturally occurring single nucleotide polymorphisms in the human V_2R gene. One such polymorphism results in the substitution of Arg for His at position 137 of the encoded receptor protein.[42] This mutation, located in the second intracellular loop of the receptor, severely affects V_2R agonist-promoted receptor desensitization and downregulation. It was initially identified as a "loss-of-function" V_2R polymorphism, strongly associated with the clinical syndrome of nephrogenic diabetes insipidus (NDI) in afflicted carriers.[94] Subsequently, at the molecular level, it was found to cause constitutive GRK-mediated receptor phosphorylation followed by unrestrained βarr-mediated V_2R desensitization, internalization, and loss of responsiveness to circulating vasopressin.[42] Thus, this molecular abnormality produces the phenotype of an inactive V_2R, which translates into the inability of the kidneys to reabsorb water and hence, uncontrollable water loss in the urine, that is, "diabetes insipidus."

Of note, another naturally occurring polymorphism, resulting in substitution of either Cys or Leu in the place of Arg137 of the human V_2R, causes a different clinical syndrome, the "nephrogenic syndrome of inappropriate antidiuresis" (NSIAD).[43] This substitution also leads to constitutive βarr-dependent V_2R endocytosis, but contrary to the Arg137His one, also (somewhat paradoxically) results in constitutive V_2R-stimulated cAMP accumulation. This impairs the kidney's ability to excrete excess free water, which most probably underlies NSIAD in afflicted carriers.[43] To sum up, βarrs can cause severe dysfunction of several naturally occurring V_2R variants in humans, simply by acting as classical desensitizers and mediators of internalization/downregulation of the V_2R, but doing so in an aberrantly excessive (constitutive) manner. Thus, βarr inhibition in the kidneys of carriers of such V_2R polymorphisms might be of therapeutic value in the treatment of renal syndromes, such as NDI and NSIAD, which can directly and detrimentally impact cardiovascular function and homeostasis in these individuals (Table 12.1).

4.3. Niacin receptor (GPR109A)

Niacin (nicotinic acid), despite recent trials challenging its clinical effectiveness,[95] has been a very useful lipid-lowering drug for over 50 years. In can, however, induce severe cutaneous flushing, a major adverse effect that significantly limits its utility in hyperlipidemic patients.[96] Both the serum free fatty acid lowering and the cutaneous blood flow/flushing-stimulating actions of niacin are exerted via activation of its receptor, GPR109A, which is a $G_{i/o}$-protein-coupled receptor.[97,98] In mice and humans, GPR109A has been shown to activate $G_{i/o}$ protein signaling, leading to a decrease in plasma triglycerides, a major risk factor for atherosclerosis and cardiovascular disease, and simultaneously to stimulate cytosolic phospholipase A_2 ($_cPLA_2$), an important proinflammatory enzyme, which results in the production and secretion of the vasodilatory prostaglandin D_2 (PGD_2).[97,98] It is this production of PGD_2 that underlies the adverse effect of flushing with niacin treatment. Recently, it was demonstrated that βarr1, although dispensable for the beneficial effects of niacin on free fatty acid lowering, promotes the niacin-related flushing, as measured by perfusion of the ventral ear of mice.[44] The underlying mechanism appears to be GPR109A-induced βarr1-mediated activation of $_{c}PLA_2$, leading to release of arachidonic acid (AA), the precursor of all prostaglandins, including PGD_2, and the dilation of cutaneous vessels that produces the unpleasant "flushing" effect.[44] Thus, βarr1 mediates signal transduction of the niacin receptor toward vasodilatation, which in this case is a negative (adverse) effect, whereas the therapeutically desirable GPR109A signaling toward lipid lowering is entirely G-protein-dependent (Table 12.1).

In this setting, βarr1 inhibition at the GPR109A receptor or a "biased" GPR109A ligand capable of activating G protein signaling without recruiting βarrs would be therapeutically desirable, as it would confer the same therapeutic effect of niacin on lipid metabolism without the adverse effect of "flushing" (Table 12.1). Such a G-protein-biased GPR109A receptor agonist might not only reduce the adverse effect of flushing, but also have the added benefit of enhancing the therapeutic effect of lipid level reduction, given that βarr1 desensitizes the beneficial G-protein-dependent signaling of the GPR109A receptor. Notably, a series of pyrazole derivatives with GPR109A receptor agonist properties have been synthesized that are devoid of the ability to internalize the GPR109A receptor and activate ERK, that is, the classic βarr-dependent signaling effects, and, indeed, they do not cause "flushing."[99] These compounds, including one with very potent serum free fatty acid-lowering properties, 3-(1*H*-tetrazol-5-yl)-1,4,5,6-tetrahydrocyclopentapyrazole (MK-0354),[100] appear superior to niacin for lipid-lowering therapy and are most probably G-protein-selective "biased" agonists at the GPR109A receptor, that is, they activate G protein signaling while blocking βarr1 activation. In conclusion, βarr1 blockade has potential as a strategy to improve the lipid-lowering treatment with niacin (Table 12.1), which despite some recent controversies,[95] is still thought to be useful for the prevention of cardiovascular diseases, such as atherosclerosis and acute coronary syndrome.

4.4. P2Y Receptors

The P2Y family of purinergic receptors (receptors for purine nucleosides and nucleotides, such as adenosine 5′-diphosphate, ADP; and adenosine 5′-triphosphate, ATP) constitutes the metabotropic purinergic receptor family, all the members of which are GPCRs.[101] $P2Y_1$ and $P2Y_{12}$ receptors for ADP are expressed in platelets and play a key role in platelet activation and thrombosis.[101] They are targeted by several antiplatelet drugs, such as clopidogrel, prasugrel, and ticagrelor.[101] Platelets activated by ADP become refractory to restimulation, a phenomenon that has been attributed to $P2Y_1$ receptor desensitization and internalization.[102] Although a substantial fraction of $P2Y_{12}$ receptors also rapidly internalize upon stimulation, it remains uncertain whether this ADP receptor type is also subject to desensitization.[103]

Very recently, it was shown in platelets derived from βarr1KO mice that βarr1 can promote thrombus formation, not by desensitizing P2Y receptors but by transducing their signaling to integrin $\alpha_{IIb}\beta_3$.[45] In fact, neither βarr appears to be involved in agonist (ADP)-induced platelet activation and P2Y receptor desensitization.[45] Platelets from either βarr1KO or βarr2KO mice responded normally to a series of agonists, and displayed similar desensitization to ADP as wild-type mouse platelets, suggesting no role for βarrs in these processes.[45] No defect in hemostasis was observed in either KO line. In βarr1KO platelets only, defects in arterial thrombosis formation attributed to defective integrin $\alpha_{IIb}\beta_3$ shape change were observed,[45] suggesting a role for βarr1, but not for βarr2, in thrombosis and P2Y receptor signaling toward integrin $\alpha_{IIb}\beta_3$. This βarr1-dependent signaling to integrin $\alpha_{IIb}\beta_3$ was Src kinase- and Akt kinase-mediated.[45] Thus, it appears that βarr1, acting as signal transducer of platelet ADP P2Y receptors, can promote arterial thrombosis (Table 12.1). Therefore, inhibition of platelet βarr1 might be of therapeutic value in thromboembolic disease treatment.

P2Y receptors are also present in arterial smooth muscle cells, wherein they induce vasoconstriction, hypertrophy, and hyperplasia, via typical coupling to the $G_{q/11}$ protein–phospholipase C–Ca^{2+} signaling pathway.[46] Using βarr isoform-specific siRNA knockdown, βarr1 was recently shown to act as a desensitizer of P2Y receptor-induced contractility in adult Wistar rat mesenteric arterial smooth muscle cells *in vitro*, in response to uridine 5′-triphosphate (UTP), another purine nucleotide.[46] βarr1 appears to be the only βarr isoform capable of conferring P2Y receptor desensitization in these cells and it does so following phosphorylation of the receptors by GRK2 only.[46] In addition, siRNA-mediated depletion of βarr1 in rat aortic smooth muscle cells *in vitro* prolongs signaling of UTP-stimulated P2Y receptors toward activation of p38 MAPK and ERKs, providing additional evidence that βarr1, but not βarr2, acts as a desensitizer of arterial smooth muscle cell P2Y receptors.[41] Thus, with regard to VSM cell-residing P2Y receptors, βarr1 appears capable of antagonizing purine nucleotide-dependent vasoconstriction and arterial smooth muscle cell migration, which suggests that enhancing its activity in VSM cells might be beneficial in treatment of hypertension and vascular stenosis/arteriosclerosis (Table 12.1).

4.5. Protease-activated receptors

Another family of GPCRs with important cardiovascular system effects is the protease-activated receptor (PAR) family, which responds to thrombin and other proteases that activate the receptor by proteolytic cleavage of its extracellular N-terminus.[104] They are divided into four distinct types (PAR_{1-4}), all of which couple primarily to the $G_{q/11}$ protein–phospholipase C–Ca^{2+} signaling pathway, although PAR_1 and PAR_2 can also couple to G_i proteins, leading to inhibition of AC and lowering of cAMP levels. They reside in membranes of endothelial cells and platelets, among other cell types, wherein they mediate blood clot formation, that is, thrombosis.[104] PARs are well-characterized βarr substrates, with βarrs acting upon them as either desensitizers/internalizers or signal transducers.[104]

A recent study in βarr1KO mouse platelets suggests that βarr1 can mediate signaling from thrombin-stimulated PAR4, in addition to its role in ADP P2Y receptor signaling in these cells. βArr1, but not βarr2, was found to transduce the agonist-activated PAR4 signal to phosphatidylinositol-3 kinase (PI3K) and subsequent Akt/PKB activation in platelets, leading to increased fibrinogen binding and thrombus formation.[47] Interestingly, the authors also found that this βarr1-dependent PAR4 signaling was potentiated by ADP P2Y receptors, as $P2Y_1$ receptor activation appears to enhance it via PKC activation and the $P2Y_{12}$ receptors appear to heterodimerize with PAR4, facilitating its interaction with βarr1.[47] These signaling effects of βarr1 on PAR4 are consistent with the reduced development of ferric chloride-stimulated thrombosis observed in βarr1KO mice.[45,47] Thus, platelet βarr1 seems to mediate not only the prothrombotic signaling of P2Y receptors but also that of PARs (Table 12.1). Taken together, these findings on P2Y receptor and PAR βarr1-dependent signaling in platelets make platelet βarr1 an attractive therapeutic target for treatment of various thromboembolic and other blood coagulation disorders.

5. CARDIOVASCULAR NON-GPCRs AND βARRS

As already described in several examples of βarr-dependent signaling by cardiac GPCRs, an important mechanism by which βarrs direct ERK1/2 signaling in the heart, regulating cardiomyocyte hypertrophy and survival/apoptosis, is transactivation of the cardiac EGFR, a receptor belonging to the receptor tyrosine kinase family.[105] In this pathway, βarrs act "upstream" of the EGFR and "downstream" of a GPCR. In some cases, βarrs are known to mediate downstream signaling of non-7TMRs, for example, the insulin-like growth factor-1 receptor and transforming growth factor beta receptor type III.[106] However, all the cardiovascular-related actions of βarrs on non-GPCRs reported to date pertain to EGFR transactivation. Thus, the VSM cell AT_1R, the cardiomyocyte βARs, and the cardiac urotensin II receptor (also a GPCR) have all been reported to induce EGFR transactivation and subsequent ERK1/2 activation via βarrs, which may contribute to VSM or cardiac hypertrophy.[36,107–109]

The precise molecular pathways underlying this type of cardiovascular βarr-dependent signaling have not been fully elucidated and may vary for each GPCR. Cardiac AT_1R-stimulated βarr-dependent EGFR transactivation opposes chronic catecholaminergic stimulation-induced dilated cardiomyopathy and cardiac apoptosis in a mouse model of HF, since mice incapable of transactivating cardiac EGFR displayed significantly less cardiac hypertrophy and improved survival compared to wild-type mice.[25] The mechanism probably involves transactivated EGFR-induced ERK1/2 activation.[109,110] Similarly, urotensin II receptor-mediated EGFR transactivation has been shown to be βarr-dependent and to reduce cardiac apoptosis in mice with chronic pressure overload compared to mice in which cardiac EGFR was blocked.[107]

However, the involvement of βarrs in cardiac GPCR-mediated EGFR transactivation, cardiomyocyte proliferation, and survival appears to be case- and GPCR-specific. AT_1R-stimulated EGFR transactivation and hypertrophy are totally dependent on $G_{q/11}$ proteins, and not at all on βarrs, in neonatal rat ventricular cardiomyocytes.[111] Conversely, ligand-independent, mechanical stretch-stimulated AT_1R-mediated EGFR transactivation in the heart, which increases survival by promoting Akt/PKB activation and lowering cardiac apoptosis, is βarr2-dependent (Table 12.1).[34] In VSM cells, AT_1R transactivates the EGFR via βarr2 (Table 12.1) leading to ERK1/2 activation and increased VSM cell DNA synthesis.[36] Of note, there is significant cross-talk between AT_1R and βAR signaling on cardiomyocyte contractility and ERK1/2 activation,[112,113] which is subject to regulation by βarrs. This could represent another mechanism by which cardiac βarrs, acting as transducers of signals from one receptor to another (either a GPCR or a non-GPCR), affect cardiac hypertrophy and survival/apoptosis. To sum up, cardiac EGFR transactivation by GPCRs, which enhances both survival and hypertrophy, can proceed either through βarrs (mainly βarr2) or through G proteins, depending on the GPCR, ligand and cellular/tissue model studied. In any case, transactivation of this non-GPCR appears to be an important signaling mechanism by which cardiovascular βarrs stimulate cardiomyocyte or VSM cell proliferation and survival.

6. BIASED AGONISM/ANTAGONISM OF CARDIOVASCULAR GPCR LIGANDS

Numerous studies have investigated the ability of ligands to "bias" GPCR activation of βarr- versus G-protein-dependent signaling pathways. Among the GPCRs studied are the β_2AR and the AT_1R, two of the most important cardiovascular GPCRs. Thus, a short discussion of the properties of ligands for these two receptors with regard to their "biased" agonism/antagonism of βarr-mediated signaling is worthwhile to have a more complete picture of the physiology and pharmacology of cardiovascular βarrs. This section gives a brief overview of what is currently known, first about βarr signaling "bias" of β_2AR ligands, followed by what is known about βarr signaling "bias" of AT_1R ligands. Some examples of βarr-biased ligands for other cardiovascular GPCRs, for example, α_2AR and GPR109A, have already been given in preceding sections.

With regard to the β_2AR, considerable bias toward βarr activation was observed for compounds containing an ethyl substitution of the α-carbon within a series of phenylethylamines.[114] Stereoisomers of fenoterol, a β_2AR-selective agonist, differentially activate G_s and G_i proteins in rat cardiomyocytes.[115] It is now known that, in addition to blockade of βAR-induced G protein activation, some β-blockers produce activation of ERKs through βarr.[116] In view of the fact that a number of β-blockers have been tested in clinical trials for therapy of congestive HF and only a few have shown beneficial effect, this ERK1/2-stimulating effect may be physiologically and therapeutically relevant. In particular, carvedilol (Fig. 12.1), a non-subtype selective β-blocker that is beneficial for the treatment of congestive HF,[63] has been shown to stimulate βarr-mediated activation of ERK1/2, while blocking G protein activation by βARs; in other words, it is a fully "biased" agonist for βarr signaling at βARs.[117] Bucindolol, another β-blocker that is very similar to carvedilol in terms of stimulating ERK1/2 while blocking β_2AR activation of G proteins,[116] has also shown some benefit in HF treatment.[118]

A recent study by the Raymond Stevens' group investigated the biophysical basis of βarr "biased" signaling.[119] Using site-specific ^{19}F (fluorine-19)-labeled β_2AR bound to a panel of structurally diverse β_2AR ligands and NMR (nuclear magnetic resonance) spectroscopy, these authors were able

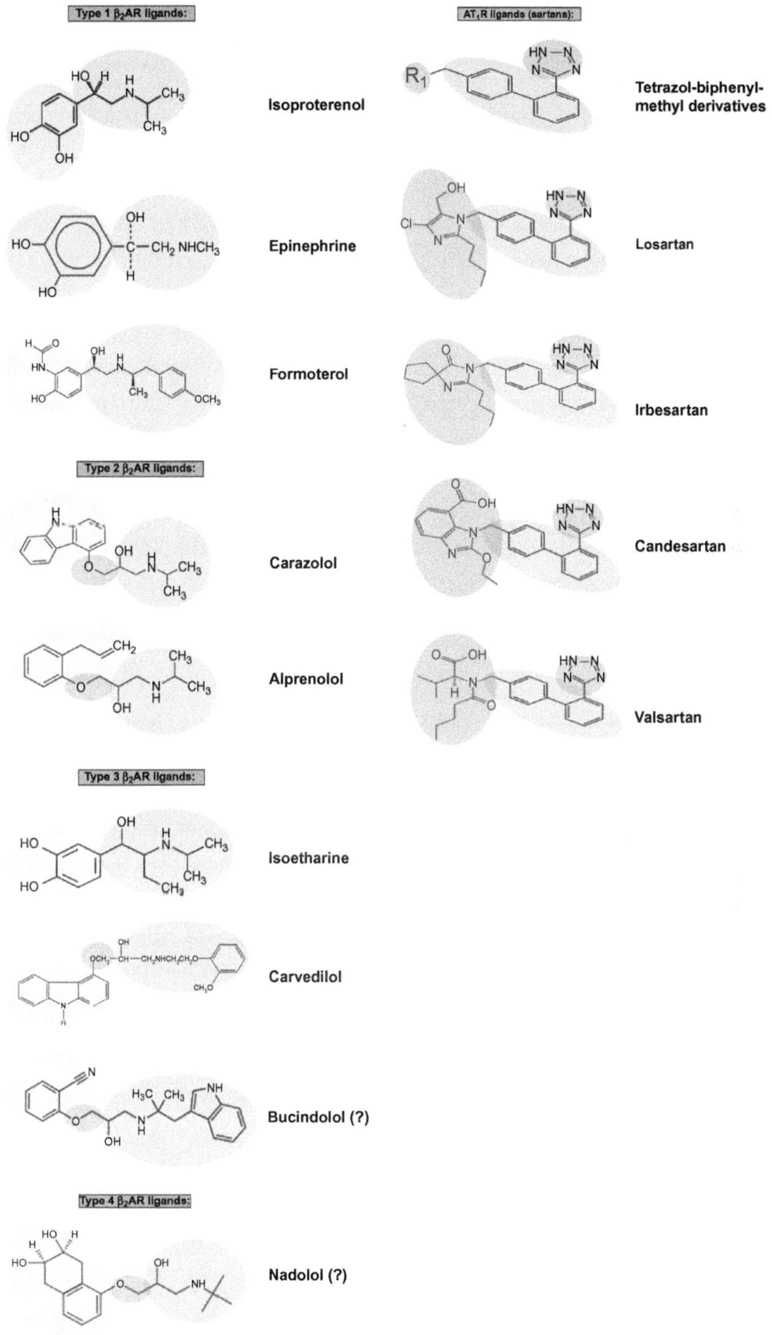

Figure 12.1 Chemical structures of the β_2AR (left column) and AT_1R (right column) ligands discussed in Section 6. Left column: light pink indicates the aromatic "head group"; light blue, the ethanolamine "tail"; and olive green, the two-atom "linker." Notice the absence of this two-atom linker from ligands that have agonistic activity at the β_2AR.

(Continued)

to relate ligand-induced conformational changes in the receptor to ligand "bias" for βarr versus G protein agonism. The β_2AR interacts with G proteins via its TM (transmembrane) helices V and VI, and intracellular loops (ICLs) ICL2 and ICL3, but TM helix VII makes no substantial contact with the G protein. Thus, the degree to which β_2AR ligands shift the receptor equilibrium toward the active state of TM helix VI dictates the resulting G protein signaling capacity/efficacy.[119] For instance, the antagonist/weak partial agonist alprenolol can induce a small shift on Cys^{265}6.27 (TM VI), but has a minimal effect on Cys^{327}7.54 (TM VII), whereas the inverse agonist carazolol induces no apparent shifts to either Cys.[119] Conversely, two βarr-biased ligands, isoetharine and carvedilol, cause large conformational shifts on Cys^{327}7.54, while producing minimal changes at Cys^{265}6.27.[119] Thus, for the β_2AR, it appears that TM helix VII conformations primarily impact βarr signaling, whereas TM helix VI conformations G protein signaling.[119]

The majority of known high-affinity β_2AR ligands have a common structural motif consisting of an aromatic "head group" and an ethanolamine "tail group" (Fig. 12.1). The head group, that is, catechol, of agonists is directly connected to the ethanolamine tail, while antagonists and inverse agonists, both of which act as competitive inhibitors to agonists and either have no impact or a negative impact on the basal receptor activity, have an additional two-atom linker between the "head" and "tail" groups. Crystal structures of the β_2AR bound to various ligands, as well as biochemical and modeling studies, have established that the aromatic "head" groups interact with TM helices V and VI, whereas the ethanolamine "tails" are anchored by ionic and polar interactions at TM helices III and VII.[119,120] Consequently, four different types of β_2AR ligands can be distinguished with regard to their βarr versus G protein signaling "bias" properties (Fig. 12.1):

Type 1. Ligands that induce conformational activation of both TM helices VI and VII and thus activate both G proteins and βarrs equally (unbiased agonists), for example, isoproterenol, epinephrine, and formoterol.

Type 2. Ligands that block conformational activation of both TM helices VI and VII and thus block both G proteins and βarrs equally

Figure 12.1 Cont'd—Bucindolol (based on its pharmacological similarity to carvedilol) and nadolol are putative type 3 and type 4 β_2AR ligands, respectively, that is, their types have not been verified as yet. Right column: light pink indicates the R_1 substitution; light blue, the biphenyl-methyl "backbone"; and olive green, the tetrazole ring group. See text for details. (See Color Insert.)

(unbiased antagonists/inverse agonists), for example, carazolol and alprenolol.

Type 3. Ligands that induce preferential conformational activation of TM helix VII over TM helix VI, and thus selectively activate βarrs while blocking G proteins (βarr-biased agonists or G-protein-biased antagonists), for example, isoetharine and carvedilol.

Type 4. Ligands that induce preferential conformational activation of TM helix VI over TM helix VII, and thus selectively activate G proteins while blocking βarrs (G-protein-biased agonists or βarr-biased antagonists), for example, MK-0354 for the niacin receptor GPR109A. Notably, no compound of this type has been identified for the β_2AR to date, but nadolol (Fig. 12.1) might be a good candidate.[117,121]

Considering βarr-biased agonism/antagonism at the AT_1R, data from our lab indicate that losartan, the prototypic AT_1R-selective antagonist (ARB), fails to suppress adrenal βarr1-dependent post-MI hyperaldosteronism,[39] a finding that prompted us to test various other structurally similar ARBs (i.e., tetrazolo-biphenyl-methyl derivatives; Fig. 12.1) for their efficacy in inhibiting the human AT_1R–βarr interaction. Our latest data indicate that irbesartan, like losartan, is a very weak βarr-biased antagonist at the AT_1R, displaying very little (if any) "bias" for βarr inhibition versus G protein inhibition.[59] In contrast, candesartan and valsartan are the most potent βarr blockers at the AT_1R and thus, the most "biased" βarr antagonists.[59] Based on a comparison of the structures of these tetrazolo-biphenyl-methyl derivatives, all of which are very potent G protein antagonists of the AT_1R, some assumptions about the structural requirements of AT_1R ligands for βarr-biased antagonism can be made (Fig. 12.1): (a) the tetrazole ring attached to the biphenyl "backbone" appears essential for βarr inhibition at the AT_1R (as it is for G protein inhibition at the AT_1R) and (b) the "bulkier" the R_1 substitution of the biphenyl-methyl "backbone," the more "biased" the ligand is toward βarr blockade.[59] Thus, candesartan and valsartan, which have the "bulkiest" R_1 substitutions, are the most "biased" βarr antagonists at the AT_1R, whereas losartan and irbesartan, which have the "smallest" R_1 substitutions, are the least "biased" βarr antagonists.[59]

As for βarr-biased agonism at the AT_1R, our data from a series of AngII analogs with various substitutions at amino acid position 5, all of which are devoid of G protein activation properties due to a sarcosine (*N*-methyl-glycine) substitution at amino acid position 1, indicate that the "bulkier" the amino acid substitution at position 5 of the AngII peptide, the more biased the ligand is toward βarr activation.[59,122] Of note, [Sar^1,Ile^4,Ile^8]-AngII, the "prototypic" βarr-biased agonist at the AT_1R, has no substitution at this position; it has a substitution at position 4 instead.[123,124] Substitutions at positions 1 and 8 are indispensable for an AngII peptide analog to lack any G protein-activating properties and thus be a pure βarr-biased agonist at the AT_1R.[59,122,123]

A final note on βarr-biased agonism/antagonism of GPCR ligands: it should always be kept in mind that ligand "bias" with regard to βarr-dependent signaling exists not only at the level of receptor interactions with βarrs and G proteins, which dictates the extent of "bias" for βarr-mediated versus G-protein-mediated signaling, but also at the level of receptor-induced βarr activation *per se*. βarrs can adopt various active conformations upon their interaction with a GPCR, which apparently dictates their downstream signaling, that is, which of the several available cellular signaling pathways will be activated.[125] Indeed, a recent study uncovered significant variation in the extent of "bias" of various AngII analogs for distinct AT_1R-bound βarr active conformations, leading to different signaling events and cellular responses.[126]

7. PERSPECTIVES AND FUTURE DIRECTIONS

Given that βarrs are ubiquitous in the cardiovascular system and regulate all the important GPCRs within its various parts, organs, and tissues, the recently discovered and increasingly expanding field of βarr-dependent GPCR signaling offers several exciting opportunities for therapeutic intervention in cardiovascular disease. Although cardiovascular βarr physiology and pharmacology are still in their infancy, with a lot of blank or shady areas awaiting elucidation, the potential advantages of targeting cardiovascular βarrs for therapeutic purposes are enormous.

Admittedly, the picture is still hazy regarding several physiological actions of βarrs in the cardiovascular system, especially when it comes to ascribing certain physiological/pathophysiological effects to specific βarr isoforms in specific tissues/organs. For instance, cardiac βarr2 reduces βAR-stimulated contractility (which is G protein-dependent), but, at the same time, appears to enhance AT_1R-dependent contractile function. This βarr isoform also appears to have beneficial effects in the heart as it increases survival and cardiomyocyte proliferation; however, it promotes atherosclerosis in the vasculature. On the other hand, βarr1 seems to have negative effects in most cardiovascular tissues/organs. It opposes cardiac βAR (and AT_1R) procontractile signaling, promotes AngII-dependent aldosterone production/secretion by the adrenal gland, and mediates niacin-induced "flushing" via the GPR109A receptor. Furthermore, since it theoretically competes with βarr2 for GPCR binding, inhibition of βarr1 would indirectly enhance βarr2 activity and its purportedly beneficial effects in the heart. What has become crystal clear from the studies on cardiovascular βarrs to date is that the two βarr isoforms are by no means redundant with regard to their cardiovascular actions; in fact, more often than not, their signaling paths end at diametrically opposite cellular responses. Consequently, targeting of βarrs for cardiovascular disease treatment might lead to unexpected, and adverse, outcomes.

It appears likely that cardiovascular tissue- and/or individual βarr isoform-specific targeting will ultimately prove to be the only realistic approach to exploiting βarr-dependent signaling in the cardiovascular system for therapeutic purposes. Thus, agents acting specifically in the cardiac muscle and selectively inhibiting βarr1 or selectively stimulating βarr2 therein might be useful for HF therapy, as might adrenal βarr1-specific inhibitors, whereas vascular-selective drugs with the reverse properties, that is, βarr2 inhibitors or βarr1 stimulators, might be useful in the treatment of atherosclerosis and vessel (re)stenosis. Future studies will hopefully clarify the roles of individual βarr isoforms *in vivo* in each part of the cardiovascular system. For the time being, synthesis and development of new cardiovascular GPCR "biased" ligands, which selectively activate one signaling pathway over others or stimulate one signaling pathway, for example, βarr, while blocking another, for example, G protein, seems the most reasonable and pragmatic way to move forward. In any case, the field of physiology of cardiovascular βarrs, regardless of whether they are viewed in their traditional role of GPCR desensitizers or in their newly emerged role of receptor signal transducers, appears to be a treasure-trove of possibilities for novel cardiovascular therapies. Thus, it is bound to flourish with tremendous amounts of new information in the foreseeable future.

8. CONCLUSIONS

Given the tremendously important roles of several GPCRs in regulation of cardiovascular physiology and function, it comes as no surprise that cardiovascular βarrs, which act upon these receptors either to desensitize/internalize them or to mediate their signal transduction to various cellular effectors, are extremely important regulators of cardiovascular biology. As more and more information about their roles in eliciting (or preventing) specific biological effects in the individual organs/tissues of the cardiovascular system keeps coming to light, their potential as therapeutic targets for cardiovascular diseases, a cluster of diseases that still carries the biggest morbidity and mortality burden in the entire modern world, will become increasingly evident and attainable.

REFERENCES

1. Piascik MT, Perez DM. Alpha1-adrenergic receptors: new insights and directions. *J Pharmacol Exp Ther.* 2001;298:403–410.
2. Jensen BC, O'Connell TD, Simpson PC. Alpha-1-adrenergic receptors: targets for agonist drugs to treat heart failure. *J Mol Cell Cardiol.* 2011;51:518–528.
3. Jensen BC, Swigart PM, De Marco T, Hoopes C, Simpson PC. {alpha}1-Adrenergic receptor subtypes in nonfailing and failing human myocardium. *Circ Heart Fail.* 2009;2:654–663.
4. Van Tassell BW, Rondina MT, Huggins F, Gilbert EM, Munger MA. Carvedilol increases blood pressure response to phenylephrine infusion in heart failure subjects with systolic dysfunction: evidence of improved vascular alpha1-adrenoreceptor signal transduction. *Am Heart J.* 2008;156:315–321.
5. Stewart JM, Patel MB, Wang J, et al. Chronic elevation of norepinephrine in conscious dogs produces hypertrophy with no loss of LV reserve. *Am J Physiol.* 2009;262: H331–H339.
6. Vecchione C, Fratta L, Rizzoni D, et al. Cardiovascular influences of alpha1b-adrenergic receptor defect in mice. *Circulation.* 2002;105:1700–1707.
7. Cotecchia S. The α1-adrenergic receptors: diversity of signaling networks and regulation. *J Recep Signal Transduct.* 2010;30:410–419.
8. Stanasila L, Abuin L, Dey J, Cotecchia S. Different internalization properties of the alpha1a- and alpha1b-adrenergic receptor subtypes: the potential role of receptor interaction with beta-arrestins and AP50. *Mol Pharmacol.* 2008;74:562–573.
9. Phillipp M, Hein L. Adrenergic receptor knockout mice: distinct functions of 9 receptor subtypes. *Pharmacol Ther.* 2004;101:65–74.
10. Small KM, Brown KM, Forbes SL, Liggett SB. Polymorphic deletion of three intracellular acidic residues of the alpha2B-adrenergic receptor decreases G protein-coupled receptor kinase-mediated phosphorylation and desensitization. *J Biol Chem.* 2001;276:4917–4922.
11. Nguyen K, Kassimatis T, Lymperopoulos A. Impaired desensitization of a human polymorphic α2B-adrenergic receptor variant enhances its sympatho-inhibitory activity in chromaffin cells. *Cell Commun Signal.* 2011;9:5.
12. Snapir A, Mikkelsson J, Perola M, Penttila A, Scheinin M, Karhunen PJ. Variation in the alpha2B-adrenoceptor gene as a risk factor for prehospital fatal myocardial infarction and sudden cardiac death. *J Am Coll Cardiol.* 2003;41:190–194.
13. Von Wowern F, Bengtsson K, Lindblad U, Rastam L, Melander O. Functional variant in the (alpha)2B adrenoceptor gene, a positional candidate on chromosome 2, associates with hypertension. *Hypertension.* 2004;43:592–597.
14. Ohlin B, Berglund G, Nilsson PM, Melander O. Job strain, decision latitude and alpha2B-adrenergic receptor polymorphism significantly interact, and associate with high blood pressures in men. *J Hypertens.* 2007;25:1613–1619.
15. Lymperopoulos A, Rengo G, Funakoshi H, Eckhart AD, Koch WJ. Adrenal GRK2 upregulation mediates sympathetic overdrive in heart failure. *Nat Med.* 2007;13: 315–323.

16. Lymperopoulos A, Rengo G, Gao E, Ebert SN, Dorn 2nd GW, Koch WJ. Reduction of sympathetic activity via adrenal-targeted GRK2 gene deletion attenuates heart failure progression and improves cardiac function after myocardial infarction. *J Biol Chem.* 2010;285:16378–16386.
17. Lymperopoulos A, Rengo G, Zincarelli C, Soltys S, Koch WJ. Modulation of adrenal catecholamine secretion by in vivo gene transfer and manipulation of G protein-coupled receptor kinase-2 activity. *Mol Ther.* 2008;16:302–307.
18. Lymperopoulos A, Rengo G, Koch WJ. Adrenal adrenoceptors in heart failure: fine-tuning cardiac stimulation. *Trends Mol Med.* 2007;13:503–511.
19. Wang Q, Zhao J, Brady AE, et al. Spinophilin blocks arrestin actions in vitro and in vivo at G protein-coupled receptors. *Science.* 2004;304:1940–1944.
20. Schmid CL, Bohn LM. Physiological and pharmacological implications of beta-arrestin regulation. *Pharmacol Ther.* 2009;121:285–293.
21. Rengo G, Lymperopoulos A, Koch WJ. Future G protein-coupled receptor targets for treatment of heart failure. *Curr Treat Options Cardiovasc Med.* 2009;11:328–338.
22. Rengo G, Lymperopoulos A, Zincarelli C, et al. Myocardial adeno-associated virus serotype 6-betaARKct gene therapy improves cardiac function and normalizes the neurohormonal axis in chronic heart failure. *Circulation.* 2009;119:89–98.
23. Rengo G, Lymperopoulos A, Leosco D, Koch WJ. GRK2 as a novel gene therapy target in heart failure. *J Mol Cell Cardiol.* 2011;50:785–792.
24. Lymperopoulos A, Rengo G, Koch WJ. GRK2 inhibition in heart failure: something old, something new. *Curr Pharm Des.* 2012;18:186–191.
25. Noma T, Lemaire A, Naga Prasad SV, et al. Beta-arrestin-mediated beta1-adrenergic receptor transactivation of the EGFR confers cardioprotection. *J Clin Invest.* 2007;117:2445–2458.
26. Tilley DG. G protein-dependent and G protein-independent signaling pathways and their impact on cardiac function. *Circ Res.* 2011;109:217–230.
27. Noor N, Patel CB, Rockman HA. β-arrestin: a signaling molecule and potential therapeutic target for heart failure. *J Mol Cell Cardiol.* 2011;51:534–541.
28. Luttrell LM, Roudabush FL, Choy EW, et al. Activation and targeting of extracellular signal-regulated kinases by beta-arrestin scaffolds. *Proc Natl Acad Sci USA.* 2001;98:2449–2454.
29. Witherow DS, Garrison TR, Miller WE, Lefkowitz RJ. Beta-arrestin inhibits NF-kappaB activity by means of its interaction with the NF-kappaB inhibitor IkappaBalpha. *Proc Natl Acad Sci USA.* 2004;101:8603–8607.
30. Gao H, Sun Y, Wu Y, et al. Identification of beta-arrestin2 as a G protein-coupled receptor-stimulated regulator of NF-kappaB pathways. *Mol Cell.* 2004;14:303–317.
31. Zhai P, Yamamoto M, Galeotti J, et al. Cardiac-specific overexpression of AT1 receptor mutant lacking G alpha q/G alpha i coupling causes hypertrophy and bradycardia in transgenic mice. *J Clin Invest.* 2005;115:3045–3056.
32. Aplin M, Christensen GL, Schneider M, et al. Differential extracellular signal-regulated kinases 1 and 2 activation by the angiotensin type 1 receptor supports distinct phenotypes of cardiac myocytes. *Basic Clin Pharmacol Toxicol.* 2007;100:296–301.
33. Rajagopal K, Whalen EJ, Violin JD, et al. Beta-arrestin2-mediated inotropic effects of the angiotensin II type 1A receptor in isolated cardiac myocytes. *Proc Natl Acad Sci USA.* 2006;103:16284–16289.
34. Rakesh K, Yoo B, Kim IM, Salazar N, Kim KS, Rockman HA. Beta-arrestin-biased agonism of the angiotensin receptor induced by mechanical stress. *Sci Signal.* 2010;3: ra46.
35. Kim J, Zhang L, Peppel K, et al. Beta-arrestins regulate atherosclerosis and neointimal hyperplasia by controlling smooth muscle cell proliferation and migration. *Circ Res.* 2008;103:70–79.

36. Kim J, Ahn S, Rajagopal K, Lefkowitz RJ. Independent beta-arrestin2 and Gq/protein kinase Czeta pathways for ERK stimulated by angiotensin type 1A receptors in vascular smooth muscle cells converge on transactivation of the epidermal growth factor receptor. *J Biol Chem.* 2009;284:11953–11962.
37. Ahn S, Kim J, Hara MR, Ren XR, Lefkowitz RJ. Beta-arrestin-2 mediates anti-apoptotic signaling through regulation of BAD phosphorylation. *J Biol Chem.* 2009;284:8855–8865.
38. Lymperopoulos A, Rengo G, Zincarelli C, Kim J, Soltys S, Koch WJ. An adrenal beta-arrestin 1-mediated signaling pathway underlies angiotensin II-induced aldosterone production in vitro and in vivo. *Proc Natl Acad Sci USA.* 2009;106:5825–5830.
39. Lymperopoulos A, Rengo G, Zincarelli C, Kim J, Koch WJ. Adrenal beta-arrestin 1 inhibition in vivo attenuates post-myocardial infarction progression to heart failure and adverse remodeling via reduction of circulating aldosterone levels. *J Am Coll Cardiol.* 2011;57:356–365.
40. Morris GE, Nelson CP, Standen NB, Challiss RA, Willets JM. Endothelin signalling in arterial smooth muscle is tightly regulated by G protein-coupled receptor kinase 2. *Cardiovasc Res.* 2010;85:424–433.
41. Morris GE, Nelson CP, Brighton PJ, Standen NB, Challiss RA, Willets JM. Arrestins 2 and 3 differentially regulate ETA and P2Y2 receptor-mediated cell signaling and migration in arterial smooth muscle. *Am J Physiol Cell Physiol.* 2012;302:C723–C734.
42. Barak LS, Oakley RH, Laporte SA, Caron MG. Constitutive arrestin-mediated desensitization of a human vasopressin receptor mutant associated with nephrogenic diabetes insipidus. *Proc Natl Acad Sci USA.* 2001;98:93–98.
43. Rochdi MD, Vargas GA, Carpentier E, et al. Functional characterization of vasopressin type 2 receptor substitutions (R137H/C/L) leading to nephrogenic diabetes insipidus and nephrogenic syndrome of inappropriate antidiuresis: implications for treatments. *Mol Pharmacol.* 2010;77:836–845.
44. Walters RW, Shukla AK, Kovacs JJ, et al. Beta-arrestin1 mediates nicotinic acid-induced flushing, but not its antilipolytic effect, in mice. *J Clin Invest.* 2009;119:1312–1321.
45. Schaff M, Receveur N, Bourdon C, et al. β-arrestin-1 participates in thrombosis and regulates integrin aIIbβ3 signalling without affecting P2Y receptors desensitisation and function. *Thromb Haemost.* 2012;107:735–748.
46. Morris GE, Nelson CP, Everitt D, et al. G protein-coupled receptor kinase 2 and arrestin2 regulate arterial smooth muscle P2Y-purinoceptor signalling. *Cardiovasc Res.* 2011;89:193–203.
47. Li D, D'Angelo L, Chavez M, Woulfe DS. Arrestin-2 differentially regulates PAR4 and ADP receptor signaling in platelets. *J Biol Chem.* 2011;286:3805–3814.
48. Kukkonen JP. Regulation of receptor-coupling to (multiple) G proteins. A challenge for basic research and drug discovery. *Receptors Channels.* 2004;10:167–183.
49. Brodde OE. Beta-adrenoceptors in cardiac disease. *Pharmacol Ther.* 1993;60:405–430.
50. Communal C, Singh K, Sawyer DB, Colucci WS. Opposing effects of β1- and β2-adrenergic receptors on cardiac myocyte apoptosis: role of a pertussis toxin-sensitive G protein. *Circulation.* 1999;100:2210–2212.
51. Chesley A, Lundberg MS, Asai T, et al. The β2-adrenergic receptor delivers an anti-apoptotic signal to cardiac myocytes through Gi-dependent coupling to phosphatidylinositol 3-kinase. *Circ Res.* 2000;87:1172–1179.
52. Zhu WZ, Zheng M, Koch WJ, Lefkowitz RJ, Kobilka BK, Xiao RP. Dual modulation of cell survival and cell death by β2-adrenergic signalling in adult mouse cardiomyocytes. *Proc Natl Acad Sci USA.* 2001;98:1607–1612.
53. Dorn II GW, Tepe NM, Lorenz JN, Koch WJ, Liggett SB. Low- and high-level transgenic expression of β2-adrenergic receptors differentially affect cardiac hypertrophy

and function in Gαq-overexpressing mice. *Proc Natl Acad Sci USA*. 1999;96: 6400–6405.
54. Liggett SB, Tepe NM, Lorenz JN, et al. Early and delayed consequences of beta(2)-adrenergic receptor overexpression in mouse hearts: critical role for expression level. *Circulation*. 2000;101:1707–1714.
55. Bristow MR, Ginsburg R, Umans V, et al. β1-and β2-adrenergic-receptor subpopulations in nonfailing and failing human ventricular myocardium: coupling of both receptor subtypes to muscle contraction and selective β1-receptor down-regulation in heart failure. *Circ Res*. 1986;59:297–309.
56. Bristow MR, Ginsburg R, Minobe W, et al. Decreased catecholamine sensitivity and β-adrenegic receptor density in failing human hearts. *N Engl J Med*. 1982;307:205–211.
57. Rockman HA, Koch WJ, Lefkowitz RJ. Seven-transmembrane-spanning receptors and heart function. *Nature*. 2002;415:206–212.
58. Nikolaev VO, Moshkov A, Lyon AR, et al. Beta2-adrenergic receptor redistribution in heart failure changes cAMP compartmentation. *Science*. 2010;327:1653–1657.
59. Lymperopoulos A. Beta-arrestin biased agonism/antagonism at cardiovascular seven transmembrane-spanning receptors. *Curr Pharm Des*. 2012;18:192–195.
60. Conner DA, Mathier MA, Mortensen RM, et al. Beta-arrestin1 knockout mice appear normal but demonstrate altered cardiac responses to beta-adrenergic stimulation. *Circ Res*. 1997;81:1021–1026.
61. Xiang Y, Kobilka BK. Myocyte adrenoceptor signaling pathways. *Science*. 2003;300:1530–1532.
62. Bristow MR. β-Adrenergic receptor blockade in chronic heart failure. *Circulation*. 2000;101:558–569.
63. Packer M, Bristow MR, Cohn JN, et al. The effect of carvedilol on morbidity and mortality in patients with chronic heart failure. *N Engl J Med*. 1996;334:1349–1355.
64. Dessy C, Balligand JL. Beta3-adrenergic receptors in cardiac and vascular tissues emerging concepts and therapeutic perspectives. *Adv Pharmacol*. 2010;59:135–163.
65. Breit A, Lagacé M, Bouvier M. Hetero-oligomerization between beta2- and beta3-adrenergic receptors generates a beta-adrenergic signaling unit with distinct functional properties. *J Biol Chem*. 2004;279:28756–28765.
66. Gros R, Benovic JL, Tan CM, Feldman RD. G-protein-coupled receptor kinase activity is increased in hypertension. *J Clin Invest*. 1997;99:2087–2093.
67. Belmonte SL, Blaxall BC. G protein coupled receptor kinases as therapeutic targets in cardiovascular disease. *Circ Res*. 2011;109:309–319.
68. de Foucart S, Champlain J, Nadeau R. Modulation by beta-adrenoceptors and angiotensin II receptors of splanchnic nerve evoked catecholamine release from the adrenal medulla. *Can J Physiol Pharmacol*. 1991;69:1–7.
69. Foucart S, Nadeau R, de Champlain J. Local modulation of adrenal catecholamines release by beta-2 adrenoceptors in the anaesthetized dog. *Naunyn Schmiedebergs Arch Pharmacol*. 1988;337:29–34.
70. Billet S, Aguilar F, Baudry C, Clauser E. Role of angiotensin II AT1 receptor activation in cardiovascular diseases. *Kidney Int*. 2008;74:1379–1384.
71. Rockman HA, Choi DJ, Rahman NU, Akhter SA, Lefkowitz RJ, Koch WJ. Receptor-specific in vivo desensitization by the G protein-coupled receptor kinase-5 in transgenic mice. *Proc Natl Acad Sci USA*. 1996;93:9954–9959.
72. Balmforth AJ, Shepherd FH, Warburton P, Ball SG. Evidence of an important and direct role for protein kinase C in agonist-induced phosphorylation leading to desensitization of the angiotensin AT1A receptor. *Br J Pharmacol*. 1997;122:1469–1477.
73. Olivares-Reyes JA, Smith RD, Hunyady L, Shah BH, Catt KJ. Agonist-induced signaling, desensitization, and internalization of a phosphorylation-deficient AT1A angiotensin receptor. *J Biol Chem*. 2001;276:37761–37768.

74. Mundell SJ, Benovic JL. Selective regulation of endogenous G protein-coupled receptors by arrestins in HEK293 cells. *J Biol Chem*. 2000;275:12900–12908.
75. Calegari VC, Bezerra RM, Torsoni MA, et al. Suppressor of cytokine signaling 3 is induced by angiotensin II in heart and isolated cardiomyocytes, and participates in desensitization. *Endocrinology*. 2003;144:4586–4596.
76. Hein L, Meinel L, Pratt RE, Dzau VJ, Kobilka BK. Intracellular trafficking of angiotensin II and its AT1 and AT2 receptors: evidence for selective sorting of receptor and ligand. *Mol Endocrinol*. 1997;11:1266–1277.
77. Thomas WG, Thekkumkara TJ, Baker KM. Cardiac effects of AII. AT1A receptor signaling, desensitization, and internalization. *Adv Exp Med Biol*. 1996;396:59–69.
78. Richard DE, Laporte SA, Bernier SG, Leduc R, Guillemette G. Desensitization of AT1 receptor-mediated cellular responses requires long term receptor down-regulation in bovine adrenal glomerulosa cells. *Endocrinology*. 1997;138:3828–3835.
79. Zidar DA, Violin JD, Whalen EJ, Lefkowitz RJ. Selective engagement of G protein coupled receptor kinases (GRKs) encodes distinct functions of biased ligands. *Proc Natl Acad Sci USA*. 2009;106:9649–9654.
80. Boerrigter G, Lark MW, Whalen EJ, Soergel DG, Violin JD, Burnett Jr JC. Cardiorenal actions of TRV120027, a novel β-arrestin-biased ligand at the angiotensin II type I receptor, in healthy and heart failure canines: a novel therapeutic strategy for acute heart failure. *Circ Heart Fail*. 2011;4:770–778.
81. De Gasparo M, Catt KJ, Inagami T, Wright JW, Unger T. International union of pharmacology. XXIII. The angiotensin II receptors. *Pharmacol Rev*. 2000;52: 415–472.
82. MacKenzie A. Endothelium-derived vasoactive agents, AT1 receptors and inflammation. *Pharmacol Ther*. 2011;131:187–203.
83. Weber KT. Aldosterone in congestive heart failure. *N Engl J Med*. 2001;345: 1689–1697.
84. Connell JM, Davies E. The new biology of aldosterone. *J Endocrinol*. 2005;186:1–20.
85. Marney AM, Brown NJ. Aldosterone and end-organ damage. *Clin Sci (Lond)*. 2007;113:267–278.
86. Richards AM. Mineralocorticoid receptor antagonists for heart failure. *Expert Opin Pharmacother*. 2011;12:2801–2815.
87. Ganguly A, Davis JS. Role of calcium and other mediators in aldosterone secretion from the adrenal glomerulosa cells. *Pharmacol Rev*. 1994;46:417–447.
88. Harvey AN, Nguyen K, Lymperopoulos A. GRK2 and beta-arrestins in cardiovascular disease: established and emerging possibilities for therapeutic targeting. *Curr Mol Pharmacol* 2012. 2011 Jun 15 [Epub ahead of print].
89. Turu G, Szidonya L, Gáborik Z, et al. Differential beta-arrestin binding of AT1 and AT2 angiotensin receptors. *FEBS Lett*. 2006;580:41–45.
90. Ivey ME, Osman N, Little PJ. Endothelin-1 signalling in vascular smooth muscle: pathways controlling cellular functions associated with atherosclerosis. *Atherosclerosis*. 2008;199:237–247.
91. Vignon-Zellweger N, Heiden S, Miyauchi T, Emoto N. Endothelin and endothelin receptors in the renal and cardiovascular systems. *Life Sci*. 2012;91:490–500.
92. Stockand JD. Vasopressin regulation of renal sodium excretion. *Kidney Int*. 2010;78:849–856.
93. Oakley RH, Laporte SA, Holt JA, Caron MG, Barak LS. Differential affinities of visual arrestin, beta arrestin1, and beta arrestin2 for G protein-coupled receptors delineate two major classes of receptors. *J Biol Chem*. 2000;275:17201–17210.
94. Lymperopoulos A, Bathgate A. Pharmacogenomics of the heptahelical receptor regulators G-protein-coupled receptor kinases and arrestins: the known and the unknown. *Pharmacogenomics*. 2012;13:323–341.

95. Lim GB. Vascular disease: no added benefit of niacin with statins. *Nat Rev Cardiol.* 2011;9:3.
96. Digby JE, Ruparelia N, Choudhury RP. Niacin in cardiovascular disease: recent preclinical and clinical developments. *Arterioscler Thromb Vasc Biol.* 2012;32:582–588.
97. Kather H, Aktories K, Schulz G, Jakobs KH. Islet-activating protein discriminates the antilipolytic mechanism of insulin from that of other antilipolytic compounds. *FEBS Lett.* 1983;161:149–152.
98. Pike NB. Flushing out the role of GPR109A (HM74A) in the clinical efficacy of nicotinic acid. *J Clin Invest.* 2005;115:3400–3403.
99. Richman JG, Kanemitsu-Parks M, Gaidarov I, et al. Nicotinic acid receptor agonists differentially activate downstream effectors. *J Biol Chem.* 2007;282:18028–18036.
100. Semple G, Skinner PJ, Gharbaoui T, et al. 3-(1H-tetrazol-5-yl)-1,4,5,6-tetrahydro-cyclopentapyrazole (MK-0354): a partial agonist of the nicotinic acid receptor, G-protein coupled receptor 109a, with antilipolytic but no vasodilatory activity in mice. *J Med Chem.* 2008;51:5101–5108.
101. Gachet C. P2 receptors, platelet function and pharmacological implications. *Thromb Haemost.* 2008;99:466–472.
102. Hardy AR, Conley PB, Luo J, Benovic JL, Poole AW, Mundell SJ. P2Y1 and P2Y12 receptors for ADP desensitize by distinct kinase-dependent mechanisms. *Blood.* 2005;105:3552–3560.
103. Baurand A, Eckly A, Hechler B, et al. Differential regulation and relocalization of the platelet P2Y receptors after activation: a way to avoid loss of hemostatic properties? *Mol Pharmacol.* 2005;67:721–733.
104. Soh UJ, Dores MR, Chen B, Trejo J. Signal transduction by protease-activated receptors. *Br J Pharmacol.* 2010;160:191–203.
105. Iwamoto R, Mekada E. ErbB and HB-EGF signaling in heart development and function. *Cell Struct Funct.* 2006;31:1–14.
106. DeWire SM, Ahn S, Lefkowitz RJ, Shenoy SK. Beta-arrestins and cell signaling. *Annu Rev Physiol.* 2007;69:483–510.
107. Esposito G, Perrino C, Cannavo A, et al. EGFR trans-activation by urotensin II receptor is mediated by beta-arrestin recruitment and confers cardioprotection in pressure overload-induced cardiac hypertrophy. *Basic Res Cardiol.* 2011;106:577–589.
108. Zhai P, Galeotti J, Liu J, et al. An angiotensin II type 1 receptor mutant lacking epidermal growth factor receptor transactivation does not induce angiotensin II-mediated cardiac hypertrophy. *Circ Res.* 2006;99:528–536.
109. Kim IM, Tilley DG, Chen J, et al. Beta-blockers alprenolol and carvedilol stimulate betaarrestin-mediated EGFR transactivation. *Proc Natl Acad Sci USA.* 2008;105:14555–14560.
110. Tilley DG, Kim IM, Patel PA, Violin JD, Rockman HA. Beta-arrestin mediates beta1-adrenergic receptor-epidermal growth factor receptor interaction and downstream signaling. *J Biol Chem.* 2009;284:20375–20386.
111. Smith NJ, Chan HW, Qian H, et al. Determination of the exact molecular requirements for type 1 angiotensin receptor epidermal growth factor receptor transactivation and cardiomyocyte hypertrophy. *Hypertension.* 2011;57:973–980.
112. Barki-Harrington L, Luttrell LM, Rockman HA. Dual inhibition of beta-adrenergic and angiotensin II receptors by a single antagonist: a functional role for receptor–receptor interaction in vivo. *Circulation.* 2003;108:1611–1618.
113. Cervantes D, Crosby C, Xiang Y. Arrestin orchestrates crosstalk between G protein-coupled receptors to modulate the spatiotemporal activation of ERK MAPK. *Circ Res.* 2010;106:79–88.
114. Drake MT, Violin JD, Whalen EJ, Wisler JW, Shenoy SK, Lefkowitz RJ. β-Arrestin-biased agonism at the β_2-adrenergic receptor. *J Biol Chem.* 2008;283:5669–5676.

115. Woo AY, Wang TB, Zeng X, et al. Stereochemistry of an agonist determines coupling preference of beta2-adrenoceptor to different G proteins in cardiomyocytes. *Mol Pharmacol*. 2009;75:158–165.
116. Azzi M, Charest PG, Angers S, et al. β-Arrestin-mediated activation of MAPK by inverse agonists reveals distinct active conformations for G protein-coupled receptors. *Proc Natl Acad Sci USA*. 2003;100:11406–11411.
117. Wisler JW, DeWire SM, Whalen EJ, et al. A unique mechanism of β-blocker action: carvedilol stimulates β-arrestin signaling. *Proc Natl Acad Sci USA*. 2007;104: 16657–16662.
118. Fiuzat M, O'Connor CM. Bucindolol: new hopes from reviewing past data. *Drugs Today (Barc)*. 2011;47:347–351.
119. Liu JJ, Horst R, Katritch V, Stevens RC, Wüthrich K. Biased signaling pathways in β_2-adrenergic receptor characterized by 19F-NMR. *Science*. 2012;335:1106–1110.
120. Rosenbaum DM, Rasmussen SG, Kobilka BK. The structure and function of G-protein-coupled receptors. *Nature*. 2009;459:356–363.
121. Peng H, Bond RA, Knoll BJ. The effects of acute and chronic nadolol treatment on β2AR signaling in HEK293 cells. *Naunyn Schmiedebergs Arch Pharmacol*. 2011;383: 209–216.
122. Lymperopoulos A, Rengo G, Magafa V, Cordopatis P, Koch WJ. Biased agonism/ antagonism of β-arrestin activation by the angiotensin II type 1 receptor: a study of sartans and angiotensin II analogs using aldosterone turnover as a readout. *J Card Fail Suppl*. 2010;16:S30–S31.
123. Holloway AC, Qian H, Pipolo L, et al. Side-chain substitutions within angiotensin II reveal different requirements for signaling, internalization, and phosphorylation of type 1A angiotensin receptors. *Mol Pharmacol*. 2002;61:768–777.
124. Wei H, Ahn S, Shenoy SK, et al. Independent beta-arrestin 2 and G protein-mediated pathways for angiotensin II activation of extracellular signal-regulated kinases 1 and 2. *Proc Natl Acad Sci USA*. 2003;100:10782–10787.
125. Shukla AK, Violin JD, Whalen EJ, Gesty-Palmer D, Shenoy SK, Lefkowitz RJ. Distinct conformational changes in beta-arrestin report biased agonism at seven-transmembrane receptors. *Proc Natl Acad Sci USA*. 2008;105:9988–9993.
126. Zimmerman B, Beautrait A, Aguila B, et al. Differential β-arrestin-dependent conformational signaling and cellular responses revealed by angiotensin analogs. *Sci Signal*. 2012;5:ra33.

CHAPTER THIRTEEN

Arrestins in Bone

Brittany N. Bohinc*, **Diane Gesty-Palmer***,†
*Department of Medicine, Division of Endocrinology and Metabolism, Duke University Medical Center, Durham, North Carolina, USA
†Parexel International, Research Triangle Park, Durham, North Carolina, USA

Contents

1. Introduction 336
2. Arrestin Signaling: A New Dimension to GCPR Signaling in Bone 337
   2.1 G protein-coupled receptors 337
   2.2 Multiple dimensions of GPCR signaling 339
   2.3 Functional selectivity 339
   2.4 Arrestins as mediators of GPCR signaling 340
3. Functional Selectivity in Bone 341
   3.1 Regulation of bone metabolism by PTH 341
   3.2 PTH receptor-mediated G protein signaling in bone 342
   3.3 PTH receptor-mediated arrestin signaling in bone 342
   3.4 Biased agonism at the PTH receptor 344
4. Arrestin Signaling Effects in Bone 345
   4.1 Influence of β-arrestin 2 on bone formation and turnover 345
   4.2 The effect of PTH arrestin signaling: Beyond desensitization 346
   4.3 Skeletal effects of an arrestin pathway-selective PTH_1R agonist 347
   4.4 hPTH(1–34) and bPTH(7–34) affect bone mass *In vivo* through distinct genomic mechanisms 349
   4.5 Arrestin pathway-selectivity as a strategy for drug design 352
5. Perspectives and Future Directions 352
6. Conclusions 353
Acknowledgments 353
References 353

Abstract

Parathyroid hormone (PTH) is the principle regulator of calcium–phosphorus metabolism and bone turnover. Because of its central role in bone remodeling, recombinant human PTH (i.e., Forteo®; PTH(1–34)) has been developed for the treatment of osteoporosis. PTH(1–34) acts principally through the type I PTH/PTH-related peptide receptor (PTH_1R), a classic seven transmembrane G protein-coupled receptor (GPCR). Intermittent treatment with PTH(1–34) promotes osteoblast and osteoclast recruitment through activation of PTH_1R with resultant net bone gain. Recent studies have demonstrated

Progress in Molecular Biology and Translational Science, Volume 118
ISSN 1877-1173
http://dx.doi.org/10.1016/B978-0-12-394440-5.00013-9

that the complex metabolic effects induced by PTH_1R stimulation are not entirely a consequence of conventional GPCR signaling. β-Arrestins, in addition to their desensitizing actions, also serve as multifunctional scaffolding proteins linking the PTH_1R to signaling molecules independent of classic G protein-mediated second messenger-dependent pathways. *In vitro*, D-Trp^{12}, Tyr^{34}-bPTH(7–34) [bPTH(7–34)], a β-arrestin-selective biased agonist for the PTH_1R, antagonizes G protein signaling but activates arrestin-dependent signaling. *In vivo*, intermittent administration of bPTH(7–34) to mice induces anabolic bone formation independent of classic G protein-coupled signaling mechanisms. While both the conventional PTH_1R agonists, PTH(1–34) and bPTH(7–34), stimulate anabolic bone formation in mice, the latter does not induce hypercalcemia nor does it increase markers of bone resorption. This newly recognized ability of β-arrestins to serve as signal transducers for the PTH_1R independent of classic GPCR signaling represents a novel paradigm with therapeutic potential. Exploitation of β-arrestin-biased agonism may offer therapeutic benefit for the treatment of metabolic bone diseases such as osteoporosis with an improved side effect profile.

ABBREVIATIONS

7TMRs seven transmembrane receptors
BMD bone mineral density
bPTH(7–34) D-Trp^{12}, Tyr^{34}-bPTH(7–34)
DPD deoxypyrodiniline
GPCR G protein-coupled receptor
MAPK mitogen-activated protein kinase
OPG osteoprotegerin
PTH parathyroid hormone
PTH_1R type I parathyroid hormone receptor
RANKL receptor activator of NFκB ligand

1. INTRODUCTION

Osteoporosis, or low bone mineral density (BMD), is a significant health threat among our aging population and is associated with increased fracture risk, mortality, and health care cost.[1] It is characterized by deterioration in bone microarchitecture with resultant need for anabolic therapy.[2] The etiology of osteoporosis is complex, representing the net imbalance between osteoblast-mediated bone formation and osteoclast-mediated bone resorption. Currently employed antiresorptive therapies, such as bisphosphonates, selective estrogen modulators, supplemental calcium and vitamin D, are not sufficient to regenerate lost trabecular bone architecture. Thus, anabolic agents that target osteoblast-mediated bone formation are needed. Ideally, these therapies would uncouple bone formation from osteoclast-mediated bone resorption and improve mineral content and bone quality without contributing to side effects. Several therapeutic agents, including fluoride,[3–5] human growth hormone,[6,7] prostaglandins,[8,9] and parathyroid hormone (PTH) analogs,[10] have been explored as prospective anabolic treatments. Of these, PTH analogs are the most efficacious anabolic therapies developed to date.

PTH acts principally through the type I PTH/PTH-related peptide receptor (PTH_1R), a classic seven transmembrane G protein-coupled receptor (GPCR). Recent studies have demonstrated that the complex metabolic effects induced by PTH_1R stimulation are not entirely a consequence of conventional GPCR signaling. β-Arrestins, in addition to their GPCR desensitizing actions, also serve as multifunctional scaffolding proteins linking the PTH_1R to signaling molecules independent of the classic G protein-mediated second messenger-dependent pathways. Presently, the conventional PTH_1R agonist, PTH(1–34) (i.e., Forteo®), is the only FDA-approved anabolic approach to stimulate bone formation. Intermittent treatment with PTH(1–34) promotes osteoblast and osteoclast recruitment through activation of PTH_1R with resultant net bone gain. Although effective in anabolic bone mineralization, its clinical utility is hampered by its propensity to activate bone resorption and predispose patients to hypercalcemia/hypercalcuria with prolonged administration. In addition, long-term exposure to PTH(1–34) in rats increases osteosarcoma risk, and evidence links Ca^{2+}-dependent protein kinase C signaling to the initiation and progression of these PTH-induced osteosarcomas.[11,12] Thus, the development of PTH receptor agonists that promote osteoblastic bone formation without stimulating bone resorption may have improved clinical utility for the treatment of osteoporosis.[13–15] A PTH analog that activates β-arrestin-mediated signaling in the absence of GPCR-dependent signaling shares this desired pharmacologic profile.[15] The recent identification of β-arrestin-biased agonists may form the basis for pharmacologic agents with enhanced therapeutic specificity and efficacy.[13–16]

2. ARRESTIN SIGNALING: A NEW DIMENSION TO GCPR SIGNALING IN BONE

2.1. G protein-coupled receptors

Heptahelical GPCRs constitute the largest and the most diverse superfamily of cell surface receptors, accounting for as many as 1000 discrete receptor proteins.[17] The superfamily of seven transmembrane receptors (7TMRs) is believed to have evolved from a common ancestor. While retaining their heptahelical structure and their functional coupling to heterotrimeric G proteins, this class of receptors has evolved into subfamilies characterized by distinct structural elements and diverse modes of ligand binding, receptor activation, and receptor regulation.[18,19] This evolutionary diversity translates into the capacity for GPCRs to detect an enormous array of extracellular stimuli. GPCR signaling is fundamental to numerous biologic functions including vision, taste, and odorant sensory detection, neurotransmission, endocrine control of physiological homeostasis and reproduction, and regulation of hemodynamics and intermediary metabolism. As a result, agents targeting GPCR signaling have been a principal focus of pharmaceutical development, and it is estimated that almost half of all drugs in clinical use target this ubiquitous family of receptors.[18,19]

Earlier models of GPCR signaling envisioned the receptor existing in equilibrium between two states: an inactive state (R) and an active state (R^*) which is stabilized by agonist binding.[20–22] This allosteric model viewed agonist binding as triggering conformational receptor changes that coupled it to its associated G proteins, thereby regulating the activity of enzymatic effectors, such as adenylate cyclases, phospholipase Cβ isoforms, and ion channels, leading to the generation of small molecule "second messengers" that altered the physiological function of the cell. In these models, the intrinsic efficacy of a ligand was simply seen as a reflection of its ability to alter the R–R^* equilibrium. *Agonists* stabilized the R^* conformation, pulling the equilibrium toward the "on" state, whereas "neutral" *antagonists* bound indiscriminately to both R and R^*, producing no physiological response but blocking the response to agonists. The eventual discovery of constitutively activating GPCR mutations allowed the detection of receptor activity in the absence of ligand and enabled the identification of an additional class of ligands, termed *inverse agonists*, that suppressed basal receptor activity.[23–25] Thus, inverse agonists appear as antagonists when basal receptor activity is low but have the added property of reducing constitutive receptor activity by binding preferentially to R and pulling the equilibrium toward the "off" state.

With the development of increasingly sophisticated biochemical and biophysical approaches for measuring ligand–receptor interactions, evidence emerged that GPCRs assume multiple "active" conformations that are capable of interacting with intracellular effectors with varying efficiency.[22] Moreover, these conformations can interact with ligands in a highly selective manner, meaning that structurally distinct ligands may elicit unique ligand-specific efficacy signatures.[16,26] Recognition that the relative activity of agonists does not always adhere to the predictions of the simple "two-state" model has led to a broader view of GPCR signaling. In turn, the potential for drugs that promote receptor conformations with distinctive intracellular signaling capacity has moved to the forefront of drug development.

2.2. Multiple dimensions of GPCR signaling

While all conventional GPCRs share the ability to function as ligand-activated guanine nucleotide exchange factors for heterotrimeric G proteins, it is increasingly apparent that GPCR signal transduction is far more complex than classically envisioned. Many GPCRs have been shown to couple to two, three, or more unrelated G protein classes, enabling a single receptor to engage multiple effectors simultaneously or activate them differentially in a tissue-selective manner.[26,27] Moreover, a host of protein–protein interactions are known to affect the specificity, selectivity, and temporal signaling of GPCRs and their effector components. These include the formation of GPCR dimers,[28] the interaction of GPCRs with receptor activity-modifying proteins,[29,30] and the binding of PDZ domain-containing and non-PDZ domain scaffold proteins to the intracellular loops and C-termini of receptors.[31–33] Prominent among the latter are the arrestins, which are now known to act as ligand-regulated scaffolds, recruiting a host of non-G protein-regulated effectors that endow GPCRs with novel signaling capacity.[34] This broader view of GPCR signaling has led to the term "pluridimensional efficacy,"[35] which recognizes that GPCRs interact with multiple effectors and that, in addition to the receptor, cellular environment plays a key role in determining the consequences of GPCR activation.

2.3. Functional selectivity

It is now evident that GPCRs adopt multiple active conformations that couple to the receptor to a diverse set of downstream effectors with varying efficiency and that structurally distinct ligands can effect receptor coupling by stabilizing these active states in different proportions than the native ligand. Thus, the ligand–receptor complex, not the receptor alone, specifies the active state, and the formation of agonist-selective active states can "bias" the coupling of the receptor to different signaling pathways.[36] Many terms have been used to describe this phenomenon, including "stimulus trafficking," "functional dissociation," "biased agonism," "biased inhibition," "differential engagement," "discrete activation of transduction," and "functional selectivity." Whatever term is applied, the implications for signal transduction are striking. To the extent that ligand binding can alter this diverse array of protein interactions, functional selectivity has the potential to modify GPCR signaling.

The concept of ligand-directed signaling raises the potential of pharmacologic agents with novel efficacy profiles possessing improved therapeutic efficacy or reduced side effects. Functional selectivity has been described for a number of GPCRs, including the PTH_1R in bone,[15,37–45] the angiotensin II type 1A receptor,[46] the β2-adrenergic receptor,[37] and the thyrotropin receptor.[47] "Functional selectivity" can range from relatively modest deviations in rank order of potency to frank reversal of efficacy, such that the characterization of a ligand as agonist, antagonist or inverse agonist becomes assay dependent. Put another way, conventional GPCR pharmacology addresses changing the *quantity* of efficacy; that is, increasing or decreasing receptor activity. In contrast, functional selectivity revolves around changing the *quality* of efficacy by amplifying desired signaling effects while diminishing other unwanted effects. Thus, development of ligands with novel GPCR stimulus trafficking characteristics is of significant pharmacologic interest in relation to drug design.[48–52] To date, however, there is little experimental evidence that biased ligands offer advantages over conventional agonists/antagonists *in vivo*.

2.4. Arrestins as mediators of GPCR signaling

The arrestins, a family of four GPCR binding proteins that regulate receptor desensitization and endocytosis, are among the most studied protein modulators of GPCR signaling. Arrestins bind tightly and specifically to agonist-occupied GPCRs that have been phosphorylated by GPCR kinases[53] and sterically inhibit the receptor from further G protein activation. The two nonvisual arrestin isoforms, β-arrestins 1 and 2, also regulate the agonist-induced internalization of desensitized GPCRs via clathrin-coated pits.[54]

The discovery that arrestins serve as adapters not only in the context of GPCR sequestration but also in linking activated receptors to additional effectors of downstream signaling[55] advanced our understanding of GPCR signal transduction. It is now recognized that a number of catalytically active proteins bind arrestins and are recruited to agonist-occupied GPCRs. Among these are Src family tyrosine kinases; components of the ERK1/2 and c-Jun N-terminal kinase 3; mitogen-activated protein kinase

(MAPK) cascades; the E3 ubiquitin ligase, Mdm2; the cAMP phosphodiesterases, PDE4D3/5; diacylglycerol kinase; the inhibitor of nuclear factor-κ B, IκB; the Ral-GDP dissociation stimulator, Ral-GDS; and the Ser/Thr protein phosphatase 2A.[55–57,34,58–61] It is through these interactions that arrestin-binding confers unique signaling properties upon agonist-occupied GPCRs, opening up a broad realm of possibilities for manipulating GPCR signal transduction.[58]

A novel feature of at least some arrestin-dependent signals is that they can be initiated independent of heterotrimeric G protein activation. Downstream signals emanating from GPCRs in the absence of G protein coupling have been demonstrated for several different 7TMRs using G protein-uncoupled receptor mutants and arrestin pathway-selective biased ligands.[37,46,62,63] In bone, the PTH_1R has been shown to produce arrestin-dependent activation of ERK1/2 that is temporally distinct from G protein-dependent mechanisms.[63] This ability to dissociate arrestin- and G protein-dependent PTH_1R signaling in bone has the potential to impact future therapeutics in osteoporosis.

3. FUNCTIONAL SELECTIVITY IN BONE

3.1. Regulation of bone metabolism by PTH

PTH is an 84-amino acid peptide hormone that serves as the primary regulator of bone and mineral metabolism. Expressed principally in the parathyroid gland, PTH is excreted in a pulsatile manner and tightly regulates calcium and phosphate homeostasis. The principle targets of PTH in the periphery are the kidney and the bone, where its actions promote a rise in serum calcium. In the kidney, PTH regulates renal tubular calcium resorption through a well characterized cAMP-dependent mechanism. It also regulates renal expression of the 1α-hydroxyase necessary to convert 25(OH)-vitamin D to its active form $1,25(OH)_2$-vitamin D, which in turn enhances intestinal calcium absorption.

The physiologic actions of PTH on bone are complex. PTH is known for both its anabolic effects as well as its catabolic/resorptive effects on bone. PTH is an important regulator of bone "remodeling," a lifelong process that is necessary to maintain structural integrity of the bone microarchitecture and maintain calcium–phosphorus homeostasis. At the cellular level, PTH directly stimulates osteoblasts to build bone by increasing osteoblast number and activity, promoting the deposition of new bone matrix, and accelerating the rate of mineralization.[64,65] At the same time, PTH stimulates bone resorption by indirectly recruiting osteoclasts. Because they lack PTH receptors, osteoclasts respond to factors, such as receptor activator of NFκB ligand (RANKL) and osteoprotegerin (OPG), secreted by osteoblasts in response to PTH. Osteoclastic activity is triggered via the osteoblasts' surface-bound RANKL activating the osteoclasts' surface-bound receptor activator of NFκB, OPG works to inhibit osteoclastogenesis by serving as a decoy receptor for RANKL.

The net effect of PTH on bone density is dependent upon the kinetics of receptor activation and the intrinsic concentrations of OPG and RANKL.[66,10] Intermittent exposure to PTH produces a net increase in bone formation, while continuous exposure produces a net decrease in bone formation and possible hypercalcemia.[64,10,67,68] The underlying molecular mechanisms contributing to these opposing physiologic effects are not well understood.

3.2. PTH receptor-mediated G protein signaling in bone

PTH acts principally through the PTH_1R, a class II GPCR that is highly expressed in the kidney and the bone. Most of its known effects are mediated by classic G protein signaling mechanisms, including G_s-mediated activation of adenylyl cyclase, resulting in cAMP production and PKA activation, and $G_{q/11}$-mediated activation of phospholipase-Cβ, leading to inositol-1,4,5-trisphosphate production, calcium mobilization, and PKC activation.[69–73] In renal tubular epithelium, PDZ domain-mediated binding of Na^+/H^+ exchanger regulatory factor 2 to the PTH_1R C-terminus permits the receptor to engage $G_{i/o}$ proteins, leading to inhibition of adenylyl cyclase while simultaneously enhancing receptor coupling to $G_{q/11}$.[74] PTH also activates the ERK1/2 MAPK cascade through both PKA and PKC in a cell-specific and G protein-dependent manner.[75–77] PTH-stimulated MAPK pathway activation has been shown to have proliferative and differentiative effects in bone.[78,79]

3.3. PTH receptor-mediated arrestin signaling in bone

Arrestin function in bone was first described in an osteoblastic cell line (UMR 106-H5), where β-arrestin 2 was found to be involved in PTH_1R desensitization.[80] Further studies by Ferrari and Bouxsein showed that the PTH_1R and β-arrestin 2 colocalized intracellularly upon receptor activation and that β-arrestin 2 promoted endocytosis of the PTH_1R and attenuated PTH-stimulated cAMP accumulation.[81] Despite finding that β-arrestin 2 was a negative regulator of PTH signaling, intermittent treatment of male β-arrestin 2 knockout (KO) mice with PTH(1–34) did not increase total body bone mass and cortical and trabecular bone parameters to the same degree as wild-type mice, suggesting that arrestin's actions helped to promote net bone formation.[82] They found that PTH significantly increased osteoclast number and surface (up to 80%) in KO, but not in WT mice, suggesting that loss of arrestin-mediated PTH_1R desensitization, and the resultant exaggeration of cAMP signaling, leads to unbalanced PTH_1R signaling and increased osteoclastogenesis.

In contrast, work in estrogen-replete female β-arrestin 2 KO mice showed that PTH(1–34) had enhanced benefit on total body BMD, trabecular bone architecture, and enhanced cortical effects in KO mice as compared to WT, suggesting that arrestins inhibit the ability of intermittent PTH to stimulate periosteal bone apposition and endosteal resorption.[83] In females, the lack of β-arrestin 2 led to altered PTH-stimulated osteoblast activity, characterized by a lower OPG/RANKL mRNA expression ratio, enhanced osteoclastogenesis, and endosteal bone resorption,[82] but also a higher bone formation at the periosteum.[83] These findings support the conclusion that PTH-mediated osteoblast–osteoclast coupling is cAMP dependent and that sustained cAMP signaling in the absence of an arrestin–PTH1R interaction promotes bone turnover.[38,82] Further *in vitro* and *in vivo* work demonstrated that β-arrestin 2 decreases PTH-dependent and -independent bone resorption by modulating RANKL-dependent osteoclastogenesis.[84]

Studies focusing on molecular mechanisms of arrestin signaling in bone suggest that even conventional PTH_1R agonists signal by coupling to arrestins.[15] PTH_1R activation by PTH(1–34) promotes translocation of both β-arrestin 1 and β-arrestin 2 to the plasma membrane, association of the receptor with β-arrestins, internalization of receptor–β-arrestin complexes, and arrestin-dependent activation of ERK1/2.[63,81,85] *In vitro*, PTH(1–34) stimulates ERK1/2 by multiple distinct mechanisms, conventional G protein-dependent pathways that involve PKA and/or PKC, and a G protein-independent pathway mediated by arrestins.[63,75,77] These discrete G protein-dependent and G protein-independent, β-arrestin-mediated signaling pathways are temporally distinct. G protein-dependent ERK1/2 activation is rapid and transient over the course of minutes, whereas the time course of G protein-independent ERK1/2 activation, mediated by β-arrestins, is delayed and sustained over the course of hours.[63] Microarray gene expression analysis of osteoblastic cells isolated from WT and β-arrestin 2 KO mice suggests that β-arrestin 2 targets prominently p38 MAPK- and NFκB-dependent signaling pathways in osteoblasts exposed to intermittent PTH.[86]

3.4. Biased agonism at the PTH receptor

The PTH_1R has long served as a model for the study of biased agonism. Through the generation of numerous PTH peptide analogs, it has been demonstrated that the downstream signaling events induced by the PTH_1R are sensitive to alterations in ligand structure. Examples of pathway-selective PTH_1R agonists are highlighted in Table 13.1. The closest mimetic of intact PTH(1–84), PTH(1–34) acts as a conventional/full agonist with respect to PTH_1R signaling, activating G_s and $G_{q/11}$ signaling and promoting β-arrestin-dependent receptor desensitization and internalization. In addition, PTH(1–34) stimulates β-arrestin-dependent downstream signaling. Shorter N-terminal fragments of the PTH peptide, for example, PTH(1–31) and [Aib1,3,M]PTH(1–15), preferentially promote G_s coupling,[88–90] while N-terminal truncations, for example, PTH(3–34), promote $G_{q/11}$ coupling while failing to activate G_s.[89,91,92] Trp[1]-PTHrp(1–36) has been shown to activate ERK1/2 exclusively through a G_s/PKA-

Table 13.1 Reported efficacy profiles of the conventional agonist, PTH(1–34), and selected biased PTH_1R agonists

Ligand	G Protein coupling	β-Arrestin coupling	References
PTH(1–34)	G_s and $G_{q/11}$	β-Arrestin 1 and β-arrestin 2	63,69–72
[Aib^{1,3},M]PTH(1–15)	G_s only	ND	43
PTH(1–31)	G_s only	ND	44,45
PTH(3–34)	G_q only	ND	40–42
PTH(28–42)	G_q only	ND	41,42
PTH(28–48)	G_q only	ND	41,42
Trp^1-PTHrp(1–36)	G_s only	Antagonist	63
Bpa^1-PTHrp(1–36)	G_s only	Antagonist	38,39
D-Trp^{12},Tyr^{34}-bPTH (7–34)	Inverse G_s agonist	β-Arrestin 1 and β-arrestin 2	15,63,87

ND, not determined.

dependent pathway that is unaffected by β-arrestin expression.[63] Similarly, Bpa^1-PTHrp(1–36) has been shown to induce sustained G_s coupling without promoting β-arrestin-dependent receptor desensitization.[38,39] More novel, (D-Trp^{12}, Tyr^{34})bPTH(7–34) (bPTH(7–34)) is a PTH analog previously identified as an inverse agonist for PTH_1R-G_s coupling that is capable of signaling via a G protein-independent, β-arrestin-mediated pathway as a β-arrestin pathway-selective biased agonist.[15,63,87]

Recent work examining biased agonism at the PTH_1R demonstrates that selective activation of G protein-independent arrestin-mediated signaling pathways elicits a physiologic response in bone distinct from that induced by the conventional PTH_1R agonist, PTH(1–34).[15,63,88] While intermittent administration of PTH(1–34) is effective in increasing bone formation, it is also associated with increases in bone resorption and a propensity to promote hypercalcemia and hypercalcuria. In contrast, the arrestin pathway-selective agonist, bPTH(7–34), induces bone formation independent of classic G protein-coupled signaling mechanisms. Unlike PTH(1–34), bPTH(7–34) appears to "uncouple" the anabolic effects of PTH_1R activation from its catabolic and calcitropic effects. Identification of β-arrestin pathway-biased agonists like bPTH(7–34) permits the investigation of β-arrestin-mediated signaling pathways *in vivo* and offers evidence that arrestin pathway-selective GPCR agonists can elicit potentially beneficial effects *in vivo* that cannot be achieved using conventional agonist or antagonist ligands.

4. ARRESTIN SIGNALING EFFECTS IN BONE

4.1. Influence of β-arrestin 2 on bone formation and turnover

Both β-arrestin 1 and β-arrestin 2 are expressed in osteoblasts. While the contributions of β-arrestin 1 to osteoblast function and skeletal metabolism are not well understood, β-arrestin 2 has been shown to affect bone remodeling and the skeletal response to endogenous PTH.[15,82–84] β-Arrestin 2 null mice lack both β-arrestin 2-dependent desensitization of PTH-stimulated G protein activation and β-arrrestin 2-mediated signaling. *In vivo*, β-arrestin 2 null mice have normal serum calcium levels and no gross alterations in skeletal morphology or size compared to congenic wild-type mice. However, the loss of β-arrestin 2 alters underlying bone metabolism. Circulating levels of endogenous PTH are suppressed in β-arrestin 2 null mice,[89] possibly a compensatory mechanism to maintain physiologic calcium homeostasis in the setting of impaired PTH_1R desensitization. β-Arrestin 2 null mice exhibit higher basal rates of bone turnover compared to wild-type mice. Osteoid surface and osteocalcin mRNA levels are increased, consistent with an overall increase in the rate of bone formation, while at the same time bone resorption is accelerated, as evidenced by increased osteoclast surface and bone turnover markers such as urine deoxypyrodiniline (DPD) excretion.[15,84]

Although the trabecular bone mineral densities and bone volume fraction of β-arrestin 2 null and wild-type mice are comparable, the KOs show microarchitectural differences, such as increased trabecular thickness and decreased trabecular number. These differences likely represent the net effect of increased bone formation that is offset by accelerated bone resorption. Collectively, these findings clearly demonstrate that β-arrestin 2 is not required for skeletal patterning and development and suggest that a major function in bone is to dampen heterotrimeric G protein signaling, consistent with its ubiquitous role in GPCR desensitization.

4.2. The effect of PTH arrestin signaling: Beyond desensitization

When one overrides compensatory physiologic mechanisms by exposing β-arrestin 2 null mice to pharmacologic levels of intermittent PTH (1–34), differences emerge that suggest arrestins in bone may play roles beyond GPCR desensitization.[13,15,82,84] In wild-type animals, intermittent administration of PTH(1–34) produces the expected increases in indices of bone formation including increased osteoblast number and osteoid surface, increased osteocalcin mRNA, and increased serum osteocalcin level. PTH (1–34) also produces the expected increase in osteoclast activity marked by increased RANKL mRNA expression, increased osteoclast surface, and increased urine DPD excretion. Together, these opposing responses reflect the PTH_1R-dependent coupling of osteoblastic bone formation to osteoclastic bone resorption. The net effect of intermittent PTH(1–34) is increased bone formation, as evidenced by an increase in trabecular bone volume, trabecular number, and trabecular thickness within cancellous bone. Additional increases in periosteal circumference and cortical thickness in the femur indicate a net increase in cortical bone formation.

The response to intermittent PTH(1–34) in β-arrestin 2 null mice is more complex. β-Arrestin 2 null mice treated with PTH(1–34) demonstrate attenuated bone formation at trabecular and endocortical bone surfaces as well as increased markers of bone resorption.[15] As previously discussed, Ferrari and

Bouxsein reported that intermittent administration of PTH(1–34) increased bone mass in female β-arrestin 2 null mice but failed to exert an anabolic effect in male animals.[82,83] The lack of effect in the male β-arrestin $2^{-/-}$ mice was attributed to the loss of β-arrestin-mediated desensitization of G protein signaling, increased and sustained cyclic AMP, and exaggerated osteoclastogenesis resulting from an increased RANKL/OPG ratio in the KO animals,[82,83] but the parallel decrease in bone formation markers suggests that the loss of β-arrestin 2 might also impair new bone formation. In all, evidence suggests that β-arrestin 2 is important not only in GCPR desensitization but that β-arrestin signaling in bone may be important in promoting bone formation.

4.3. Skeletal effects of an arrestin pathway-selective PTH_1R agonist

β-Arrestin 2 null mice are impaired in both arrestin-mediated PTH_1R desensitization and arrestin-mediated signaling. As the two effects cannot be independently reconstituted *in vivo*, it is not possible to determine their contributions to the actions of PTH on bone by comparing PTH(1–34) effects in wild-type and β-arrestin 2 null mice. However, the identification of arrestin pathway-selective biased agonists for the PTH_1R has provided a way to examine the contribution of arrestin-dependent signaling to bone remodeling *in vivo*, independent of its role in the desensitization of PTH_1R-mediated G protein activation. Administering an arrestin pathway-selective PTH analog to wild-type mice allows one to examine the contribution of β-arrestin-mediated signaling to skeletal metabolism in the absence of pharmacologic activation of G protein pathways. The analogous experiment performed in β-arrestin 2 null mice would reveal skeletal effects that might arise from the transient *inhibition* of G protein signaling resulting from competitive antagonism of endogenous PTH signaling. Conversely, administering PTH(1–34) to β-arrestin 2 null animals allows separation of the effects of G protein signaling from β-arrestin signaling because PTH(1–34) activates both pathways in wild-type animals, but only G protein signaling in the KO.

The results of such an experiment,[13] performed in congenic male β-arrestin 2 null mice and wild-type C57BL/6 controls, are summarized in Fig. 13.1. Despite the antagonism of G protein signaling, wild-type animals treated with bPTH(7–34) exhibited increases in bone formation, associated with increased osteoblast number, osteocalcin mRNA expression and serum osteocalcin level, increased trabecular number and thickness, and greater bone volume fraction as compared to wild-type animals treated with PTH(1–34). In β-arrestin 2 null mice, bPTH(7–34) had no significant effect on bone formation markers, indicating that the changes seen in

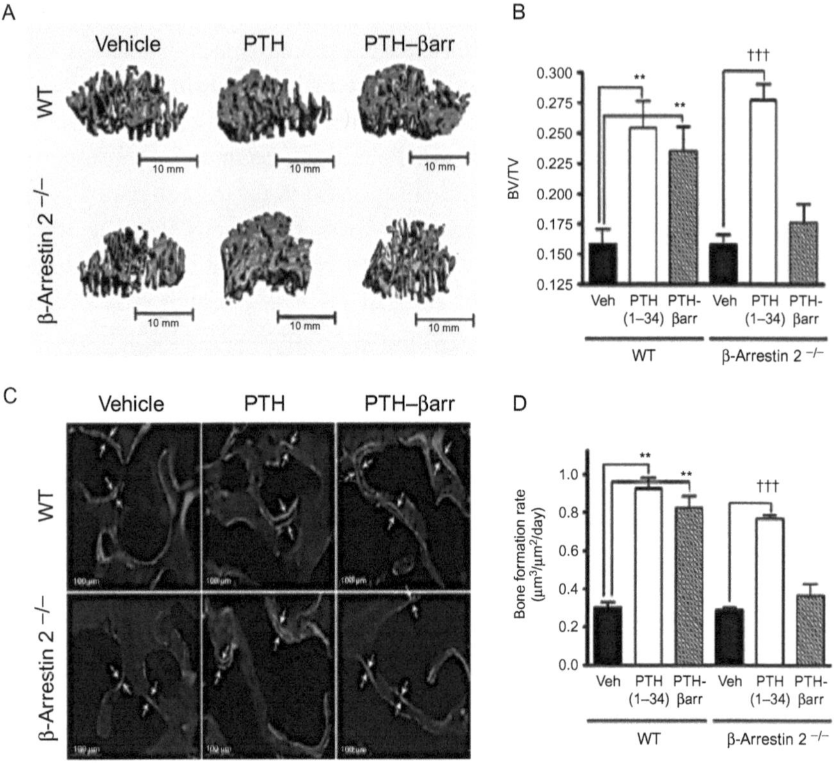

Figure 13.1 Arrestin pathway-selective biased agonism at the PTH_1R promotes bone formation independent of G protein activation. (A) Representative quantitative micro CT (qCT) of proximal tibia from male wild-type (WT) and β-arrestin $2^{-/-}$ mice treated for 8 weeks with daily injections of vehicle, or (40 μg/kg/day) of PTH(1–34) or D-Trp^{12}, Tyr^{34}-bPTH(7–34) [PTH-βarr]. Scale bar = 1.0 mm. (B) qCT of proximal tibia was used to determine the effects of intermittent PTH(1–34) or bPTH(7–34) on trabecular bone (Tb) volume fraction (BV/TV). Data represent the mean ± SEM of measurements taken from at least seven male mice. (C) Representative calcein double-labeled, non-decalcified, 10-μm sections of lumbar vertebrae from male WT and β-arrestin 2 null mice were treated for 8 weeks with either vehicle, PTH(1–34) or bPTH(7–34). Scale bar = 100 μm. Bone formation rates are determined by calcein-double labeling (arrows). (D) Quantitation of bone formation rates from calcein-labeled trabecular bone. Data represent the mean ± SEM of measurements from four mice. $^{*}P < 0.05$, $^{**}P < 0.01$, $^{***}P < 0.001$ compared with vehicle-treated mice. $^{\dagger}P < 0.05$, $^{\dagger\dagger}P < 0.01$, $^{\dagger\dagger\dagger}P < 0.001$ compared with vehicle-treated β-arrestin 2 null mice. Significance determined by one-way ANOVA with Bonferroni correction. *Adapted from data originally published in Ref. 15.*

wild-type animals resulted from arrestin signaling and not from inhibition of PTH_1R-G protein signaling.

Despite surprising similarity to the anabolic response to hPTH(1–34) in wild-type animals, it is clear that the arrestin-selective PTH analog does not elicit the full PTH_1R signaling response in bone. Unlike PTH(1–34), the anabolic effect of bPTH(7–34) appeared to be confined to the trabecular bone compartment. Moreover, selective activation of the β-arrestin 2 pathway by bPTH(7–34) did not significantly increase any indices of osteoclastic bone resorption. In wild-type animals, PTH(1–34) stimulated osteoblast/osteoclast coupling and bone resorption, as evidenced by increases in RANKL mRNA expression, osteoclast number, and markers of bone resorption, including urinary calcium excretion and urinary DPD. β-Arrestin 2 null animals treated with hPTH(1–34) exhibited an exaggerated increase in osteoclast number and urine DPD, supporting the conclusion that PTH_1R-mediated bone resorption is principally mediated via G protein-dependent signaling pathways that are not activated by the arrestin-selective PTH analog. These differential effects of PTH(1–34) and bPTH(7–34) on markers of osteoblast-mediated bone formation and osteoclast-mediated bone resorption are summarized in Fig. 13.2.

4.4. hPTH(1–34) and bPTH(7–34) affect bone mass *in vivo* through distinct genomic mechanisms

Although PTH(1–34) and bPTH(7–34) have markedly different *in vitro* efficacy, both stimulate anabolic bone formation *in vivo*. To understand how two PTH_1R ligands with such different efficacy profiles could both increase bone mass, a recent study compared their mechanisms of action at the tissue level using functional genomics.[88] In this study, wild-type and β-arrestin 2 null mice were treated with PTH(1–34) or bPTH(7–34) for 8 weeks, and gene expression profiles from calvarial bone were analyzed by geneset enrichment analysis to identify signaling pathway and biological process gene clusters that were significantly different from vehicle-treated animals. As shown in Fig. 13.3, bPTH(7–34) produced a genomic signature with limited overlap with that produced by the conventional agonist, suggesting that the two agents have distinctly different mechanisms of action at the tissue level. As expected, PTH(1–34)-treated mice exhibited genome level changes in pathways classically associated with bone development, remodeling, and differentiation, notably Wnt/β-catenin, bone morphogenic protein, transforming growth factor-beta (TGF-β), phosphoinositide 3-kinase/AKT (PI3K/AKT), and MAPK/ERK.[90–92] At the biological

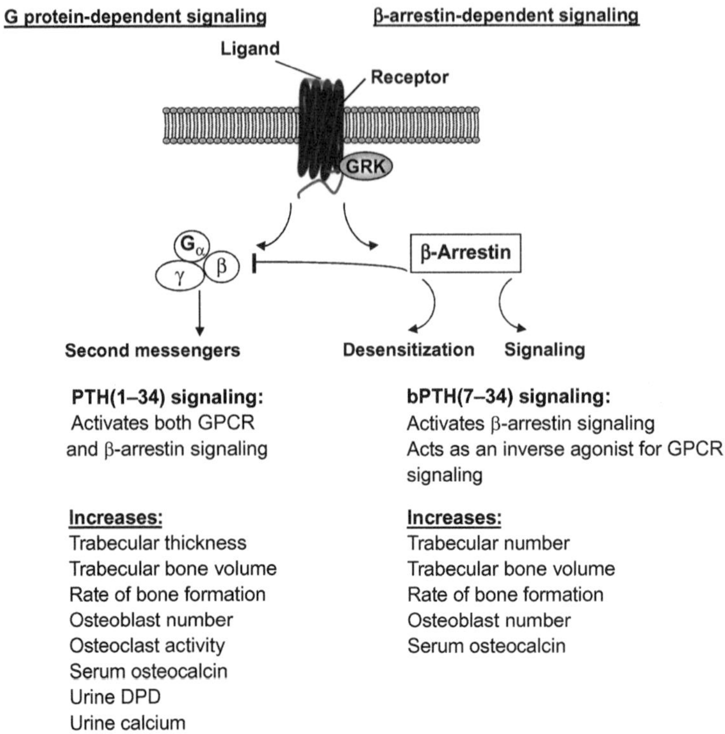

Figure 13.2 Effects of conventional and arrestin pathway-selective biased PTH_1R agonists on bone metabolism in WT and β-arrestin 2 null mice. Stimulation of the PTH_1R results in the activation of two distinct signaling pathways: one G protein-mediated and the other β-arrestin-mediated. Concomitantly, β-arrestins desensitize the G protein-activated response. The binding of the conventional PTH_1R agonist, hPTH(1–34), results in the activation of G protein- and β-arrestin-dependent signaling, whereas bPTH(7–34) activates only the β-arrestin-dependent pathway. bPTH(7–34) stimulates anabolic bone formation through a β-arrestin-dependent mechanism independent of G protein activation. Summarized are the effects of PTH(1–34) and bPTH(7–34) on markers of osteoblast-mediated bone formation and osteoclast-mediated bone resorption.

process level, this translated into regulation of collagen synthesis, matrix mineralization, and morphogenic pathways involved in skeletal patterning and development. On the other hand, bPTH(7–34) primarily affected cell survival, cell cycle progression, and migration through regulation of gene clusters associated with PI3K/AKT, p53, phosphatase and tensin homolog, and ataxia telangiectasia mutated kinase signaling (Fig. 13.3). Confirmatory *in vitro* experiments performed using primary calvarial osteoblasts demonstrated that bPTH(7–34) had direct effects on osteoblast proliferation, anti-apoptosis, and migration that were absent in β-arrestin 2 null osteoblasts, suggesting that these responses are dependent on β-arrestin 2 signaling.

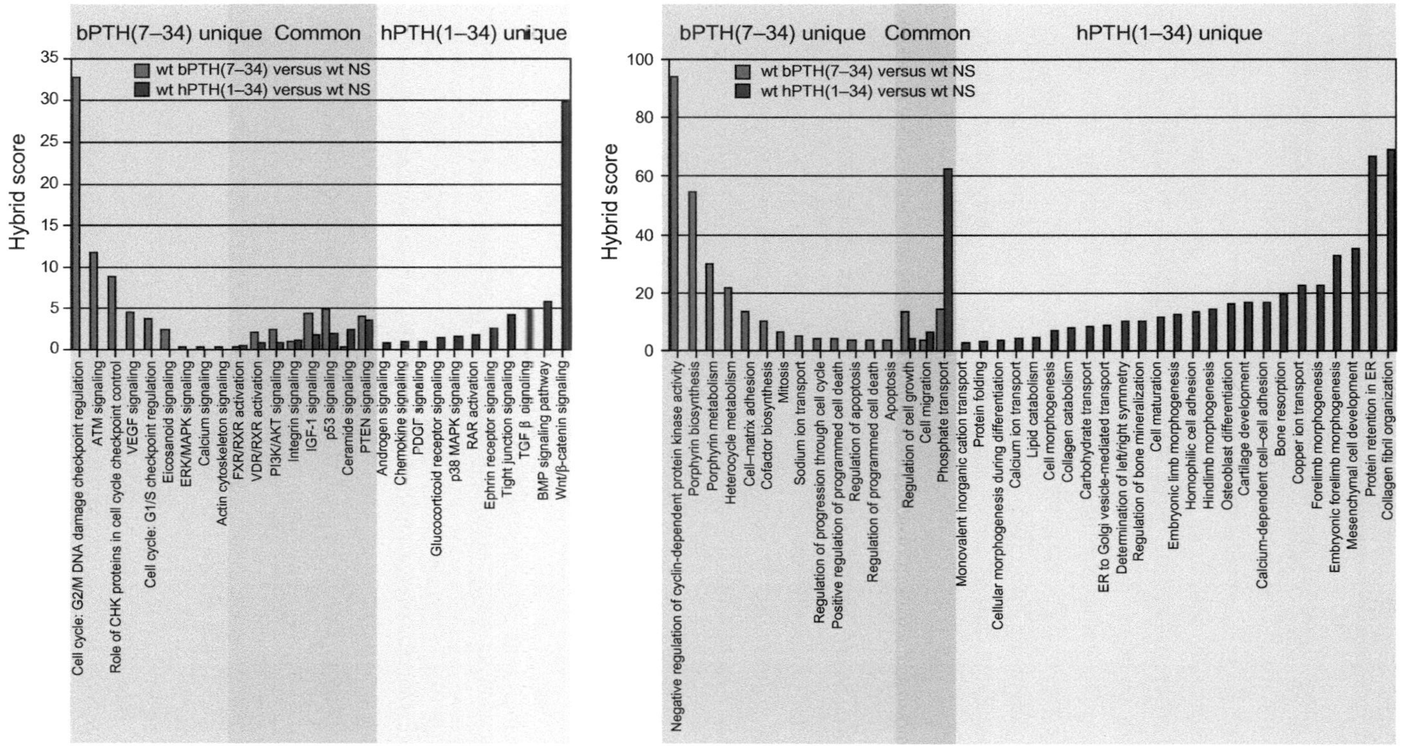

Figure 13.3 PTH(1–34) and bPTH(7–34) produce distinct genomic signatures in calvarial bone. Microarray genesets consisting of calvarial transcripts with significantly different expression between wild-type (wt) mice treated with vehicle (NS), hPTH(1–34), or bPTH(7–34) were analyzed by parametric geneset enrichment analysis. The *left* panel compares signaling pathway gene clusters identified using the wt bPTH(7–34) versus wt NS and wt hPTH(1–34) versus wt NS genesets. Signaling pathways corresponding to signal transduction, growth factor signaling, nuclear receptor signaling, and cell cycle control are shown. The *right* panel depicts an identical comparison derived by querying the GObp database. In each panel, hybrid scores ($-\log_{10}(p) \times$ pathway enrichment ratio) reflect the statistical probability that the observed differences did not occur by chance. All signaling pathway and GObp terms shown exceed a minimal threshold significance of $P \leq 0.05$. *Adapted from Ref. 88.* (See Color Insert.)

Consistent with previous studies,[13,38,83] PTH(1–34) produced larger increases in RANKL mRNA expression than bPTH(7–34). As β-arrestin-2 null mice have higher osteoclast numbers and activity and an exaggerated osteoclastic response to PTH(1–34) compared to wild-type mice, these results confirm that PTH_1R-mediated bone resorption requires G protein signaling that is normally restrained by arrestin-mediated desensitization. Thus, both facets of arrestin action, desensitization and signaling, contribute to PTH_1R actions in bone. The data suggest that the arrestin pathway-specific ligand increases matrix synthesis and mineralization *in vivo* by promoting osteoblast survival and expansion of the osteoblast pool without simultaneously accelerating bone turnover, whereas the conventional agonist stimulates osteoblast differentiation and function that is coupled with recruitment of osteoclasts. Both ligands increase bone mass, but the anabolic effects are achieved through substantially different mechanisms of action.

4.5. Arrestin pathway-selectivity as a strategy for drug design

The finding that bPTH(7–34) can uncouple bone formation from bone resorption offers compelling evidence that arrestin pathway-selective GPCR ligands may possess clinically useful properties distinct from conventional agonists or antagonists. The phenomenon of functional selectivity presents the opportunity to develop drugs that target GPCRs with unique biologic actions as well as improved specificity and efficacy. Ligands that direct signaling toward individual G protein pathways may prove useful in a variety of settings. G protein-selective ligands that signal without producing arrestin-dependent desensitization have seemingly obvious applications, for example, targeting opioid receptors for the management of chronic pain.[93] Settings where arrestin pathway-selectivity could prove valuable are perhaps less obvious, as our present understanding of the physiological roles of arrestin signaling *in vivo* is incomplete.[34] Regardless, current data suggest that functional selectivity can be exploited to change the quality of GPCR efficacy *in vivo* and may prove a useful strategy in future drug development.

5. PERSPECTIVES AND FUTURE DIRECTIONS

The true scope of physiologically relevant arrestin signaling remains largely unknown. However, a growing literature supports the concept that arrestin-bound effectors perform numerous functions, among them enhancing second messenger degradation, regulating cytoskeletal dynamics controlling GPCR endocytosis, postendocytic receptor trafficking, vesicle exocytosis, and cell migration, and impacting more distant processes such as protein translation and gene transcription.[34] The demonstration that a biased PTH_1R agonist selective for the β-arrestin pathway can promote anabolic bone formation without inducing bone resorption or hypercalcemia offers evidence that biased activators of G protein-independent signaling can produce *in vivo* biological responses that cannot be attained with conventional ligands.[15] While no currently used pharmaceuticals are known to possess unique clinical efficacy based on their ability to bias GPCR signaling in favor of arrestin-specific transduction, the opportunities for drug development in osteoporosis are promising.

6. CONCLUSIONS

Pathway-selective PTH analogs have proved to be valuable tools for determining the contribution of different PTH_1R signaling pathways to bone metabolism both *in vitro* and *in vivo*. Recent work using an arrestin pathway-selective PTH_1R agonist *in vivo* suggests that activation of arrestin signaling is sufficient to promote bone formation but is unable to stimulate bone resorption, meaning that it uncouples the bone forming effects of PTH on osteoblasts from its previously nondissociable effects on osteoclastic bone resorption. Although considerable additional work will be required to fully understand the mechanistic basis of β-arrestin signaling in bone and determine whether arrestin pathway-selective ligands are effective in preclinical models of metabolic bone disease, the demonstration that an arrestin pathway-selective biased agonist of the PTH_1R can accelerate bone formation *in vivo* offers the best evidence to date that biased activators of G protein-independent signaling can achieve biological responses that cannot be attained with nonselective agonists or antagonists.

ACKNOWLEDGMENTS

This work was supported by National Institutes of Health Grants DK64353, and HD043446; and the Arthritis Foundation.

REFERENCES

1. Orsini LS, Rousculp MD, Long SR, Wang S. Health care utilization and expenditures in the United States: a study of osteoporosis-related fractures. *Osteoporos Int.* 2005;16:359–371.
2. Kanis JA, Melton LJ, Christiansen C, Johnston CC, Khaltaev N. The diagnosis of osteoporosis. *J Bone Miner Res.* 1994;9:1137–1141.

3. Caverzasio J, Palmer G, Suzuki A, Bonjour J. Mechanism of the mitogenic effect of fluoride on osteoblast-like cells: evidences for a G protein-dependent tyrosine phosphorylation process. *J Bone Miner Res.* 1997;12:1975–1983.
4. Lau KHW, David JB. Molecular mechanism of action of fluoride on bone cells. *J Bone Miner Res.* 1998;13:1660–1667.
5. Vestergaard P, Jorgensen N, Schwarz P, Mosekilde L. Effects of treatment with fluoride on bone mineral density and fracture risk—a meta-analysis. *Osteoporos Int.* 2008;19: 257–268.
6. Giustina A, Mazziotti G, Canalis E. Growth hormone, insulin-like growth factors, and the skeleton. *Endocr Rev.* 2008;29:535–559.
7. Sugimoto T, Kaji H, Nakaoka D, et al. Effect of low-dose of recombinant human growth hormone on bone metabolism in elderly women with osteoporosis. *Eur J Endocrinol.* 2002;147:339–348.
8. Quarles LD, Haupt DM, Davidai G, Middleton JP. Prostaglandin F2 alpha-induced mitogenesis in MC3T3-E1 osteoblasts: role of protein kinase-C-mediated tyrosine phosphorylation. *Endocrinology.* 1993;132:1505–1513.
9. Soper DL, Milbank JBJ, Mieling GE, et al. Synthesis and biological evaluation of prostaglandin-F alkylphosphinic acid derivatives as bone anabolic agents for the treatment of osteoporosis. *J Med Chem.* 2001;44:4157–4169.
10. Qin L, Raggatt LJ, Partridge NC. Parathyroid hormone: a double-edged sword for bone metabolism. *Trends Endocrinol Metab.* 2004;15:60–65.
11. Shukla VK. Treating osteoporosis with teriparatide: many unknowns? *Issues Emerg Health Technol.* 2003;51:1–4.
12. Mackay HJ, Twelves CJ. Targeting the protein kinase C family: are we there yet? *Nat Rev Cancer.* 2007;7:554–562.
13. Bouxsein ML, Ferrari SL. Coupling PTH and arrestins to uncouple bone formation from resorption: a new road to osteoporosis anabolic therapy? *IBMS BoneKEy.* 2009;12: 470–476.
14. Ferrari SL, Bouxsein ML. Beta-arrestin-biased parathyroid hormone ligands: a new approach to the development of agents that stimulate bone formation. *Sci Transl Med.* 2009;1:1ps1.
15. Gesty-Palmer D, Flannery P, Yuan L, et al. A beta-arrestin biased agonist of the parathyroid hormone receptor (PTH1R) promotes bone formation independent of G protein activation. *Sci Transl Med.* 2009;1:1ra1.
16. Kenakin T, Miller LJ. Seven transmembrane receptors as shapeshifting proteins: the impact of allosteric modulation and functional selectivity on new drug discovery. *Pharmacol Rev.* 2010;62:265–304.
17. Lagerström MC, Schiöth HB. Structural diversity of G protein-coupled receptors and significance for drug discovery. *Nat Rev Drug Discov.* 2008;7:339–357.
18. Fredriksson R, Lagerström MC, Lundin L-G, Schiöth HB. The G-protein-coupled receptors in the human genome form five main families. Phylogenetic analysis, paralogon groups, and fingerprints. *Mol Pharmacol.* 2003;63:1256–1272.
19. Römpler H, Stäubert C, Thor D, Schulz A, Hofreiter M, Schöneberg T. G protein-coupled time travel: evolutionary aspects of GPCR research. *Mol Interv.* 2007;7:17–25.
20. Karlin A. On the application of "a plausible model" of allosteric proteins to the receptor for acetylcholine. *J Theor Biol.* 1967;16:306–320.
21. Thron CD. On the analysis of pharmacological experiments in terms of an allosteric receptor model. *Mol Pharmacol.* 1973;9:1–9.
22. Kenakin T. Receptor conformational induction versus selection: all part of the same energy landscape: agonists can differentially stabilize multiple active states of receptors. *Trends Pharmacol Sci.* 1996;17:190–191.

23. Costa T, Herz A. Antagonists with negative intrinsic activity at delta opioid receptors coupled to GTP-binding proteins. *Proc Natl Acad Sci USA*. 1989;86(19):7321–7325.
24. Samama P, Cotecchia S, Costa T, Lefkowitz RJ. A mutation-induced activated state of the beta 2-adrenergic receptor. Extending the ternary complex model. *J Biol Chem*. 1993;268:4625–4636.
25. Weiss JM, Morgan PH, Lutz MW, Kenakin TP. The cubic ternary complex receptor-occupancy model. III. Resurrecting efficacy. *J Theor Biol*. 1996;181:381–397.
26. Maudsley S, Martin B, Luttrell LM. The origins of diversity and specificity in G protein-coupled receptor signaling. *J Pharmacol Exp Ther*. 2005;314:485–494.
27. Laugwitz KL, Allgeier A, Offermanns S, et al. The human thyrotropin receptor: a heptahelical receptor capable of stimulating members of all four G protein families. *Proc Natl Acad Sci USA*. 1996;93:116–120.
28. Angers S, Salahpour A, Joly E, et al. Detection of beta 2-adrenergic receptor dimerization in living cells using bioluminescence resonance energy transfer (BRET). *Proc Natl Acad Sci USA*. 2000;97:3684–3689.
29. Hay DL, Poyner DR, Sexton PM. GPCR modulation by RAMPs. *Pharmacol Ther*. 2006;109:173–197.
30. Foord SM, Marshall FH. RAMPs: accessory proteins for seven transmembrane domain receptors. *Trends Pharmacol Sci*. 1999;20:184–187.
31. Brady AE, Limbird LE. G protein-coupled receptor interacting proteins: emerging roles in localization and signal transduction. *Cell Signal*. 2002;14:297–309.
32. Hall RA, Lefkowitz RJ. Regulation of G protein-coupled receptor signaling by scaffold proteins. *Circ Res*. 2002;91:672–680.
33. Bockaert J, Marin P, Dumuis A, Fagni L. The 'magic tail' of G protein-coupled receptors: an anchorage for functional protein networks. *FEBS Lett*. 2003;546:65–72.
34. Luttrell LM, Gesty-Palmer D. Beyond desensitization: physiological relevance of arrestin-dependent signaling. *Pharmacol Rev*. 2010;62:305–330.
35. Galandrin S, Bouvier M. Distinct signaling profiles of beta1 and beta2 adrenergic receptor ligands toward adenylyl cyclase and mitogen-activated protein kinase reveals the pluridimensionality of efficacy. *Mol Pharmacol*. 2006;70:1575–1584.
36. Kenakin T. Agonist-receptor efficacy II: agonist trafficking of receptor signals. *Trends Pharmacol Sci*. 1995;16:232–238.
37. Wisler JW, DeWire SM, Whalen EJ, et al. A unique mechanism of beta-blocker action: carvedilol stimulates beta-arrestin signaling. *Proc Natl Acad Sci USA*. 2007;104:16657–16662.
38. Bisello A, Chorev M, Rosenblatt M, Monticelli L, Mierke DF, Ferrari SL. Selective ligand-induced stabilization of active and desensitized parathyroid hormone type 1 receptor conformations. *J Biol Chem*. 2002;277:38524–38530.
39. Bisello A, Manen D, Pierroz DD, Usdin TB, Rizzoli R, Ferrari SL. Agonist-specific regulation of parathyroid hormone (PTH) receptor type 2 activity: structural and functional analysis of PTH- and tuberoinfundibular peptide (TIP) 39-stimulated desensitization and internalization. *Mol Endocrinol*. 2004;18:1486–1498.
40. Jouishomme H, Whitfield JF, Chakravarthy B, et al. The protein kinase-C activation domain of the parathyroid hormone. *Endocrinology*. 1992;130:53–60.
41. Azarani A, Goltzman D, Orlowski J. Structurally diverse N-terminal peptides of parathyroid hormone (PTH) and PTH-related peptide (PTHRP) inhibit the Na+/H+ exchanger NHE3 isoform by binding to the PTH/PTHRP receptor type I and activating distinct signaling pathways. *J Biol Chem*. 1996;271:14931–14936.
42. Takasu H, Gardella TJ, Luck MD, Potts JT, Bringhurst FR. Amino-terminal modifications of human parathyroid hormone (PTH) selectively alter phospholipase C signaling via the type 1 PTH receptor: implications for design of signal-specific PTH ligands. *Biochemistry*. 1999;38:13453–13460.

43. Dean T, Linglart A, Mahon MJ, et al. Mechanisms of ligand binding to the parathyroid hormone (PTH)/PTH-related protein receptor: selectivity of a modified PTH(1–15) radioligand for G{alpha}S-coupled receptor conformations. *Mol Endocrinol.* 2006;20:931–943.
44. Jouishomme H, Whitfield JF, Gagnon L, et al. Further definition of the protein kinase C activation domain of the parathyroid hormone. *J Bone Miner Res.* 1994;9:943–949.
45. Whitfield JF, Morley P. Small bone-building fragments of parathyroid hormone: new therapeutic agents for osteoporosis. *Trends Pharmacol Sci.* 1995;16:382–386.
46. Wei H, Ahn S, Shenoy SK, et al. Independent beta-arrestin 2 and G protein-mediated pathways for angiotensin II activation of extracellular signal-regulated kinases 1 and 2. *Proc Natl Acad Sci USA.* 2003;100:10782–10787.
47. Vassart G, Dumont JE. The thyrotropin receptor and the regulation of thyrocyte function and growth. *Endocr Rev.* 1992;13:596–611.
48. Kenakin T. Inverse, protean, and ligand-selective agonism: matters of receptor conformation. *FASEB J.* 2001;15:598–611.
49. Brzostowski JA, Kimmel AR. Signaling at zero G: G-protein-independent functions for 7-TM receptors. *Trends Biochem Sci.* 2001;26:291–297.
50. Kenakin T. Functional selectivity through protean and biased agonism: who steers the ship? *Mol Pharmacol.* 2007;72:1393–1401.
51. Gesty-Palmer D, Luttrell LM. Heptahelical terpsichory. Who calls the tune? *J Recept Signal Transduct Res.* 2008;28:39–58.
52. Violin JD, Lefkowitz RJ. [beta]-Arrestin-biased ligands at seven transmembrane receptors. *Trends Pharmacol Sci.* 2007;28:416–422.
53. Lefkowitz RJ. G protein-coupled receptor kinases. *Cell.* 1993;74:409–412.
54. Ferguson SS. Evolving concepts in G protein-coupled receptor endocytosis: the role in receptor desensitization and signaling. *Pharmacol Rev.* 2001;53:1–24.
55. Luttrell LM, Ferguson SS, Daaka Y, et al. Beta-arrestin-dependent formation of beta2 adrenergic receptor-Src protein kinase complexes. *Science.* 1999;283:655–661.
56. DeFea KA, Zalevsky J, Thoma MS, Dery O, Mullins RD, Bunnett NW. Beta-arrestin-dependent endocytosis of proteinase-activated receptor 2 is required for intracellular targeting of activated ERK1/2. *J Cell Biol.* 2000;148:1267–1281.
57. DeWire SM, Ahn S, Lefkowitz RJ, Shenoy SK. beta-Arrestins and cell signaling. *Annu Rev Physiol.* 2007;69:483–510.
58. Luttrell LM, Roudabush FL, Choy EW, et al. Activation and targeting of extracellular signal-regulated kinases by beta-arrestin scaffolds. *Proc Natl Acad Sci USA.* 2001;98: 2449–2454.
59. McDonald PH, Chow CW, Miller WE, et al. Beta-arrestin 2: a receptor-regulated MAPK scaffold for the activation of JNK3. *Science.* 2000;290:1574–1577.
60. Rey A, Manen D, Rizzoli R, Caverzasio J, Ferrari SL. Proline-rich motifs in the parathyroid hormone (PTH)/PTH-related protein receptor C terminus mediate scaffolding of c-Src with beta-Arrestin2 for ERK1/2 activation. *J Biol Chem.* 2006;281: 38181–38188.
61. Shenoy SK, Lefkowitz RJ. Seven-transmembrane receptor signaling through {beta}-arrestin. *Sci STKE.* 2005;308:cm10.
62. Azzi M, Charest PG, Angers S, et al. Beta-arrestin-mediated activation of MAPK by inverse agonists reveals distinct active conformations for G protein-coupled receptors. *Proc Natl Acad Sci USA.* 2003;100:11406–11411.
63. Gesty-Palmer D, Chen M, Reiter E, et al. Distinct beta-arrestin- and G protein-dependent pathways for parathyroid hormone receptor-stimulated ERK1/2 activation. *J Biol Chem.* 2006;281:10856–10864.
64. Dobnig H, Turner RT. Evidence that intermittent treatment with parathyroid hormone increases bone formation in adult rats by activation of bone lining cells. *Endocrinology.* 1995;136:3632–3638.

65. Schmidt I, Dobnig H, Turner R. Intermittent parathyroid hormone treatment increases osteoblast number, steady state messenger ribonucleic acid levels for osteocalcin, and bone formation in tibial metaphysis of hypophysectomized female rats. *Endocrinology.* 1995;136:5127–5134.
66. Kroll M. Parathyroid hormone temporal effects on bone formation and resorption. *Bull Math Biol.* 2000;62:163–188.
67. Tam CS, Heersche JN, Murray TM, Parsons JA. Parathyroid hormone stimulates the bone apposition rate independently of its resorptive action: differential effects of intermittent and continuous administration. *Endocrinology.* 1982;110:506–512.
68. Hock JM, Gera I. Effects of continuous and intermittent administration and inhibition of resorption on the anabolic response of bone to parathyroid hormone. *J Bone Miner Res.* 1992;7:65–72.
69. Abou-Samra AB, Juppner H, Force T, et al. Expression cloning of a common receptor for parathyroid hormone and parathyroid hormone-related peptide from rat osteoblast-like cells: a single receptor stimulates intracellular accumulation of both cAMP and inositol trisphosphates and increases intracellular free calcium. *Proc Natl Acad Sci USA.* 1992;89:2732–2736.
70. Juppner H, Abou-Samra AB, Freeman M, et al. A G protein-linked receptor for parathyroid hormone and parathyroid hormone-related peptide. *Science.* 1991;254: 1024–1026.
71. Bringhurst FR, Juppner H, Guo J, et al. Cloned, stably expressed parathyroid hormone (PTH)/PTH-related peptide receptors activate multiple messenger signals and biological responses in LLC-PK1 kidney cells. *Endocrinology.* 1993;132:2090–2098.
72. Iida-Klein A, Guo J, Takemura M, et al. Mutations in the second cytoplasmic loop of the rat parathyroid hormone (PTH)/PTH-related protein receptor result in selective loss of PTH-stimulated phospholipase C activity. *J Biol Chem.* 1997;272:6882–6889.
73. Koh AJ, Beecher CA, Rosol TJ, McCauley LK. 3′,5′-Cyclic adenosine monophosphate activation in osteoblastic cells: effects on parathyroid hormone-1 receptors and osteoblastic differentiation in vitro. *Endocrinology.* 1999;140:3154–3162.
74. Mahon MJ, Cole JA, Lederer ED, Segre GV. Na+/H+ exchanger-regulatory factor 1 mediates inhibition of phosphate transport by parathyroid hormone and second messengers by acting at multiple sites in opossum kidney cells. *Mol Endocrinol.* 2003;17: 2355–2364.
75. Cole JA. Parathyroid hormone activates mitogen-activated protein kinase in opossum kidney cells. *Endocrinology.* 1999;140:5771–5779.
76. Lederer ED, Sohi SS, McLeish KR. Parathyroid hormone stimulates extracellular signal-regulated kinase (ERK) activity through two independent signal transduction pathways: role of ERK in sodium-phosphate cotransport. *J Am Soc Nephrol.* 2000;11:222–231.
77. Verheijen MH, Defize LH. Parathyroid hormone activates mitogen-activated protein kinase via a cAMP-mediated pathway independent of Ras. *J Biol Chem.* 1997;272: 3423–3429.
78. Garcia-Ocana A, Gomez-Casero E, Penaranda C, Esbrit P. Parathyroid hormone-related protein increases DNA synthesis in proximal tubule cells by cyclic AMP- and protein kinase C-dependent pathways. *Life Sci.* 1998;62:2267–2273.
79. Swarthout JT, Doggett TA, Lemker JL, Partridge NC. Stimulation of extracellular signal-regulated kinases and proliferation in rat osteoblastic cells by parathyroid hormone is protein kinase C-dependent. *J Biol Chem.* 2001;276:7586–7592.
80. Bliziotes M, Murtagh J, Wiren K. Beta-adrendergic receptor kinase-like activity and beta-arrestin are expressed in osteoblastic cells. *J Bone Miner Res.* 1996;11:820–826.
81. Ferrari SL, Behar V, Chorev M, Rosenblatt M, Bisello A. Endocytosis of ligand-human parathyroid hormone receptor 1 complexes is protein kinase C dependent and involves beta-arrestin 2. Real-time monitoring by fluorescence microscopy. *J Biol Chem.* 1999;274:29968–29975.

82. Ferrari SL, Pierroz DD, Glatt V, et al. Bone response to intermittent parathyroid hormone is altered in mice null for β-arrestin 2. *Endocrinology*. 2005;146:1854–1862.
83. Bouxsein ML, Pierroz DD, Glatt V, et al. Beta-arrestin2 regulates the differential response of cortical and trabecular bone to intermittent PTH in female mice. *J Bone Miner Res*. 2005;20:635–643.
84. Pierroz DD, Rufo A, Bianchi EN, et al. β-Arrestin2 regulates RANKL and ephrins gene expression in response to bone remodeling in mice. *J Bone Miner Res*. 2009;24:775–784.
85. Vilardaga JP, Krasel C, Chauvin S, Bambino T, Lohse MJ, Nissenson RA. Internalization determinants of the parathyroid hormone receptor differentially regulate beta-arrestin/receptor association. *J Biol Chem*. 2002;277:8121–8129.
86. Bianchi EN, Ferrari SL. Beta-arrestin2 regulates parathyroid hormone effects on a p38 MAPK and NfkappaB gene expression network in osteoblasts. *Bone*. 2009;45:716–725.
87. Gardella TJ, Luck MD, Jensen GS, Schipani E, Potts Jr JT, Juppner H. Inverse agonism of amino-terminally truncated parathyroid hormone (PTH) and PTH-related peptide (PTHrP) analogs revealed with constitutively active mutant PTH/PTHrP receptors. *Endocrinology*. 1996;137:3936–3941.
88. Gesty-Palmer D, Yuan L, Martin B, et al. β-Arrestin selective G protein-coupled receptor agonists engender unique biological efficacy in vivo. *Mol Endocrinol*. 2013;27(2): 296–314.
89. Pi M, Oakley RH, Gesty-Palmer D, et al. {beta}-Arrestin- and G protein receptor kinase-mediated calcium-sensing receptor desensitization. *Mol Endocrinol*. 2005;19: 1078–1087.
90. Deng ZL, Sharff KA, Tang N, et al. Regulation of osteogenic differentiation during skeletal development. *Front Biosci*. 2008;13:2001–2021.
91. Lian JB, Stein GS, Javed A, et al. Networks and hubs for the transcriptional control of osteoblastogenesis. *Rev Endocr Metab Disord*. 2006;7:1–16.
92. Guo X, Wang XF. Signaling cross-talk between TGF-beta/BMP and other pathways. *Cell Res*. 2009;19:71–88.
93. Bohn LM, Lefkowitz RJ, Gainetdinov RR, Peppel K, Caron MG, Lin FT. Enhanced morphine analgesia in mice lacking beta-arrestin 2. *Science*. 1999;286:2495–2498.

CHAPTER FOURTEEN

β-Arrestins in the Immune System

Dianhua Jiang, Ting Xie, Jiurong Liang, Paul W. Noble
Division of Pulmonary, Allergy, and Critical Care Medicine, Department of Medicine, Duke University School of Medicine, Durham, North Carolina, USA

Contents

1. Introduction 360
2. β-Arrestins in Innate Immunity 361
   2.1 β-Arrestins and macrophages 361
   2.2 β-Arrestins and polymorphonuclear leukocytes 362
   2.3 β-Arrestins and natural killer cells 363
   2.4 β-Arrestins and mast cells 363
   2.5 β-Arrestins and complement 363
3. β-Arrestins in Adaptive Immunity 364
   3.1 β-Arrestins and lymphocytes 364
   3.2 T-cell activation 365
   3.3 Migration and chemotaxis 365
   3.4 Lymphocyte survival and apoptosis 365
   3.5 TH1, TH2, and TH17 cells 366
4. β-Arrestins and Structural Cells 367
5. β-Arrestins Regulate Immune Signaling Pathways 367
   5.1 β-Arrestins and chemokine receptors 367
   5.2 β-Arrestins and other GPCRs in inflammation 369
   5.3 β-Arrestins and TLR signaling 371
   5.4 Other non-GPCR signaling pathways 373
6. Role of β-Arrestins in Human Diseases 374
   6.1 Pathogen recognition and clearance 374
   6.2 Fibrosis 377
   6.3 Arthritis 379
   6.4 Multiple sclerosis 380
   6.5 Encephalomyelitis 381
   6.6 Acute rejection in organ transplant 381
   6.7 Asthma 381
   6.8 Cystic fibrosis 382
   6.9 Cardioprotection 382
7. β-Arrestins in Therapeutic Development for Inflammatory Diseases 383
References 384

Progress in Molecular Biology and Translational Science, Volume 118
ISSN 1877-1173
http://dx.doi.org/10.1016/B978-0-12-394440-5.00014-0

Abstract

β-Arrestins regulate G protein-coupled receptors through receptor desensitization while also acting as signaling scaffolds to facilitate numerous effector pathways. Recent studies have provided evidence that β-arrestins play a key role in inflammatory responses. Here, we summarize these advances on the roles of β-arrestins in immune regulation and inflammatory responses under physiological and pathological conditions, with an emphasis on translational implications of β-arrestins on human diseases.

1. INTRODUCTION

β-Arrestins are classically known to regulate G protein-coupled receptor (GPCR) signaling through receptor desensitization and internalization.[1] The structure and molecular characterization of β-arrestins have been established.[2–4] Collaborating with GPCR kinases (GRKs),[5] β-arrestins are known for their role in desensitizing the β2-adrenergic receptor (β2AR).[6] Much of the investigation on the role of β-arrestins has been associated with the deactivation and desensitization of the GPCRs. GPCR internalization is believed to occur through clathrin-mediated endocytosis, which consists of three processes: receptor desensitization, sequestration of receptors to clathrin-coated pits, and receptor internalization. These steps are each regulated by distinct signaling events, and β-arrestins have a role in these processes. In addition, β-arrestins have been shown to mediate non-GPCR signaling pathways, such as through TGF-β receptor III.[7]

Reports have also shown that β-arrestins facilitate the activation of numerous effector pathways, such as the mitogen-activated protein kinases (MAPKs) and Akt.[8] β-Arrestin 2 serves as a scaffold to enhance MAPK kinase kinase (cRaf-1) and MAPK kinase (MEK)-dependent activation of extracellular signal-regulated kinase 2 (ERK2[9]) (reviewed in Refs. 10,11). Furthermore, evidence suggests that β-arrestins mediate signaling pathways through transcriptional regulation (reviewed in Refs. 12). For example, β-arrestin 1 is able to translocate from the cytosol to the nucleus and recruit histone acetyltransferase p300, leading to enhanced local histone H4 acetylation and transcription of p27 and c-fos.[13]

Extensive research demonstrates that β-arrestins have functions in development,[14,15] cancer,[16,17] and chemotaxis.[18,19] During the past several years, roles of β-arrestins in innate and adaptive immunity, as well as in inflammatory responses, have been reported. Studies in mice with targeted deletion of β-arrestins have identified important roles of these adaptor molecules in immune responses, and regulation of the pathogenesis of several disease processes has begun to be explored.

The immune system consists of innate and adaptive responses. Innate immune cells, receptor systems, and related signaling pathways defend the host from infections such as bacteria and viruses, as well as from noninfectious tissue injuries. Defense by the innate immune system includes chemotaxis of inflammatory cells, release of cytokines, pathogen clearance, as well as the activation of the adaptive immune system through antigen presentation. The adaptive immune system, on the other hand, consists of highly specialized cells (such as T lymphocytes and B lymphocytes), proteins, and processes that aim to further eliminate pathogens. In addition to their role in pathogen clearance, both systems have a role in tissue injury, inflammation, and repair processes. Expression of β-arrestins has been found in innate immune cells, such as macrophages and polymorphonuclear leukocytes (PMNs), as well as in adaptive immune cells, including T lymphocytes and B cells. β-Arrestins have been suggested to have a role in both systems. In addition, β-arrestins may have a role in mammalian hematopoiesis as β-arrestin 1-deficient mice demonstrate basal hematologic defects.[20] Similarly in zebrafish, β-arrestin 1 relieves polycomb group-mediated repression of the cdx4–hox pathway, regulating hematopoietic lineage specification during primitive hematopoiesis.[20]

2. β-ARRESTINS IN INNATE IMMUNITY

2.1. β-Arrestins and macrophages

β-Arrestins are expressed in macrophages. Both β-arrestins 1 and 2 mRNA and protein expression are found in human and mouse mononuclear cells and macrophages.[21,22] Lipopolysaccharides downregulate β-arrestin 2 expression in RAW264 macrophages.[23] Furthermore, β-arrestin 1 expression can be downregulated by the activation of TLR2 and TLR4 in primary mouse macrophages, and the regulation is both transcriptional and posttranslational.[24] Expression of β-arrestins 1 and 2 by macrophages can be regulated by cytokines. The cytokine, granulocyte–macrophage colony-stimulating factor-1, increased β-arrestin 1 expression associated with the downregulation of surface chemokine receptor CXCR4 expression in human primary monocytes–macrophages.[25] Lectin phytoemagglutinin reduces β-arrestin 1 expression in mononuclear leukocytes, and interferon β-1a can quench the effect of phytoemagglutinin on the expression of β-arrestin 1.[26] β-Arrestin 2 mRNA and protein levels were significantly elevated in peripheral blood mononuclear cells of cryptococcal meningitis patients.[27] β-Arrestin 1 and 2 expression is reduced in blood leukocytes during acute kidney transplantation.[28] The modulation of β-arrestin expression by cytokines is involved in regulating related effector pathways.

β-Arrestins regulate macrophage chemotaxis induced by chemokines and serine proteases. Chemokine CCL4 (MIP-1β) binding to CCR5 induces the translocation of β-arrestins, PI3K, and Pyk2 to the membrane and forms a multikinase signaling complex mediating macrophage migration.[29] Knockdown of β-arrestins 1 and 2 by small interfering RNA (siRNA) impairs complex formation and inhibits macrophage chemotaxis toward CCL4.[29] Chemokine CCL2 (MCP-1) binds with the CCR2 receptor, leading to the formation of a CCR2/β-arrestin/GRK2 complex in human monocytes/macrophages. This process promotes the rapid desensitization of CCL2-induced calcium flux responses, which are essential for chemotaxis in macrophages.[30] Protease-activated receptor-2 (PAR-2) activation has been shown to enhance β-arrestin and ERK1/2-regulated assembly of the actin cytoskeleton in pseudopodia, promoting polarized pseudopodia extension and facilitating macrophage chemotaxis.[31] Collectively, these studies demonstrate that β-arrestins are an integrated regulator of macrophage-mediated innate immune responses.

2.2. β-Arrestins and polymorphonuclear leukocytes

PMNs are essential for host defense against infection, and the activation of PMNs is pivotal in the process of innate immunity.[32,33] β-Arrestin 1 is highly expressed in peripheral blood leukocytes including PMNs.[22] β-Arrestin 1 and 2 are critical modulators of inflammatory responses in PMNs, including PMN cytokine production, recruitment,[34] and granule release.[35] β-Arrestins and GRK2 associated with the CCR2 receptor play an important role in CCL2 (MCP-1)-regulated leukocyte activation and migration following CCL2 binding.[30] PMNs from β-arrestin 2-deficient mice show enhanced release of inflammatory cytokines, including TNF-α and IL-6, compared with WT PMNs.[34] β-Arrestin 2 deficiency results in augmented PMN recruitment with increased expression of CD18 and CD62L and enhanced PMN infiltration in the lungs after cecal ligation and puncture (CLP[34]). Furthermore, β-arrestins regulate interleukin (IL)-8-induced granule exocytosis in PMNs, which is critical for the innate immune response against infectious bacteria.[35] β-Arrestin 2 plays a negative role in CXCR2 signaling in neutrophils. Neutrophils isolated from peritoneal cavity of β-arrestin 2-deficient mice show enhanced calcium mobilization, superoxide anion production, and GTPase activity, but decreased receptor internalization in CXCR2 signaling.[36] β-Arrestin 2-deficient mice also demonstrate increased recruitment of PMNs in response to CXCL1 in the air pouch model and in excisional wound beds, leading to faster wound reepithelialization.[36]

2.3. β-Arrestins and natural killer cells

Natural killer (NK) cells constitute a major component of the innate immune system and play a significant role in tumor rejection and viral clearance. NK cell receptors express activating NK receptors such as NKG2D and inhibitory receptors such as killer-cell immunoglobulin-like receptors. β-Arrestin 2 has been reported to interact with the inhibitory receptor KIR2DL1 and mediate the binding of tyrosine phosphatases to KIR2DL1 to facilitate the inhibitory signaling of NK cells.[37] A mutated KIR2DL1 with an arginine residue at position 245 in its transmembrane domain recruited more β-arrestin 2 and Src-homology-2 domain-containing protein tyrosine phosphatase 2.[38] NK cells isolated from β-arrestin 2-deficient mice demonstrated higher cytotoxicity than WT mice, while NK cells from β-arrestin 2 transgenic mice showed reduced cytotoxicity which could be restored by β-arrestin 2 knockdown.[37] Thus, β-arrestin 2 is a negative regulator in NK cell cytotoxicity.

2.4. β-Arrestins and mast cells

Mast cells contain granules rich in histamine and heparin, and play a role in allergy and anaphylaxis, as well as in tissue injury.[7] Mast cell degranulation can be triggered by injury, cross-linking of Immunoglobulin E receptors, and complement proteins.[7] In the intestine, mast cells release tryptase to activate PAR2 in colonocytes. PAR2 then associates with β-arrestins to activate ERK1/2, which increases epithelial permeability by regulating the assembly of perijunctional F-actin.[39] β-Arrestins may play a role in the increased epithelial permeability of the intestine during stress and inflammation through the mediation of mast cell activation.

2.5. β-Arrestins and complement

The complement system contains more than 30 proteins, which regulate tissue injury and inflammation.[40] The complement system can be activated as part of the innate immune, as well as the adaptive immune, response. Complement fragments bind to specific receptors to induce cellular responses that trigger inflammatory and immune responses. β-Arrestins 1 and 2 are involved in complement receptor C3aR desensitization, internalization, degranulation, NF-κB activation, and chemokine generation in mast cells.[41] Stable knockdown of β-arrrestin 2 expression attenuated C3aR desensitization, internalization, and NF-κB activity, as well as chemokine generation in human mast cell lines that endogenously express C3aR.[41] However, silencing β-arrestin 1 decreased C3a-induced mast cell degranulation, without affecting C3aR desensitization. Knockdown of β-arrestin 1, β-arrestin 2, or both enhanced the early response to C3a but inhibited G protein-dependent ERK1/2 phosphorylation.[41] Other studies suggested that although C3aR interacts with β-arrestins, they do not appear to be involved in C3a-induced ERK activation in a leukemia cell line.[42] Complement C5a binds to the C5a receptor (C5aR), a GPCR, and the C5L2 receptor, a non-GPCR, to elicit its proinflammatory functions. A recent study suggested that C5a binding to C5aR recruits C5L2 and β-arrestin 2.[43] Activation of C5L2 results in the inhibition of C5aR-β-arrestin-mediated ERK1/2 activation, with no apparent alteration of G protein-mediated functions such as calcium influx and receptor endocytosis.[43] Bone marrow-derived macrophages devoid of β-arrestin 2 showed decreased complement C1q gene expression and enhanced factor-independent survival of CSF-1 through JNK/ERK signaling.[44] Therefore, β-arrestins appear to regulate complement functions through receptor desensitization as well as ERK signaling.

3. β-ARRESTINS IN ADAPTIVE IMMUNITY

3.1. β-Arrestins and lymphocytes

Both β-arrestins 1 and 2 are expressed in lymphocytes.[45] In thymocytes, β-arrestin 1 expression was very low compared with splenocytes or lymphocytes isolated from lymph nodes.[46] Compared with B cells, T cells contain substantially more β-arrestin 2.[45] For both naïve and activated cells, $CD4^+$ T cells had much more nuclear β-arrestin 1 than $CD8^+$ T cells.[46] $CD4^+$ T lymphocytes of mice with allergic asthma expressed higher levels of β-arrestin 2 at both the mRNA and protein levels compared with normal mice.[47] β-Arrestin expression levels are dependent upon cell type, activation status, and disease states.

3.2. T-cell activation

T lymphocytes are activated by the binding of the T-cell receptor (TCR) to antigens bound by major histocompatibility complex molecules on antigen-presenting cells. The binding of costimulatory molecules to the CD28 receptor is also required for full T-cell activation. In addition, the activation is mediated by kinases, phosphatases, and transcription factors. Engagement of the TCR is followed by rapid cyclic AMP (cAMP) production in lipid rafts and activation of the cAMP–protein kinase A (PKA)–Csk pathway inhibiting proximal T-cell signaling.[48,49] It appears that β-arrestins do not regulate the cAMP–PKA–Csk pathway.[48,49] However, β-arrestins participate in CD28 receptor coligation signaling.[49] Upon TCR and CD28 coligation, β-arrestin was recruited to lipid rafts and augmented TCR/CD28-stimulated cytokine production.[48,50] β-Arrestins form complexes with cAMP phosphodiesterase 4 and mediate cross talk between phosphatidylinositol 3-kinase and cAMP–PKA signaling pathways.[51] A PDE4/β-arrestin complex attenuates PKA phosphorylation of Csk,[48,52] allowing full T-cell activation to proceed.

3.3. Migration and chemotaxis

The migration of T cells is important in adaptive immune responses. T cells must migrate to sites where antigen is found, because T cells respond to pathogens only on direct contact with pathogen-derived antigen.[53] β-Arrestins function as fundamental adaptors connecting receptors to cell trafficking machinery.[48] β-Arrestin-2-mediated activation of the p38 MAPK cascade is required for CXCR4-mediated migration of human embryonic kidney 293 cells to CXCL12 (SDF-1α[54]). It has been reported that lymphocytes devoid of β-arrestin-2 demonstrate impaired migration toward the chemotactic factor CXCL12 *in vitro*.[45] A subsequent study showed that β-arrestin-2-deficient mice have defective macrophage-derived chemokine-mediated $CD4^+$ T-cell migration to the lungs,[55] and T cells in β-arrestin-2-deficient mice exhibit migration defects in a model of metastatic tumor growth.[17]

3.4. Lymphocyte survival and apoptosis

Several studies indicate that β-arrestins play a critical role in preventing apoptosis.[56,57] For example, neuropeptide substance P inhibits apoptosis via a β-arrestin scaffolding complex,[58] and β-arrestins play a critical role in the suppression of apoptosis mediated by stimulated GPCRs, as demonstrated using β-arrestin 1/2-knockout cells.[57] IGF1-mediated antiapoptosis was diminished by knockdown of β-arrestin 1/2 and was restored by transfection of β-arrestin 1.[56] β-Arrestin 1 epigenetically regulates Bcl2 expression, regulating the survival and homeostasis of $CD4^+$ T cells.[46] β-Arrestin 1 promotes acetylation of histone H4[59] at the Bcl2 locus and Bcl2 expression in both naïve and activated $CD4^+$ T cells.[46] Furthermore, β-arrestin 1 is highly expressed in $CD4^+$ T cells from multiple sclerosis patients, and RNAi knockdown of β-arrestin 1 in those cells increased apoptosis after induction by cytokine withdrawal.[46] β-Arrestin 1 did not affect the homeostasis of $CD8^+$ T cells,[46] possibly due to the difference in the subcellular distribution of β-arrestin 1 in $CD4^+$ and $CD8^+$ T cells. Moreover, β-arrestin 2 overexpression and interference can decrease and increase, respectively, the level of gp120/morphine-induced lymphocyte apoptosis.[60] Collectively, these studies support a role for β-arrestins in regulating lymphocyte survival.

3.5. TH1, TH2, and TH17 cells

Classically, helper T cells include Th1 and Th2 T cells. T-cell antigen receptor-mediated activation of the Ras/MAP kinase pathway controls IL-4 receptor function and Th2 cell differentiation.[61] Although the direct demonstration of a role for β-arrestins in Th1 and Th2 skewing is lacking, it seems that β-arrestins play a role in both conditions. For example, β-arrestin-2-deficient mice showed reduced $CD4^+$ T-cell accumulation in the airway in an asthma model.[55] This may be a Th2-dependent event as β-arrestin-2-deficient mice showed a normal inflammatory response in an LPS model.[55] Knocking down β-arrestin 2 with RNAi decreased the expression of the Th2 cytokine IL-4, as well as the T-cell-specific transcription factor GATA3 in $CD4^+$ T lymphocytes after terbutaline stimulation.[47] On the other hand, overexpression of β-arrestin 1 increased production of the Th1 cytokine, interferon-γ.[59] The stress-induced suppression of Th1 cytokines and the increased production of Th2 cytokines were greatly enhanced in β-arrestin 2-deficient mice compared with wild-type mice.[62] These studies suggest that β-arrestins may not have a role in T-cell differentiation but have a role in Th1 and Th2 responses during inflammation.

The involvement of β-arrestins in Th17 cell function has not been reported. Blocking CCL2–CCR2 interactions with a fusion peptide failed to recruit β-arrestin 2, leading to blockade of IL-17 production *in vitro*.[63]

Silencing β-arrestin 2 with specific siRNA reduced IL-17 expression.[64] Overexpression of β-arrestin 1 increased IL-17 expression, possibly by acetylation of histone H4 in the promoter region of IL-17.[59] It would be interesting to determine if β-arrestins have any role in activation of Th17 cells.

4. β-ARRESTINS AND STRUCTURAL CELLS

β-Arrestins are expressed in epithelial cells,[65,66] endothelial cells,[22,67,68] and smooth muscle.[22] β-Arrestin 1 has been found to be expressed in human endothelial cells and smooth muscle cells.[22] In these cells, β-arrestin 1 expression can be upregulated by cAMP-inducing agents such as cholera toxin, forskolin, iloprost, and isoproterenol.[22] β-Arrestins mediate the infections of *Streptococcus pneumonia*[67] and *Neisseria meningitidis*[69] through interactions with endothelial cells. β-Arrestin 2 regulates thrombin–PAR1 signaling in endothelial cell adhesion.[68]

β-Arrestin influences CXCL12-mediated epithelial cell migration. CXCL12-stimulated cell migration is enhanced by CXCR7/CXCR4 coexpression in a β-arrestin-dependent manner.[65] In a CXCR4/CXCR7-expressing MDA-MB-231 breast cancer epithelial cell line, inhibition of β-arrestin using either siRNA knockdown or a dominant negative mutant suppressed CXCL12-induced cell migration.[65]

5. β-ARRESTINS REGULATE IMMUNE SIGNALING PATHWAYS

5.1. β-Arrestins and chemokine receptors

Chemokines are a family of small cytokines secreted by immune cells and inflammatory cells. Receptors for chemokines are typically seven-transmembrane GPCRs. Binding of chemokines and their cognate receptors regulate chemotaxis of leukocytes into inflammatory sites[70] and cytokine release. β-Arrestins are recruited to the receptors and mediate receptor desensitization and internalization (reviewed in Refs. 11,18,21,71).

CXC chemokines are one of the main subfamilies of chemokines. β-Arrestins have been shown to mediate the internalization and desensitization of IL-8-induced CXCR1,[72,73] CXCR2,[36,73,74] and CXCL9- and CXCL10-induced CXCR3.[75] β-Arrestins regulate CXCR4 internalization in response to the specific ligand CXCL12/SDF-1.[76] CXCR4 can also form a heteromeric complex with the CXCR7 chemokine receptor upon in response to CXCL12.[65] Defects in chemokine receptor–β-arrestin interactions may promote human disease states. WHIM (warts, hypogammaglobulinemia, infections, and myelokatexis) syndrome is an immunodeficiency syndrome linked to heterozygous mutations of the chemokine receptor CXCR4 resulting in truncations of its cytoplasmic tail. WHIM-associated mutant CXCR4 was unable to recruit β-arrestin 2, leading to the impaired CXCR4 internalization and enhanced chemotaxis in response to CXCL12.[77]

CC chemokines are a second group of chemokines. Reports show that β-arrestins regulate CC chemokine-induced CC chemokine receptor activation by desensitization and internalization. CCL2 (MCP-1) binds to CCR2 inducing the activation of CCR2 and, shortly after stimulation, recruits a protein complex including GRK2 and β-arrestin.[30] In addition, the actin binding protein filamin A binds CCR2B[78] and together with β-arrestin regulates the internalization of the receptor complex. The chemokine CCL5 (RANTES) binding to the CCR5 induces the phosphorylation of CCR5. β-Arrestin 1 is able to desensitize and internalize homo- and hetero-oligomers of CCR5.[79,80] β-Arrestins thus regulate immune and inflammatory responses, in part, by desensitization and internalization of activated chemokine receptors.

β-Arrestins also regulate chemokine signaling via noncanonical G protein pathways. β-Arrestin 2 and GRK6 play positive regulatory roles, mediating the chemotactic responses of T and B lymphocytes to CXCL12.[45] β-Arrestin 2 strengthened CXCR4 activation through the ASK1–p38MAPK pathway.[54] In addition to the formation of a CXCR4 homodimer, CXCL12 also binds to CXCR7[81,82] and induces CXCR4–CXCR7 heterodimers.[65] Upon CXCL12 binding, the CXCR4–CXCR7 heterodimer complex recruits β-arrestin, resulting in the preferential activation of β-arrestin-linked signaling pathways over canonical G protein pathways.[65] This activation was dependent on β-arrestin-mediated cell signaling pathways, such as ERK1/2, p38 MAPK, and SAPK, leading to enhanced cell migration in response to CXCL12 stimulation.[65] The CC chemokine CCL4 (MIP-1β) binds to CCR5-regulating macrophage migration. Chemotaxis upon CCR5 stimulation by CCL4 requires activation of Pyk2, PI3K p85, Lyn, and ERK. β-Arrestins regulate the receptor complex formation and enhance macrophage chemotaxis toward the ligand.[29]

5.2. β-Arrestins and other GPCRs in inflammation

5.2.1 Angiotensin II receptor type 1A

Angiotensin II is a potent vasopressor peptide hormone. In addition to its role as an important effector controlling blood pressure, it binds its receptor, the angiotensin II type 1A receptor (AGTR1, AT1R, or AT1AR), to mediate the formation of reactive oxygen species and inflammatory responses.[83] For example, in liver, angiotensin II-dependent NF-κB activation leads to proinflammatory cytokine release from hepatocytes and stellate cells that is implicated in the development of hepatic fibrosis.[84] Studies suggested that β-arrestins promote desensitization and sequestration of AGTR1.[85] Both β-arrestin 1 and β-arrestin 2 showed similar roles in agonist-stimulated receptor desensitization, while the β-arrestin 2 may have more profound effects on angiotensin II-stimulated internalization of the AGTR1.[85] Furthermore, β-arrestin 2 also mediates chemotaxis through an AGTR1-dependent and G protein-independent mechanism.[86] β-Arrestins also facilitate GPCR-mediated ERK activation but inhibit ERK-dependent transcription by targeting activated ERK1/2 to nonnuclear substrates following angiotensin AT1A receptor stimulation.[87] Furthermore, β-arrestin 2 binds JNK and mediates ASK1 and MKK4 activity.[88]

The angiotensin II receptor has been shown to play a role in immune-mediated renal injury.[89,90] Deletion of AGTR1 in mice markedly attenuated glomerular expression of CCL2 (MCP-1), proteinuria, and tissue damage in an antiglomerular basement membrane nephritis model.[89] Angiotensin II binds its receptors on immune cells, triggering the proliferation of splenic lymphocytes and leading to the potency of cellular alloimmune responses.[90] The absence of AGTR1 signaling accentuates the immunosuppressive effects of the calcineurin inhibitor, cyclosporine.[90] The manipulation of β-arrestin-mediated renin angiotensin signaling can be used as a way to regulate immune responses.

5.2.2 β2-Adrenergic receptor

β-Arrestin attenuates agonist-induced β2AR signaling by GRK-mediated desensitization[91] and by clathrin-mediated receptor internalization.[92] It is reported that both isoforms of β-arrestins promote desensitization of β2ARs[85]; however, deficiency in β-arrestin 2, not in β-arrestin 1, significantly compromised the sequestration of β2ARs.[85] A cAMP phosphodiesterase PDE4 is recruited to the receptor–β-arrestin complex, and PDE4 plays a role in the regulation of G protein switching by the β2AR in cardiac myocytes.[93]

5.2.3 Lipid signaling receptors

Prostaglandins are lipid compounds derived from fatty acids. Prostaglandins bind to their respective receptors to elicit functions including inflammatory regulation. Prostaglandin E2 receptor (EP4) is a typical GPCR. Studies suggested that β-arrestins have both positive and negative effects on the EP4 receptor. β-Arrestin 1 is necessary for maximal agonist-induced internalization through the interaction with a Ser/Thr cluster in the C-terminal domain of EP4 receptor.[94]

Leukotrienes are arachidonic acid metabolites, binding to their respective receptors to elicit their role in inflammatory responses. β-Arrestin 1 is required for internalization of the leukotriene B4-activated leukotriene B4 receptor.[95]

Lysophosphatidylcholines are phospholipids that bind to G2A receptors on PMNs recruiting clathrin, β-arrestin-1, and GRK6 for receptor desensitization. Phospholipids cause acute lung injury[96] and are involved in transfusion-related acute lung injury.[97] β-Arrestin 1 both propagates and terminates inflammatory signals.[97]

Lysophosphatidic acid binds to lysophosphatidic acid receptors, including LPAR1, and mediates diverse biologic roles such as proliferation, platelet aggregation, smooth muscle contraction, chemotaxis, fibroblast migration, and tumor cell invasion. β-Arrestin 1 and GRK2 mediate lysophosphatidic acid receptor LPAR1 desensitization.[98] Furthermore, β-arrestin 2 promotes a distinct subset of ERK1/2-mediated responses to LPA receptor activation[99] and lysophosphatidic acid-induced NF-κB activation.[100] Suppression of β-arrestin 2 expression using siRNA abolished chemotaxis induced by lysophosphatidic acid.[86] β-Arrestin signaling regulates lysophosphatidic acid-mediated migration and invasion of human breast tumor cells.[101]

The phospholipid, platelet-activating factor (PAF), is an effective chemoattractant that primes PMN oxidases. The binding of PAF with the PAF receptor (PAFR or PTAFT) recruits β-arrestin 1 and desensitizes the receptor.[102] β-Arrestin recruitment blocks both NF-κB activation and Ca2+/calcineurin-dependent signaling pathways, leading to reduced cytokine production.[103] Platelet-activating factor-induced clathrin-mediated endocytosis requires a β-arrestin 1–MKK3–p38 MAPK complex and p38 MAPK activation.[104] Finally, the same scaffolding complex regulates actin rearrangement in human neutrophils.[104,105]

5.2.4 Protease-activated receptors

Both protease-activated receptor-1 (PAR-1) and PAR-2 mediate leukocyte recruitment to sites of injury and infection. β-Arrestins may play a dual role in both PAR-1 and PAR-2 desensitization and scaffolding. A report suggested that only β-arrestin 1 is capable of PAR-1 receptor desensitization, but receptor internalization is independent of either β-arrestin.[106] In contrast, β-arrestins function as scaffolds for PAR-2 activation. A recent study reported that upon PAR-2 activation, β-arrestin scaffolding aggregates the intracellular actin-modulating protein cofilin and its activator chronophin to facilitate the localized generation of free actin barbed ends, leading to the reorganization of the actin cytoskeleton and chemotaxis.[107] The formation of the signaling complex of β-arrestins and activated ERK1/2 is required to mediate PAR-2 signaling.[108]

5.3. β-Arrestins and TLR signaling

Toll-like receptors (TLRs) are pattern recognition receptors that recognize pathogen-associated molecular patterns such as bacterial coat components[109] as well as endogenous ligands.[110] TLRs belong to the IL-1 receptor/Toll-like receptor superfamily that play a key role in the innate immune system.[109] β-Arrestins are implicated in TLR4 signaling in several studies.[111–114] β-Arrestin 2 attenuates TLR4-initiated apoptosis by controlling the activation and inactivation of GSK-3β.[115] β-Arrestin 2 modulates stress-induced immune responses through TLR2-mediated signaling.[116]

Studies suggest that endosomal trafficking of the LPS–TLR receptor complex is essential for signal termination and LPS-associated antigen presentation.[117] Reports have shown that β-arrestins may have a direct role in regulating TLR sensitization and internalization. β-Arrestin 2 directly interacts with TRAF6 in response to LPS or IL-1β stimulation, preventing TRAF6 autoubiquitination and oligomerization.[111] β-Arrestin 2 also inhibits AP-1 and NF-κB activation and inhibits LPS- and IL-1-induced inflammatory cytokine production.[111] β-Arrestin 2 and GRK5 have been reported to negatively regulate TLR4 signaling through NF-κB and ERK phosphorylation.[118] Conversely, another study suggested that β-arrestins regulate cytokine production following LPS-induced chromatin modification, and not MAPK and IκBα–NFκB pathways.[112] In *Drosophila*, the β-arrestin, Kurtz, inhibits MAPK and Toll signaling during development.[119]

β-Arrestins may have a role in regulating cross talk between TLRs and other receptor families (Fig. 14.1). A recent study showed that there was cross talk between the β2AR and TLR4, and that the suppression of β-arrestin 2 eliminated the antiinflammatory effects of β2AR activation in monocytes.[120] β-Arrestin-2 mediated redistribution of CD14 and the TLR4/CD14 complex.[120] The β2AR regulates TLR4-induced late-phase NF-κB activation.[121] Furthermore, recent studies suggest there is cross talk between TLRs and chemokine receptors. TLR2 activation leads to β-arrestin-mediated endocytosis of CCR1, CCR2, and CCR5 on human blood monocytes, providing a molecular mechanism for inhibiting monocyte migration after pathogen recognition.[122] Moreover, it seems that TLRs also directly regulate β-arrestins. For instance, activation of TLR2 and TLR4 significantly decreased β-arrestin 2 protein expression in macrophages.[24]

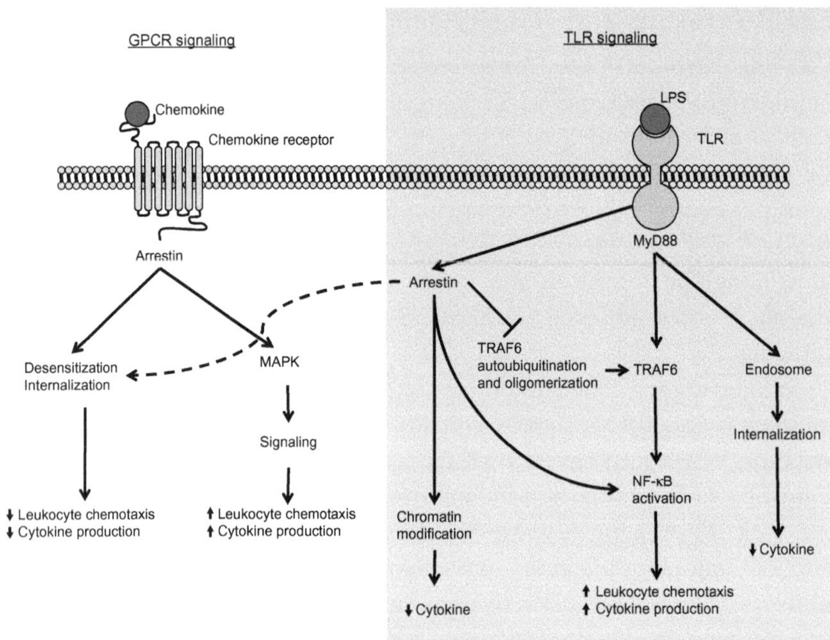

Figure 14.1 β-Arrestins regulate GPCR and TLR signaling during inflammation. Upon binding of chemokines to their respective receptors, β-arrestins are recruited to the receptors and regulate chemotaxis of leukocytes into inflammatory sites and cytokine release in two ways: mediating receptor desensitization and internalization, and acting as signaling scaffolds to facilitate numerous effector pathways. Classically, LPS–TLR interaction induces signaling cascades including TRAF6 and NF-κB activation, leading to an inflammatory response. LPS–TLR binding also recruits β-arrestins and the latter mediate TLR signaling through regulating TRAF6 and NF-κB, as well as chromatin modification. Furthermore, β-arrestins are able to mediate the cross talk to chemokine receptors.

5.4. Other non-GPCR signaling pathways

Peroxisome proliferator-activated receptors are nuclear receptor proteins, regulating metabolism and inflammation. β-Arrestin 1 binds peroxisome proliferator-activated receptor-γ (PPARγ) to elicit the repression of PPARγ/9-*cis*-retinoic acid receptor-α function and to promote PPARγ/nuclear receptor corepressor function in PPARγ-mediated adipogenesis and inflammatory cytokine expression.[123]

TNFα binds to its receptor, a type I transmembrane protein, recruiting death domain-containing proteins and leading to activation of apoptotic pathways. As TNFα is such a pleiotropic cytokine, the interaction of TNF and its cognate receptor regulates an array of biological functions including apoptosis, immune regulation, and inflammatory responses. β-Arrestin 1 can function as a signaling molecule in the TNFα action cascade that stimulates ERK activation and lipolysis, mediates phosphatidylinositol 3-kinase activation and inflammatory gene expression.[60] β-Arrestins 1 and 2 also bind and prevent degradation of IκBα, inhibiting NF-κB activation[124,125] and leading to the suppression of TNFα-induced proinflammatory cytokines. A subsequent report confirmed that although TNF receptor-1 is a non-GPCR, it functions through a Gαq/11 signaling complex and ERK activation, mediating TNFα-induced phosphatidylinositol 3-kinase activation and inflammatory gene expression.[126] β-Arrestin 1 was found to interact with STAT1 upon interferon-γ stimulation.[127] β-Arrestin 1 accelerates STAT1 tyrosine dephosphorylation by recruiting T-cell protein tyrosine phosphatase (TC45) and negatively regulates interferon-γ-induced gene transcription.[127] Interestingly, β-arrestin 2 did not have a similar effect in regulating interferon-γ signaling, although β-arrestin 2 also binds to STAT1.

TGF-β binds to TGF-β receptors, regulating cell growth, differentiation, and immune responses. β-Arrestin 2, but not β-arrestin 1, binds to the phosphorylated type III TGF-β receptor and mediates endocytosis of the type II TGF-β receptor/type III TGF-β receptor complex.[128] Further study suggested that β-arrestin 2 modulates the association of type III TGF-β receptor with ALK6 and ALK3 and enhances ALK6-mediated bone morphogenetic protein 2 signaling.[129] TGF-β signaling mediates immune regulation, inflammatory responses, and tissue fibrosis. However, functional studies on β-arrestin-mediating TGF-β signaling are largely lacking. Interaction of β-arrestin 2 and type III TGF-β receptors regulates epithelial cell migration through activation of Cdc42.[130] In a tissue fibrosis model, our initial findings suggest that deficiency in β-arrestins may not affect TGF-β-induced matrix production.[131] Further analysis, such as prolonged treatment with TGF-β, may shed some light on the role of TGF-β and β-arrestins in inflammatory and fibrotic diseases. In addition, β-arrestins have been shown to regulate the orexin-1 receptor[132] and the TCR/CD28 coreceptor complex.[133] Through interactions with these signaling pathways, β-arrestins regulate inflammatory responses.

6. ROLE OF β-ARRESTINS IN HUMAN DISEASES

The involvement of β-arrestins in a number of inflammatory diseases has been investigated using human tissues and animal models. Table 14.1 summarizes the animal models of inflammatory disease that have been studied using β-arrestin 1 or β-arrestin 2 knockout mice.

6.1. Pathogen recognition and clearance

β-Arrestins play a role in bacterial-induced inflammation and clearance, such as in sepsis,[34,134,135] and *Cryptococcus* infections.[27] For example, mice deficient in β-arrestin-2 showed decreased mortality and reduced infections. Therefore, inhibition of β-arrestins may be used to limit bacterial and fungal infections in some instances.[27,69]

6.1.1 Neisseria meningitidis

N. meningitidis, a commensal bacterium causing cerebrospinal meningitis, is an intracellular human-specific pathogen that functions through Type IV pilus-mediated adhesion to human brain endothelial cells, leading to the opening of intercellular junctions and allowing meningeal colonization.[136] The signaling receptor activated by the pathogen may be the β2AR.[69] Studies have suggested that the β2AR/β-arrestin signaling pathway may mediate bacterial colonization.[136,137] Type IV pili of *N. meningitidis* are able to activate the β2AR/β-arrestin signaling pathway, thus recruiting the polarity complex and the cell junction proteins and allowing the opening of a paracellular route in endothelial cells.[138] By this process, *N. meningitidis* is able to cross the blood–brain barrier and affect the central nervous system.[138] Activation of β2AR endocytosis with specific agonists prevents signaling events downstream of *N. meningitidis* adhesion and inhibits bacterial crossing of the endothelial barrier.[69] Therefore, targeting the β2AR/β-arrestin signaling pathway may be used for treatment and prevention of meningococcal infection.[69]

Table 14.1 Disease models with β-arrestin 1- and β-arrestin 2-deficient mice

Disease	Disease model	Mouse model	Key phenotype	References
Lung fibrosis	Bleomycin-induced lung fibrosis	$\beta arr1^{-/-}$ or $\beta arr2^{-/-}$	Reduced susceptibility and lung fibrosis	131
Asthma	Ovalbumin sensitization model of asthma	$\beta arr2^{-/-}$	Reduced airway hyperresponsiveness	55,66
Infection	Cytomegalovirus infection	$\beta arr2^{-/-}$	Reduced cytomegalovirus infection	37
Sepsis	Galactosamine-D sensitized model of endotoxic shock	$\beta arr2^{-/-}$	More susceptible to endotoxic shock	111
Sepsis	LPS injection or cecal ligation and puncture	$\beta arr1^{-/-}$ or $\beta arr2^{-/-}$	Protected from endotoxic shock and lethality	112,135
Wound healing	Skin air pouch model of inflammation	$\beta arr2^{-/-}$	Increased reepithelialization	36
Encephalomyelitis	Experimental autoimmune encephalomyelitis	$\beta arr1^{-/-}$	Resistant to experimental autoimmune encephalomyelitis	46
Stress	Chronic physical restraint	$\beta arr2^{-/-}$	Augments stress-induced immune suppression	62,116
Cutaneous flushing	Nicotinic acid-induced cutaneous flushing	$\beta arr1^{-/-}$	Reduced cutaneous flushing in response to nicotinic acid	171
Lung cancer	Heterotopic murine Lewis lung cancer and tail vein metastasis tumor model	$\beta arr2^{-/-}$	Increase in Lewis lung cancer tumor growth and metastasis	17
Rheumatoid arthritis	Collagen-induced arthritis	$\beta arr2^{-/-}$	More severe arthritis in collagen-induced arthritis	153

6.1.2 *Streptococcus pneumoniae*

β-Arrestin 1 has been reported to be involved in pneumococcal infection. A report showed that upon infection with *S. pneumoniae*, β-arrestin 1 moved to the plasma membrane of endothelial cells.[67] Pneumococcal invasion was enhanced in a cell line by transfection of β-arrestin 1. β-Arrestin 1 was shown to interact with the PAF receptor and contribute to successful infection by *S. pneumoniae*.[67]

6.1.3 *Polymicrobial sepsis*

β-Arrestin 2 protein expression was reduced in a murine model of CLP-induced polymicrobial sepsis.[134] β-Arrestin-2-deficient mice exhibited decreased mortality, increased IL-6 production, and elevated cecal myeloperoxidase in a mouse model of polymicrobial sepsis by CLP. Splenocytes devoid of β-arrestin 2 produced higher levels of TNF-α, IL-6, and IL-10.[135] This is reminiscent of another study showing that CLP-induced myeloperoxidase was increased in β-arrestin-2-deficient mouse lung.[34] β-Arrestin 1 and β-arrestin-2-deficient mice were protected from TLR4-mediated endotoxic shock and lethality.[112] These studies suggest that β-arrestin 2 is a negative regulator in polymicrobial sepsis. Therefore, treatment with flavocoxid preserved β-arrestin 2 expression and reduced cytokine production, leading to protection against tissue damage induced by CLP, presumably by decreasing myeloperoxidase activity in the lung and the liver.[134]

As chemokine receptors play key roles in pathogen recognition and β-arrestins regulate chemokine receptor internalization, it is not surprising that β-arrestins regulate chemokine receptor-mediated pathogen clearance. For example, during *Staphylococcus aureus* associated lipoteichoic acid induced pathogen recognition and monocyte migration, β-arrestins mediate endocytosis of CCR5 into recycling endosomes, thus facilitating the TLR2-negative regulation of CCR1, CCR2, and CCR5 on monocytes.[122]

6.1.4 *Cryptococcal meningitis*

β-Arrestin 2 mRNA and protein levels were significantly elevated in peripheral blood mononuclear cells of patients with cryptococcal meningitis. The increased β-arrestin 2 levels were concurrent with reduced cytotoxic activity against *Cryptococcus*, increased IL-10 levels, and reduced IFN-γ expression in patient serum.[27] This study suggests that increasing IFN-γ by the inhibition of β-arrestin 2 could be a therapeutic strategy in *Cryptococcus* infections.[27]

6.1.5 Viral responses

Reports suggest that β-arrestins mediate antiviral responses. β-Arrestin 1 accelerates STAT1 tyrosine dephosphorylation by recruiting T-cell protein tyrosine phosphatase (TC45) and negatively regulates IFN-γ-induced gene transcription in IFN-γ-induced cellular antiviral responses.[127] β-Arrestin 1 knockdown promotes the antiviral immune response in vesicular stomatitis virus-infected cells.[127] β-Arrestin 2 plays a negative role in HIV-1 gp120 and opioid-induced lymphocyte cell death.[60] Increased or decreased β-arrestin 2 expression inhibited or enhanced gp120/morphine-induced apoptosis, respectively.[60] β-Arrestins 1 and 2 can function as both positive and negative regulators of adenovirus-induced innate immune responses both *in vivo* and *in vitro*.[114]

6.2. Fibrosis

The molecular mechanisms that control progressive tissue fibrosis remain poorly understood. Idiopathic pulmonary fibrosis (IPF) is a progressive disease and a major cause of morbidity and mortality. Progressive pulmonary fibrosis results from numerous microinjuries to the alveolar epithelia that lead to excessive fibroblast activity. Lung epithelial cell apoptosis is believed to be responsible for the initiation of fibrogenesis. β-Arrestins are known to regulate cell apoptosis (reviewed in Refs. 19,57,139). Some GPCRs are reported to have a role in lung epithelial apoptosis. For instance, angiotensin II receptors are expressed by lung alveolar epithelial cells, and the interaction of angiotensin II with its receptors mediates alveolar epithelial cell apoptosis.[140,141] However, it is unclear whether β-arrestins mediate type II epithelial cell apoptosis during lung fibrosis. During tissue injury, TGF-β is released by fibroblasts and the epithelium. There is abundant evidence that TGF-β is critical for the progression of pulmonary fibrosis due to its role in regulating collagen synthesis, fibroblast proliferation, apoptosis, and myofibroblast differentiation.[142–144] β-Arrestin 2, but not β-arrestin 1, mediates endocytosis of the type II TGF-β receptor/type III TGF-β receptor complex and downregulates TGF-β signaling[128] and bone morphogenetic protein 2 signaling.[129] In a lung fibrosis model, our initial findings suggest that deficiency in β-arrestins may not affect TGF-β-induced matrix production.[131] Further detailed studies, such as prolonged treatment with TGF-β, may shed some light on the role of β-arrestins in TGF-β signaling in inflammatory and fibrotic diseases (Fig. 14.2).

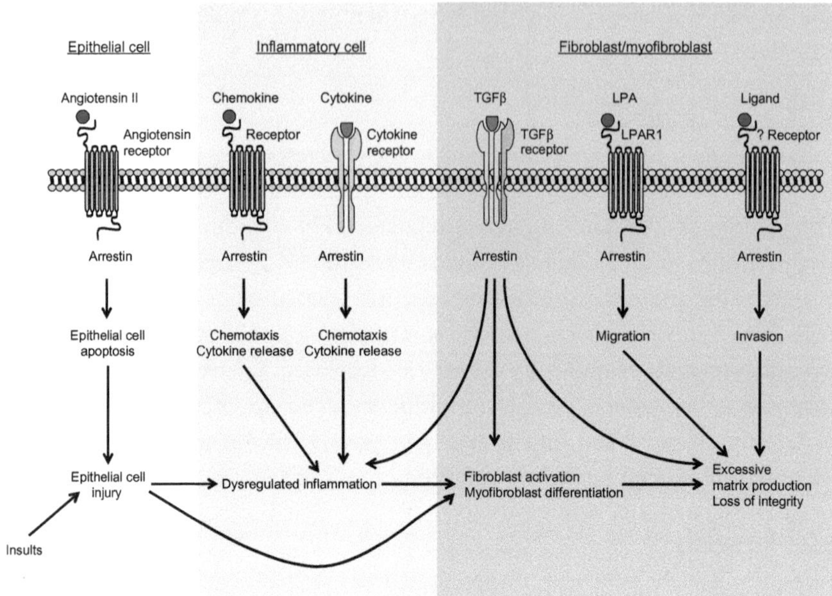

Figure 14.2 β-Arrestins mediate signaling pathways during fibrogenesis. A simplified model of tissue fibrosis proposes that epithelial cell injury and apoptosis lead to dysregulated inflammatory responses and fibroblast/myofibroblast activation, resulting in excessive matrix production and eventually fibrosis. β-Arrestins regulate signaling pathways mediating cell injury and apoptosis in epithelial cells, inflammatory responses in immune cells, and fibroblast migration and invasiveness, matrix production in effector cells.

Fibroblast and myofibroblast migration is an important process during fibrogenesis. Chemokines and chemokine receptors regulate fibroblast migration.[145,146] Furthermore, lysophosphatidic acid in bronchoalveolar lavage fluid interacts with its receptor LPAR1, regulating fibroblast migration during fibrogenesis.[147] β-Arrestins regulate chemokine signaling as well as LPAR1 signaling. The role of β-arrestins in fibroblast migration has not been explored. We have shown that deletion of β-arrestins did not significantly affect fibroblast migration.[131]

The destruction of alveolar basement membrane was observed in experimental lung fibrosis and the biopsies of patients with lung fibrosis.[148] Moreover, fibroblasts and myofibroblasts from patients with progressive lung fibrosis have been shown to have the ability to invade extracellular matrix in the manner of metastatic cancer cells.[148] β-Arrestins have been shown to play a role in tumor metastasis.[16,17] For example, β-arrestin signaling mediates lysophosphatidic acid-induced cancer cell migration and invasion.[101] We found that fibroblasts from bleomycin-treated β-arrestin-deficient mice and human IPF fibroblasts with β-arrestin 2 knockdown failed to invade extracellular matrix.[131] Furthermore, we showed that loss of β-arrestin 1 or β-arrestin 2 in primary lung fibroblasts resulted in altered expression of genes involved in matrix production, basement membrane degradation, and cell adhesion.[131] Therefore, this study suggests that inhibiting β-arrestin expression in lung fibroblasts would be a potential therapy for IPF and other types of pulmonary fibrosis.

β-Arrestin 1 expression is increased in T cells from primary biliary cirrhosis patients. There is also a positive correlation between the levels of β-arrestin 1 mRNA in peripheral blood mononuclear cells and progression of cirrhosis.[59] β-Arrestin 1 was overexpressed in autoreactive T cells from cirrhotic patients and augmented cell proliferation, increased interferon release, and regulated autoimmune-related gene expression.[59] The study suggests that β-arrestin 1 overexpression contributes to the pathogenesis of primary biliary cirrhosis. Angiotensin II binds the angiotensin II type 1A receptor to mediate inflammatory and fibrotic responses.[83] β-Arrestins regulate desensitization and sequestration of AGTR1.[85,149] In cirrhosis, impaired vascular reactivity leads to vasodilation and contributes to portal hypertension. Vascular hyporesponsiveness to angiotensin II in CCL4-treated rats is due to enhanced interaction of the AGTR1 with β-arrestin 2 and alterations in receptor function.[150] Silencing β-arrestins might be a therapeutic strategy for cirrhosis.

6.3. Arthritis

Rheumatoid arthritis is an autoimmune disease characterized by joint inflammation and destruction. GPCRs, such as chemokine receptors, G-protein-coupled receptor kinases, and β-arrestin1 and 2 are involved in the pathogenesis of rheumatoid arthritis. Reduction in GRK activity with a decreased level of GRK-2 and GRK-6 protein expression was found in patients with rheumatoid arthritis, while β-arrestin-1 expression was unchanged between rheumatoid arthritis patients and controls.[151] In a rat adjuvant arthritis model, the expression of β-arrestin 1 in splenocytes was increased at the most severe period of the disease.[152] Using collagen-induced arthritis and human TNFα transgenic mouse models, another group demonstrated that expression of β-arrestin 1 and 2 was elevated in joint tissue.[153] Furthermore, β-arrestins differentially regulated inflammatory cytokine production by fibroblast-like synoviocytes, in which β-arrestin 1 increased

TNFα and IL-6 production, while β-arrestin decreased their production by synoviocytes.[153] Finally, mice deficient in β-arrestin 2 exhibited more severe arthritis in the collagen-induced arthritis model.[153]

The role of chemokine receptor CCR2 and β-arrestins has been investigated in inflammatory arthritis, with some studies suggesting that blocking CCR2 signaling may ameliorate arthritis in animal models. A fusion protein using GM-CSF and an N-terminal truncated version of CCL2 binding to CCR2 led to altered conformational changes in the CCR2 homodimer and failed to induce the recruitment of β-arrestin 2 to the receptor.[63] The fusion peptide was capable of blocking IL-17 production *in vitro* and attenuated inflammatory arthritis *in vivo* in mice.[63] This study suggests that suppressing expression of β-arrestins may protect joints from rheumatoid arthritis.

Thus, strategies that regulate β-arrestin expression or activity may be developed for patients with rheumatoid arthritis. For example, Paeoniflorin, a monoterpene glucoside that decreased expression of β-arrestin 1 and β-arrestin 2, and increased β2AR and GRK2 in mesenteric lymph node lymphocytes, significantly attenuated arthritis scores and reversed the changes of cytokines.[154]

6.4. Multiple sclerosis

Multiple sclerosis is characterized by the presence of plaques of demyelination throughout the central nervous system. The destruction of myelin is believed to involve autoimmune mechanisms. β-Arrestin 1 autoantibodies have been found in sera from multiple sclerosis patients.[155] Multiple sclerosis patients had a greater prevalence of positive T-cell proliferative responses to β-arrestin than healthy controls.[156] These studies suggest a role for β-arrestin 1 in the pathogenesis of multiple sclerosis. β-Arrestin 1 expression was increased in brains of multiple sclerosis patients relative to nonmultiple sclerosis brains.[157] This suggests that β-arrestin 1 may be a negative regulator in multiple sclerosis. In contrast, another study showed that β-arrestin 1 mRNA levels were reduced by phytohemagglutinin, but increased in interferon β-1a-treated mononuclear leukocytes.[26] As IFNβ-1a is known to ameliorate the course of multiple sclerosis, β-arrestin 1 may be a downstream mediator in multiple sclerosis. Additional studies suggest that β-arrestin 1 is critical for $CD4^+$ T-cell survival and may be a factor in susceptibility to autoimmunity.[46]

6.5. Encephalomyelitis

β-Arrestin protein expression was significantly increased in splenocytes following myelin oligodendrocyte glycoprotein-induced experimental autoimmune encephalomyelitis (EAE), while GRK2, GRK6, and A1 adenosine receptors were decreased during the disease.[157,158] Mice deficient in β-arrestin 1 were more resistant to EAE, whereas overexpression of β-arrestin 1 increased susceptibility to the disease.[46] Glucocorticoid treatment eliminated EAE-induced neuroinflammation and neurobehavioral deficits, and also reduced β-arrestin 1 expression and enhanced A1 adenosine receptor expression.[157] These results suggest that β-arrestins may play a pathogenic role in EAE.

6.6. Acute rejection in organ transplant

β-Arrestins 1 and 2 are expressed by intravascular renal graft monocytes and T cells in a rat model of kidney transplantation, and β-arrestin 1 and 2 expression is decreased in monocytes during rejection. Decreased β-arrestin 2 expression was concurrent with depression of I-κB and augmentation of NF-κB.[28] These data suggest that the upregulation of β-arrestin 2 in renal isografts may suppress the activation of blood monocytes via NF-κB activation. Ischemic injury following heart transplantation reduced β-arrestin 1 expression.[159] The immunomodulatory peptide α-melanocyte-stimulating hormone was able to preserve β-arrestin 1 protein levels[159] and reduced damage in transplanted heart grafts.

6.7. Asthma

Asthma is characterized by chronic airway inflammation and airway hyperresponsiveness. Studies have showed that Th2 cytokines such as IL-4 and IL-13 play a key role in the pathogenesis. β-Arrestin 2 regulates T-cell function in human asthma and animal models of allergic asthma. β-Arrestin 2 knockout mice are protected from OVA-induced allergic asthmatic inflammation, but not LPS-mediated nonallergic inflammation, suggesting that neutrophil recruitment is not dependent on β-arrestin 2. β-Arrestin 2-deficient mice are competent in their ability to present antigen, to generate antigen-specific T-cell responses, and to undergo Ig isotype switching, but they have defective macrophage-derived chemokine-mediated $CD4^+$ T lymphocytes migration to sites of inflammation.[55] A subsequent study determined the cell types required for β-arrestin 2-dependent allergic pathology and inflammation. By using bone marrow chimera mice, the same group was able to show expression of β-arrestin 2 in eosinophilic and lymphocytic cells contributes to airway inflammation, while expression of β-arrestin 2 in structural cells contributes to airway hyperresponsiveness.[66] Another study investigated the role of β-arrestin 2 in $CD4^+$ T lymphocytes and found that β-arrestin 2 regulated IL-4 production and GATA3 expression.[47]

β-Arrestins can regulate β-agonist-induced desensitization of airway smooth muscle β2ARs, which is the cause of loss of bronchoprotective effect and deterioration of asthma control by β-agonists. β-Arrestin-2-deficient mice show augmented β-agonist-mediated airway smooth muscle relaxation and increased β-agonist-stimulated cyclic adenosine monophosphate production. Knockdown of β-arrestins in cultured human airway smooth muscle cells produced increased β-agonist-stimulated cyclic adenosine monophosphate production. Collectively, these studies suggest that β-arrestins regulate β2AR signaling and function in airway smooth muscle.[160–162]

IL-17 is a proinflammatory cytokine that plays a critical role in the pathogenesis of allergic asthma. A recent study found that β-arrestin 2 and IL-17 expression in $CD4^+$ T cells from a murine asthma model were increased compared with those from wild-type mice.[64] Moreover, β-arrestin 2 stimulated IL-17 production and expression of $CD4^+$ T lymphocytes in a murine asthma model.[64]

6.8. Cystic fibrosis

β-Arrestin 2 protein level is increased in cystic fibrosis cell models, *Cftr*$^{-/-}$ mouse nasal epithelia, and nasal scrapes obtained from cystic fibrosis patients.[163] Elevated β-arrestin 2 expression leads to increased cAMP response element-binding protein (CREB) activation and ERK activation in βarr2-GFP expressing tracheal epithelial cells. Inhibition of ERK resulted in diminished cAMP response element activation in both cells with and without βarr2–GFP expression. Nasal epithelium excised from *Cftr* and β-arrestin 2 double knockout mice exhibit reduced pCREB and pERK levels compared with *Cftr*$^{-/-}$ mice, but similar to WT mice. Thus, β-arrestin 2 directly regulates cystic fibrosis-induced CREB activation through the ERK signaling pathway.[164]

6.9. Cardioprotection

In addition to regulating the PKA-apoptosis pathway, β1AR activation also signals through the β-arrestin–Src–MMP pathway, leading to transactivation of the EGF receptor and promoting cardiomyocyte survival.[165] Both β-arrestin 1 and β-arrestin 2 are required for β1AR transactivation of the EGF receptor.[165] Mechanical stretch in the heart triggers activation of AT1AR and β-arrestin recruitment to the receptor in the absence of ligand.[166] β-Arrestin-deficient hearts failed to induce a cardioprotective response to mechanical stretch and demonstrated enhanced myocyte apoptosis,[166] suggesting that the heart responds to mechanical stress by activating β-arrestin-mediated cell survival signals.[167,168]

7. β-ARRESTINS IN THERAPEUTIC DEVELOPMENT FOR INFLAMMATORY DISEASES

Understanding β-arrestin biology offers the potential to accelerate drug development in targeting GPCR functions (reviewed in Refs. 169,170). Several assay systems have been proposed and tested to target β-arrestins. For example, β-arrestin conformational changes were used to develop a biased agonist to the GPR109A receptor that retains the therapeutic lipolytic effect, but is devoid of cutaneous side effects.[171]

RNA interference targeting β-arrestins has been used to demonstrate the role of arrestins in GPCR desensitization, internalization, and signaling functions.[125,172] Furthermore, these specific siRNAs have been used in *in vivo* as well as *in vitro* settings. Silencing β-arrestin 2 with RNA interference in allergic asthmatic mice reduced Th2 cytokines such as IL-4.[47] Silencing β-arrestin 2 reduced fibroblast invasiveness.[131] Suppression of β-arrestin 2 expression using siRNA eliminated AT(1A)R- and LPAR1-mediated chemotaxis.[86]

There are few reports exploring the interplay between microRNAs and β-arrestins. For example, β-arrestin 2-mediated ERK phosphorylation is required for the downregulation of miR-190.[173] Expression of miR-326 is decreased in human glioblastomas and correlated with decreased expression of the host gene β-arrestin 1.[174] Although there are no reports identifying β-arrestins as targets of any microRNAs thus far, microRNAs regulating β-arrestin expression may prove to be a powerful strategy for therapeutic development.

Therapeutic targeting of β-arrestins has been challenging thus far. β-Arrestins have been shown to interact with an array of receptors as well as downstream proteins. For example, by proteomic analysis, there are over a hundred proteins interacting with either β-arrestin 1 or β-arrestin 2 when the AT1AR is activated.[175] Furthermore, there are 171 proteins with increased phosphorylation, and 53 with decreased phosphorylation, which represent a β-arrestin-mediated kinase-substrate network involved in a number of cellular pathways.[176] However, the protein impact of β-arrestins in the pathobiology of a variety of disease processes suggests efforts at developing agonists or antagonists will continue.

REFERENCES

1. Luttrell LM, Lefkowitz RJ. The role of beta-arrestins in the termination and transduction of G-protein-coupled receptor signals. *J Cell Sci*. 2002;115:455–465.
2. Han M, Gurevich VV, Vishnivetskiy SA, Sigler PB, Schubert C. Crystal structure of beta-arrestin at 1.9 A: possible mechanism of receptor binding and membrane translocation. *Structure*. 2001;9:869–880.
3. Xiao K, Shenoy SK, Nobles K, Lefkowitz RJ. Activation-dependent conformational changes in {beta}-arrestin 2. *J Biol Chem*. 2004;279:55744–55753.
4. Vishnivetskiy SA, Schubert C, Climaco GC, Gurevich YV, Velez MG, Gurevich VV. An additional phosphate-binding element in arrestin molecule. Implications for the mechanism of arrestin activation. *J Biol Chem*. 2000;275:41049–41057.
5. Pitcher JA, Freedman NJ, Lefkowitz RJ. G protein-coupled receptor kinases. *Annu Rev Biochem*. 1998;67:653–692.
6. Lohse MJ, Benovic JL, Codina J, Caron MG, Lefkowitz RJ. beta-Arrestin: a protein that regulates beta-adrenergic receptor function. *Science*. 1990;248:1547–1550.
7. Galli SJ, Nakae S, Tsai M. Mast cells in the development of adaptive immune responses. *Nat Immunol*. 2005;6:135–142.
8. McDonald PH, Chow CW, Miller WE, et al. Beta-arrestin 2: a receptor-regulated MAPK scaffold for the activation of JNK3. *Science*. 2000;290:1574–1577.
9. Luttrell LM, Roudabush FL, Choy EW, et al. Activation and targeting of extracellular signal-regulated kinases by beta-arrestin scaffolds. *Proc Natl Acad Sci USA*. 2001;98:2449–2454.
10. Pierce KL, Lefkowitz RJ. Classical and new roles of beta-arrestins in the regulation of G-protein-coupled receptors. *Nat Rev Neurosci*. 2001;2:727–733.
11. Miller WE, Lefkowitz RJ. Expanding roles for beta-arrestins as scaffolds and adapters in GPCR signaling and trafficking. *Curr Opin Cell Biol*. 2001;13:139–145.
12. Ma L, Pei G. Beta-arrestin signaling and regulation of transcription. *J Cell Sci*. 2007;120:213–218.
13. Kang J, Shi Y, Xiang B, et al. A nuclear function of beta-arrestin1 in GPCR signaling: regulation of histone acetylation and gene transcription. *Cell*. 2005;123:833–847.
14. Chen W, ten Berge D, Brown J, et al. Dishevelled 2 recruits beta-arrestin 2 to mediate Wnt5A-stimulated endocytosis of Frizzled 4. *Science*. 2003;301:1391–1394.
15. Chen W, Ren XR, Nelson CD, et al. Activity-dependent internalization of smoothened mediated by beta-arrestin 2 and GRK2. *Science*. 2004;306:2257–2260.
16. Buchanan FG, Gorden DL, Matta P, Shi Q, Matrisian LM, DuBois RN. Role of beta-arrestin 1 in the metastatic progression of colorectal cancer. *Proc Natl Acad Sci USA*. 2006;103:1492–1497.
17. Raghuwanshi SK, Nasser MW, Chen X, Strieter RM, Richardson RM. Depletion of beta-arrestin-2 promotes tumor growth and angiogenesis in a murine model of lung cancer. *J Immunol*. 2008;180:5699–5706.
18. DeFea KA. Stop that cell! Beta-arrestin-dependent chemotaxis: a tale of localized actin assembly and receptor desensitization. *Annu Rev Physiol*. 2007;69:535–560.

19. DeWire SM, Ahn S, Lefkowitz RJ, Shenoy SK. Beta-arrestins and cell signaling. *Annu Rev Physiol.* 2007;69:483–510.
20. Yue R, Kang J, Zhao C, et al. Beta-arrestin1 regulates zebrafish hematopoiesis through binding to YY1 and relieving polycomb group repression. *Cell.* 2009;139:535–546.
21. Lattin J, Zidar DA, Schroder K, Kellie S, Hume DA, Sweet MJ. G-protein-coupled receptor expression, function, and signaling in macrophages. *J Leukoc Biol.* 2007;82:16–32.
22. Parruti G, Peracchia F, Sallese M, et al. Molecular analysis of human beta-arrestin-1: cloning, tissue distribution, and regulation of expression. Identification of two isoforms generated by alternative splicing. *J Biol Chem.* 1993;268:9753–9761.
23. Kizaki T, Izawa T, Sakurai T, et al. Beta2-adrenergic receptor regulates Toll-like receptor-4-induced nuclear factor-kappaB activation through beta-arrestin 2. *Immunology.* 2008;124:348–356.
24. Loniewski K, Shi Y, Pestka J, Parameswaran N. Toll-like receptors differentially regulate GPCR kinases and arrestins in primary macrophages. *Mol Immunol.* 2008;45:2312–2322.
25. Wang J, Guan E, Roderiquez G, Calvert V, Alvarez R, Norcross MA. Role of tyrosine phosphorylation in ligand-independent sequestration of CXCR4 in human primary monocytes-macrophages. *J Biol Chem.* 2001;276:49236–49243.
26. Giorelli M, Livrea P, Defazio G, et al. Interferon beta-1a counteracts effects of activation on the expression of G-protein-coupled receptor kinases 2 and 3, beta-arrestin-1, and regulators of G-protein signalling 2 and 16 in human mononuclear leukocytes. *Cell Signal.* 2002;14:673–678.
27. Xia R, Hu Z, Sun Y, et al. Overexpression of beta-arrestin 2 in peripheral blood mononuclear cells of patients with cryptococcal meningitis. *J Interferon Cytokine Res.* 2010;30:155–162.
28. Zakrzewicz A, Krasteva G, Wilhelm J, et al. Reduced expression of arrestin beta 2 by graft monocytes during acute rejection of rat kidneys. *Immunobiology.* 2011; 216:854–861.
29. Cheung R, Malik M, Ravyn V, Tomkowicz B, Ptasznik A, Collman RG. An arrestin-dependent multi-kinase signaling complex mediates MIP-1beta/CCL4 signaling and chemotaxis of primary human macrophages. *J Leukoc Biol.* 2009;86:833–845.
30. Aragay AM, Mellado M, Frade JM, et al. Monocyte chemoattractant protein 1 induced CCR2B receptor desensitization mediated by the G protein-coupled receptor kinase 2. *Proc Natl Acad Sci USA.* 1998;95:2985–2990.
31. Ge L, Ly Y, Hollenberg M, DeFea K. A beta-arrestin-dependent scaffold is associated with prolonged MAPK activation in pseudopodia during protease-activated receptor-2-induced chemotaxis. *J Biol Chem.* 2003;278:34418–34426.
32. Faurschou M, Borregaard N. Neutrophil granules and secretory vesicles in inflammation. *Microbes Infect.* 2003;5:1317–1327.
33. Mizgerd JP. Molecular mechanisms of neutrophil recruitment elicited by bacteria in the lungs. *Semin Immunol.* 2002;14:123–132.
34. Basher F, Fan H, Zingarelli B, et al. beta-Arrestin 2: a negative regulator of inflammatory responses in polymorphonuclear leukocytes. *Int J Clin Exp Med.* 2008; 1:32–41.
35. Barlic J, Andrews JD, Kelvin AA, et al. Regulation of tyrosine kinase activation and granule release through beta-arrestin by CXCRI. *Nat Immunol.* 2000;1:227–233.
36. Su Y, Raghuwanshi SK, Yu Y, Nanney LB, Richardson RM, Richmond A. Altered CXCR2 signaling in beta-arrestin-2-deficient mouse models. *J Immunol.* 2005;175: 5396–5402.
37. Yu MC, Su LL, Zou L, et al. An essential function for beta-arrestin 2 in the inhibitory signaling of natural killer cells. *Nat Immunol.* 2008;9:898–907.

38. Bari R, Bell T, Leung WH, et al. Significant functional heterogeneity among KIR2DL1 alleles and a pivotal role of arginine 245. *Blood.* 2009;114:5182–5190.
39. Jacob C, Yang PC, Darmoul D, et al. Mast cell tryptase controls paracellular permeability of the intestine. Role of protease-activated receptor 2 and beta-arrestins. *J Biol Chem.* 2005;280:31936–31948.
40. Wagner E, Frank MM. Therapeutic potential of complement modulation. *Nat Rev Drug Discov.* 2010;9:43–56.
41. Vibhuti A, Gupta K, Subramanian H, Guo Q, Ali H. Distinct and shared roles of beta-arrestin-1 and beta-arrestin-2 on the regulation of C3a receptor signaling in human mast cells. *PLoS One.* 2011;6:e19585.
42. Ahamed J, Haribabu B, Ali H. Cutting edge: differential regulation of chemoattractant receptor-induced degranulation and chemokine production by receptor phosphorylation. *J Immunol.* 2001;167:3559–3563.
43. Bamberg CE, Mackay CR, Lee H, et al. The C5a receptor (C5aR) C5L2 is a modulator of C5aR-mediated signal transduction. *J Biol Chem.* 2010;285:7633–7644.
44. Lattin JE, Greenwood KP, Daly NL, et al. Beta-arrestin 2 is required for complement C1q expression in macrophages and constrains factor-independent survival. *Mol Immunol.* 2009;47:340–347.
45. Fong AM, Premont RT, Richardson RM, Yu YR, Lefkowitz RJ, Patel DD. Defective lymphocyte chemotaxis in beta-arrestin2- and GRK6-deficient mice. *Proc Natl Acad Sci USA.* 2002;99:7478–7483.
46. Shi Y, Feng Y, Kang J, et al. Critical regulation of CD4+ T cell survival and autoimmunity by beta-arrestin 1. *Nat Immunol.* 2007;8:817–824.
47. Wang G, Liu Y, Yang M, et al. Effects of beta-arrestin 2 on cytokine production of CD4+ T lymphocytes of mice with allergic asthma. *Indian J Exp Biol.* 2011;49:585–593.
48. Bjorgo E, Solheim SA, Abrahamsen H, et al. Cross talk between phosphatidylinositol 3-kinase and cyclic AMP (cAMP)-protein kinase a signaling pathways at the level of a protein kinase B/beta-arrestin/cAMP phosphodiesterase 4 complex. *Mol Cell Biol.* 2010;30:1660–1672.
49. Bjorgo E, Moltu K, Tasken K. Phosphodiesterases as targets for modulating T-cell responses. *Handb Exp Pharmacol.* 2011;204:345–363.
50. Abrahamsen H, Baillie G, Ngai J, et al. TCR- and CD28-mediated recruitment of phosphodiesterase 4 to lipid rafts potentiates TCR signaling. *J Immunol.* 2004; 173:4847–4858.
51. Bjorgo E, Tasken K. Role of cAMP phosphodiesterase 4 in regulation of T-cell function. *Crit Rev Immunol.* 2006;26:443–451.
52. Baillie GS, Houslay MD. Arrestin times for compartmentalised cAMP signalling and phosphodiesterase-4 enzymes. *Curr Opin Cell Biol.* 2005;17:129–134.
53. von Andrian UH, Mackay CR. T-cell function and migration. Two sides of the same coin. *N Engl J Med.* 2000;343:1020–1034.
54. Sun Y, Cheng Z, Ma L, Pei G. Beta-arrestin2 is critically involved in CXCR4-mediated chemotaxis, and this is mediated by its enhancement of p38 MAPK activation. *J Biol Chem.* 2002;277:49212–49219.
55. Walker JK, Fong AM, Lawson BL, et al. Beta-arrestin-2 regulates the development of allergic asthma. *J Clin Invest.* 2003;112:566–574.
56. Povsic TJ, Kohout TA, Lefkowitz RJ. Beta-arrestin1 mediates insulin-like growth factor 1 (IGF-1) activation of phosphatidylinositol 3-kinase (PI3K) and anti-apoptosis. *J Biol Chem.* 2003;278:51334–51339.
57. Revankar CM, Vines CM, Cimino DF, Prossnitz ER. Arrestins block G protein-coupled receptor-mediated apoptosis. *J Biol Chem.* 2004;279:24578–24584.
58. DeFea KA, Vaughn ZD, O'Bryan EM, Nishijima D, Dery O, Bunnett NW. The proliferative and antiapoptotic effects of substance P are facilitated by formation of a

beta-arrestin-dependent scaffolding complex. *Proc Natl Acad Sci USA*. 2000;97: 11086–11091.
59. Hu Z, Huang Y, Liu Y, et al. beta-Arrestin 1 modulates functions of autoimmune T cells from primary biliary cirrhosis patients. *J Clin Immunol*. 2011;31:346–355.
60. Moorman J, Zhang Y, Liu B, et al. HIV-1 gp120 primes lymphocytes for opioid-induced, beta-arrestin 2-dependent apoptosis. *Biochim Biophys Acta*. 2009;1793: 1366–1371.
61. Yamashita M, Kimura M, Kubo M, et al. T cell antigen receptor-mediated activation of the Ras/mitogen-activated protein kinase pathway controls interleukin 4 receptor function and type-2 helper T cell differentiation. *Proc Natl Acad Sci USA*. 1999; 96:1024–1029.
62. Li H, Smalligan DA, Xie N, et al. beta-Arrestin 2-mediated immune suppression induced by chronic stress. *Neuroimmunomodulation*. 2011;18:142–149.
63. Rafei M, Berchiche YA, Birman E, et al. An engineered GM-CSF-CCL2 fusokine is a potent inhibitor of CCR2-driven inflammation as demonstrated in a murine model of inflammatory arthritis. *J Immunol*. 2009;183:1759–1766.
64. Liu Y, Wang GY, Liu SK, et al. beta-Arrestin2 stimulates interleukin-17 production and expression of CD4+ T lymphocytes in a murine asthma model. *Iran J Allergy Asthma Immunol*. 2011;10:171–182.
65. Decaillot FM, Kazmi MA, Lin Y, Ray-Saha S, Sakmar TP, Sachdev P. CXCR7/ CXCR4 heterodimer constitutively recruits beta-arrestin to enhance cell migration. *J Biol Chem*. 2011;286:32188–32197.
66. Hollingsworth JW, Theriot BS, Li Z, et al. Both hematopoietic-derived and non-hematopoietic-derived {beta}-arrestin-2 regulates murine allergic airway disease. *Am J Respir Cell Mol Biol*. 2010;43:269–275.
67. Radin JN, Orihuela CJ, Murti G, Guglielmo C, Murray PJ, Tuomanen EI. beta-Arrestin 1 participates in platelet-activating factor receptor-mediated endocytosis of Streptococcus pneumoniae. *Infect Immun*. 2005;73:7827–7835.
68. Delekta PC, Apel IJ, Gu S, et al. Thrombin-dependent NF-{kappa}B activation and monocyte/endothelial adhesion are mediated by the CARMA3.Bcl10.MALT1 signalosome. *J Biol Chem*. 2010;285:41432–41442.
69. Coureuil M, Lecuyer H, Scott MG, et al. Meningococcus hijacks a beta2-adrenoceptor/beta-arrestin pathway to cross brain microvasculature endothelium. *Cell*. 2010;143:1149–1160.
70. Neptune ER, Bourne HR. Receptors induce chemotaxis by releasing the betagamma subunit of Gi, not by activating Gq or Gs. *Proc Natl Acad Sci USA*. 1997; 94:14489–14494.
71. Oppermann M. Chemokine receptor CCR5: insights into structure, function, and regulation. *Cell Signal*. 2004;16:1201–1210.
72. Barlic J, Khandaker MH, Mahon E, et al. beta-Arrestins regulate interleukin-8-induced CXCR1 internalization. *J Biol Chem*. 1999;274:16287–16294.
73. Richardson RM, Marjoram RJ, Barak LS, Snyderman R. Role of the cytoplasmic tails of CXCR1 and CXCR2 in mediating leukocyte migration, activation, and regulation. *J Immunol*. 2003;170:2904–2911.
74. Fan GH, Yang W, Wang XJ, Qian Q, Richmond A. Identification of a motif in the carboxyl terminus of CXCR2 that is involved in adaptin 2 binding and receptor internalization. *Biochemistry*. 2001;40:791–800.
75. Colvin RA, Campanella GS, Sun J, Luster AD. Intracellular domains of CXCR3 that mediate CXCL9, CXCL10, and CXCL11 function. *J Biol Chem*. 2004;279: 30219–30227.
76. Orsini MJ, Parent JL, Mundell SJ, Marchese A, Benovic JL. Trafficking of the HIV coreceptor CXCR4. Role of arrestins and identification of residues in the c-terminal tail that mediate receptor internalization. *J Biol Chem*. 1999;274:31076–31086.

77. McCormick PJ, Segarra M, Gasperini P, Gulino AV, Tosato G. Impaired recruitment of Grk6 and beta-arrestin 2 causes delayed internalization and desensitization of a WHIM syndrome-associated CXCR4 mutant receptor. *PLoS One*. 2009;4:e8102.
78. Minsaas L, Planaguma J, Madziva M, et al. Filamin a binds to CCR2B and regulates its internalization. *PLoS One*. 2010;5:e12212.
79. Huttenrauch F, Nitzki A, Lin FT, Honing S, Oppermann M. Beta-arrestin binding to CC chemokine receptor 5 requires multiple C-terminal receptor phosphorylation sites and involves a conserved Asp-Arg-Tyr sequence motif. *J Biol Chem*. 2002;277: 30769–30777.
80. Huttenrauch F, Pollok-Kopp B, Oppermann M. G protein-coupled receptor kinases promote phosphorylation and beta-arrestin-mediated internalization of CCR5 homo- and hetero-oligomers. *J Biol Chem*. 2005;280:37503–37515.
81. Burns JM, Summers BC, Wang Y, et al. A novel chemokine receptor for SDF-1 and I-TAC involved in cell survival, cell adhesion, and tumor development. *J Exp Med*. 2006;203:2201–2213.
82. Boldajipour B, Mahabaleshwar H, Kardash E, et al. Control of chemokine-guided cell migration by ligand sequestration. *Cell*. 2008;132:463–473.
83. Hunyady L, Catt KJ. Pleiotropic AT1 receptor signaling pathways mediating physiological and pathogenic actions of angiotensin II. *Mol Endocrinol*. 2006;20:953–970.
84. McAllister-Lucas LM, Ruland J, Siu K, et al. CARMA3/Bcl10/MALT1-dependent NF-kappaB activation mediates angiotensin II-responsive inflammatory signaling in nonimmune cells. *Proc Natl Acad Sci USA*. 2007;104:139–144.
85. Kohout TA, Lin FS, Perry SJ, Conner DA, Lefkowitz RJ. beta-Arrestin 1 and 2 differentially regulate heptahelical receptor signaling and trafficking. *Proc Natl Acad Sci USA*. 2001;98:1601–1606.
86. Hunton DL, Barnes WG, Kim J, et al. Beta-arrestin 2-dependent angiotensin II type 1A receptor-mediated pathway of chemotaxis. *Mol Pharmacol*. 2005;67:1229–1236.
87. Tohgo A, Pierce KL, Choy EW, Lefkowitz RJ, Luttrell LM. beta-Arrestin scaffolding of the ERK cascade enhances cytosolic ERK activity but inhibits ERK-mediated transcription following angiotensin AT1a receptor stimulation. *J Biol Chem*. 2002; 277:9429–9436.
88. Willoughby EA, Collins MK. Dynamic interaction between the dual specificity phosphatase MKP7 and the JNK3 scaffold protein beta-arrestin 2. *J Biol Chem*. 2005;280:25651–25658.
89. Hisada Y, Sugaya T, Yamanouchi M, et al. Angiotensin II plays a pathogenic role in immune-mediated renal injury in mice. *J Clin Invest*. 1999;103:627–635.
90. Nataraj C, Oliverio MI, Mannon RB, et al. Angiotensin II regulates cellular immune responses through a calcineurin-dependent pathway. *J Clin Invest*. 1999;104: 1693–1701.
91. Premont RT, Inglese J, Lefkowitz RJ. Protein kinases that phosphorylate activated G protein-coupled receptors. *FASEB J*. 1995;9:175–182.
92. Goodman Jr OB, Krupnick JG, Santini F, et al. Beta-arrestin acts as a clathrin adaptor in endocytosis of the beta2-adrenergic receptor. *Nature*. 1996;383:447–450.
93. Baillie GS, Sood A, McPhee I, et al. beta-Arrestin-mediated PDE4 cAMP phosphodiesterase recruitment regulates beta-adrenoceptor switching from Gs to Gi. *Proc Natl Acad Sci USA*. 2003;100:940–945.
94. Neuschafer-Rube F, Hermosilla R, Rehwald M, et al. Identification of a Ser/Thr cluster in the C-terminal domain of the human prostaglandin receptor EP4 that is essential for agonist-induced beta-arrestin1 recruitment but differs from the apparent principal phosphorylation site. *Biochem J*. 2004;379:573–585.

95. Jala VR, Shao WH, Haribabu B. Phosphorylation-independent beta-arrestin translocation and internalization of leukotriene B4 receptors. *J Biol Chem.* 2005;280: 4880–4887.
96. Silliman CC, Voelkel NF, Allard JD, et al. Plasma and lipids from stored packed red blood cells cause acute lung injury in an animal model. *J Clin Invest.* 1998;101: 1458–1467.
97. Khan SY, McLaughlin NJ, Kelher MR, et al. Lysophosphatidylcholines activate G2A inducing G(alphai)-/G(alphaq/)-Ca(2)(+) flux, G(betagamma)-Hck activation and clathrin/beta-arrestin-1/GRK6 recruitment in PMNs. *Biochem J.* 2010;432:35–45.
98. Iacovelli L, Capobianco L, D'Ancona GM, Picascia A, De Blasi A. Regulation of lysophosphatidic acid receptor-stimulated response by G-protein-coupled receptor kinase-2 and beta-arrestin1 in FRTL-5 rat thyroid cells. *J Endocrinol.* 2002;174: 103–110.
99. Gesty-Palmer D, El Shewy H, Kohout TA, Luttrell LM. beta-Arrestin 2 expression determines the transcriptional response to lysophosphatidic acid stimulation in murine embryo fibroblasts. *J Biol Chem.* 2005;280:32157–32167.
100. Sun J, Lin X. Beta-arrestin 2 is required for lysophosphatidic acid-induced NF-kappaB activation. *Proc Natl Acad Sci USA.* 2008;105:17085–17090.
101. Li TT, Alemayehu M, Aziziyeh AI, et al. Beta-arrestin/Ral signaling regulates lysophosphatidic acid-mediated migration and invasion of human breast tumor cells. *Mol Cancer Res.* 2009;7:1064–1077.
102. Chen Z, Dupre DJ, Le Gouill C, Rola-Pleszczynski M, Stankova J. Agonist-induced internalization of the platelet-activating factor receptor is dependent on arrestins but independent of G-protein activation. Role of the C terminus and the (D/N)PXXY motif. *J Biol Chem.* 2002;277:7356–7362.
103. Venkatesha RT, Ahamed J, Nuesch C, Zaidi AK, Ali H. Platelet-activating factor-induced chemokine gene expression requires NF-kappaB activation and Ca2+/calcineurin signaling pathways. Inhibition by receptor phosphorylation and beta-arrestin recruitment. *J Biol Chem.* 2004;279:44606–44612.
104. McLaughlin NJ, Banerjee A, Kelher MR, et al. Platelet-activating factor-induced clathrin-mediated endocytosis requires beta-arrestin-1 recruitment and activation of the p38 MAPK signalosome at the plasma membrane for actin bundle formation. *J Immunol.* 2006;176:7039–7050
105. McLaughlin NJ, Banerjee A, Khan SY, et al. Platelet-activating factor-mediated endosome formation causes membrane translocation of p67phox and p40phox that requires recruitment and activation of p38 MAPK, Rab5a, and phosphatidylinositol 3-kinase in human neutrophils. *J Immunol.* 2008;180:8192–8203.
106. Paing MM, Stutts AB, Kohout TA, Lefkowitz RJ, Trejo J. beta-Arrestins regulate protease-activated receptor-1 desensitization but not internalization or down-regulation. *J Biol Chem.* 2002;277:1292–1300.
107. Zoudilova M, Min J, Richards HL, Carter D, Huang T, DeFea KA. beta-Arrestins scaffold cofilin with chronophin to direct localized actin filament severing and membrane protrusions downstream of protease-activated receptor-2. *J Biol Chem.* 2010;285: 14318–14329.
108. DeFea KA, Zalevsky J, Thoma MS, Dery O, Mullins RD, Bunnett NW. beta-Arrestin-dependent endocytosis of proteinase-activated receptor 2 is required for intracellular targeting of activated ERK1/2. *J Cell Biol.* 2000;148:1267–1281.
109. Medzhitov R. Toll-like receptors and innate immunity. *Nat Rev Immunol.* 2001;1:135–145.
110. Jiang D, Liang J, Fan J, et al. Regulation of lung injury and repair by Toll-like receptors and hyaluronan. *Nat Med.* 2005;11:1173–1179.

111. Wang Y, Tang Y, Teng L, Wu Y, Zhao X, Pei G. Association of beta-arrestin and TRAF6 negatively regulates Toll-like receptor-interleukin 1 receptor signaling. *Nat Immunol.* 2006;7:139–147.
112. Porter KJ, Gonipeta B, Parvataneni S, et al. Regulation of lipopolysaccharide-induced inflammatory response and endotoxemia by beta-arrestins. *J Cell Physiol.* 2010;225:406–416.
113. Fan H, Luttrell LM, Tempel GE, Senn JJ, Halushka PV, Cook JA. Beta-arrestins 1 and 2 differentially regulate LPS-induced signaling and pro-inflammatory gene expression. *Mol Immunol.* 2007;44:3092–3099.
114. Seregin SS, Appledorn DM, Patial S, et al. beta-Arrestins modulate Adenovirus-vector-induced innate immune responses: differential regulation by beta-arrestin-1 and beta-arrestin-2. *Virus Res.* 2010;147:123–134.
115. Li H, Sun X, LeSage G, et al. β-Arrestin 2 regulates Toll-like receptor 4-mediated apoptotic signalling through glycogen synthase kinase-3beta. *Immunology.* 2010;130: 556–563.
116. Li H, Chen L, Zhang Y, et al. Chronic stress promotes lymphocyte reduction through TLR2 mediated PI3K signaling in a beta-arrestin 2 dependent manner. *J Neuroimmunol.* 2011;233:73–79.
117. Husebye H, Halaas O, Stenmark H, et al. Endocytic pathways regulate Toll-like receptor 4 signaling and link innate and adaptive immunity. *EMBO J.* 2006;25:683–692.
118. Parameswaran N, Pao CS, Leonhard KS, et al. Arrestin-2 and G protein-coupled receptor kinase 5 interact with NFkappaB1 p105 and negatively regulate lipopolysaccharide-stimulated ERK1/2 activation in macrophages. *J Biol Chem.* 2006;281:34159–34170.
119. Tipping M, Kim Y, Kyriakakis P, Tong M, Shvartsman SY, Veraksa A. β-arrestin Kurtz inhibits MAPK and Toll signalling in Drosophila development. *EMBO J.* 2010;29:3222–3235.
120. Wang W, Xu M, Zhang YY, He B. Fenoterol, a beta(2)-adrenoceptor agonist, inhibits LPS-induced membrane-bound CD14, TLR4/CD14 complex, and inflammatory cytokines production through beta-arrestin-2 in THP-1 cell line. *Acta Pharmacol Sin.* 2009;30:1522–1528.
121. Kizaki T, Shirato K, Sakurai T, et al. Beta2-adrenergic receptor regulate Toll-like receptor 4-induced late-phase NF-kappaB activation. *Mol Immunol.* 2009;46: 1195–1203.
122. Fox JM, Letellier E, Oliphant CJ, Signoret N. TLR2-dependent pathway of heterologous down-modulation for the CC chemokine receptors 1, 2, and 5 in human blood monocytes. *Blood.* 2011;117:1851–1860.
123. Zhuang LN, Hu WX, Xin SM, Zhao J, Pei G. Beta-arrestin-1 protein represses adipogenesis and inflammatory responses through its interaction with peroxisome proliferator-activated receptor-gamma (PPARgamma). *J Biol Chem.* 2011;286: 28403–28413.
124. Gao H, Sun Y, Wu Y, et al. Identification of beta-arrestin2 as a G protein-coupled receptor-stimulated regulator of NF-kappaB pathways. *Mol Cell.* 2004;14:303–317.
125. Witherow DS, Garrison TR, Miller WE, Lefkowitz RJ. beta-Arrestin inhibits NF-kappaB activity by means of its interaction with the NF-kappaB inhibitor IkappaBalpha. *Proc Natl Acad Sci USA.* 2004;101:8603–8607.
126. Kawamata Y, Imamura T, Babendure JL, Lu JC, Yoshizaki T, Olefsky JM. Tumor necrosis factor receptor-1 can function through a G alpha q/11-beta-arrestin-1 signaling complex. *J Biol Chem.* 2007;282:28549–28556.
127. Mo W, Zhang L, Yang G, et al. Nuclear beta-arrestin1 functions as a scaffold for the dephosphorylation of STAT1 and moderates the antiviral activity of IFN-gamma. *Mol Cell.* 2008;31:695–707.

128. Chen W, Kirkbride KC, How T, et al. Beta-arrestin 2 mediates endocytosis of type III TGF-beta receptor and down-regulation of its signaling. *Science*. 2003;301:1394–1397.
129. Lee NY, Kirkbride KC, Sheu RD, Blobe GC. The transforming growth factor-beta type III receptor mediates distinct subcellular trafficking and downstream signaling of activin-like kinase (ALK)3 and ALK6 receptors. *Mol Biol Cell*. 2009;20: 4362–4370.
130. Mythreye K, Blobe GC. The type III TGF-beta receptor regulates epithelial and cancer cell migration through beta-arrestin2-mediated activation of Cdc42. *Proc Natl Acad Sci USA*. 2009;106:8221–8226.
131. Lovgren AK, Kovacs JJ, Xie T, et al. beta-Arrestin deficiency protects against pulmonary fibrosis in mice and prevents fibroblast invasion of extracellular matrix. *Sci Transl Med*. 2011;3:74ra23.
132. Milasta S, Evans NA, Ormiston L, Wilson S, Lefkowitz RJ, Milligan G. The sustainability of interactions between the orexin-1 receptor and beta-arrestin-2 is defined by a single C-terminal cluster of hydroxy amino acids and modulates the kinetics of ERK MAPK regulation. *Biochem J*. 2005;387:573–584.
133. Tasken K, Stokka AJ. The molecular machinery for cAMP-dependent immunomodulation in T-cells. *Biochem Soc Trans*. 2006;34:476–479.
134. Bitto A, Minutoli L, David A, et al. Flavocoxid, a dual inhibitor of COX-2 and 5-LOX of natural origin, attenuates the inflammatory response and protects mice from sepsis. *Crit Care*. 2012;16:R32.
135. Fan H, Bitto A, Zingarelli B, et al. Beta-arrestin 2 negatively regulates sepsis-induced inflammation. *Immunology*. 2010;130:344–351.
136. Coureuil M, Mikaty G, Miller F, et al. Meningococcal type IV pili recruit the polarity complex to cross the brain endothelium. *Science*. 2009;325:83–87.
137. Tourret J, Finlay BB. A receptor for meningococcus: eliciting beta-arrestin signaling for barrier breaching. *Dev Cell*. 2011;20:7–8.
138. Lecuyer H, Nassif X, Coureuil M. Two strikingly different signaling pathways are induced by meningococcal type IV pili on endothelial and epithelial cells. *Infect Immun*. 2012;80:175–186.
139. Lefkowitz RJ, Shenoy SK. Transduction of receptor signals by beta-arrestins. *Science*. 2005;308:512–517.
140. Papp M, Li X, Zhuang J, Wang R, Uhal BD. Angiotensin receptor subtype AT(1) mediates alveolar epithelial cell apoptosis in response to ANG II. *Am J Physiol Lung Cell Mol Physiol*. 2002;282:L713–L718.
141. Uhal BD, Li X, Xue A, Gao X, Abdul-Hafez A. Regulation of alveolar epithelial cell survival by the ACE-2/angiotensin 1-7/Mas axis. *Am J Physiol Lung Cell Mol Physiol*. 2011;301:L269–L274.
142. Sime PJ, Xing Z, Graham FL, Csaky KG, Gauldie J. Adenovector-mediated gene transfer of active transforming growth factor-beta1 induces prolonged severe fibrosis in rat lung. *J Clin Invest*. 1997;100:768–776.
143. Lee CG, Cho SJ, Kang MJ, et al. Early growth response gene 1-mediated apoptosis is essential for transforming growth factor beta1-induced pulmonary fibrosis. *J Exp Med*. 2004;200:377–389.
144. Munger JS, Huang X, Kawakatsu H, et al. The integrin alpha v beta 6 binds and activates latent TGF beta 1: a mechanism for regulating pulmonary inflammation and fibrosis. *Cell*. 1999;96:319–328.
145. Tager AM, Kradin RL, LaCamera P, et al. Inhibition of pulmonary fibrosis by the chemokine IP-10/CXCL10. *Am J Respir Cell Mol Biol*. 2004;31:395–404.
146. Jiang D, Liang J, Campanella GS, et al. Inhibition of pulmonary fibrosis in mice by CXCL10 requires glycosaminoglycan binding and syndecan-4. *J Clin Invest*. 2010;120:2049–2057.

147. Tager AM, LaCamera P, Shea BS, et al. The lysophosphatidic acid receptor LPA1 links pulmonary fibrosis to lung injury by mediating fibroblast recruitment and vascular leak. *Nat Med*. 2008;14:45–54.
148. White ES, Lazar MH, Thannickal VJ. Pathogenetic mechanisms in usual interstitial pneumonia/idiopathic pulmonary fibrosis. *J Pathol*. 2003;201:343–354.
149. Hennenberg M, Trebicka J, Biecker E, Schepke M, Sauerbruch T, Heller J. Vascular dysfunction in human and rat cirrhosis: role of receptor-desensitizing and calcium-sensitizing proteins. *Hepatology*. 2007;45:495–506.
150. Hennenberg M, Trebicka J, Kohistani AZ, Heller J, Sauerbruch T. Vascular hyporesponsiveness to angiotensin II in rats with CCl(4)-induced liver cirrhosis. *Eur J Clin Invest*. 2009;39:906–913.
151. Lombardi MS, Kavelaars A, Schedlowski M, et al. Decreased expression and activity of G-protein-coupled receptor kinases in peripheral blood mononuclear cells of patients with rheumatoid arthritis. *FASEB J*. 1999;13:715–725.
152. Lombardi MS, Kavelaars A, Cobelens PM, Schmidt RE, Schedlowski M, Heijnen CJ. Adjuvant arthritis induces down-regulation of G protein-coupled receptor kinases in the immune system. *J Immunol*. 2001;166:1635–1640.
153. Li P, Cook JA, Gilkeson GS, et al. Increased expression of beta-arrestin 1 and 2 in murine models of rheumatoid arthritis: isoform specific regulation of inflammation. *Mol Immunol*. 2011;49:64–74.
154. Wu H, Wei W, Song L, Zhang L, Chen Y, Hu X. Paeoniflorin induced immune tolerance of mesenteric lymph node lymphocytes via enhancing beta 2-adrenergic receptor desensitization in rats with adjuvant arthritis. *Int Immunopharmacol*. 2007;7:662–673.
155. Ohguro H, Chiba S, Igarashi Y, Matsumoto H, Akino T, Palczewski K. Beta-arrestin and arrestin are recognized by autoantibodies in sera from multiple sclerosis patients. *Proc Natl Acad Sci USA*. 1993;90:3241–3245.
156. Forooghian F, Cheung RK, Smith WC, O'Connor P, Dosch HM. Enolase and arrestin are novel nonmyelin autoantigens in multiple sclerosis. *J Clin Immunol*. 2007;27: 388–396.
157. Tsutsui S, Vergote D, Shariat N, Warren K, Ferguson SS, Power C. Glucocorticoids regulate innate immunity in a model of multiple sclerosis: reciprocal interactions between the A1 adenosine receptor and beta-arrestin-1 in monocytoid cells. *FASEB J*. 2008;22:786–796.
158. Vroon A, Lombardi MS, Kavelaars A, Heijnen CJ. Changes in the G-protein-coupled receptor desensitization machinery during relapsing-progressive experimental allergic encephalomyelitis. *J Neuroimmunol*. 2003;137:79–86.
159. Colombo G, Sordi A, Lonati C, et al. Treatment with alpha-melanocyte stimulating hormone preserves calcium regulatory proteins in rat heart allografts. *Brain Behav Immun*. 2008;22:817–823.
160. Deshpande DA, Theriot BS, Penn RB, Walker JK. Beta-arrestins specifically constrain beta2-adrenergic receptor signaling and function in airway smooth muscle. *FASEB J*. 2008;22:2134–2141.
161. Wang WC, Mihlbachler KA, Brunnett AC, Liggett SB. Targeted transgenesis reveals discrete attenuator functions of GRK and PKA in airway beta2-adrenergic receptor physiologic signaling. *Proc Natl Acad Sci USA*. 2009;106:15007–15012.
162. Dickey BF, Walker JK, Hanania NA, Bond RA. beta-Adrenoceptor inverse agonists in asthma. *Curr Opin Pharmacol*. 2010;10:254–259.
163. Manson ME, Corey DA, White NM, Kelley TJ. cAMP-mediated regulation of cholesterol accumulation in cystic fibrosis and Niemann-Pick type C cells. *Am J Physiol Lung Cell Mol Physiol*. 2008;295:L809–L819.
164. Manson ME, Corey DA, Rymut SM, Kelley TJ. beta-Arrestin-2 regulation of the cAMP response element binding protein. *Biochemistry*. 2011;50:6022–6029.

165. Noma T, Lemaire A, Naga Prasad SV, et al. Beta-arrestin-mediated beta1-adrenergic receptor transactivation of the EGFR confers cardioprotection. *J Clin Invest.* 2007;117:2445–2458.
166. Rakesh K, Yoo B, Kim IM, Salazar N, Kim KS, Rockman HA. beta-Arrestin-biased agonism of the angiotensin receptor induced by mechanical stress. *Sci Signal.* 2010; 3:ra46.
167. Engelhardt S. Alternative signaling: cardiomyocyte beta1-adrenergic receptors signal through EGFRs. *J Clin Invest.* 2007;117:2396–2398.
168. Noor N, Patel CB, Rockman HA. Beta-arrestin: a signaling molecule and potential therapeutic target for heart failure. *J Mol Cell Cardiol.* 2011;51:534–541.
169. Rajagopal S, Rajagopal K, Lefkowitz RJ. Teaching old receptors new tricks: biasing seven-transmembrane receptors. *Nat Rev Drug Discov.* 2010;9:373–386.
170. Schmid CL, Bohn LM. Physiological and pharmacological implications of beta-arrestin regulation. *Pharmacol Ther.* 2009;121:285–293.
171. Walters RW, Shukla AK, Kovacs JJ, et al. beta-Arrestin1 mediates nicotinic acid-induced flushing, but not its antilipolytic effect, in mice. *J Clin Invest.* 2009;119:1312–1321.
172. Ahn S, Nelson CD, Garrison TR, Miller WE, Lefkowitz RJ. Desensitization, internalization, and signaling functions of beta arrestins demonstrated by RNA interference. *Proc Natl Acad Sci USA.* 2003;100:1740–1744.
173. Zheng H, Zeng Y, Zhang X, Chu J, Loh HH, Law PY. mu-Opioid receptor agonists differentially regulate the expression of miR-190 and NeuroD. *Mol Pharmacol.* 2010;77:102–109.
174. Kefas B, Comeau L, Floyd DH, et al. The neuronal microRNA miR-326 acts in a feedback loop with notch and has therapeutic potential against brain tumors. *J Neurosci.* 2009;29:15161–15168.
175. Xiao K, McClatchy DB, Shukla AK, et al. Functional specialization of beta-arrestin interactions revealed by proteomic analysis. *Proc Natl Acad Sci USA.* 2007; 104:12011–12016.
176. Xiao K, Sun J, Kim J, et al. Global phosphorylation analysis of beta-arrestin-mediated signaling downstream of a seven transmembrane receptor (7TMR). *Proc Natl Acad Sci USA.* 2010;107:15299–15304.

CHAPTER FIFTEEN

The Role of β-Arrestins in Cancer

Philip Michael Sobolesky, Omar Moussa
Department of Pathology and Laboratory Medicine, Medical University of South Carolina, Charleston, South Carolina, USA

Contents

1. Introduction 396
2. ARRBs and Cancer-Associated Cell Phenotypes 397
   2.1 Cell proliferation 397
   2.2 Cell migration and invasion 398
   2.3 Apoptosis and survival 399
3. The Role of ARRBs in Cancer 400
   3.1 Breast cancer 400
   3.2 Colorectal cancer 402
   3.3 Lung cancer 402
   3.4 Bladder cancer 403
   3.5 Hematological malignancies 404
   3.6 Prostate cancer 405
   3.7 Pancreatic cancer 405
   3.8 Glioblastoma 406
4. Conclusions 407
Acknowledgments 407
References 407

Abstract

Beyond their well-characterized roles in G protein-coupled receptor desensitization and trafficking, β-arrestins (ARRBs) have been implicated in the regulation of several basic cellular functions, including cell cycle regulation, cell migration, and apoptotic signaling. Nowhere are the data supporting a physiologically relevant role for these arrestin-mediated responses stronger than in cancer. *In vitro*, ARRBs regulate cell proliferation, promote migration, and transmit anti-apoptotic survival signals by scaffolding cytosolic signaling protein networks and even translocating to the nucleus to directly regulate gene expression. In animal models, ARRB expression affects tumor initiation time, growth rate, vascularization, survival under hypoxic conditions, invasiveness, and metastatic potential. Studies in human cancer patients have demonstrated that dysregulation of ARRB expression, localization, or phosphorylation is associated with more aggressive cancer phenotypes and poorer outcomes in malignancies involving

Progress in Molecular Biology and Translational Science, Volume 118
ISSN 1877-1173
http://dx.doi.org/10.1016/B978-0-12-394440-5.00015-2

the breast, lung, prostate, brain, and hematologic system. Collectively, these data build a strong case that ARRB-dependent signaling contributes to the cancer phenotype and that the ARRBs may represent novel therapeutic targets in certain malignancies.

1. INTRODUCTION

The nonvisual arrestins, β-arrestin (ARRB) 1 and 2, play important roles in regulating cellular signaling processes including several that alter cell phenotype and lead to the progression of cancer. ARRBs were initially reported to mediate desensitization of G protein-coupled receptor (GPCR) signaling by regulating receptor internalization.[1] Subsequently, major progress in our understanding of ARRB function has revealed additional roles in cell signal regulation as ARRBs have been shown to scaffold proteins within the cytoplasm and translocate to the nucleus to regulate several cellular processes.[2] ARRB-regulated cellular processes are commonly associated with cancer cell phenotypes such as cell proliferation,[3] cancer cell migration, invasion, and metastasis.[4,5] Recent advances have further implicated ARRB in novel roles related to cancer progression, including the inhibition of cell death, activation of tyrosine kinase pathways, and the regulation of gene transcription. ARRBs have a role in the regulation of cellular apoptosis demonstrated in a variety of cancer and noncancerous cell lines, contributing to uncontrolled growth.[6–14] Extracellular signal-regulated kinase (ERK) signaling can also contribute to abnormal cell proliferation and cancer progression. ARRBs mediate ERK1/2 activation both directly and indirectly via interaction with the cellular proto-oncogene c-SRC.[15] ARRBs can also directly interact with c-SRC and promote migration and invasion independent of ERK1/2. Furthermore, malignant cells often take hostage of GPCRs to migrate and invade into surrounding tissues. ARRBs facilitate this process by recruiting a large number of other nonreceptor tyrosine kinases to various GPCRs.[16] The role of ARRBs in regulating gene transcription was first described in 2001; ARRB1 regulated the activation of the lymphoid enhancer transcription factor through its interaction with phosphorylated dishevelled 1 and 2.[17] Following nuclear translocation, ARRB1 is able to interact with histone acetyltransferase p300 and cAMP response element-binding protein to regulate histone acetylation and gene transcription.[18]

Major advances in our understanding of the molecular mechanisms of ARRB signaling over the past decade have been made possible by an integrated approach combining the strengths of three systems: cell culture models provide an assay system for cellular and molecular phenotypes, the mouse provides an *in vivo* model, and cancer patients provide clinical and epidemiologic evidence for biological relevance. This chapter reviews the mechanisms of β-arrestin 1 and 2 signaling in cancer and their roles in multiple aspects of tumor development and disease progression.

2. ARRBs AND CANCER-ASSOCIATED CELL PHENOTYPES

ARRBs are proteins that interact with almost all the GPCRs and serve multiple purposes.[19] ARRB2, also known as arrestin3 and β-arrestin2,[20] was first discovered by its ability to enhance desensitization and phosphorylation of GPCRs and was subsequently found to have additional effects. Both ARRB1 and ARRB2 interact with GPCRs to act as scaffolds for the recruitment of cytoplasmic signaling proteins.[21] This enables ARRBs to mediate agonist-induced signaling, for example, ERK activation, in a non-G protein-dependent manner. Signaling via ARRBs can result in activation of ERKs, MAPKs, tyrosine kinases, AKT, PI3 kinases, and inhibition of the NFκB pathway.[21] ARRB1 may play an important role in prostaglandin E_2 (PGE_2) activation of *Src* and transactivation of epidermal growth factor (EGF) receptors in colorectal cancer cells.[5] Activation of these signaling molecules was found to inhibit apoptosis and increase migration of cancer cells.[21] In addition, ARRB1 enhances the metastatic potential of colorectal cancer cells *in vivo*. The thromboxane receptor β isoform (TPβ), but not TPα receptors, has been found to interact with ARRB2,[22–25] raising the possibility that the TP isoform-specific interaction with ARRB2 leads to the pathophysiologic effects seen in cancer cells. Published data demonstrate that ARRB2-RNA interference results in decreased TP receptor agonist (U46619)-dependent cell migration and invasion. This led to the hypothesis that TPβ activation results in enhanced cell proliferation, migration, and invasion mediated by an ARRB2-dependent mechanism indicating a novel role for ARRB2 in the malignant transformation induced by TPβ.

2.1. Cell proliferation

The role of the activated ERK1/2 and PI3K/AKT signaling pathways in tumor initiation and progression is well characterized.[26–32] In the past decade, new roles for ARRBs in regulating the intracellular signaling network including the MAPKs have been identified. The first study that described the role of ARRBs in cellular proliferation through ERK activation was published over a decade ago by Pierce *et al.*[33] Several studies confirmed the role of ARRBs in the regulation of cellular proliferation through their interaction with both G protein-dependent and -independent signaling pathways.[34] ARRB signaling has been shown to regulate the activity of PI3K following stimulation of the insulin-like growth factor type 1 (IGF-1) receptor tyrosine kinase.[11] In non-small-cell lung cancer (NSCLC), ARRB1 has been shown to regulate nicotine-induced epithelial cell proliferation by activating Src and Rb–Raf-1 pathways. The binding of the nicotine to nicotinic acetylcholine receptors (nAChRs) causes the recruitment of ARRBs and Src and results in the activation of the MAPK and Rb–Raf-1 pathways. These events subsequently cause the recruitment of E2F1, Rb, and Raf-1 to E2F-responsive proliferative promoters. The inactivation of Rb by Raf-1 facilitates the dissociation of Rb from the promoters, which combined with the increased binding of E2F1, causes transcription of S-phase genes and cell cycle progression. Ablation of ARRBs using RNA interference blocked nicotine-induced proliferation of NSCLCs, blocked activation of Src, suppressed levels of phosphorylated ERK, and abrogated Rb–Raf-1 binding in response to nicotine.[35]

2.2. Cell migration and invasion

The ARRB proteins play a significant role in mediating signals for cell migration and invasion of normal and malignant cells. The role of ARRBs in regulating cell migration was first demonstrated in a study that found impaired CXCR-mediated cell motility in the lymphocytes obtained from ARRB2 knockout mice, compared to their wild-type litter mates.[36] Additional studies found that the CXCR-mediated cell migration was ARRB dependent, resulting in increased activation of p38, MAPK, SAPK, and ERK1/2 pathways.[37] Knocking down ARRB expression with siRNA or introduction of a dominant negative ARRB attenuated the increased migration and associated signaling.[37] Suppression of ARRB2 was associated with decreased stromal cell-derived factor 1α-induced cell migration via decreased p38 MAPK activation in HEK293 cells.[38] The IGF-1 receptor is involved in the cell cycle regulation of normal and malignant cells.[39] Stimulation of the IGF-1 receptor causes it to associate with ARRB1 resulting in both the formation of a complex with Mdm2, leading to its proteosomal destruction, and ERK1/2 activation.[40]

Recent studies have demonstrated the importance of ARRBs in migration and invasion of various types of cancers. In transgenic mice, overexpressing HA-tagged ARRB1 increased vascular endothelial growth factor (VEGF) expression, MMP9 activity, and overall growth, and decreased tumor initiation time when subcutaneously injected with tumor cells.[41] The lipid mediator lysophosphatidic acid (LPA) receptor is a GPCR that associates with ARRB proteins to mediate tumor cell invasion and migration, and both are found overexpressed in advanced stages of breast cancer.[42] Breast carcinoma cells under hypoxic conditions expressed elevated levels of vascular endothelial growth factor-A (VEGF-A) as a result of stabilization of its transcriptional regulator HIF-1α by ARRB1. This stabilization of HIF-1α by ARRB1 in the nucleus was essential for mediating VEGF cell survival signals in hypoxic conditions.[43] Another study in breast cancer cells demonstrated that protease-activated receptor-2 (PAR-2) requires the recruitment and scaffolding of both ARRB proteins to activate ERK1/2 and promote tumor cell migration.[44] Further studies determined that PAR-2 stimulation resulted in ARRB1 binding to PI3K, which inhibited its activity.[45] In colorectal cancer, ARRB1 is necessary for the formation of a signaling complex facilitating in the interaction of the PGE_2 receptor with c-Src, resulting in transactivation of the EGF receptor and downstream Akt signaling.[5] Blocking the interaction between ARRB1 and c-Src decreased migration and invasiveness of colorectal carcinoma cells, as well as the number of metastatic liver lesions *in vivo*.[5] In ovarian cancer cells, the interaction between ARRB and c-Src following endothelin-A receptor stimulation was found to be crucial for EGF receptor transactivation, inactivation of glycogen synthase kinase-3β (GSK-3β), and activation of the β-catenin signaling pathways.[4] Similar to colorectal carcinoma metastases, ovarian cancer cells that expressed a mutated ARRB1 protein that prevented its interaction with c-Src exhibited a reduced metastatic rate.[4] Thus, the scaffolding ability of the ARRB proteins appears to determine its involvement in invasion and migration of normal and malignant cells.

2.3. Apoptosis and survival

A novel role of ARRBs has been identified in regulating GPCR-mediated apoptosis. The first study demonstrated that the anti-apoptotic effects of the neuropeptide substance P (SP) on neurokinin-1 receptor (NK1R) expressing endothelial cells involved ERK1/2 activation following formation of a complex consisting of NK1R–ARRB–Src–ERK. Disrupting the formation of this complex by expression of a dominant negative ARRB inhibited SP-induced anti-apoptotic effects.[46] Another study demonstrated that stimulation of various GPCRs (*N*-formyl peptide receptor, AT_1AR, V_2R, and CXCR2) was able to initiate apoptosis in ARRB1 and ARRB2 knockout MEFs and restoring expression of ARRBs in these deficient cells attenuated GPCR-stimulated cell death.[47,48] Studies expressing truncated versions of ARRB2 in MEFs deficient in both ARRBs found sequences in the C-terminal tail of ARRB2 that are essential to preventing apoptosis.[7] Interestingly, the binding of ARRB2 alone to the GPCR was insufficient to prevent apoptosis in MEFs. Preventing GPCR-mediated apoptosis was dependent on the interaction between the GPCR-bound ARRB and adaptor protein-2.[6,7]

The anti-apoptotic properties of ARRBs are the result of its ability to act as a scaffold following receptor stimulation and subsequently mediate downstream anti-apoptotic signals. A recent study showed the apoptotic signaling promoted by Toll-like receptor 4 following serum deprivation is through activation of GSK-3β, and expression of ARRB2 greatly attenuated GSK-3β activation.[49] A similar study performed in endometrial cancer cells found that apoptosis induced by resveratrol treatment was negatively regulated by ARRB2 expression by affecting the Akt/GSK-3β pathways.[13] ARRB2 has been shown to mediate anti-apoptotic signaling through MAPK-RSK and PI3K–AKT activation upon stimulation of the angiotensin II type 1A (AT_{1A}) receptor by phosphorylation of pro-apoptotic protein BAD.[50] Conversely, one study demonstrated that a nonphosphorylated form of ARRB2 promoted apoptosis following exposure to UV through the stabilization of IκBα which binds to and prevents NFκB nuclear translocation, consequently suppressing its anti-apoptotic signaling.[51] The role of ARRB proteins in apoptosis remains to be fully elucidated, but it is clear that both ARRB proteins are important in mediating anti-apoptotic signals.

3. THE ROLE OF ARRBs IN CANCER

3.1. Breast cancer

The role of ARRBs in breast cancer is not well understood. Very few studies have explored the impact of ARRBs in breast cancer cell models. The earliest report indicated that both ARRBs are critical for constitutive PAR-2-mediated cell migration in the metastatic breast cancer cell line MDA MB-231.[44] Signaling pathways for the coagulation cascade involving PAR-2 and tissue factor cytoplasmic domain signaling have been linked to the metastatic potential of breast cancer cells.[52] PAR-2 expression is linked to cancer cell migration by G protein-independent pathways through ARRB scaffolding. When activated by trypsin, PAR-2 increases breast cancer cell migration through ERK1/2 activation. This process has been found to be dependent on ARRBs.[44] Knockdown of ARRBs by RNA interference, or the use of a MEK-1 inhibitor, decreased breast cancer cell migration.[44] These data demonstrate a requirement for both ARRBs in PAR-2-mediated motility and suggest that autocrine activation of PAR-2 by secreted proteases may contribute to the migration of metastatic tumor cells through ARRB-dependent ERK1/2 activation.[44] The significance of this study was that it was one of the earliest reports to indicate the novel functions of ARRBs other than their well-characterized role as G-protein signal terminators.

More recent studies further support the role of ARRBs in breast cancer. ARRBs have been shown to be involved in LPA receptor signaling. LPA is a biologically active phospholipid that can regulate several cellular processes by the activation of specific GPCRs. LPA plays a key role in *de novo* lipid synthesis and also regulates cellular phenotypes such as cell migration, cytoskeletal reorganization, and survival. The role of LPA in breast cancer cell migration, invasion, and metastases is well documented,[53–56] and evidence suggests that ARRBs mediate the effects of LPA on cancer cells migration through a Ral GTPase-dependent pathway.[42]

The majority of the studies that have explored the role of the ARRBs in cancer were performed in cell models. Fewer studies have been designed to explore the clinical relevance of ARRBs expression in disease outcomes such as response to chemotherapeutic agents. Using a real-time PCR-based assay, Li *et al.*[42] detected high levels of both ARRBs in advanced stage breast cancer tissues compared to early stage disease. In the same study, they examined the mRNA expression of the two molecules in a breast cancer cell line MDA MB-231 and the normal mammary epithelial line MCF 10A. While the protein levels of both ARRBs were increased compared to the MCF 10A cells, only the mRNA of the ARRB1 was elevated. The mechanism by which the ARRB2 protein increases but the mRNA does not increase is yet to be characterized.

A single recent study investigated the clinical relevance of ARRB1 expression in tumor and stromal cells and correlated the levels of tissue expression detected by immunohistochemical analysis with disease outcomes such as histological grade, tumor size, and node status in two large cohorts of breast cancer patients. In the first cohort of 179 pre- and postmenopausal patients, higher expression of ARRB1 in the stroma, but not the tumor cells, correlated with reduced patient survival. In the second cohort, the expression of ARRB1 was evaluated in 500 premenopausal patients randomized for receiving Tamoxifen as neo-adjuvant therapy or no-adjuvant therapy. Interestingly, patients with low or moderate levels of stromal ARRB1 protein expression did not benefit from the neo-adjuvant therapy.[86]

3.2. Colorectal cancer

One of the earliest reports investigated the role of ARRB1 in the metastatic progression of colon cancer.[5] PGE_2-induced transactivation of the EGF receptor is mediated by the association of ARRB1 with c-Src.[5] Activation of this pathway was found to be dependent on ARRB1 and resulted in increased colon cancer cell proliferation and survival.[5] The activity of ARRB1 is regulated by phosphorylation at the Ser-412 position, which controls its association with c-Src.[5,57]

A more recent study examined the role of ARRBs in mice with the truncated tumor suppressor gene, adenomatous polyposis coli (APC). The disruption of APC induces intestinal tumor growth via an activated Wnt signaling pathway. Targeted knockdown of ARRB1 did not result in any significant change in tumor formation, but depletion of ARRB2 levels by targeted disruption or using RNA interference reduced intestinal tumor initiation in 67% of the animals.[57] The remaining 33% of animals formed tumors that were similar in size, characteristics, and gene expression profile to the control group mice with wild-type ARRB2. In an *ex vivo* model using cells isolated from $APC^{\Delta 14/+}$ mouse tumors, downregulation of ARRB2 decreased Wnt pathway activity, indicating that the tumor formation in these mice was dependent on Wnt pathway activity. In spite of the absence of any other reports evaluating the role of ARRBs in intestinal neoplasms, these two important studies point to the differential biological role of the two ARRB proteins and their impact on cancer signaling pathways. While ARRB1 may play an essential role in the cancer-associated c-Src signaling pathway,[5] it might not be involved in other pathways such as Wnt signaling, which have been shown to be dependent on ARRB2.[57]

3.3. Lung cancer

Lung cancer is the leading cause of cancer death for men and women in the United States.[58] There are two major forms of lung cancer; small-cell lung cancer and NSCLC. Smoking is a significant contributing factor involved in the initiation and progression of lung cancer. Nicotine, an addictive substance found in tobacco, was found to stimulate angiogenesis and promote endothelial cell tumor growth through nAChR cross talk with ARRB1.[59] It was shown that siRNA knockdown of ARRB1 in NSCLC cells treated with nicotine prevented the activation of Src and Rb–Raf-1 pathways and blocked nicotine-induced proliferation.[35] A later study found that nicotine stimulation of normal and NSCLC cells promoted the translocation of ARRB1 to the nucleus where it bound to E2F transcription factors. This interaction resulted in the regulation of the proliferative genes, CDC6 and TYMS, as well as pro-survival genes like BIRC5.[3] The ability of ARRB1 proteins to translocate to the nucleus, and directly bind and regulate the activity of transcription factors involved in tumor growth and survival, validates future studies on determining the potential of inhibiting ARRB proteins as therapeutic targets.

Tissue microarray analysis revealed that human NSCLC tumors displayed increased levels of nuclear ARRB1 compared to normal lung tissue.[3] A recent study investigating the expression levels of ARRB2 in NSCLC by ELISA found the serum levels of ARRB2 were significantly lower in the NSCLC patients compared to the healthy control patients. The decreased serum levels of ARRB2 were also associated with a poorer prognosis in NSCLC patients.[60]

The prostaglandins, that is, PGD_2, PGE_2, PGF_2, and TxA_2, are commonly dysregulated in cancer due to the overexpression of the cyclooxygenase proteins (COX-1 and -2), resulting in increased tumor cell survival, growth, migration, and invasion.[61,62] The expression of COX-2 protein is increased in lung cancer, as well as in numerous other cancers.[63,64] The increased expression results in enhanced PG synthesis. Treatment of lung cancer cells with PGE_2 resulted in an ARRB-dependent increase in cell migration.[65] In lung cancer cells, PGE_2 bound to the EP4 receptor causing ARRB1 recruitment and mediating the activation of c-Src. Knocking down expression of ARRB1 with shRNA decreased PGE_2-mediated c-Src phosphorylation and migration independent of ERK activation.[65] This makes targeting the inhibition of ARRB proteins an appealing therapeutic strategy, as it may reduce tumor cell migration while avoiding the serious adverse side effects that have been associated with pharmacologic inhibitors of COX-2.

3.4. Bladder cancer

Bladder cancer is the fifth most frequent type of cancer in the United States, and the National Cancer Institute estimates 73,510 new cases, and 14,880 deaths will occur from bladder cancer in 2012.[58] Bladder cancer is initiated in the epithelial cell layer lining the bladder lumen. The majority of bladder cancer patients present with the superficial form. Unfortunately, recurrence, invasion, and metastasis, even after seemingly successful treatment at very early stages, are characteristic of bladder cancer. The need for life-long treatment and surveillance makes it the most expensive cancer to treat on a per patient basis.[66] We found that the β-isoform of the thromboxane receptor (TPβ) can cause malignant transformation in immortalized nontransformed urothelial cells.[67] TPβ, but not TPα, receptors have been found to interact with ARRB2,[22–24] thus raising the possibility that the TPβ isoform-specific interaction with ARRB2 leads to the pathophysiologic effects seen in cancer cells. ARRB2-RNA interference treatment of bladder cancer cells decreased TP receptor agonist (U46619) dependent increases in cell migration and invasion. ARRB1-RNA interference knockdown did not alter agonist-dependent stimulation of cell migration and invasion. These data represent evidence of a role for ARRB2, but not ARRB1, in the malignant transformation and maintenance of the malignant phenotypes induced by TPβ.[67]

3.5. Hematological malignancies

Although the critical role of ARRBs in maintaining the transformed status of many epithelial cells has been demonstrated in several studies, their role in the initiation and progression of hematological malignancies is not well studied. Acute lymphoblastic leukemia (ALL) is the main subtype of childhood leukemia. Standard treatment protocols used for ALL involve risk stratification and risk-based therapy. These protocols ensure the use of low toxicity protocols for low-risk patients and more aggressive treatment for high-risk individuals. The risk stratification process utilizes several clinical and laboratory features such as patient age, white blood cell count, and immunologic subtype of the disease.[68] Genetic-based stratification of diseases is proving increasingly more effective. Recently, two published reports supported a potential role of ARRBs in hematological malignancies. The first report studied the mRNA expression of ARRB1 in 155 newly diagnosed children with ALL and correlated the expression with 51 controls.[69] The expression of ARRB1 was significantly elevated in the ALL patients and its levels were correlated with white blood cell count and risk classification. Interestingly, the levels of ARRB1 also inversely correlated with Notch1, which is essential for T-cells and developing hematological cells. The second study investigated the role of ARRB1 and ARRB2 in the establishment and propagation of chronic myelogenous leukemia (CML). Using knockout mouse models, loss of ARRB2, but not ARRB1, led to severe impairment in CML initiation and progression. The effect of ARRB2 on CML was mediated in part through the Wnt/β-catenin signaling pathway.[70]

3.6. Prostate cancer

Prostate cancer often manifests as an androgen-dependent disease that can be managed initially through inhibition of the androgen receptor (AR), yet many patients with seemingly successful treatment will develop reoccurrence of a more severe androgen-independent, hormone-refractory disease.[71] The ARRB2 protein in prostate cancer was found to repress AR activity by functioning as a scaffold and forming a complex with the AR and Mdm2 resulting in ubiquitylation and downregulation of the AR.[72] Also, an inverse relationship between PSA protein expression and ARRB2 protein expression was observed in prostate cancer tissue samples.[71] Interestingly, the staining of ARRB2 in the cancer samples was localized to the epithelial cells and not the stromal cells, whereas in normal prostate tissue, ARRB2 staining was only observed in the stromal cells.[71,73] The type III TGF-β receptor (TβRIII) expression was found decreased or lost in the majority of human prostate cancers at the protein and mRNA level, and restoring expression decreased cell motility, cell invasion, and tumorigenicity.[74] A follow-up study identified ARRB2 as the essential scaffolding protein that allowed TβRIII to activate Cdc42 and inhibit cell migration.[75] Conversely, the β2 adrenergic receptor (β_2AR) is a GPCR associated with increased activity levels in prostate cancer patients.[76] A recent study using prostate cancer cells demonstrated that the increased cell proliferation induced by β_2AR activation was mediated by an ARRB2 complex with c-Src.[76] More studies are needed to differentiate the role of ARRB2 in prostate cancer migration, invasion, and tumorigenicity.

3.7. Pancreatic cancer

It has been shown that the chemokine receptors, CXCR4 and CXCR7 are highly overexpressed in pancreatic cancer.[77] Following CXCL12 stimulation, both receptors signal through ARRB2 to activate the MAPK pathway, which promotes increased cellular proliferation of pancreatic cells.[77,78] Another signaling pathway dysregulated in pancreatic cancer is the hedgehog (Hh) pathway. Binding of the ligand sonic hedgehog (Shh) to Patched (Ptc) relieves the inhibitory control of Ptc over

Smoothened (Smo) resulting in its activation and promotion of downstream Hh targets involved in cell division and the suppression of apoptosis.[79] Shh stimulation induces Smo to interact with ARRB2, resulting in Smo internalization, while treatment with a Smo antagonist (cyclopamine) prevents the interaction and internalization.[80] In pancreatic cancer-derived cell lines and tissue samples, Shh expression was found significantly increased.[79] Inhibition of Smo by cyclopamine treatment resulted in decreased cell proliferation and increased apoptosis.[79] While the involvement of ARRB proteins on Hh signaling has not been specifically shown in a pancreatic model, it is likely that Smo internalization and activation is ARRB dependent. Thus, future studies confirming the role of ARRB proteins in Hh signaling may identify additional therapeutic targets for pancreatic cancer treatment.

3.8. Glioblastoma

Few studies have been published specifically examining the role of ARRBs in glioblastoma (GBM) formation and progression. One study performed immunohistochemistry on GBM samples and identified a strong correlation between grades II and III GBM and ARRB1 phosphorylation at Ser412, such that GBMs expressed less phosphorylated ARRB1 than non-neoplastic brain.[81] Survival curves found that GBM patients negative for phosphorylated ARRB1 at Ser412 had significantly decreased survival.[81] The total and phosphorylated expression of ARRB2 was not examined in GBM samples and would be worth investigating in the future to determine if there is any differential expression between the two proteins, similar to that found in breast cancer.[82] GBM, as well as medulloblastoma, has been found to overexpress CXCR4 and its ligand CXCL12, and as mentioned before, increased tumor cell migration by CXCR4 signaling is mediated through ARRB proteins.[37,83] While CXCR4 antagonists (AMD3100 and TC14012) initially reduce the growth of primary brain tumors, the same antagonists activate CXCR7 signaling mediated through ARRB.[83,84] In astrocytes, CXCR7 activation of ERK and Akt by CXCL12 was abolished by pertussis toxin and thus is mediated through $G_{i/o}$ proteins.[85] But in astrocytes stimulated with the CXCR7 ligand, I-TAC/CXCL11, ERK and Akt activation was ARRB dependent and not pertussis toxin sensitive.[85] More studies need to be done on the effects of ARRB pathway activation to improve current therapies that target CXCR4 and CXCR7 signaling.

4. CONCLUSIONS

The essential role played by the ARRBs as key modulators of tumor initiation and progression is well documented in several studies reviewed in this chapter. Although many of these studies support the notion that ARRB proteins are necessary for a malignant phenotype, none of the current studies have demonstrated that their altered expression is sufficient for malignant transformation. The observed differences in ARRB1 and ARRB2, as well as the effect of overexpression or knockdown of either one, are striking and seem to be organ and site dependent. While some studies indicate that ARRB2, not ARRB1, expression is critical for maintaining malignant phenotypes, other studies support the role of ARRB1 in tumorigenesis. Further work is needed to address the interplay between the two forms and their downstream targets in cancer progression.

ACKNOWLEDGMENTS

The authors would like to thank Perry V. Halushka, M.D., Ph.D. for critical review of the chapter. The authors' work is supported by NIH Grants R01 CA127905 (O. M.).

REFERENCES

1. Barak LS, Ferguson SS, Zhang J, Caron MG. A beta-arrestin/green fluorescent protein biosensor for detecting G protein-coupled receptor activation. *J Biol Chem*. 1997;272 (44):27497–27500.
2. Beaulieu JM, Caron MG. Beta-arrestin goes nuclear. *Cell*. 2005;123(5):755–757.
3. Dasgupta P, Rizwani W, Pillai S, et al. ARRB1-mediated regulation of E2F target genes in nicotine-induced growth of lung tumors. *J Natl Cancer Inst*. 2011;103(4):317–333.
4. Rosano L, Cianfrocca R, Masi S, et al. Beta-arrestin links endothelin A receptor to beta-catenin signaling to induce ovarian cancer cell invasion and metastasis. *Proc Natl Acad Sci USA*. 2009;106(8):2806–2811.
5. Buchanan FG, Gorden DL, Matta P, Shi Q, Matrisian LM, DuBois RN. Role of beta-arrestin 1 in the metastatic progression of colorectal cancer. *Proc Natl Acad Sci USA*. 2006;103(5):1492–1497.
6. Vines CM, Revankar CM, Maestas DC, et al. N-formyl peptide receptors internalize but do not recycle in the absence of arrestins. *J Biol Chem*. 2003;278(43):41581–41584.
7. Wagener BM, Marjon NA, Revankar CM, Prossnitz ER. Adaptor protein-2 interaction with arrestin regulates GPCR recycling and apoptosis. *Traffic*. 2009;10(9):1286–1300.
8. Khan MS, Devaraj H, Devaraj N. Chrysin abrogates early hepatocarcinogenesis and induces apoptosis in N-nitrosodiethylamine-induced preneoplastic nodules in rats. *Toxicol Appl Pharmacol*. 2011;251(1):85–94.
9. Li Y, Sun X, Zhang Y, et al. Morphine promotes apoptosis via TLR2, and this is negatively regulated by beta-arrestin 2. *Biochem Biophys Res Commun*. 2009;378(4):857–861.
10. Moorman J, Zhang Y, Liu B, et al. HIV-1 gp120 primes lymphocytes for opioid-induced, beta-arrestin 2-dependent apoptosis. *Biochim Biophys Acta*. 2009;1793(8): 1366–1371.

11. Povsic TJ, Kohout TA, Lefkowitz RJ. Beta-arrestin1 mediates insulin-like growth factor 1 (IGF-1) activation of phosphatidylinositol 3-kinase (PI3K) and anti-apoptosis. *J Biol Chem.* 2003;278(51):51334–51339.
12. Shi Q, Hou Y, Yang Y, Bai G. Protective effects of glycyrrhizin against beta(2)-adrenergic receptor agonist-induced receptor internalization and cell apoptosis. *Biol Pharm Bull.* 2011;34(5):609–617.
13. Sun X, Zhang Y, Wang J, et al. Beta-arrestin 2 modulates resveratrol-induced apoptosis and regulation of Akt/GSK3ss pathways. *Biochim Biophys Acta.* 2010;1800(9):912–918.
14. Zhang Z, Hao J, Zhao Z, et al. beta-Arrestins facilitate ubiquitin-dependent degradation of apoptosis signal-regulating kinase 1 (ASK1) and attenuate H_2O_2-induced apoptosis. *Cell Signal.* 2009;21(7):1195–1206.
15. Reiter E, Lefkowitz RJ. GRKs and beta-arrestins: roles in receptor silencing, trafficking and signaling. *Trends Endocrinol Metab.* 2006;17(4):159–165.
16. Penela P, Elorza A, Sarnago S, Mayor Jr F. Beta-arrestin- and c-Src-dependent degradation of G-protein-coupled receptor kinase 2. *EMBO J.* 2001;20(18):5129–5138.
17. Chen W, Hu LA, Semenov MV, et al. beta-Arrestin1 modulates lymphoid enhancer factor transcriptional activity through interaction with phosphorylated dishevelled proteins. *Proc Natl Acad Sci USA.* 2001;98(26):14889–14894.
18. Kang J, Shi Y, Xiang B, et al. A nuclear function of beta-arrestin1 in GPCR signaling: regulation of histone acetylation and gene transcription. *Cell.* 2005;123(5):833–847.
19. Laporte SA, Oakley RH, Holt JA, Barak LS, Caron MG. The interaction of beta-arrestin with the AP-2 adaptor is required for the clustering of beta 2-adrenergic receptor into clathrin-coated pits. *J Biol Chem.* 2000;275(30):23120–23126.
20. Pierce KL, Lefkowitz RJ. Classical and new roles of beta-arrestins in the regulation of G-protein-coupled receptors. *Nat Rev Neurosci.* 2001;2(10):727–733.
21. Lefkowitz RJ, Shenoy SK. Transduction of receptor signals by beta-arrestins. *Science.* 2005;308(5721):512–517.
22. Parent JL, Labrecque P, Orsini MJ, Benovic JL. Internalization of the TXA2 receptor alpha and beta isoforms. Role of the differentially spliced cooh terminus in agonist-promoted receptor internalization. *J Biol Chem.* 1999;274(13):8941–8948.
23. Parent JL, Labrecque P, Driss Rochdi M, Benovic JL. Role of the differentially spliced carboxyl terminus in thromboxane A2 receptor trafficking: identification of a distinct motif for tonic internalization. *J Biol Chem.* 2001;276(10):7079–7085.
24. Laroche G, Rochdi MD, Laporte SA, Parent JL. Involvement of actin in agonist-induced endocytosis of the G protein-coupled receptor for thromboxane A2: overcoming of actin disruption by arrestin-3 but not arrestin-2. *J Biol Chem.* 2005;280(24):23215–23224.
25. Kelley-Hickie LP, Kinsella BT. Homologous desensitization of signalling by the beta (beta) isoform of the human thromboxane A2 receptor. *Biochim Biophys Acta.* 2006;1761(9):1114–1131.
26. Park KS, Jeon SH, Oh JW, Choi KY. p21Cip/WAF1 activation is an important factor for the ERK pathway dependent anti-proliferation of colorectal cancer cells. *Exp Mol Med.* 2004;36(6):557–562.
27. Park KS, Ahn Y, Kim JA, Yun MS, Seong BL, Choi KY. Extracellular zinc stimulates ERK-dependent activation of p21(Cip/WAF1) and inhibits proliferation of colorectal cancer cells. *Br J Pharmacol.* 2002;137(5):597–607.
28. Ding XZ, Adrian TE. MEK/ERK-mediated proliferation is negatively regulated by P38 map kinase in the human pancreatic cancer cell line, PANC-1. *Biochem Biophys Res Commun.* 2001;282(2):447–453.
29. Xu CL, Lu XL, Yan XN, Wang HL, Chen SQ. Effects of PI3K/Akt/NF-kappaB signal pathway on FSH facilitation on cell proliferation and invasion by human epithelial ovarian cancer. *Zhonghua fu chan ke za zhi.* 2012;47(2):134–138.

30. Wei Y, Zhang Z, Liao H, et al. Nuclear estrogen receptor-mediated Notch signaling and GPR30-mediated PI3K/AKT signaling in the regulation of endometrial cancer cell proliferation. *Oncol Rep*. 2012;27(2):504–510.
31. Qu JL, Qu XJ, Zhao MF, et al. Gastric cancer exosomes promote tumour cell proliferation through PI3K/Akt and MAPK/ERK activation. *Dig Liver Dis*. 2009;41 (12):875–880.
32. Meng Q, Xia C, Fang J, Rojanasakul Y, Jiang BH. Role of PI3K and AKT specific isoforms in ovarian cancer cell migration, invasion and proliferation through the p70S6K1 pathway. *Cell Signal*. 2006;18(12):2262–2271.
33. Pierce KL, Luttrell LM, Lefkowitz RJ. New mechanisms in heptahelical receptor signaling to mitogen activated protein kinase cascades. *Oncogene*. 2001;20(13):1532–1539.
34. Chun KS, Lao HC, Langenbach R. The prostaglandin E2 receptor, EP2, stimulates keratinocyte proliferation in mouse skin by G protein-dependent and {beta}-arrestin1-dependent signaling pathways. *J Biol Chem*. 2010;285(51):39672–39681.
35. Dasgupta P, Rastogi S, Pillai S, et al. Nicotine induces cell proliferation by beta-arrestin-mediated activation of Src and Rb-Raf-1 pathways. *J Clin Invest*. 2006;116 (8):2208–2217.
36. Fong AM, Premont RT, Richardson RM, Yu YR, Lefkowitz RJ, Patel DD. Defective lymphocyte chemotaxis in beta arrestin2- and GRK6-deficient mice. *Proc Natl Acad Sci USA*. 2002;99(11):7478–7483.
37. Decaillot FM, Kazmi MA, Lin Y, Ray-Saha S, Sakmar TP, Sachdev P. CXCR7/CXCR4 heterodimer constitutively recruits beta-arrestin to enhance cell migration. *J Biol Chem*. 2011;286(37):32188–32197.
38. Sun Y, Cheng Z, Ma L, Pei G. Beta-arrestin2 is critically involved in CXCR4-mediated chemotaxis, and this is mediated by its enhancement of p38 MAPK activation. *J Biol Chem*. 2002;277(51):49212–49219.
39. Girnita L, Shenoy SK, Sehat B, et al. {beta}-Arrestin is crucial for ubiquitination and down-regulation of the insulin-like growth factor-1 receptor by acting as adaptor for the MDM2 E3 ligase. *J Biol Chem*. 2005;280(26):24412–24419.
40. Girnita L, Shenoy SK, Sehat B, et al. Beta-arrestin and Mdm2 mediate IGF-1 receptor-stimulated ERK activation and cell cycle progression. *J Biol Chem*. 2007;282 (15):11329–11338.
41. Zou L, Yang R, Chai J, Pei G. Rapid xenograft tumor progression in beta-arrestin1 transgenic mice due to enhanced tumor angiogenesis. *FASEB J*. 2008;22(2):355–364.
42. Li TT, Alemayehu M, Aziziyeh AI, et al. Beta-arrestin/Ral signaling regulates lysophosphatidic acid-mediated migration and invasion of human breast tumor cells. *Mol Cancer Res*. 2009;7(7):1064–1077.
43. Shenoy SK, Han S, Zhao YL, et al. Beta-arrestin1 mediates metastatic growth of breast cancer cells by facilitating HIF-1-dependent VEGF expression. *Oncogene*. 2012;31 (3):282–292.
44. Ge L, Shenoy SK, Lefkowitz RJ, DeFea K. Constitutive protease-activated receptor-2-mediated migration of MDA MB-231 breast cancer cells requires both beta-arrestin-1 and -2. *J Biol Chem*. 2004;279(53):55419–55424.
45. Wang P, DeFea KA. Protease-activated receptor-2 simultaneously directs beta-arrestin-1-dependent inhibition and Galphaq-dependent activation of phosphatidylinositol 3-kinase. *Biochemistry*. 2006;45(31):9374–9385.
46. DeFea KA, Vaughn ZD, O'Bryan EM, Nishijima D, Dery O, Bunnett NW. The proliferative and antiapoptotic effects of substance P are facilitated by formation of a beta-arrestin-dependent scaffolding complex. *Proc Natl Acad Sci USA*. 2000;97(20): 11086–11091.
47. Revankar CM, Vines CM, Cimino DF, Prossnitz ER. Arrestins block G protein-coupled receptor-mediated apoptosis. *J Biol Chem*. 2004;279(23):24578–24584.

48. Lefkowitz RJ, Rajagopal K, Whalen EJ. New roles for beta-arrestins in cell signaling: not just for seven-transmembrane receptors. *Mol Cell.* 2006;24(5):643–652.
49. Li H, Sun X, LeSage G, et al. beta-Arrestin 2 regulates Toll-like receptor 4-mediated apoptotic signalling through glycogen synthase kinase-3beta. *Immunology.* 2010;130 (4):556–563.
50. Ahn S, Kim J, Hara MR, Ren XR, Lefkowitz RJ. {beta}-Arrestin-2 mediates anti-apoptotic signaling through regulation of BAD phosphorylation. *J Biol Chem.* 2009;284(13):8855–8865.
51. Luan B, Zhang Z, Wu Y, Kang J, Pei G. Beta-arrestin2 functions as a phosphorylation-regulated suppressor of UV-induced NF-kappaB activation. *EMBO J.* 2005;24 (24):4237–4246.
52. Schaffner F, Ruf W. Tissue factor and PAR2 signaling in the tumor microenvironment. *Arterioscler Thromb Vasc Biol.* 2009;29(12):1999–2004.
53. Boucharaba A, Guillet B, Menaa F, et al. Bioactive lipids lysophosphatidic acid and sphingosine 1-phosphate mediate breast cancer cell biological functions through distinct mechanisms. *Oncol Res.* 2009;18(4):173–184.
54. Boucharaba A, Serre CM, Guglielmi J, Bordet JC, Clezardin P, Peyruchaud O. The type 1 lysophosphatidic acid receptor is a target for therapy in bone metastases. *Proc Natl Acad Sci USA.* 2006;103(25):9643–9648.
55. Peyruchaud O, Boucharaba A, Saulnier-Blache JS, Clezardin P. Lysophosphatidic acid: a new link between blood platelets and bone metastasis. *Med Sci.* 2005;21(4):353–355.
56. Boucharaba A, Serre CM, Gres S, et al. Platelet-derived lysophosphatidic acid supports the progression of osteolytic bone metastases in breast cancer. *J Clin Invest.* 2004;114 (12):1714–1725.
57. Bonnans C, Flaceliere M, Grillet F, et al. Essential requirement for beta-arrestin2 in mouse intestinal tumors with elevated Wnt signaling. *Proc Natl Acad Sci USA.* 2012;109(8):3047–3052.
58. Siegel R, Naishadham D, Jemal A. Cancer statistics, 2012. *CA Cancer J Clin.* 2012;62 (1):10–29.
59. Heeschen C, Jang JJ, Weis M, et al. Nicotine stimulates angiogenesis and promotes tumor growth and atherosclerosis. *Nat Med.* 2001;7(7):833–839.
60. Wu Z, Tong W, Tan Z, Wang S, Lin P. The clinical significance of beta-arrestin 2 expression in the serum of non-small cell lung cancer patients. *Zhongguo Fei Ai Za Zhi.* 2011;14(6):497–501.
61. Wang MT, Honn KV, Nie D. Cyclooxygenases, prostanoids, and tumor progression. *Cancer Metastasis Rev.* 2007;26(3–4):525–534.
62. Schneider C, Pozzi A. Cyclooxygenases and lipoxygenases in cancer. *Cancer Metastasis Rev.* 2011;30(3–4):277–294.
63. Achiwa H, Yatabe Y, Hida T, et al. Prognostic significance of elevated cyclooxygenase 2 expression in primary, resected lung adenocarcinomas. *Clin Cancer Res.* 1999;5 (5):1001–1005.
64. Ristimaki A, Sivula A, Lundin J, et al. Prognostic significance of elevated cyclooxygenase-2 expression in breast cancer. *Cancer Res.* 2002;62(3):632–635.
65. Kim JI, Lakshmikanthan V, Frilot N, Daaka Y. Prostaglandin E2 promotes lung cancer cell migration via EP4-betaArrestin1-c-Src signalsome. *Mol Cancer Res.* 2010;8 (4):569–577.
66. Avritscher EB, Cooksley CD, Grossman HB, et al. Clinical model of lifetime cost of treating bladder cancer and associated complications. *Urology.* 2006;68(3):549–553.
67. Moussa O, Ashton AW, Fraig M, et al. Novel role of thromboxane receptors beta isoform in bladder cancer pathogenesis. *Cancer Res.* 2008;68(11):4097–4104.

68. Schultz KR, Bowman WP, Aledo A, et al. Improved early event-free survival with imatinib in Philadelphia chromosome-positive acute lymphoblastic leukemia: a children's oncology group study. *J Clin Oncol*. 2009;27(31):5175–5181.
69. Liu H, Long J, Zhang PH, et al. Elevated beta-arrestin1 expression correlated with risk stratification in acute lymphoblastic leukemia. *Int J Hematol*. 2011;93(4):494–501.
70. Fereshteh M, Ito T, Kovacs JJ, et al. beta-Arrestin2 mediates the initiation and progression of myeloid leukemia. *Proc Natl Acad Sci USA*. 2012;109(31):12532–12537.
71. Feldman BJ, Feldman D. The development of androgen-independent prostate cancer. *Nat Rev Cancer*. 2001;1(1):34–45.
72. Lakshmikanthan V, Zou L, Kim JI, et al. Identification of betaArrestin2 as a corepressor of androgen receptor signaling in prostate cancer. *Proc Natl Acad Sci USA*. 2009;106 (23):9379–9384.
73. Hennenberg M, Schlenker B, Roosen A, et al. beta-Arrestin-2 is expressed in human prostate smooth muscle and a binding partner of alpha1A-adrenoceptors. *World J Urol*. 2011;29(2):157–163.
74. Turley RS, Finger EC, Hempel N, How T, Fields TA, Blobe GC. The type III transforming growth factor-beta receptor as a novel tumor suppressor gene in prostate cancer. *Cancer Res*. 2007;67(3):1090–1098.
75. Mythreye K, Blobe GC. The type III TGF-beta receptor regulates epithelial and cancer cell migration through beta-arrestin2-mediated activation of Cdc42. *Proc Natl Acad Sci USA*. 2009;106(20):8221–8226.
76. Zhang P, He X, Tan J, Zhou X, Zou L. beta-Arrestin2 mediates beta-2 adrenergic receptor signaling inducing prostate cancer cell progression. *Oncol Rep*. 2011;26 (6):1471–1477.
77. Heinrich EL, Lee W, Lu J, Lowy AM, Kim J. Chemokine CXCL12 activates dual CXCR4 and CXCR7-mediated signaling pathways in pancreatic cancer cells. *J Transl Med*. 2012;10:68.
78. Shen X, Artinyan A, Jackson D, Thomas RM, Lowy AM, Kim J. Chemokine receptor CXCR4 enhances proliferation in pancreatic cancer cells through AKT and ERK dependent pathways. *Pancreas*. 2010;39(1):81–87.
79. Thayer SP, di Magliano MP, Heiser PW, et al. Hedgehog is an early and late mediator of pancreatic cancer tumorigenesis. *Nature*. 2003;425(6960):851–856.
80. Chen W, Ren XR, Nelson CD, et al. Activity dependent internalization of smoothened mediated by beta-arrestin 2 and GRK2. *Science*. 2004;306(5705):2257–2260.
81. Mandell JW, Glass G, Gianchandani EP, et al. Dephosphorylation of beta-arrestin 1 in glioblastomas. *J Neuropathol Exp Neurol*. 2009;68(5):535–541.
82. Michal AM, Peck AR, Tran TH, et al. Differential expression of arrestins is a predictor of breast cancer progression and survival. *Breast Cancer Res Treat*. 2011;130(3):791–807.
83. Rubin JB, Kung AL, Klein RS, et al. A small-molecule antagonist of CXCR4 inhibits intracranial growth of primary brain tumors. *Proc Natl Acad Sci USA*. 2003;100 (23):13513–13518.
84. Gravel S, Malouf C, Boulais PE, et al. The peptidomimetic CXCR4 antagonist TC14012 recruits beta-arrestin to CXCR7: roles of receptor domains. *J Biol Chem*. 2010;285(49):37939–37943.
85. Odemis V, Lipfert J, Kraft R, et al. The presumed atypical chemokine receptor CXCR7 signals through G(i/o) proteins in primary rodent astrocytes and human glioma cells. *Glia*. 2012;60(3):372–381.
86. Lundgren K, Tobin NP, Lehn S, et al. Stromal expression of beta-arrestin-1 predicts clinical outcome and tamoxifen response in breast cancer. *J Mol Diagn*. May 2011;13 (3):340–351.

CHAPTER SIXTEEN

Arrestins in Metabolic Regulation

Jian Zhao*, Gang Pei*,†
*State Key Laboratory of Cell Biology, Institute of Biochemistry and Cell Biology, Shanghai Institutes for Biological Sciences, Chinese Academy of Sciences, Shanghai, PR China
†School of Life Science and Technology, Tongji University, Shanghai, PR China

Contents

1. A Short Introduction to Metabolic Regulation 413
2. Functional Roles of β-Arrestins in Regulation of Whole-Body Energy Balance and Body Weight Control 415
   2.1 Hypothalamic signals to peripheral tissues 415
   2.2 The peripheral signals 416
3. Functional Role of β-Arrestins in Regulation of Carbohydrate and Lipid Homeostasis 417
   3.1 Insulin signaling 417
   3.2 Signals mediating insulin secretion 419
   3.3 Glucagon and other GPCR signals 420
   3.4 Peroxisome proliferator-activated receptors 421
4. Conclusions 423
References 424

Abstract

This review summarizes the regulatory roles of β-arrestins in whole-body energy balance, body weight control, and carbohydrate and lipid homeostasis. Much research has pointed in the direction of the functions of β-arrestins in mediating desensitization and endocytosis of G protein-coupled receptors as well as in activating the receptor/β-arrestin/ERK signaling pathway being crucial for metabolic regulation. Furthermore, β-arrestins form diverse signal complexes for the activation of the downstream cassettes for the body's metabolic reactions. However, further studies are required to fully address the emerging roles of β-arrestins in metabolic regulation and related diseases.

1. A SHORT INTRODUCTION TO METABOLIC REGULATION

Metabolic regulation is the physiological mechanism by which the body takes in nutrients and delivers energy as required. Metabolic regulation

Progress in Molecular Biology and Translational Science, Volume 118
ISSN 1877-1173
http://dx.doi.org/10.1016/B978-0-12-394440-5.00016-4

works ultimately at a molecular level, mainly by modulation of enzyme activities that function together as a whole system to sense the balance of energy coming in and energy required. The different organs in the body have their own characteristic patterns of metabolism according to their functions in the body. Thus, it is critical that metabolic pathways interact in a dynamic sense, in the entire organism. Furthermore, the endocrine and nervous systems need to precisely coordinate to control the flow of energy within the body.

Much of the metabolic regulation is governed by hormones that are delivered through the bloodstream and act through specific cellular receptors. Both the cell-surface receptors (that usually bind peptide hormones) and the nuclear receptors (that bind thyroid hormones, steroid hormones, and other membrane-permeant ligands) play critical roles in metabolic regulation. Hormones acting through cell-surface receptors are involved in rapid metabolic adjustments. These receptors signal via the small molecule cyclic adenosine 3,5-monophosphate (cyclic AMP or cAMP) and the membrane lipid phosphatidylinositol (3,4,5) trisphosphate. Following this, the activities of downstream metabolic enzymes are regulated by covalent modification, especially phosphorylation and dephosphorylation, and/or translocation of enzymes within the cell. Activation of the nuclear receptors by their ligands, on the other hand, directly controls the transcription of metabolic genes and leads to long-term metabolic regulation.

Studies carried out with several different families of receptors have pointed out that β-arrestins determine the specificity, spatiality, and temporality of cellular signals as well as the intracellular movement of receptors and other signal complexes.[1–4] Binding of β-arrestins to ligand-bound G protein-coupled receptors (GPCRs) physically uncouples the G protein from the receptor and effectively terminates G protein-mediated signaling. By coupling to a ligand-activated receptor, β-arrestins also initiate GPCR signaling in a G protein-independent manner. Furthermore, β-arrestins scaffold diverse signal complexes, thereby linking activated receptors with distinct sets of accessory and effector proteins. Considering the intricate metabolic regulatory network composed of a variety of hormones and their specific receptors, it should not be a surprise that proper functioning of β-arrestins is indispensable for the body's metabolic function. This chapter summarizes the function of β-arrestins in metabolic regulation and also discusses their association with metabolic syndromes including insulin resistance, type 2 diabetes, and obesity.

2. FUNCTIONAL ROLES OF β-ARRESTINS IN REGULATION OF WHOLE-BODY ENERGY BALANCE AND BODY WEIGHT CONTROL

2.1. Hypothalamic signals to peripheral tissues

In response to the body energy storage level, hypothalamic nuclei signal to different peripheral tissues to ensure a well-orchestrated response to food intake and energy expenditure thereby maintaining a stable body weight. The key hypothalamic neuroendocrine peptides include proopiomelanocortin (POMC), neuropeptide Y (NPY), agouti-related peptide (AgRP),[5] orexin, melanin-concentrating hormone (MCH), alpha-melanocyte stimulating hormone (α-MSH), and cocaine- and amphetamine-regulated transcript (CART).[6] These peptides bind with their GPCRs on cell membranes to influence energy homeostasis. Studies show not only that β-arrestins influence these peptide signals by mediating receptor desensitization and endocytosis but also that receptor/β-arrestin/ubiquitin complexes are intrinsic to activation of the downstream pathways for organismal metabolic reactions.

The 36-amino acid peptide neurotransmitter NPY family peptides [NPY, peptide YY (PYY), and pancreatic polypeptide] bind to four GPCRs (Y1, Y2, Y4, and Y5). Due to their profound central effects on energy intake, mood, and reproduction, the pharmacological profiles and distribution of NPYs and their receptors have been studied extensively.[7] NPYs bind to Y receptors with different affinities and lead to activation of $G_{i/o}$ protein and inhibition of cAMP synthesis. Desensitization and internalization of the Y receptors are controlled by β-arrestin-2.[8] Interestingly, the receptor that is known to internalize rapidly (Y1) exhibits the most rapid association with β-arrestin-2, whereas the receptor that is known to internalize slowly, or not at all, (Y2) associates slowly with β-arrestin-2, suggesting that the association of β-arrestin-2 with the activated receptors is a critical step for Y receptor desensitization.[9–11]

Melanocortin receptors (MC3R and MC4R) expressed in the arcuate nucleus (MC3R) and in the nucleus paraventricular (MC4R) play important roles in energy homeostasis.[12] α-Melanocyte-stimulating hormone (α-MSH), generated by processing of hypothalamic expressed POMC, binds to MC3R and MC4R and leads to cAMP accumulation and inhibition of food intake.[13] Agouti and AgRP act as natural inverse agonists to diminish signaling of MC3R and MC4R and control body weight.[14] First, Agouti-related peptide exerts its physiological effects by antagonizing the actions of α-MSH on MCRs. Furthermore, binding of Agouti-related peptide to MC3R and MC4R decreases basal, agonist-independent, receptor signaling. Studies using bioluminescence resonance energy transfer assays showed that binding of Agouti-related peptide promotes the interaction of β-arrestins with MC3R and MC4R and initiates receptor endocytosis,[15] pointing out a new mechanism by which the inverse agonist Agouti-related peptide reduces MCR density at the cell surface and fine tunes the response of MCR-expressing cells to melanocortins.

Orexin-A (OxA) and -B (OxB), also called hypocretins, are neurotransmitters produced by neurons in the hypothalamus that regulate arousal, wakefulness, and appetite.[16] Orexins increase the craving for food. OxA stimulates glucose uptake in 3T3-L1 adipocytes, and increases energy uptake and lipogenesis.[17] Orexin-knockout mice showed impaired brown adipose tissue thermogenesis.[18] OxA binds to both orexin receptors 1 and 2 (OxR1 and OxR2) with similar affinity, whereas OxB has less affinity for OxR1 compared with OxR2.[19] Both OxR1 and OxR2 interact with β-arrestin-1 and -2 in an agonist dose-dependent manner, but with a two orders of magnitude lower affinity than observed for G protein activation.[20,21] A potency shift of OxA-induced ERK1/2 phosphorylation between early (2 min; predominantly G protein-mediated) and late (90 min; predominantly β-arrestin-mediated) signaling responses has been observed. Furthermore, the OxA-induced receptor–arrestin–ubiquitin complex differs between OxR subtypes. Compared with OxR2, OxR1 dissociates from either form of β-arrestin more rapidly, with a stronger initial interaction and more rapid recycling back to the plasma membrane. Consistently, OxA-dependent ERK1/2 phosphorylation in the presence of OxR2 is more sustained compared with that of OxR1. Temporal profiling of OxA-dependent receptor-β-arrestin and β-arrestin-ubiquitin proximity suggests tangible subtype-specific differences in receptor trafficking and function. Importantly, these delicate differences in protein complex formation, stability, and duration might be useful in targeting these receptors for therapeutic purposes.

2.2. The peripheral signals

In response to different nutrients, the gastrointestinal tract produces over 20 enteroendocrine hormones that act locally, peripherally, and centrally to affect food intake and regulate the digestion, absorption, distribution, and metabolism of nutrients. The enteroendocrine hormones include serotonin (5-hydroxytryptamine; 5-HT), ghrelin, glucagon-like peptide (GLP)-1 and -2,

PYY, glucose-dependent insulinotropic polypeptide (GIP), CCK, and others. Besides the enteroendocrine hormones, adipocytes secrete leptin as a signal of total body fat mass. β-Arrestins actively participate in the signaling of these orexigenic and anorexigenic peptides.

Ghrelin is an acylated 28 amino acid peptide secreted by the stomach just before an expected meal, and strongly promotes food intake. Ghrelin is the endogenous ligand for the growth hormone (GH) secretagogue-receptor (GHS-R). Ghrelin inhibits insulin release in mice, rats, and humans. Low plasma ghrelin levels are associated with elevated fasting insulin levels and insulin resistance in humans. Camiña *et al.* found that besides the $Ca(^{2+})$-dependent PKC/Src and phosphoinositide 3-kinase (PI3K)/PKC/Src pathways, the ghrelin receptor GHS-R1a also activates ERK1/2 via β-arrestin-mediated pathways.[22] Binding of β-arrestin-1 and -2 to the receptor leads to the formation of a multiprotein complex containing the β-arrestins, Src, Raf-1, and ERK 1/2, indicating that β-arrestins act as scaffolding proteins and signal transducers for GHS-R-activated ERK1/2 signaling. Damian *et al.* showed that ghrelin binding induces a specific GHS-R1a conformational change leading to agonist-dependent β-arrestin-2 recruitment to the isolated GHS-R in lipid discs, whereas the interaction of GHS-R with μ-AP2 is independent of ghrelin.[23] The distinct ligand requirements for the interaction of the purified GHS-R1a with β-arrestin or AP2 provide a new mechanism for the differential regulation of basal- and agonist-induced receptor internalization in cells.

3. FUNCTIONAL ROLE OF β-ARRESTINS IN REGULATION OF CARBOHYDRATE AND LIPID HOMEOSTASIS

Glucose metabolism in the liver (glycolysis and gluconeogenesis) is regulated by insulin and glucagon via several transcription factors responding to cAMP concentrations (cAMP-responsive element binding proteins; CREBs). Key regulators of glucose metabolism include the forkhead transcription factor, FOXO1, a CREB-regulated transcription coactivator 2, GCN5 acetyltransferase, and ATF4.

3.1. Insulin signaling

The insulin signaling pathway is pivotal in maintaining metabolic homeostasis. Of most importance, glucose homeostasis is maintained by the fine orchestration of insulin secretion and insulin action to promote glucose transport into muscle and adipocytes. Normally, the insulin receptor

(IR), stimulated by insulin, recruits and phosphorylates insulin receptor substrate (IRS) protiens, which leads to activation of the phosphatidylinositol 3-kinase (PI3K)/Akt pathway. The PI3K/Akt pathway then mediates metabolic actions of insulin. Deficits in the insulin signal pathway lead to insulin resistance and, eventually, type 2 diabetes mellitus. Insulin resistance is characterized by a diminished IR response to insulin stimulation which leads to decreased capacity of circulating insulin to regulate glucose and lipid metabolism in adipose tissue, liver, and skeletal muscle.[24]

The prevailing view on insulin signaling considers mainly the recruitment of adaptor proteins such as IRS, GAP, Grb2, and Shc to IR. Interestingly, β-arrestins are also able to interact with IR. β-Arrestin-1 associates with IR in a ligand-dependent manner.[25–27] Binding with β-arrestin-1 promotes receptor endocytosis, however, neither glucose metabolic action nor mitogenic signaling by insulin is dependent on β-arrestin-1. Surprisingly, β-arrestin-2, which shares a high similarity with β-arrestin-1 and functions redundantly to β-arrestin-1 in most circumstances, participates in insulin signaling as well as in downstream glucose metabolic actions. We discovered that β-arrestin-2 promotes Akt activation in IR signaling through scaffolding an IR/Akt/β-arrestin-2/Src signal complex.[28] The formation of this protein complex depends on the stimulation of IR by insulin, indicating that the β-arrestin-2-mediated signal complex is an intrinsic component of IR signaling pathway. β-Arrestin-2 mediates Akt activation through Src, but not PI3K, following insulin stimulation. Thus, in addition to the classical PI3K/Akt pathway, modulation of IR activation by β-arrestin-2 provides a novel regulatory mechanism for hormonal regulation of insulin sensitivity. Further, the IR/Akt/β-arrestin-2/Src signal complex is essential for proper insulin signaling and whole-body insulin action. Downregulation of β-arrestin-2 expression or mutation of β-arrestin-2 leads to the disruption of the Akt/Src interaction, Src-dependent Akt activation, and insulin actions. This, in turn, increases insulin resistance and type 2 diabetes phenotypes *in vivo*.

It is well known that chronic insulin treatment leads to desensitization of insulin signaling. This process involves insulin-induced ubiquitination and degradation of IRS proteins. Although β-arrestin-1 does not appear to mediate insulin responses, it does alter insulin signaling in the long term. Insulin stimulates the association of IRS-1 and Mdm2, an E3 ubiquitin ligase, which leads to the degradation of IRS-1. Overexpression of β-arrestin-1 inhibited the formation of the IRS-1/Mdm2 complex and thereby attenuated insulin-induced degradation of IRS-1, leading to an increased insulin signaling downstream of IRS-1.[29] When endogenous β-arrestin-1 was knocked down by β-arrestin-1 siRNA, insulin-induced IRS-1 degradation was enhanced. Thus, β-arrestin-1 participates in long-term and β-arrestin-2 in short-term insulin signaling.

3.2. Signals mediating insulin secretion

Normal islet function is a prerequisite for normal glucose homeostasis. In fact, islet dysfunction is a key event underlying development of type 2 diabetes, as manifested by impaired insulin secretion.[30] Islet function is regulated by a number of different signals including hormones, neurotransmitters, and nutrients. Several of these factors signal through GPCRs for regulation of insulin secretion and glucose homeostasis.[31] Current studies show that β-arrestin-1 functions to mediate GPCR signaling in islets.[32]

GLP-1 is a polypeptide hormone secreted from enteroendocrine L cells and potentiates glucose-dependent insulin secretion in pancreatic cells. Recently, the GLP-1 receptor (GLP-1R) has been the focus of a new form of antidiabetic therapy, with the introduction of GLP-1 analogs and dipeptidyl peptidase type IV inhibitors. This has stimulated additional interest in the mechanisms of GLP-1 signaling.[33] β-Arrestin-1 mediates the effects of GLP-1 to stimulate cAMP production and insulin secretion in cells.[34] Using a coimmunoprecipitation technique, Sonoda *et al.* found a physical association between the GLP-1R and β-arrestin-1 in cultured INS-1 pancreatic cells. Knockdown of β-arrestin-1 broadly attenuated GLP-1 signaling, causing decreased ERK and CREB activation, and IRS-2 expression, as well as reduced cAMP levels and impaired insulin secretion. However, β-arrestin-1 knockdown did not affect GLP-1R surface expression or ligand-induced GLP-1R internalization/desensitization. Further, Queyer *et al.* found that in MIN6 β-cells, GLP-1 stimulates the phosphorylation of the proapoptotic protein BAD through the β-arrestin-1-dependent ERK1/2-p90RSK cascade.[35] Thus, β-arrestin-1 seems to play important roles in mediating GLP-1-stimulated insulin secretion as well as β-cell survival.

The pituitary adenylate cyclase-activating polypeptide (PACAP) is expressed in islet parasympathetic nerve terminals and strongly potentiates insulin secretion in a glucose-dependent manner. PACAP functions through three GPCRs (PAC1, VPAC1, and VPAC2) via the Gαs/cAMP/PKA pathway.[36] Broca *et al.* found that, in the rat pancreatic INS-1E cell line, PACAP via PAC1R activates ERK1/2 by two spatiotemporally separated signaling pathways.[37] PACAP alone induces transient ERK1/2 activation via PAC1R through PKA- and Src-dependent, but β-arrestin-1-independent pathways, and promotes the nuclear translocation of a fraction of phosphorylated ERK1/2 (p-ERK1/2).[38] In contrast, in the presence of glucose, PACAP potentiates the glucose induced long-lasting ERK1/2 activation via a β-arrestin-1-dependent pathway. In addition, activation of ERK1/2 through the β-arrestin-1-dependent pathway is involved in PACAP-induced IRS-2 protein expression, which is crucial for β-cell function and survival.

The islets express two families of purinergic GPCRs, the P1 receptors (activated by adenosine) and P2 receptors (activated by ATP and ADP). The P2 receptor P2Y is expressed in islets.[39] Furthermore, there are six different P2Y receptor subtypes. Studies showed that in clonal β cells, P2Y1 and P2Y6 receptors are expressed in islets and are functionally active in stimulating insulin secretion in a Ca^{2+}- and cAMP-dependent manner. In contrast to P2Y, P1 receptor activation inhibits insulin secretion through inhibition of adenylate cyclase activity. Hoffmann *et al.* found that the P2Y1 receptor binds β-arrestin-2 with higher affinity than β-arrestin-1.[40] Reiner *et al.* analyzed the role of potential phosphorylation sites in the third intracellular loop and the C-terminus of the human P2Y1 receptor for β-arrestin binding and receptor internalization.[41] The mutation of potential phosphorylation sites in the distal C-terminus of P2Y1 receptor fully prevented β-arrestin-2 translocation, whereas truncation of the C-terminus adjacent to the important sites (P2Y1-363) had no influence on internalization.[42] siRNA-mediated depletion of β-arrestin-1 almost completely attenuated UTP-stimulated P2Y2-receptor desensitization, whereas knockdown of β-arrestin-2 had no effect on P2Y2 signaling.

GIP is one of the major mediators in the regulation of nutrient-dependent insulin release from the pancreas.[43,44] The GIP receptor is a member of the class II GPCRs.[45] In bTC3 cells, expression of β-arrestin-1 attenuated GIP-induced insulin release and cAMP production, whereas glucose-stimulated insulin secretion was not affected.[46] These results suggest a potential role for β-arrestin-1 in modulating GIP-mediated insulin secretion in pancreatic islet cells but the exact mechanism is not clear.

3.3. Glucagon and other GPCR signals

Glucagon is a hormone secreted by the pancreas that raises blood glucose levels. Its effect is opposite to that of insulin, which lowers blood glucose levels. Glucagon is synthesized and secreted from the α cells of the islets, which are located adjacent to β cells in the endocrine portion of the pancreas. Merlen *et al.* studied the functional relationship between glucagon receptor endocytosis, phosphorylation, and coupling to the adenylate cyclase pathway.[47] They found that following administration of a saturating dose of glucagon to rats, a rapid internalization of glucagon receptor was observed coincident with its serine phosphorylation both at the plasma membrane and within endosomes. Coincident with glucagon receptor endocytosis, a massive internalization of both the 45- and 47-kDa Gsα proteins were also observed. In contrast, no change in the subcellular distribution of adenylate cyclase or β-arrestin-1 and -2 were observed. Krilov *et al.* found that the glucagon receptor recycles through Rab4- and Rab11-positive vesicles. The actin cytoskeleton, β-arrestin-1, β-arrestin-2, and the receptor's carboxyl terminus are involved in recycling.[48]

Both insulin and endothelin type A (ETA) can stimulate the activation of ETA receptor, leading to $G_{q/11}$ protein and PI3K activation, subsequently promoting glucose transport by enhancing the translocation of GLUT4, the predominant glucose transporter in skeletal muscle, and white and brown adipocytes.[49] β-Arrestin-1 promoted ETA-stimulated, but not insulin-stimulated, GLUT4 translocation. β-Arrestin-1 associated with the ETA receptor in an agonist-dependent manner and recruited Src kinase to the ETA receptor thus forming a signal complex.[50] It was concluded that β-arrestin-1 and Src kinase form a molecular complex with the ETA receptor to mediate ETA signaling to $G_{q/11}$ with subsequent stimulation of glucose transport.

The pineal hormone melatonin is well known to affect carbohydrate metabolism, but it also inhibits insulin secretion through the GPCRs mT1 (also known as mTNr1A) and mT2 (also known as mTNr1B), which are expressed in islets and clonal β-cells.[51] mT1 and mT2 signal through G_i proteins to inhibit adenylate cyclase and reduce cAMP. Bondi *et al.* used coimmunoprecipitation analysis and found that in the presence of melatonin, β-arrestin-2 is required for the formation of MT1 receptor/G protein/MEK1/2/ERK1/2 complexes.[52] β-Arrestin-2 mediates the endocytosis of mT1 receptor and forms β-arrestin-2/MEK1/2/ERK1/2 complexes upon the activation of mT1.

3.4. Peroxisome proliferator-activated receptors

Peroxisome proliferator-activated receptors (PPARs) are a group of nuclear receptor proteins that function as master transcription factors to regulate gene expression according to the availability of fatty acids.[53] The endogenous ligands for PPARs are fatty acids or compounds derived from fatty acids, such as certain prostaglandins. There are three major isoforms of PPAR (α, β/δ, and γ). Upregulated PPARα in the liver increases fatty acid oxidation and overall fatty acid availability. Fatty acid storage as triacylglycerol in adipose tissue is also upregulated by activated PPARγ. As PPAR ligands, the *n*-3 series polyunsaturated fatty acids are more potent than the *n*-6 series or the saturated fatty acids. The PPAR protein, with its ligand bound, binds to the Retinoid X Receptor (RXR), whose ligand is all-*cis*-retinoic acid, a derivative of vitamin A. PPARγ is considered as one of the master regulators of adipogenesis and macrophage function.[53,54] Activated PPARγ/RXR heterodimers bind to peroxisome proliferator response elements (PPREs) and trigger the expression of PPARγ-targeted genes.[55–57] The coordinated activation of these PPARγ-targeted adipogenic genes leads to a flux of fatty acids from the circulation and other tissues into adipocytes. PPARγ transcriptional activity is also modulated by interaction with a number of different coactivators or corepressors, such as NCoR and SMRT.[54,58,59] The PPARγ-repressive complexes possess histone deacetylase or methyltransferase activities through which they remodel the chromatin structure and repress gene transcription.[54] In activated macrophages, PPARγ-repressive complexes function to mediate the expression of inflammatory genes for immune responses.[54] In these ways, PPARγ governs the function of adipocytes and macrophages and helps to achieve whole-body energy balance, and has become a central focus of obesity and diabetes research.

The PPAR system has aroused intense interest because of its potential for pharmacological manipulation. Thiazolidinediones, synthetic ligands for PPARγ, have been widely in use for the treatment of diabetes. These ligands act by increasing the ability of adipose tissue to store excess fatty acids as triacylglycerol. Thus, circulating fatty acid concentrations are reduced, and tissues such as skeletal muscle are able to utilize more glucose because of reduced substrate competition. More recently, drugs have been designed to activate PPARδ. Although the physiological role of PPARδ has not been clear, studies have shown that it can markedly increase the muscle's oxidative capacity.[60] Furthermore, PPARδ agonists have been shown to improve HDL-cholesterol concentrations and reduce insulin resistance in monkeys and humans.[61]

Our previous studies showed that in the nucleus, β-arrestin-1 regulates histone modification and gene transcription through its interaction with p300, the same cofactor necessary for hypoxia inducible factor (HIF)-mediated transcription.[62] Recently, we reported that β-arrestin-1 modifies recruitment of PPARγ coregulators through a β-arrestin-1/PPARγ interaction.[63,64] β-Arrestin-1 exhibits antiobesity effects through inhibition of the PPARγ/RXRα heterodimer formation and promotion of PPARγ/NCoR repressive complex, thus repressing PPARγ-mediated adipogenesis and inflammatory response. It is still unclear which upstream signal may regulate the association or disassociation of β-arrestin-1 with PPARγ.

Interestingly, we found that knockout of β-arrestin-1 led to an increased expression of adipogenic proteins such as PPARγ, Fabp4, CD36, Lpl, and Fasn, without significant effects on the expression of PPARδ, Acox1, Acadm, RXRα, C/EBPα, and Glut4, suggesting that β-arrestin-1 might mediate PPARγ-dependent expression of adipogenic proteins, lipid metabolic proteins, and adipokines differently. It is worth noting that we found that rosiglitazone-stimulation decreased the β-arrestin-1/PPARγ association without influencing PPARγ/RXRα heterodimer formation. Recently, two elegant studies have added very strong evidence that the PPARγ ligands can be subcategorized based on their ability to maintain PPARγ-controlled transcription (for adipogenesis) and to antagonize the Cdk5-mediated phosphorylation of PPARγ (for insulin sensitization).[65,66] Compounds that block the obesity-linked phosphorylation of PPARγ by Cdk5 without reducing its classical transcriptional activation property have potent antidiabetic activity while not causing the fluid retention, weight gain, or interference with bone formation that are serious side effects of the thiazolidinedione class of PPAR ligands. Thus, it would be of interest to investigate whether β-arrestin-1 affects Cdk5-phosphorylation of PPARγ and whether mediation of PPARγ activity by either β-arrestin-1 or Cdk5 shares the same or similar underlying mechanisms. Nevertheless, all these data illustrate that new classes of drugs specifically modulating PPARγ activities could be developed against metabolic diseases.

4. CONCLUSIONS

The evidence summarized in this chapter has clearly shown that β-arrestins are actively involved in the regulation of pathways that are crucial for metabolic regulation, not only through the classical mechanisms of regulating GPCRs that are known to be critical in whole-body energy balance, body weight control, carbohydrate and lipid homeostasis but also through the novel mechanisms related to the so-called β-arrestin-biased signaling. It is worth pointing out that β-arrestins also regulate other non GPCR receptors and transcription factors, mostly via direct interaction or by forming diverse signal complexes, to influence body metabolism. Furthermore, any defects or disturbances of β-arrestin-mediated regulation would likely lead to metabolic disorders such as insulin resistance, type 2 diabetes, and obesity, indicating the diagnostic and therapeutic potential of modulating β-arrestin-mediated pathways.

REFERENCES

1. Whalen EJ, Rajagopal S, Lefkowitz RJ. Therapeutic potential of beta-arrestin- and G protein-biased agonists. *Trends Mol Med.* 2011;17:126–139.
2. Shenoy SK, Lefkowitz RJ. β-Arrestin-mediated receptor trafficking and signal transduction. *Trends Pharmacol Sci.* 2011;32:521–533.
3. DeWire SM, Ahn S, Lefkowitz RJ, Shenoy SK. β-Arrestins and cell signaling. *Annu Rev Physiol.* 2007;69:483–510.
4. Shukla AK, Xiao K, Lefkowitz RJ. Emerging paradigms of β-arrestin-dependent seven transmembrane receptor signaling. *Trends Biochem Sci.* 2011;36:457–469.
5. Biebermann H, Kuhnen P, Kleinau G, Krude H. The neuroendocrine circuitry controlled by POMC, MSH, and AGRP. *Handb Exp Pharmacol.* 2012;209:47–75.
6. Harrold JA, Halford JC. The hypothalamus and obesity. *Recent Pat CNS Drug Discov.* 2006;1:305–314.
7. Zhang L, Bijker MS, Herzog H. The neuropeptide Y system: pathophysiological and therapeutic implications in obesity and cancer. *Pharmacol Ther.* 2011;131:91–113.
8. Holliday ND, Lam CW, Tough IR, Cox HM. Role of the C terminus in neuropeptide Y Y1 receptor desensitization and internalization. *Mol Pharmacol.* 2005;67:655–664.
9. Berglund MM, Schober DA, Statnick MA, McDonald PH, Gehlert DR. The use of bioluminescence resonance energy transfer 2 to study neuropeptide Y receptor agonist-induced beta-arrestin 2 interaction. *J Pharmacol Exp Ther.* 2003;306:147–156.
10. Kilpatrick LE, Briddon SJ, Hill SJ, Holliday ND. Quantitative analysis of neuropeptide Y receptor association with beta-arrestin2 measured by bimolecular fluorescence complementation. *Br J Pharmacol.* 2010;160:892–906.
11. Kilpatrick LE, Briddon SJ, Holliday ND. Fluorescence correlation spectroscopy, combined with bimolecular fluorescence complementation, reveals the effects of beta-arrestin complexes and endocytic targeting on the membrane mobility of neuropeptide Y receptors. *Biochim Biophys Acta.* 2012;1823:1068–1081.
12. Tao YX. Molecular mechanisms of the neural melanocortin receptor dysfunction in severe early onset obesity. *Mol Cell Endocrinol.* 2005;239:1–14.
13. Fan ZC, Sartin JL, Tao YX. Molecular cloning and pharmacological characterization of porcine melanocortin-3 receptor. *J Endocrinol.* 2008;196:139–148.
14. Yang YK, Thompson DA, Dickinson CJ, et al. Characterization of Agouti-related protein binding to melanocortin receptors. *Mol Endocrinol.* 1999;13:148–155.
15. Breit A, Wolff K, Kalwa H, Jarry H, Buch T, Gudermann T. The natural inverse agonist agouti-related protein induces arrestin-mediated endocytosis of melanocortin-3 and -4 receptors. *J Biol Chem.* 2006;281:37447–37456.
16. Nixon JP, Kotz CM, Novak CM, Billington CJ, Teske JA. Neuropeptides controlling energy balance: orexins and neuromedins. *Handb Exp Pharmacol.* 2012;209:77–109.
17. Skrzypski M, T Le T, Kaczmarek P, et al. Orexin A stimulates glucose uptake, lipid accumulation and adiponectin secretion from 3T3-L1 adipocytes and isolated primary rat adipocytes. *Diabetologia.* 2011;54:1841–1852.
18. Sellayah D, Bharaj P, Sikder D. Orexin is required for brown adipose tissue development, differentiation, and function. *Cell Metab.* 2011;14:478–490.

19. Sakurai T, Amemiya A, Ishii M, et al. Orexins and orexin receptors: a family of hypothalamic neuropeptides and G protein-coupled receptors that regulate feeding behavior. *Cell.* 1998;92:573–585.
20. Evans NA, Groarke DA, Warrack J, et al. Visualizing differences in ligand-induced beta-arrestin-GFP interactions and trafficking between three recently characterized G protein-coupled receptors. *J Neurochem.* 2001;77:476–485.
21. Dalrymple MB, Jaeger WC, Eidne KA, Pfleger KD. Temporal profiling of orexin receptor-arrestin-ubiquitin complexes reveals differences between receptor subtypes. *J Biol Chem.* 2011;286:16726–16733.
22. Camina JP, Lodeiro M, Ischenko O, Martini AC, Casanueva FF. Stimulation by ghrelin of p42/p44 mitogen-activated protein kinase through the GHS-R1a receptor: role of G-proteins and beta-arrestins. *J Cell Physiol.* 2007;213:187–200.
23. Damian M, Marie J, Leyris JP, et al. High constitutive activity is an intrinsic feature of ghrelin receptor protein: a study with a functional monomeric GHS-R1a receptor reconstituted in lipid discs. *J Biol Chem.* 2012;287:3630–3641.
24. Matthaei S, Stumvoll M, Kellerer M, Haring HU. Pathophysiology and pharmacological treatment of insulin resistance. *Endocr Rev.* 2000;21:585–618.
25. Lin FT, Daaka Y, Lefkowitz RJ. beta-arrestins regulate mitogenic signaling and clathrin-mediated endocytosis of the insulin-like growth factor I receptor. *J Biol Chem.* 1998;273:31640–31643.
26. Dalle S, Ricketts W, Imamura T, Vollenweider P, Olefsky JM. Insulin and insulin-like growth factor I receptors utilize different G protein signaling components. *J Biol Chem.* 2001;276:15688–15695.
27. Dalle S, Imamura T, Rose DW, et al. Insulin induces heterologous desensitization of G-protein-coupled receptor and insulin-like growth factor I signaling by down-regulating beta-arrestin-1. *Mol Cell Biol.* 2002;22:6272–6285.
28. Luan B, Zhao J, Wu H, et al. Deficiency of a beta-arrestin-2 signal complex contributes to insulin resistance. *Nature.* 2009;457:1146–1149.
29. Usui I, Imamura T, Huang J, et al. beta-arrestin-1 competitively inhibits insulin-induced ubiquitination and degradation of insulin receptor substrate 1. *Mol Cell Biol.* 2004;24:8929–8937.
30. Baggio LL, Drucker DJ. Therapeutic approaches to preserve islet mass in type 2 diabetes. *Annu Rev Med.* 2006;57:265–281.
31. Winzell MS, Ahren B. G-protein-coupled receptors and islet function-implications for treatment of type 2 diabetes. *Pharmacol Ther.* 2007;116:437–448.
32. Dalle S, Ravier MA, Bertrand G. Emerging roles for beta-arrestin-1 in the control of the pancreatic beta-cell function and mass: new therapeutic strategies and consequences for drug screening. *Cell Signal.* 2011;23:522–528.
33. D'Alessio D. The role of dysregulated glucagon secretion in type 2 diabetes. *Diabetes Obes Metab.* 2011;13:126–132.
34. Sonoda N, Imamura T, Yoshizaki T, Babendure JL, Lu JC, Olefsky JM. Beta-arrestin-1 mediates glucagon-like peptide-1 signaling to insulin secretion in cultured pancreatic beta cells. *Proc Natl Acad Sci USA.* 2008;105:6614–6619.
35. Quoyer J, Longuet C, Broca C, et al. GLP-1 mediates antiapoptotic effect by phosphorylating bad through a beta-arrestin 1-mediated ERK1/2 activation in pancreatic beta-cells. *J Biol Chem.* 2010;285:1989–2002.
36. Ahren B. Role of pituitary adenylate cyclase-activating polypeptide in the pancreatic endocrine system. *Ann N Y Acad Sci.* 2008;1144:28–35.
37. Broca C, Quoyer J, Costes S, et al. beta-Arrestin 1 is required for PAC1 receptor-mediated potentiation of long-lasting ERK1/2 activation by glucose in pancreatic beta-cells. *J Biol Chem.* 2009;284:4332–4342.

38. Leech CA, Holz GG, Habener JF. Signal transduction of PACAP and GLP-1 in pancreatic beta cells. *Ann N Y Acad Sci.* 1996;805:81–92.
39. Burnstock G, Williams M. P2 purinergic receptors: modulation of cell function and therapeutic potential. *J Pharmacol Exp Ther.* 2000;295:862–869.
40. Hoffmann C, Ziegler N, Reiner S, Krasel C, Lohse MJ. Agonist-selective, receptor-specific interaction of human P2Y receptors with beta-arrestin-1 and -2. *J Biol Chem.* 2008;283:30933–30941.
41. Reiner S, Ziegler N, Leon C, et al. beta-Arrestin-2 interaction and internalization of the human P2Y1 receptor are dependent on C-terminal phosphorylation sites. *Mol Pharmacol.* 2009;76:1162–1171.
42. Morris GE, Nelson CP, Everitt D, et al. G protein-coupled receptor kinase 2 and arrestin2 regulate arterial smooth muscle P2Y-purinoceptor signalling. *Cardiovasc Res.* 2011;89:193–203.
43. Pederson RA, Schubert HE, Brown JC. Gastric inhibitory polypeptide. Its physiologic release and insulinotropic action in the dog. *Diabetes.* 1975;24:1050–1056.
44. Pederson RA, Brown JC. The insulinotropic action of gastric inhibitory polypeptide in the perfused isolated rat pancreas. *Endocrinology.* 1976;99:780–785.
45. Tseng CC, Boylan MO, Jarboe LA, Usdin TB, Wolfe MM. Chronic desensitization of the glucose-dependent insulinotropic polypeptide receptor in diabetic rats. *Am J Physiol.* 1996;270:E661–E666.
46. Tseng CC, Zhang XY. Role of G protein-coupled receptor kinases in glucose-dependent insulinotropic polypeptide receptor signaling. *Endocrinology.* 2000;141: 947–952.
47. Merlen C, Fabrega S, Desbuquois B, Unson CG, Authier F. Glucagon-mediated internalization of serine-phosphorylated glucagon receptor and Gsalpha in rat liver. *FEBS Lett.* 2006;580:5697–5704.
48. Krilov L, Nguyen A, Miyazaki T, Unson CG, Bouscarel B. Glucagon receptor recycling: role of carboxyl terminus, beta-arrestins, and cytoskeleton. *Am J Physiol Cell Physiol.* 2008;295:C1230–C1237.
49. Imamura T, Ishibashi K, Dalle S, Ugi S, Olefsky JM. Endothelin-1-induced GLUT4 translocation is mediated via Galpha(q/11) protein and phosphatidylinositol 3-kinase in 3T3-L1 adipocytes. *J Biol Chem.* 1999;274:33691–33695.
50. Imamura T, Huang J, Dalle S, et al. Beta-arrestin-mediated recruitment of the Src family kinase Yes mediates endothelin-1-stimulated glucose transport. *J Biol Chem.* 2001;276:43663–43667.
51. Dubocovich ML, Markowska M. Functional MT1 and MT2 melatonin receptors in mammals. *Endocrine.* 2005;27:101–110.
52. Bondi CD, McKeon RM, Bennett JM, et al. MT1 melatonin receptor internalization underlies melatonin-induced morphologic changes in Chinese hamster ovary cells and these processes are dependent on Gi proteins, MEK 1/2 and microtubule modulation. *J Pineal Res.* 2008;44:288–298.
53. Tontonoz P, Spiegelman BM. Fat and beyond: the diverse biology of PPARgamma. *Annu Rev Biochem.* 2008;77:289–312.
54. Pascual G, Fong AL, Ogawa S, et al. A SUMOylation-dependent pathway mediates transrepression of inflammatory response genes by PPAR-gamma. *Nature.* 2005;437:759–763.
55. Schoonjans K, Peinado-Onsurbe J, Lefebvre AM, et al. PPARalpha and PPARgamma activators direct a distinct tissue-specific transcriptional response via a PPRE in the lipoprotein lipase gene. *EMBO J.* 1996;15:5336–5348.
56. Chandra V, Huang P, Hamuro Y, et al. Structure of the intact PPAR-gamma-RXR-alpha nuclear receptor complex on DNA. *Nature.* 2008;456:350–356.

57. Evans RM, Barish GD, Wang YX. PPARs and the complex journey to obesity. *Nat Med*. 2004;10:355–361.
58. Nofsinger RR, Li P, Hong SH, et al. SMRT repression of nuclear receptors controls the adipogenic set point and metabolic homeostasis. *Proc Natl Acad Sci USA*. 2008;105: 20021–20026.
59. Chen JD, Umesono K, Evans RM. SMRT isoforms mediate repression and anti-repression of nuclear receptor heterodimers. *Proc Natl Acad Sci USA*. 1996;93: 7567–7571.
60. Tanaka T, Yamamoto J, Iwasaki S, et al. Activation of peroxisome proliferator-activated receptor delta induces fatty acid beta-oxidation in skeletal muscle and attenuates metabolic syndrome. *Proc Natl Acad Sci USA*. 2003;100:15924–15929.
61. Oliver Jr WR, Shenk JL, Snaith MR, et al. A selective peroxisome proliferator-activated receptor delta agonist promotes reverse cholesterol transport. *Proc Natl Acad Sci USA*. 2001;98:5306–5311.
62. Kang J, Shi Y, Xiang B, et al. A nuclear function of beta-arrestin1 in GPCR signaling: regulation of histone acetylation and gene transcription. *Cell*. 2005;123:833–847.
63. Zhuang LN, Hu WX, Zhang ML, et al. β-Arrestin-1 protein represses diet-induced obesity. *J Biol Chem*. 2011;286:28396–28402.
64. Zhuang LN, Hu WX, Xin SM, Zhao J, Pei G. β-Arrestin-1 protein represses adipogenesis and inflammatory responses through its interaction with peroxisome proliferator-activated receptor-γ (PPARγ). *J Biol Chem*. 2011;286:28403–28413.
65. Choi JH, Banks AS, Estall JL, et al. Anti-diabetic drugs inhibit obesity-linked phosphorylation of PPARγ by Cdk5. *Nature*. 2010;466:451–456.
66. Choi JH, Banks AS, Kamenecka TM, et al. Antidiabetic actions of a non-agonist PPARγ ligand blocking Cdk5-mediated phosphorylation. *Nature*. 2011;477:477–481.

PART IV

The Future—The Potential for Arrestin-based Therapeutics

CHAPTER SEVENTEEN

Systems Analysis of Arrestin Pathway Functions

Stuart Maudsley*, Sana Siddiqui*, Bronwen Martin†
*Receptor Pharmacology Unit, National Institute on Aging, Baltimore, Maryland, USA
†Metabolism Unit, National Institute on Aging, Baltimore, Maryland, USA

Contents

1. G Protein-Coupled Receptor Signaling Activity 432
2. Signaling Diversity Among GPCRs 433
   2.1 Accessory protein association 434
   2.2 Arrestin-associated GPCR signaling 435
   2.3 Signal integration and encryption 436
3. Systems Analysis of Receptor Signaling Systems 438
   3.1 Transcriptomic analysis 440
   3.2 Global proteomic analysis 443
   3.3 Posttranslational proteomic analysis 446
   3.4 Physical interactomic investigation 449
   3.5 Bioinformatic interpretation 451
4. Functional Analyses of Arrestin Signaling Paradigms 457
5. Conclusions 459
Acknowledgments 459
References 459

Abstract

To fully appreciate the diversity and specificity of complex cellular signaling events, such as arrestin-mediated signaling from G protein-coupled receptor activation, a complex systems-level investigation currently appears to be the best option. A rational combination of transcriptomics, proteomics, and interactomics, all coherently integrated with applied next-generation bioinformatics, is vital for the future understanding of the development, translation, and expression of GPCR-mediated arrestin signaling events in physiological contexts. Through a more nuanced, systems-level appreciation of arrestin-mediated signaling, the creation of arrestin-specific molecular response "signatures" should be made simple and ultimately amenable to drug discovery processes. Arrestin-based signaling paradigms possess important aspects, such as its specific temporal kinetics and ability to strongly affect transcriptional activity, that make it an ideal test bed for next-generation of drug discovery bioinformatic approaches such as multi-parallel dose–response analysis, data texturization, and latent semantic indexing-based natural language data processing and feature extraction.

Progress in Molecular Biology and Translational Science, Volume 118
ISSN 1877-1173
http://dx.doi.org/10.1016/B978-0-12-394440-5.00017-6

1. G PROTEIN-COUPLED RECEPTOR SIGNALING ACTIVITY

G protein-coupled receptor (GPCR)-mediated signaling represents one of the most fundamental physiological processes. GPCR signaling has been demonstrated to regulate almost all forms of cellular communication, while utilizing an incredibly diverse series of functional mechanisms to do so.[1] As GPCR biology regulates the vast majority of physiological processes, it is unsurprising that nearly half of all current pharmacotherapeutics target this class of proteins.[2] Historically, therapeutic agents that effectively modulate GPCR biology have been described as agonists, partial agonists, or antagonists, based on a "two-state" model of receptor function embodied in the ternary complex model.[3–5] The primary theoretical mechanisms of GPCR activity involved an appreciation of the receptor system as almost a binary process. In this simple linear model, the GPCR agonist binds to and promotes the transition of the receptor from an "off" to an "on" state. The "on" state is then capable of engaging heterotrimeric ($G_{\alpha\beta\gamma}$) guanine nucleotide-binding proteins, whose dissociated G_{α} and $G_{\beta\gamma}$ subunits in turn activate or inhibit various downstream effector molecules. Partial agonists exert a similar function to agonists, but do not possess as great an efficacy as the "full" agonist, while antagonists were considered simply able to unproductively occupy the agonist-binding site.[6] With further developments in receptor dynamic theory, for example, the extended and cubic ternary complex models, the appreciation of a ligand's nature and eventual "efficacy" necessitated redefinition.[7,8] These more advanced concepts expanded the theoretical range of receptor functions beyond the purely G protein-focused realm. GPCR agonists were initially considered to exert similarly efficacious effects upon all forms of subsequent receptor activity, for example, G protein coupling, receptor desensitization, internalization, or trafficking. Considerable experimental evidence has, however, now demonstrated the existence of multiple qualitatively distinct forms of ligand efficacy for a single GPCR type.[9–12] Such data have dramatically altered our appreciation of ligand type, to such an extent that now almost all GPCR-interacting ligands are considered differentially "agonistic" across multiple functional efficacies. The etiology of this functional diversity is likely associated with the physical *speciation* of the GPCRs at the plasma membrane into various stable *receptorsome* structures.[1,13,14] This functional diversity of

GPCRs may be largely controlled by idiosyncratic stable protein–protein interactions that the GPCR engages in to generate its eventual downstream signaling spectrum. The nature and stoichiometry of these *de facto* "distinct" GPCR isoforms represent the most important new frontier of GPCR therapeutic research. Creation of the GPCR isoform *ensemble* in cells or tissues is likely to be microdomain, cell/tissue type, and temporally sensitive. The creation of these plastic and diverse signaling GPCR isoforms, therefore, opens up the potential for the development of drug compounds with tailored and nuanced forms of pharmacological efficacy profiles.[15,16,12,17] Therefore, while we could consider each GPCR-interacting ligand with a generic phrase, such as "agonist," the true pharmacological nature of the efficacy of these "agonists" is in fact multidimensional and contextually sensitive.

2. SIGNALING DIVERSITY AMONG GPCRs

Aided by advances in technology and receptor theory development, our current appreciation of the multiple molecular nuances in GPCR signaling has expanded dramatically.[18–20] Despite their G protein-biased description, their resultant spectrum of signaling output is unlikely to be solely G protein dependent. Multiple avenues of investigation have shown that GPCR cellular signaling exhibits considerably greater diversity and *texture* than previously appreciated.[17,21,22] As we have previously described, a considerable degree of GPCR signaling diversity arises from the ability of receptors to adopt multiple "active" states that are created by a wide range of stable juxtamembrane protein–protein interactions. These GPCR signaling isoforms can: exhibit distinct effector-coupling profiles; describe receptor dimers that exhibit unique pharmacology, signaling, and trafficking; reveal the dissociation of receptor "activation" from desensitization and internalization; and facilitate the discovery that non-G protein effectors mediate some aspects of GPCR signaling.[23–26] While GPCR signaling *texture* is a function of the temporal nature of ligand stimulation as well as the rates of signaling pathway activation/inactivation, in this chapter, we shall discuss the role of accessory protein association, specifically the arrestin family, in regulating diverse modes of GPCR signaling. We shall also outline how the application of novel systems-level analytical approaches may be valuable in the isolation of unique arrestin signaling functionalities, as well as for arrestin-dependent signaling discovery paradigms.

2.1. Accessory protein association

GPCRs demonstrate stable interactions with multiple juxta- or intra-membrane accessory proteins that serve to regulate the *parent* receptor structure, confer ligand interaction specificity, and regulate the subsequent association with signaling cascades. The pharmacological structure of the receptor with its specific accessory protein coterie forms a *de facto* novel receptor structure, that is, the *receptorsome*. Receptorsome creation facilitates multiple aspects of signal preorganization as well as the maintenance of transduction efficiency, specificity, kinetic modality, and dynamic feedback regulation. In addition, specific receptorsome entities are differentially associated with various targeting mechanisms to subcellular compartments (endoplasmic reticulum, Golgi apparatus) as well as diverse endocytic, recycling, and degradative fates of the *parent* GPCR.[27–29] While many of these accessory protein associations can add *texture* to GPCR-G protein-mediated signal transduction, a small subset of the accessory proteins can also engender the creation of G protein-independent signaling events from the *parent* GPCR. One of the most important functional protein constituents of many *receptorsomes* is the arrestins. A considerable body of data now exists to suggest that arrestin-mediated GPCR signaling is one of the most important G protein-independent receptor signaling modalities.[30–33] The demonstration of discrete receptorsome structures, linked to unique signaling effects, now strongly reinforces the previously developed concept of agonist-selective trafficking of receptor signaling. In this agonist trafficking paradigm, specific ligands may select between the *de facto* variants of active receptorsome conformations and induce only a subset of the possible total response profile. Such diversity in agonistic behavior presents the opportunity to develop drugs that change the quality, as well as the quantity, of response efficacy. As a more comprehensive understanding of the complexity of GPCR signaling is developed, the rational design of ligands possessing increased specific efficacy and attenuated side effects may become the standard mode of drug development. This challenge is currently being spearheaded with the *de novo* development of biased GPCR ligands that target specifically arrestin-associated receptorsome entities.[18,26,34] Given the potential therapeutic importance of these advances, we shall focus in the next section upon the specific qualitative and quantitative modalities of arrestin-mediated GPCR signaling paradigms.

2.2. Arrestin-associated GPCR signaling

There are four known genes, arrestins 1–4, encoding GPCR-associated arrestins. The expression of arrestins 1 (visual) and 4 (cone) is largely restricted to visual sensory tissue. These two arrestins are expressed in the rod and cone visual cells in the retina where they regulate the signaling activity of the rhodopsin photon receptor.[35,36] Arrestins 2 and 3 (also known as β-arrestin 1 and 2, respectively) are expressed in most somatic tissues and are largely responsible for the formation of arrestin–GPCR complexes. All four arrestins possess considerable sequence and structural homology. Arrestins possess two receptor interaction domains, an N-terminal domain and a C-terminal domain connected by a polar hinge region. It is thought that when disengaged from GPCRs, the β-arrestin C-terminus interacts with the polar core, allowing the two interaction domains to remain cryptic.[37] Ligand-activated GPCRs can recruit β-arrestins via their association with phosphorylated intracellular GPCR loop regions. Classically, these associations were considered to merely interdict subsequent G protein interaction with the receptor and to serve as a platform for endocytic protein complex assembly.[38] The discovery of the ability of β-arrestin–GPCR complexes to elicit a divergent signaling cascade from the G protein paradigm ushered in a new era of signal transduction.[30] Therefore, rather than simply terminating signal transduction, the association with β-arrestin induced a transformation of signaling activity. This finding suggested that throughout its biophysical lifetime, GPCRs may indeed possess a consistent and important series of differential signaling activities. Considerable experimental data now suggest that GPCR-β-arrestin signaling has distinct functional and physiological consequences compared to G protein-mediated GPCR signal transduction.[26,39] After the initial discovery by Luttrell and colleagues,[30] it has now been demonstrated that many GPCR types can induce multiple cellular responses via the formation of signaling complexes scaffolded by β-arrestins.[38–41] Stable β-arrestin coupling to GPCRs facilitates the productive association of a wide variety of non-G protein signaling molecules with the receptor. For example, β-arrestin-mediated scaffolding of multiple kinase systems, in addition to the original c-Src,[30] to GPCRs has now been shown, including: Akt/glycogen synthase kinase[42,43]; mitogen-activated protein kinases[44,45]; phosphoinositide 3-kinase[46]; and epidermal growth factor receptor kinase.[47,48] In addition to linking kinase systems to GPCRs, β-arrestins can facilitate the receptor-mediated cellular localization of solute carriers (Slc24a/GLUT4[49]) as well as play a vital role in chemotactic[50–52] and apoptotic[53–55] mechanisms. Along with these multiple "intermediary metabolic" actions of GPCR-mediated β-arrestin signaling, recent evidence has also demonstrated that β-arrestin signaling engenders strong transcriptional and translational signaling behavior.[26,56–59] While considerable information concerning the existence of distinctive β-arrestin signaling modalities now exists, further and more nuanced investigation of this new signaling paradigm are required. Such a "next-level" analytical investigation is required, as a large percentage of the current analysis has been performed using *in vitro* cell systems and global congenital β-arrestin knock-out mouse models. The latter murine knock-out mice for β-arrestins are likely to provide GPCR signaling data that are highly skewed, as loss of this vital molecule will impact an enormous range of collateral physiological processes. The ability to dissect specific β-arrestin signaling functionalities from knock-out mice becomes more and more unlikely as we begin to better appreciate the multiple and often paradoxical effects[60,61] upon protein and gene function that genomic ablation of vital *keystone* factors such as β-arrestin can engender. It is the multidimensional functional capacity of β-arrestins to condition multiple forms of GPCR signaling that makes their biological role so important. We shall discuss the role of such keystone molecules in signal integration and encryption in the following section.

2.3. Signal integration and encryption

It can be simply considered that the primary role of GPCRs is to facilitate transmembrane cellular signal transduction for extracellular ligands of all varieties. While the relationship between stimulating ligand and recipient GPCR is considered tight and specific, the nature of the downstream GPCR-associated signaling pathways is in contrast thought to be generic and highly flexible. For cells to allow the discrete transmission of specific ligand-based signaling information to the cell interior, the functional message needs to be efficiently encrypted at the signal transduction level.[62] Intracellular signaling pathways were initially thought to consist of linear and isolated proteins such as kinases, adapters, and subsequent targets. Evidence gathered over recent years has uncovered the fact that signaling pathways can be highly branched, and transduction modules formerly thought to operate independently participate in a substantial degree of preorganized signaling functionality.[63,64] This preorganization of signaling proteins allows for the faithful transmission, without signal dilution or disruption, of the GPCR stimulus to the correct intracellular microdomain. The molecular integration and organization of distinct multiple signals emanating from discrete GPCRs likely occurs at the level of the structured assembly and interaction of their associated signal transduction proteins. Multiple coherently associated components of these signaling cascades have been shown to be preassembled into higher order structures by molecular scaffolding proteins, including POSH (protein kinase A anchoring protein),[65] POSH (plenty of SH3 domains),[66] JIP (c-Jun amino-terminal kinase-interacting protein),[67] 14-3-3 proteins,[68] and perhaps most importantly β-arrestins.[69] These integrating proteins compartmentalize signaling pathways in the cell, enhance specificity of target–substrate interactions, and improve the speed and efficiency of signal transduction. Thus, there is a channeling of the complex and diverse signaling inputs from ligands and their specific target GPCRs at the plasma membrane into higher order multiprotein signaling scaffolds, attached either to cytoskeletal proteins or the plasma membrane itself. The enormous diversity of GPCR–ligand interactions, coupled with a complex series of potential signaling scaffolds, allows specific GPCR signaling patterns at the plasma membrane surface to be functionally encrypted in a myriad of combinations to ensure faithful transmission of virtually any type of initial stimulus. Many cell surface receptors utilize the same types of signaling proteins, so, therefore, the intrinsic individual nature of those downstream signaling molecules cannot be specific enough to solely encrypt the unique signaling "information" of the specific stimulating ligand–receptor interaction. However, a simple and unique stoichiometric combination of common downstream signaling proteins would allow the generation of a virtually limitless number of *encryptons* (multiprotein signaling quanta), all able to convey information from a wide range of inputs that impinge upon the cell in a plethora of combinations, temporal sequences, and concentrations. The enormous functional diversity of GPCR signaling traffic is, therefore, facilitated and mediated through the connection between the ligand-bound receptors and the complex signaling quanta (encryptons). These encryptons are likely to be composed of multiple signaling molecules along with their associated scaffold and trafficking proteins.[70,71] These functional microcomplexes, therefore, assemble in a manner that allows the functional encryption of the exact qualitative and quantitative nature of the initial input ligands. The effective clustering of signaling molecules into multiprotein signaling complexes eliminates delays that would otherwise occur as a result of random signaling protein diffusion in the cytoplasmic space. In addition, as many cellular stimuli are likely to simultaneously impinge upon cells, rather than individually as in most contrived *in vitro* cellular experiments, this form of signaling architecture organization is crucial for eliciting the appropriate physiological response during such times of multiple ligand input. Organizing an efficient and discrete signaling architecture requires the accurate selection of effector molecules for regulated activation and deactivation, often by phosphorylation and dephosphorylation events. A principal strategy in achieving this selection specificity is compartmentalization of signaling enzymes and trafficking adapters.[72] As we have previously discussed, β-arrestin molecules appear to play a multidimensional role in organizing the interaction and clustering of multiple enzymes, their effectors, and their associated targets. In addition to the multiple kinase systems that are recruited to GPCRs via β-arrestin, β-arrestins are also able to orchestrate the creation of signaling *encryptons* including diverse enzymes such as c-Raf,[73,74] c-Jun N-terminal kinase 3,[75] apoptosis signal regulating kinase 1,[76] phosphodiesterase 4D5,[77] the E3 ubiquitin ligase transformed mouse 3T3 cell double minute 2,[78] cAMP-regulated guanine nucleotide exchange factor I,[79] and LIM domain-containing protein kinase.[80] Therefore, as β-arrestins possess both the ability to interact with multiple cell surface GPCRs as well as a huge variety of signaling enzymes and subcellular targeting adapters, one could easily consider that β-arrestins must constitute one of the most important signaling traffic mediators and *encrypton* components. The near ubiquitous nature of β-arrestin expression, therefore, may represent one of the universal mechanisms for GPCR signal conditioning and regulation. β-arrestins are therefore likely to regulate the temporal and spatial signal encryption of the original GPCRs functional input. An understanding of how multiple inputs are integrated, or differentiated, by β-arrestin-based *encrypton* complexes will likely provide a better appreciation of how GPCRs mediate intracellular signaling events in real physiological settings. A more in-depth understanding of β-arrestin-conditioned GPCR signaling thus may facilitate the discovery of β-arrestin signaling-biased GPCR ligands. We shall consider in later sections how the molecular investigation of encrypton structure may aid this drug discovery process.

3. SYSTEMS ANALYSIS OF RECEPTOR SIGNALING SYSTEMS

The field of GPCR signal transduction over the past decade has begun to efficiently and productively interact with the mass data analytical realms of genomic, proteomic, and metabolomic analyses.[20] The consideration of the nature of signal transduction systems has moved away from linear signaling cascades, controlled by "Brownian" modes of motion of individual signaling factors, to preorganized *encrypton*-mediated signaling paradigms.[81] We, therefore, consider that GPCR signaling systems possess far more functional components than those identified in the early *two-state* models of receptor signaling. The ability to identify the multiple protein components of these increasingly complex signaling paradigms is vital for the future of translational GPCR research. Hence GPCRs are thought now to exist, in a functional sense, as higher-order receptor structures grouped into a variety of substrate-specific isoforms that are coupled to preassembled signal transduction cascades consisting of GPCR subtype-specific, stable multiprotein signaling complexes that possess distinct subcellular targeting mechanisms.[1,82] As we have previously described, these associated protein signaling complexes, for example, β-arrestin-centered *encryptons*, can describe and coordinate a specific signaling modality that targets a selective transcriptional or translational target.[26,56–59,83] The β-arrestin- versus G - protein-mediated signal bifurcation from the *parent* GPCR represents a simple mechanism to assess the now diverse and highly complex downstream signaling network. G protein signaling is strongly associated with molecular processes that regulate intermediary cell metabolism, for example, calcium mobilization, or cAMP accumulation. This rapid form of signaling, with transient levels of stimulus and reflex tachyphylaxis, is well suited to the G protein paradigm as the temporal nature of the signaling output is reflected by the transient nature of the GPCR-G protein inter action kinetics.[84] In contrast, the interaction kinetics of β-arrestins with GPCRs are slower and therefore indicate a preference for the regulation of long-term signaling paradigms that involve more profound cellular effects, for example, genomic transcription and protein translation. This development in the appreciation of GPCR signaling activity has therefore necessitated an appreciation of GPCR signaling at a functional genomic and proteomic level. With this conceptual advance, there is therefore an inherent increase in the potential complexity of GPCR signaling systems. While increases in systems complexity are often seen as potential hindrances to pharmacological research, quite the reverse is true. Indeed, the more subtle our appreciation of the intricate nature of GPCR signaling mechanisms and their contextual variety, then the more selective and specific rationally designed pharmacotherapies may become.[1,17,85,86] Thus, investigation of signaling paradigms with mass analytical genomic/proteomic technologies has forced us to consider that physiological processes/

pharmacotherapeutic effects are mediated by alterations in whole gene/protein networks, as opposed to simple activation or inhibition of linear signal transduction pathways. A common phrase often used to describe this shift in molecular biology is that "pathways no longer exist, there are only networks." This view does not disregard the data collected from many years of prior signal transduction research but suggests perhaps that the delineation of discrete signaling pathways potentially represents an abstraction of the true, hypercomplex, signaling network, due to our previous deficiencies in analytical technologies.

In the following sections, we discuss the present variety of efficient and sensitive techniques which an investigator can use to assess genomic or proteomic differences in distinct pharmacological scenarios, including fluorescent transcriptomic array analysis, genome-wide association screening and massive parallel sequencing, antibody arrays, protein-binding microarrays, quantitative mass spectrometry (MS), and MS-physical interactomics. These era-changing mass analytical technologies, however, often cause experimenters considerable issues concerning the choice of the best analytical mechanisms to allow the fullest appreciation of such a surplus of functional data. The application of biologically relevant mathematical processes to divine the eventual physiological meaning of these datasets will also be discussed. The analytical tools and processes described will be applicable to both genomic and proteomic data and will hopefully facilitate a deeper understanding of the creation and eventual pharmacological targeting of signal transduction networks associated with β-arrestin functions. The primary goal of these bioinformatic analytical tools is the rational and biologically relevant condensation of mass data lists into outputs that predict the functional activities of the genes/proteins modulated between the control and test datasets. The clustering of gene/protein factors into functional groups, signaling pathways or even three-dimensional objects will help to categorize idiosyncratic gene/protein sets for future diagnostic and therapeutic use. Therefore, in the future, patient diagnosis, drug development, testing, and design may all take place initially at the signaling network level rather than at the single gene/protein measurement index level.

3.1. Transcriptomic analysis

With the demonstration that GPCR systems can routinely control transcriptional pathways, the importance of measuring the genomic actions of receptor ligands is evident.[71,84,87] Most GPCR systems have now been demonstrated to exert significant effects upon both early and long-term gene transcription, thereby mimicking the more classically "transcriptionally oriented" growth factor receptor systems. The presence of well-established transcriptomic analytical technologies therefore allows for the facile analysis of GPCR-mediated transcriptional signatures. As we have previously discussed, it appears that the β-arrestin-mediated GPCR signaling paradigm represents one of the primary GPCR-controlled transcriptional regulatory mechanisms.[26] While gene array technologies were originally developed to create profiles of distinctly regulated transcripts, often created through long-term experimental paradigms,[88,89] they are also highly suited to investigation of both short- and long-term effects of receptor stimulation.[26,60] For a simple analysis of transcriptional activity induced by receptor stimulation, multiple (at least in triplicate) control and test condition samples are analyzed in a direct comparative manner. Upon the creation of the first series of reproducible arrays, methods such as primary data selection, statistical analysis, and quality control processes were applied to the datasets generated.[90] Many of first transcriptomic array technologies employed radioactive detection methods, but due to vagaries of spot detection and reproducibility, these were quickly superseded by differential fluorescent platforms. Employing a differential-sample fluorescent dye labeling process (Cy3/Cy5), quantitative changes in mRNA expression are now easily obtainable on both a large (>22,000 genes) or small (100 genes) scale.[91,92] As with most technologies based upon fluorescent labeling, one of the most problematic issues before data can be extracted is the inability to completely separate unbound dye from that associated with the sample. Subtraction of such background intensity is achieved by statistically computing the average background intensity and using the standard deviation among this intensity to calculate a confidence interval, the upper limit of which is used for the subsequent background correction. As array samples should be assessed with high numbers of replicates, it may be necessary to assess GPCR transcriptomic activity across multiple array chips. The comparison of multiple gene regulation profiles between microarray chips therefore requires the application of efficient dataset normalization. Current normalization processes, however, often rely upon the use of "housekeeping" genes, whose expression profiles are assumed to be largely resistant to experimental variation. However, with respect to our current thinking of physiological response/signal transduction networks, the concept of a truly nonchanging transcript on the array unfortunately becomes less and less likely.[20] Internal spotted standards of a control transcript, for example, serum albumin, can provide an adequate control for the assay instead of using an experimental sample. However, this merely controls for the experimental detection process itself and not the differential transcript data *per se*. An alternative approach though is the more reliable use of whole-array normalization. Typically, whole-array normalization is performed using linear or logarithmic regression techniques.[93–95] Whole-array normalization relies upon a potentially flawed assumption, however, that is, the majority of genes on the array are nondifferentially expressed between the experimental states and that varying genes are not solely associated with one of the fluorescent labels. The latter assumption can be checked easily by dye-swapping paradigms in which fluorescent labels are reversed and experimental data obtained again. As mentioned previously, the assumption that there is only a minimal perturbation of the majority of the genes on the array constructively reinforces our old concept of linear, discrete signaling pathways.

To further prepare microarray data for eventual functional analysis, it is typical to apply a log transformation to the fluorescent data to make numerical manipulation more acceptable. Parametric tests used for statistical analysis of the transcript variation are the most commonly utilized, as these tests are much more sensitive and require the data to be normally distributed. This is usually achieved by using log transformation of the spot intensities to achieve a Gaussian distribution of the data. To extract the actual differential expression profile of genetic factors from microarray data, a ratio of intensity (as a measure of expression level: z-ratio) between two samples is used. As with all biological experiments, replicates of array data are required if a fold-change cut-off of z-ratios is used as the primary data set filter. Several model-based techniques have been developed that facilitate the assumption of multiplicative noise and eliminate statistically significant outliers from the data.[96] The typical parametric analytical methods applied to primary gene array data management include maximum likelihood analysis, F-statistic, analysis of variance, and t-tests. As an alternative, nonparametric tests used to analyze microarray data include Kruskal–Williams rank analysis[97] and Mann–Whitney tests.[98] The primary goal of the initial statistical analysis of the array data is the calculation of significance values for gene expression. P-values, either fixed to 0.05 or 0.01 are therefore employed to reduce the dataset to significantly regulated gene lists before z-ratio/fold-change cut-offs are applied (typically ± 1.5), as well as provisions for false data creation which are highly likely when large transcription arrays are used. Protocols for the elucidation of random false results calculate the overall chance that at least one gene is a false positive or negative, that is, the familywise error rate.[99] A Bonferroni approach can also be applied to reduce error rate. This technique multiplies the uncorrected p-value by the number of genes tested, treating each gene as an individual test. This method increases significant data specificity by reducing the number of false positives identified. However, this benefit is at the cost of reduced array sensitivity as the number of potential false negatives increases. A modification of the Bonferroni approach, the false discovery rate (FDR), uses a random permutation while assuming each gene is an independent test. In addition, bootstrapping approaches can improve significantly on the Bonferroni approach, as they are less stringent.[100] Resampling-based FDR-controlling procedures can also be used.[101] These data extraction protocols can be generically applied easily to other mass analytical array platforms, for example, antibody/protein arrays.[102] However, one caveat is of course required in cases of protein/antibody arrays, that is, the likelihood of high logarithmic increases in protein expression is unlikely as even modest changes of protein expression[103] may be sufficient to generate profound signaling actions, especially if the protein possesses enzymatic or scaffolding activity, such as β-arrestin. As stated, many of these analytical techniques can be easily transferred between genomic and proteomic platforms; however, while most transcriptomic platforms are standardized, MS detection of proteins from a complex sample is actually a random discovery process. The eventual bioinformatic annotation of these massive datasets provides an invaluable approach for elucidation of the physiological interpretation of the data. The application of statistical informatic annotation is especially important for MS proteomics as this provides a vital support, via functional protein clustering, for the unavoidable variability of protein detection between experiments. This important aspect of functional annotation of proteomic data will be expanded upon in subsequent sections.

3.2. Global proteomic analysis

Global proteomic analysis of molecular signaling systems or disease processes is rapidly becoming a standardized laboratory technique for the investigation of complex biological systems.[61,19,104,105] As GPCR-based β-arrestin signaling pathways have been demonstrated to control both genomic transcription and eventual protein translation,[26,56–59] a combined approach to integrate both proteomic and transcriptomic analysis of β-arrestin signaling paradigms seems prudent. However, there are several considerations that are required when employing such an integrated analytical approach. The primary contrast between global proteomic datasets and those from array experiments is the routine and repeatable expectation of the inclusion of certain datapoints, that is, protein identities. Gene arrays provide a reproducible experimental platform, while the recovery and mass spectrometric identification of the same protein between different experiments is often unlikely. The use therefore of signaling pathway bioinformatics, which can infer function from a variety of related proteins rather than individual protein identities, in such experiments may be paramount for the eventual application of proteomic workflows for GPCR signaling analysis.

In contrast to gene arrays, the primary concern for MS-based workflows is the speed and consistency of protein detection and identification. Using the two most commonly used MS platforms, time-of-flight (TOF) or linear ion-trap tandem MS (LC–MS^2), the identification of proteins in the sample is based upon the fragmentation ion spectrum (MS^2-spectrum) of a specific peptide ion that is broken down into its constituent components in a gas filled collision cell. Due to the enormous complexity of peptides composed of 20 amino acids, a large number of MS^2 spectra do not contain sufficient identity information to allow definitive peptide identification. To minimize false peptide identification, strict filtering criteria are required, which can be enforced, for example, by searching retrieved MS^2 spectra against a composite of both "target" and "decoy" (often reverse peptide alignments) sequence databases.[106] The correct correlation and attribution of an MS^2 spectrum to its originating peptide sequence, followed by eventual protein matching and identification, is the first and central step in proteomic data processing. Numerous computational approaches and software tools have been developed to automatically assign candidate peptide sequences to fragment ion spectra, for example, SEQUEST, MASCOT, ProteinProspector, or ProbID.[107–110] These computational approaches involve database interrogation, where peptide sequences are identified by correlating acquired fragment ion spectra with theoretical spectra predicted for each peptide contained in a protein sequence database or by correlating acquired fragment ion spectra with libraries of experimental MS^2 spectra identified in previous curated experimental datasets.

Initially, MS-based proteomic analyses of biological samples were restricted to qualitative analyses. However, with the advent of multiple labeling or label-free quantitation platforms, cellular signaling proteomics has been converted to a primarily quantitative process. Perhaps, the most commonly used quantitative MS workflows are facilitated by peptide or protein mass labeling. Mass tag labeling techniques, for example, iTRAQ (isobaric tag for relative and absolute quantitation) or isotope-coded affinity tags, as well as protein labeling approaches such as SILAC (stable incorporation of labeled amino acids in culture) and SILAM (stable incorporation of labeled amino acids in mammals) allow the rapid ratiometric analysis of peptides/proteins in a complex biological sample. In both of these processes, peptides are usually separated by multidimensional cation-exchange liquid chromatography (LC) and eventually are identified with either TOF or LC–MS^2 MS, with modified dissociation techniques such as pulsed Q collision-induced dissociation[111] or higher energy collisional dissociation[112]. These diverse protein quantitation techniques provide an ability to generate an unbiased appreciation of multiple ratiometric proteomic alterations occurring within multiple samples. The ability to multiplex several diverse experimental samples within a specific MS run, when using various iTRAQ mass tag label sizes (eight distinct mass tag labels are currently available), provides two important advances compared to nonquantitative workflows. First, multiplexing allows direct intraexperiment sample comparison and also reduces the run-to-run MS identification variability. iTRAQ workflows are designed to provide relative quantitation of as many proteins from the input sample as possible. This global detection process is assisted by multiple levels of prefractionation and orthogonal separation as well as the employment of MS scanning modes designed to detect as wide an array of ionized peptides as possible. However, when a specific protein identity is sought from a complex sample, alternative quantitation procedures can be applied, for example, selective reaction monitoring (SRM[113]). SRM workflows restrict the analytical MS scanning parameters to those that will likely detect the most characteristic peptide fragments for a prespecified protein. SRM can then be converted into a quantitative process if the target protein is readily available in a purified form.[114] Such specific monitoring modes of MS may considerably slow down the rate of data retrieval and may only be suitable for experiments in which high levels of starting extract material are available. In contrast to array technology though, the detection through SRM is still dependent on the ability of the MS to physically detect the specified peptides. This detection reliability is often more likely to demonstrate experiment-to-experiment variability than gene array platforms.

While the accurate and unbiased collection of quantitative protein data is paramount, one important caveat, with respect to data retrieval, is the need to physically retain both significantly and nonsignificantly regulated protein data. The nature of the "nonsignificantly regulated" data may yet yield significance when the coexistence of related proteins is analyzed using functional annotation-based bioinformatic strategies. Often, subtle differences between experimental conditions may be missed as no individual dramatically modulated factors may present themselves. If, however, we consider the posit that GPCR signaling functions are indeed composed of multiple interlaced network activities, the appreciation and functionally relevant correlation of these small changes with each other may illuminate a more realistic view of cellular physiology and receptor signaling.

3.3. Posttranslational proteomic analysis

The ability to identify multiple quantitative proteomic effects in response to GPCR-associated activities, using mass tag or protein labeling approaches, represents a significant functional advance upon transcriptional profiling, as the protein gene products are more closely associated with the specific biological actions studied. However, the true functional status of any protein is not simply dictated by its expression level but also by the posttranslational modification (PTM) status of that protein. PTM of proteins creates an extra functional spectrum to the basic nature of the unmodified protein. While genomes, transcriptomes, and simple "expression" proteomes in higher species (mice, humans) may comprise between 20,000 and 40,000 transcripts/proteins, the posttranslationally modified proteome could approach easily 1×10^6 "unique" proteins. Therefore, for each protein, nearly hundreds of differentially modified analogs may exist, each with a subtly varied functionality. The abundance and potential variety of PTM events that can be applied to proteins (often more than one per protein) generates this enormous functional range.

Approximately, three hundred different types of PTMs have been reported to date. It is highly likely, however, that additional forms are yet to be fully characterized.[115] While the basic expression profile of a protein may control its general functional "potential," the chemical PTM of a protein tends to induce a specificity of its function. Almost all proteins undergo one or other type of PTM during or after their synthesis in a well-defined cellular location. Within any specific peptide or protein, almost all of its constituent amino acids, except Leu, Ile, Val, Ala, and Phe, are available for some form of reversible chemical modification.[116] The diversity of chemical modifications that encompass protein PTMs is extremely high, for example, PTMs include phosphorylation, glycosylation, oxidation, ubiquitination, methylation, acetylation, nitrosylation, sulfation, sumoylation, myristoylation, and palmitoylation. Proteins can also be proteolytically modified after translation, although this is often considered to be functionally separate from chemical addition of groups to the protein backbone or side chains. The chemical PTMs can either be rapidly reversible (e.g., phosphorylation) or long lasting (e.g., ubiquitination). This varied nature of protein PTMs underlies their importance in cellular signaling pathways, as these modifications allow the generation of multiple diverse GPCR ligand controllable alterations in protein signaling networks. Detection of the diverse, subtle, and dynamic protein PTM changes that occur in GPCR signaling networks poses a significant challenge to even the most advanced mass spectrometers. As we have seen in the previous section, even the reliable mass spectrometric detection of the same protein between samples is problematical, let alone its specific diverse intact PTM state. As MS techniques assess peptide identities via a combination of their mass to charge ratio (m/z), the variety of masses or charges of a single peptide amino acid sequence can be hugely variant with different PTM events. For a single analyzed peptide, phosphorylated versions may possess the same peptide backbone but distinct ionization profiles (neutral mass loss), while palmitoylated peptides may be 238 Da heavier than the same sequence non-modified peptide.[117] As the range of protein PTMs is extremely diverse, this has resulted in the general inability to create a single unified workflow that allows the simultaneous identification of multiple PTM types in a complex sample. Therefore, multiple technical approaches need to be applied to extract as much PTM information from any starting material. Employing multiple simultaneous approaches reduces MS workflow speed and also may require hugely different amounts of input sample, depending on the relative abundance in the sample of each PTM under investigation. Affinity-based enrichments, immunopurification, two-dimensional differential in-gel electrophoresis, and metal affinity chromatography are commonly used strategies for the purification of proteins containing specific PTMs. Immobilized metal affinity chromatography (IMAC) purification is a common chemical affinity strategy for the enrichment of phosphoproteins, whereby immobilized metal ions selectively bind to the phosphorylated peptides.[16,118,119] The isolation of phosphopeptide species perhaps represents the lead technological platform at the present time for proteomic PTM analysis of GPCR signaling networks. Protein phosphorylation is one of the most important and dynamic PTM required for the regulation of most intermediary metabolism signal transduction cascades.[120] An enormous range of protein kinases and phosphatases, many of which are associated with β-arrestin-dependent cascades, controls the phosphorylation status of proteins that result in changes of protein–protein interactions, regulation of enzymatic activities, and control of dynamic subcellular localization. These multiple aspects of cellular protein biochemistry underpin almost all biological functions. Rapid and comprehensive identification of protein phosphorylation sites is, therefore, essential for understanding cellular biology and signal transduction connectivity.[62]

Global profiling of phosphoproteomes in cell signaling paradigms has, however, proved a great challenge due to the relatively low stoichiometry of protein phosphorylation (often $<1\%$ of total protein) and poor ionization efficiency in mass spectrometers. Effective, physiologically relevant, phosphoproteome research, therefore, greatly relies upon the rapid and efficient phosphopeptide enrichment from complex samples. The basic chemical principle of phosphopeptide enrichment relies upon the fact that phosphate groups in phosphorylated peptides possess lower pK_a values compared with functional moieties in other amino acids, so much so that they can sustain their negative charge even in an acidic environment. To achieve efficient and selective capture of phosphorylated peptides, solid matrices comprising immobilized metal (Fe^{3+} or Ti^{2+}), with extant positive charges under acidic conditions, are routinely used. However, the binding of nonphosphorylated peptides with multiple acidic residues, such as glutamic acid or aspartic acid, can hinder the efficiency of the enrichment step and subsequently the number of phosphopeptides identified via MS^2 analysis. TiO_2 particles have been demonstrated to possess a capacity to engender a highly specific enrichment of phosphopeptides.[121] TiO_2-based techniques have subsequently been further improved by the use the nonphosphopeptide excluders that reduce nonspecific TiO_2 binding, such as 2,5-dihydroxybezoic acid, phthalic acid, and glycolic acid.[122] While IMAC/TiO_2 chromatography can greatly assist selective phosphopeptide enrichment, the complexity of resultant enriched samples is often still high, suggesting that further separation of enriched phosphopeptides is required. This can be simply achieved, however, by using a pH-gradient elution technique for enhanced phosphopeptide identification. Using this process, it is possible to generate a highly specific separation of phosphopeptides by an applied pH elution gradient between 9.2 and 11.3. Investigation of the recovered phosphopeptide sequences identified in distinct pH fractions has demonstrated that this pH chromatography can effectively separate distinct phosphopeptide kinase substrate sequences. The extraction of such nuanced functional data from cellular signaling experiments will greatly assist the elucidation of specific subtypes of GPCR signaling profiles.[119] Such phosphopeptide enrichment methods represent one of the most important breakthroughs for phosphoproteomics and GPCR cellular signaling analysis. The coherent combination of this technique with quantitative tools has significantly assisted the global understanding of phosphoproteomic signaling patterns.[123,124]

It is likely that without significant development of new, radically advanced, analytical MS-based approaches, the systematic investigation of multiple GPCR-mediated parallel protein PTM networks remains a distant target. For the present time, such investigations will be hampered by slow workflow rates and a lack of direct comparison or association of various PTM signaling patterns with either G protein or β-arrestin-mediated signaling. However, potential solutions to this issue may yet present themselves via the application of advanced bioinformatic approaches that will be discussed in subsequent sections.

3.4. Physical interactomic investigation

While assessments of the PTM status of proteins add an extra level of functional information concerning protein function, this level of investigation may still be physically removed from the true functional crux of the specific protein. It is likely that the true physiological ramifications of GPCR-mediated changes in protein activity may only be fully appreciated by the combined measurement of expression, PTM, and the eventual subcellular appreciation of the protein–protein-binding interactions of those proteins. While the expression profile of proteins forms a global proteome, the direct functional actions of proteins are probably mediated by the formation of small complexes of functionally related proteins possessing a specific PTM state, in a particular subcellular microdomain.[125,70] The physical protein–protein interaction constellation of any specific protein can be referred to as its physical *interactome*. The interactome of a given protein will depend on the presence in its primary amino acid sequence of classical protein–protein-binding modules, PTM sites, and biophysical interaction domains. Specific protein interactomes may vary from 10 to 100 proteins; however, accurate numerical estimates of the interactome size for most proteins have yet to be effectively determined. Mirroring the flexibility and variety of GPCR stable isoforms, there are likely to be, for each protein's interactome, a wide diversity of interactome compositions. As with GPCR *receptorsomes*, the direct functional interactome composition of any specific protein may therefore be a function of the cellular type as well as the temporal and pathophysiological status of the cell or tissue.

MS-based functional interactomic analysis could yield an enormously important body of information for the elucidation of GPCR signaling networks; however, the necessary specific isolation of small protein complexes will likely yield a minimal mass of starting protein. Unfortunately, MS-based protein identification techniques possess no ability to amplify peptide signals; therefore, there are significant issues of appropriate experimental scale to deal with when narrowing proteomic analyses down to the physical interactomic level. Clearly, the primary focus of most interactomic studies is the initial physical interactome enrichment process. Current interactomic studies rely upon antibody purification processes, in low detergent *near-native* conditions, for the initial target protein isolation.[126] One major consideration for antibody-based interactome purification and analysis is the likelihood that the antigenic epitope of the target protein could be idiosyncratically presented to the purifying antibody among distinct interactomic protein composition populations. Therefore, for a single protein physical interactome, biologically mediated changes in both the qualitative or stoichiometric protein composition are likely to affect the global charge/hydrophobicity profile of the purified entity. Identification of the biologically relevant changes in functional interactome composition can be monitored using microfluidic separation devices with subsequent SILAC-based protein identification. The physical interaction of an antibody-isolated interactome (composed of multiple diverse proteins) with an analytical fluidic phase will be related to the expression of hydrophilic or hydrophobic interfaces on the accessible regions of the isolated interactome. In a manner similar to ion-mobility MS,[127] microfluidic separation and immunopurification devices that can alter both the ambient electrical potential and solvent conditions[128] may represent a facile mechanism for the selective isolation of specific interactome species from any immunopurified heterogenous pool. The ability to distinguish diverse, biologically relevant, interactome variants will no doubt represent a major step forward in the understanding of cellular signaling architecture.

As we have stated, the functional direct interactome of a specific signaling protein, for example, β-arrestin, is likely to be highly variable and dynamically coordinated. To assist the analysis of these rapid interactome variations, the SILAC labeling approach can be used to great effect. To observe treatment-induced alterations in interactomes, cellular proteins can be initially labeled with nonradioactive (nitrogen-15) Arg or Lys isotopic variants, and then interactomes can be isolated from distinct isotopically labeled cellular pools (native nitrogen-14 or nitrogen-15) and then mixed together in equal protein amounts for MS analysis. Treatment-induced changes in interactome content/stoichiometry will, therefore, be indicated by either presence or absence of a specific protein, or a significant change in the expression ratio of the various proteins in the complex between the isotopic or nonisotopically labeled samples.[129,130] With the systematic application of such procedures for a greater number of proteins, it should be feasible in the future to construct functional protein complex maps that associate physical interactome composition and subcellular distribution to cell signaling or pathophysiological events. To this end, multiple currently available databases[131,132] (PSIMAP, http://metadatabase.org/wiki/PSIMAP; Human Interaction Database, http://interactome.dfci.harvard.edu/H_sapiens/) and interactomic resources[133,134] (MAPPIT, http://www.crl-mappit.be/mappit_toolbox/; PhosphoPOINT, http://kinase.bioinformatics.tw/) are now forming the basis of such an eventual cell signaling "roadmap."

3.5. Bioinformatic interpretation

In this chapter, we have previously discussed the various modes of systematic analysis that can be applied to the investigation of GPCR-related biological functions, for example, transcriptomic, proteomic, and interactomic. Considerable technical effort has been exerted to facilitate the maximal level of accurate and appropriate data extraction from these distinct experimental processes. However, while being able to generate extensive volumes of data, this output often reaches the bottleneck of effective and productive bioinformatic analysis. In this section, we describe the use of standardized as well as more advanced modes of bioinformatic analysis that can effect a meaningful and practical interpretation of mass analytical cell signaling data.

The simplest, and unfortunately still the most common, mechanism of informatic interpretation of mass analytical data is to simply perform scientific literature searches upon the highest and lowest regulated genes/proteins from the primary data. This approach, while yielding some actionable data, is now criticized for ignoring the correlated biological relevance of the multiple factors (genes/proteins) in the rest of the dataset that do not individually demonstrate either large or significant differential regulation. If we apply the concept that complete knowledge of all the gene–protein interactions does not exist, it is perhaps more prudent to assume that all interactions could occur and then only disregard analytical data for genes/proteins that one can empirically disprove. If we consider that functional signaling responses or physiological states are the effective composite of multiple linked gene/protein networks, then an appreciation of the entire dataset, in a global inclusive manner, is needed. One of the first mechanisms derived to assist in this global appreciation of signaling function was the process of looking for biological similarity between components (genes or proteins) in a dataset. In this context, biological similarity can be assessed by comparing functional annotations (derived from empirical data) linked to the gene/protein. One of the earliest developed processes that allowed efficient classification of gene/protein function was Gene Ontology (GO) (http://www.geneontology.org/index.shtml) analysis. GO analysis attempts to create a rational and physiologically/pharmacologically relevant appreciation of large datasets via the identification of gene/protein clusters within the main data *corpus* that are related to each other either by function, linkage in a metabolic process, or by subcellular localization.[135] The number of these associations and the strength of observing multiple factors possessing the same associations within a large dataset provide the first level of "contextual" relevance of the mass dataset. The three main GO classification categories commonly used to cluster genes/proteins into related and biologically relevant groups are biological process, molecular function, and cellular component. Biological process, molecular function, and cellular component are all attributes of genes, gene products, or gene product groups. Each of these may be assigned independently to factors (genes or proteins) in a dataset. The relationships between a given factor and the biological process, molecular function, and cellular component are *one-to-many*. This reflects the biological reality that a particular protein, for example, β-arrestin, may exert a function in multiple processes, contain domains that carry out diverse molecular functions, and participate in several alternative interactions with other proteins, organelles, or locations in the cell. Currently employed gene/protein ontological structures themselves reflect the current representation of biological knowledge and therefore should be considered highly plastic. GO annotation of datasets has been demonstrated to be vital for a variety of applications, for example, genome sequencing, network modeling, text data mining, and for applied clinical situations.[136–139] The association of the appropriate GO terms to a dataset corpus of significant factors is the first step in the process by which the statistical elucidation of the most likely clustering of the factors to a specific group of GO terms can predict biologically relevant actions. To facilitate this computational process, there are now a plethora of excellent mathematical applications to achieve this first level of dataset functional analysis.[20] Clustering of functionally correlated factors into common GO term groups can be used to infer which specific signaling functions the genes/proteins may be creating. The coexpression of these fac tors and the most common similarities in their functional GO term annota tion can demonstrate a potential predictive output of the dataset. Therefore, the main goal of mass analytical experimentation is the generation of differ ential datasets that, with experimental variable isolation, can be linked to a biochemical function, physiological response, or even an organismal pheno type. The creation of this functional signaling *signature* of the dataset allows the correlation of factor expression to resultant function, with the most pro foundly enriched factor clusters in the dataset being more reliably linked to the resultant output. The *signature* of a dataset is often determined by which GO terms are represented differently, in a significant fashion more or less often than expected by chance within the dataset, compared to their expres sion in a reference whole genome/protein set.[140,141] The most commonly applied approach for this is the calculation of "enrichment" for each GO term. Hence dataset enrichment is demonstrated by the presence of signifi cantly populated GO term clusters, generated by the presence of groups of related genes/proteins occurring in the experimental dataset at a fre quency (scaled according to relative dataset sizes) greater than that in the ref erence dataset. GO analysis can be considered to be a relatively simple two-way analysis in which a simple gene/protein-to-GO term association is used. GO term annotation therefore generates a series of simple gene/protein-associated functional statements. To improve upon this, the generation of more signaling pathway-focused output pathway enrichment analysis mech anisms has been developed.

Gene set enrichment analysis (GSEA) is a statistical approach that has been used to generate a statistically significant representation of a functional pathway class within a selection of factors from a heterogeneous gene/protein dataset corpus. There are multiple freely available pathway databases and facile calculation programs now able to facilitate these computational issues for molecular biologists.[20] As with most technological application fields, subsequent iterations and developments can quickly surpass the previous techniques. In recent years, the use of GSEA has been largely replaced by a parametric version of this process (PAGE, parametric analysis of gene set enrichment[142]). GSEA employs a distribution-free, nonparametric approach to the analysis of the significance of population (two factors in each pathway are typically required for effective "population" of that pathway) of signaling pathways by the input dataset. PAGE and other parametric GSEA tools use a *Central Limit Theorem*, which states that "when the sampling size is large enough, distribution of an average of sampled observations is normal regardless of the nature of parent distribution." Statistical PAGE analysis intentionally directs the analysis of predefined signaling pathways in datasets rather than of individual factors. To generate easily appreciable data, with respect to differential signaling states, PAGE uses the fold change (gene transcript or protein) between the control and experimental groups to calculate *Z*-scores of the predefined datasets and a normal distribution to assign statistical significance to the experimental data.[142] Traditional mass dataset analysis requires that individual genes/proteins have significantly different expression levels in order for them to be considered differentially regulated. PAGE specifically takes into account that factors can be both coregulated and coexistent, to help populate discrete signaling pathways. Therefore, it is possible that factors individually may not be significantly regulated above or below the baseline, but significant regulation of pathways can be generated by such factors by grouping them significantly into the predefined signaling sets.[143] PAGE and GSEA are especially powerful for the interpretation of large datasets that will possess an increased likelihood of retrieved factor identity variation between experiments (especially the case for MS-based proteomics), or when there are subtle differences between control and experimental paradigms. Therefore, the predictive functional output of the analytical system is not wholly reliant upon a small number of individual factors but on the coexpression and coherent regulation of these factors, reflecting the coordinated, interconnected nature of receptor signaling pathways themselves.

The combined employment of mass data collection and signaling pathway analytical tools is likely to revolutionize signal transduction research in the next decade. The ability to accurately appreciate and perhaps predict a global cellular impact of pharmacological signaling actions may create a greater understanding of disease etiology and eventual drug control of disease at the level of the factor *network* rather than the linear signaling pathway level. The appreciation of a network hypothesis for biological activity presents many important new avenues for signal transduction and pharmacological research. For example, the ability to identify molecular *keystone* factors that exert the most profound actions upon the state of a given pathological network may facilitate the creation of collateral pharmacological strategies.[102] Computational platforms are currently being developed using advanced bioinformatic retrieval processes such as latent semantic indexing (LSI), to facilitate the discovery of these network-regulating factors. LSI is a data extraction process recently applied to biomedical science that employs singular value decomposition to elucidate patterns of correlation between information terms and user-defined input interrogator concepts within an unstructured collection of text. LSI relies upon the potentially meaningful association of words/items within a body of text. With the creation of a matrix of scientific (e.g., PubMed abstracts) text-to-word associations, it is possible to identify and quantitate even noncurated/experimentally derived links between input concepts and a specific gene or protein.[141,144,145] The application of LSI-based techniques to biological systems will greatly assist the discovery process for signaling network analyses and has recently been demonstrated to effectively synergize with existing human-curated KEGG (Kyoto Encyclopedia of Genes and Genomes, http://www.genome.jp/kegg/) pathway analytical workflows.[102] Using this novel *combinatorial informatics* approach, the methodological dissection of highly complex protein networks to reveal functional keystone factors is effi ciently achieved. This form of practical informatics development represents the next wave of biological interpretation of datasets of ever increasing volume. In addition to tackling individual datasets of ever increasing volume, the need for advanced data separation and classification of multiple simultaneous datasets has also driven the creation of advanced platforms such as higher-order Venn diagram analyzers.[16] The recently developed VENNTURE application (http://www.nia.nih.gov) facilitates the ability of experimenters to manage and investigate up to six parallel large datasets (Fig. 17.1A). To distinguish spe cific β-arrestin signaling, it may be vital to correlate specific gene/protein expression signatures in multiple diverse cell or tissue environments. The abil ity to perform this over at least six different data streams may greatly accelerate the discovery of a generic β-arrestin-specific signaling paradigm. Applications such as VENNTURE may be able to revolutionize functional signaling anal ysis by developing basic scientific approaches, that is, Venn diagram analysis. Analytical advances are also currently under way at the other end of the com plexity spectrum of data analysis, for example, the recently developed field of *data texturization*. Data texturization applications such as the NIH *Omnimorph* (http://www.ott.nih.gov/Technologies/abstractDetails.aspx?RefNo=2118; Fig. 17.1B) and *Iris* (http://www.ayasdi.com/index.php/iris/) attempt to explore previously cryptic nuances in highly complex datasets by creating three-dimensional proxies of the data possessing various structural or chromatic properties. Data texturization may represent one of the ultimate forms of signaling data analysis that may allow the effective combination of human user investigation and machine-based learning algorithms to uncover idiosyncratic and discriminatory patterns within hypercomplex cell signaling data.

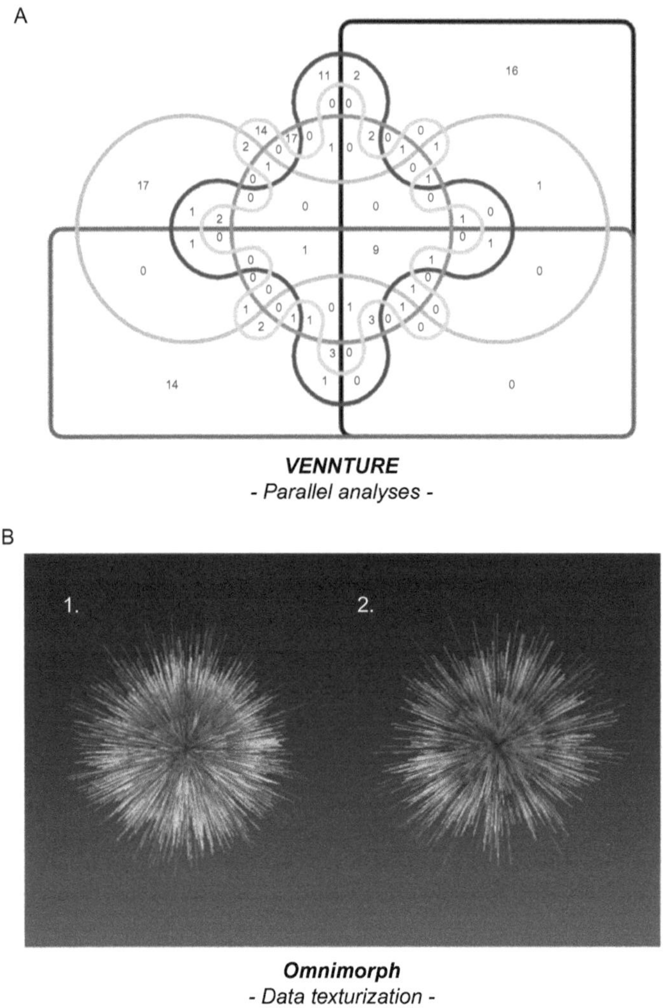

Figure 17.1 Systems-level cellular signaling analysis requires the generation of advanced informatic resources that are able to manage, organize, and facilitate an in-depth interpretation of highly complex large datasets. When assessing the generic nature of signaling systems, it may be necessary to perform multiple simultaneous analyses of complex datasets, therefore, simple Venn analysis tools, for example, *VENNTURE* (A) that are capable of handling six and even seven sets of data may be a vital addition to the scientific technological community. On the other end of the analytical complexity scale, data texturizers such as *Omnimorph* (B) may facilitate through higher-order data organization, novel, and highly nuanced mechanisms for differential (e.g., dataset 1 vs. 2) complex data analysis. (See Color Insert.)

4. FUNCTIONAL ANALYSES OF ARRESTIN SIGNALING PARADIGMS

We have so far described how the initial discovery of β-arrestin-mediated GPCR signaling activity was able to uncover an entirely new era of receptor signaling.[30] The demonstration that the "desensitized" GPCR was indeed still active revealed the presence of a previously unseen spectrum of GPCR signaling activity. As with many paradigm-creating discoveries, it has taken nearly a decade for this concept to reach the level of common acceptance and translational therapeutic development.[26,83] At the present time, β-arrestin signaling from GPCRs is the subject of standardized analysis and industrial therapeutic creation. For example, several studies have demonstrated the existence of idiosyncratic GPCR-associated β-arrestin signaling paradigms that appear to control bone density regulation,[26,83] cell proliferation,[146] and antipsychotic activity.[147] These physiological findings have been supported by mechanistic studies, at the mass analytical level using tissue-based transcriptomics,[26] phosphoproteomics,[148] MS-based direct β-arrestin interactomics,[149] or bioluminescence resonance energy transfer-based network analysis.[150] These experiments have therefore reinforced the existence of unique signaling activities emanating from β-arrestin-coupled GPCRs. With the demonstration of important and distinct physiological effects of β-arrestin signaling, via unique molecular mechanisms, the logical progression from this has been the development of GPCR ligands that demonstrate selective *in vitro* and *in vivo* β-arrestin signaling-dependent efficacies.[26,83,151–155]

In the previous sections, we have outlined many, but not all, of the potentially productive analytical methodologies that can be applied to biological systems to increase the depth of our appreciation of the GPCR-based β-arrestin signaling at a systems level. The term *systems biology* is often employed in a random manner that may or may not be entirely appropriate for the specific study in question. A systems-level appreciation of any biological process often will seem antithetical to the majority of molecular biologists, as in its truest sense "systems" approaches do not place supreme importance upon a single gene or protein. Systems-level signaling approaches are inherently nonlinear and unbiased and therefore directly at odds with the many years of reductionist one-gene/protein-centric peer-reviewed science. Understanding biological processes at a systems-level is therefore often considered nebulous and undirected compared to dogmatic *hypothesis-driven* linear science. The emergence of a need for this systems-level appreciation of biological processes, for example, β-arrestin signaling profiles, was created by the technological advances made in mass data collection and analytical approaches we have discussed previously. As systems-level approaches generate far larger datasets than reductionist linear science, the *data-dominance* of individual genes/proteins has been superseded by the prominence of coherent groups and networks of genes/proteins. Within systems-level scale datasets, there are still computational hierarchies of relevant genes/proteins in these systems, but the singular reliance on *cause-and-effect* monolithic linear signaling paradigms within the actual network has hopefully been relegated to historical science.

Our emerging appreciation of the coherent connectivity between multiple physiological systems has generated the need to empirically develop the systems biology "higher-order" level of understanding of the integration of these systems. This concept of network or "systems" biology implies that biological functions are mediated by strongly or weakly connected groups of genes/proteins rather than simple linear signaling pathways. Resistance to such concepts is related to the opinion that network-based analyses are too diffuse and non-specific to yield actionable data for biomedical or pharmaceutical science. Advances in the mathematical modeling of "small-world" networks have demonstrated that, in most cases, complex network systems are not connected in an equitable and homogenous manner but rather typically consist of at least two levels of connectivity. Therefore, networks may contain small, tightly connected, "subnetworks" which are then collected together into larger constellations of groups of multiple subnetworks.[156] Within a biological framework, it is simple to analogize the "subnetworks" to biological programs such as kinase signaling cascades (e.g., mitogen-activated protein kinase cascades) or receptor signaling systems (e.g., β-arrestin-mediated GPCR signaling) in different tissues, while endocrine or neuronal axes could represent the constellations of these groups of smaller subnetworks. While the functional output of any given gene/protein network may be an eventual function of the activity of all of the constituent components, the relative contribution (to the eventual functional output) of each gene/protein in this network is highly unlikely to be perfectly equal.[157,158] Therefore, within networks of functionally related genes/proteins, there are likely to exist specific nodes that consist of genes/proteins that form the most important bridges, or "*loose connections,*" between the smaller functional programs (subnetworks) contained within the global system. Such genes/proteins within a functional network are effective *keystones*. These keystones profoundly enhance and facilitate rapid and robust connections between disparate parts of a network constellation and, as such, can be considered as functional "shortcuts" in the complex system.[156] It has been demonstrated, using mathematical modeling of graph and network theories, that even in networks containing thousands to millions of nodes, surprisingly few (5–10) "shortcuts" (keystones) are required to facilitate rapid transfer across even the largest of systems.[156] Considering the outlined functional roles and activities of β-arrestin signaling paradigms, the ability of this molecule to acts as a functional keystone in GPCR cell signaling networks may be fully appreciated in the near future with the diligent application of a systems-level conceptual approach.

5. CONCLUSIONS

The demonstration that β-arrestin-regulated signaling pathways emanate from GPCRs represented a significant shift, not just for cell signaling research, but for our understanding of the diversity of signaling systems and for the rational design of selective, discrete therapeutics. Rather than considering signaling systems as linear pathways composed of a simple sequence of kinases and substrates, applying a systems-level investigation has led to the appreciation that GPCR signaling engenders the creation of idiosyncratic "patterns" of signaling activity instead. The β-arrestin GPCR signaling paradigm is currently one of the most prominent physiological processes that has been subjected to this novel form of systems-level analysis. It is hoped that the eventual results of the research carried out in this emerging field will eventually lead to the creation of a new β-arrestin-biased, effective, and unique GPCR pharmacopeia.

ACKNOWLEDGMENTS

This work was supported by the Intramural Program of the National Institute on Aging and National Institutes of Health.

REFERENCES

1. Maudsley S, Martin B, Luttrell LM. The origins of diversity and specificity in G protein-coupled receptor signaling. *J Pharmacol Exp Ther.* 2005;314:485–494.
2. Brink CB, Harvey BH, Bodenstein J, Venter DP, Oliver DW. Recent advances in drug action and therapeutics: relevance of novel concepts in G-protein-coupled receptor and signal transduction pharmacology. *Br J Clin Pharmacol.* 2004;57:373–387.
3. Cheney BV, Lahti RA, Barsuhn C, Gay DD. An analysis of binding at the opioid receptor based upon an agonist/antagonist two-state model. *Mol Pharmacol.* 1982; 22:349–359.

4. Davies AO, Lefkowitz RJ. Agonist-promoted high affinity state of the beta-adrenergic receptor in human neutrophils: modulation by corticosteroids. *J Clin Endocrinol Metab.* 1981;53:703–708.
5. DeLean A, Stadel JM, Lefkowitz RJ. A ternary complex model explains the agonist-specific binding properties of the adenylate cyclase-coupled beta-adrenergic receptor. *J Biol Chem.* 1980;255:7108–7117.
6. Kenakin T. Drugs and receptors. An overview of the current state of knowledge. *Drugs.* 1990;40:666–687.
7. Samama P, Cotecchia S, Costa T, Lefkowitz RJ. A mutation-induced activated state of the beta 2-adrenergic receptor. Extending the ternary complex model. *J Biol Chem.* 1993;268:4625–4636.
8. Weiss JM, Morgan PH, Lutz MW, Kenakin TP. The cubic ternary complex receptor-occupancy model. III. Resurrecting efficacy. *J Theor Biol.* 1996;181:381–397.
9. Berg KA, Maayani S, Goldfarb J, Scaramellini C, Leff P, Clarke WP. Effector pathway-dependent relative efficacy at serotonin type 2A and 2C receptors: evidence for agonist-directed trafficking of receptor stimulus. *Mol Pharmacol.* 1988;54:94–104.
10. Sagan S, Chassaing G, Pradier L, Lavielle S. Tachykinin peptides affect differently the second messenger pathways after binding to CHO-expressed human NK-1 receptors. *J Pharmacol Exp Ther.* 1996;276:1039–1048.
11. Kohout TA, Nicholas SL, Perry SJ, Reinhart G, Junger S, Struthers RS. Differential desensitization, receptor phosphorylation, beta-arrestin recruitment, and ERK1/2 activation by the two endogenous ligands for the CC chemokine receptor 7. *J Biol Chem.* 2004;279:23214–23222.
12. Maudsley S, Davidson L, Pawson AJ, Chan R, López de Maturana R, Millar RP. Gonadotropin-releasing hormone (GnRH) antagonists promote proapoptotic signaling in peripheral reproductive tumor cells by activating a Galphai-coupling state of the type I GnRH receptor. *Cancer Res.* 2004;64:7533–7544.
13. Caunt CJ, Finch AR, Sedgley KR, Oakley L, Luttrell LM, McArdle CA. Arrestin-mediated ERK activation by gonadotropin-releasing hormone receptors: receptor-specific activation mechanisms and compartmentalization. *J Biol Chem.* 2006;281: 2701–2710.
14. Bockaert J, Dumuis A, Fagni L, Marin P. GPCR-GIP networks: a first step in the discovery of new therapeutic drugs? *Curr Opin Drug Discov Dev.* 2004;7:649–657.
15. Kenakin T, Miller LJ. Seven transmembrane receptors as shapeshifting proteins: the impact of allosteric modulation and functional selectivity on new drug discovery. *Pharmacol Rev.* 2010;62:265–304.
16. Martin B, Chadwick W, Yi T, et al. VENNTURE—a novel Venn diagram investigational tool for multiple pharmacological dataset analysis. *PLoS One.* 2012;7:e36911.
17. Gesty-Palmer D, Luttrell LM. Refining efficacy: exploiting functional selectivity for drug discovery. *Adv Pharmacol.* 2011;62:79–107.
18. Maudsley S, Patel SA, Park SS, Luttrell LM, Martin B. Functional signaling biases in G protein-coupled receptors: game theory and receptor dynamics. *Mini Rev Med Chem.* 2012;12:831–840.
19. Wu WW, Wang G, Insel PA, et al. Discovery- and target-based protein quantification using iTRAQ and pulsed Q collision induced dissociation (PQD). *J Proteomics.* 2012;75:2480–2487.
20. Maudsley S, Chadwick W, Wang L, Zhou Y, Martin B, Park SS. Bioinformatic approaches to metabolic pathways analysis. *Methods Mol Biol.* 2011;756:99–130.
21. Rajagopal S, Rajagopal K, Lefkowitz RJ. Teaching old receptors new tricks: biasing seven-transmembrane receptors. *Nat Rev Drug Discov.* 2012;9:373–386.
22. Wang L, Martin B, Brenneman R, Luttrell LM, Maudsley S. Allosteric modulators of G protein-coupled receptors: future therapeutics for complex physiological disorders. *J Pharmacol Exp Ther.* 2009;331:340–348.

23. Maudsley S, Gent JP, Findlay JB, Donnelly D. The relationship between the agonist-induced activation and desensitization of the human tachykinin NK2 receptor expressed in Xenopus oocytes. *Br J Pharmacol.* 1998;124:675–684.
24. Donnelly D, Maudsley S, Gent JP, Moser RN, Hurrell CR, Findlay JB. Conserved polar residues in the transmembrane domain of the human tachykinin NK2 receptor: functional roles and structural implications. *Biochem J.* 1999;339:55–61.
25. Franco R, Ferré S, Agnati L, et al. Evidence for adenosine/dopamine receptor interactions: indications for heteromerization. *Neuropsychopharmacology.* 2000;23:50–59.
26. Gesty-Palmer D, Yuan L, Martin B, et al. β-Arrestin-selective G protein-coupled receptor agonists engender unique biological efficacy in vivo. *Mol Endocrinol.* 2013;27:296–314.
27. Shaw AS, Filbert EL. Scaffold proteins and immune-cell signaling. *Nat Rev Immunol.* 2009;9:47–56.
28. Wong W, Scott JD. AKAP signaling complexes: focal points in space and time. *Nat Rev Mol Cell Biol.* 2004;5:959–970.
29. Bockaert J, Perroy J, Bécamel C, Marin P, Fagni L. GPCR interacting proteins (GIPs) in the nervous system: roles in physiology and pathologies. *Annu Rev Pharmacol Toxicol.* 2012;50:89–109.
30. Luttrell LM, Ferguson SS, Daaka Y, et al. Beta-arrestin-dependent formation of beta2 adrenergic receptor-Src protein kinase complexes. *Science.* 1999;283:655–661.
31. Hall RA, Premont RT, Lefkowitz RJ. Heptahelical receptor signaling: beyond the G protein paradigm. *J Cell Biol.* 1999;145:927–932.
32. Lefkowitz RJ, Shenoy SK. Transduction of receptor signals by beta-arrestins. *Science.* 2005;308:512–517.
33. Rajagopal K, Lefkowitz RJ, Rockman HA. When 7 transmembrane receptors are not G protein coupled receptors. *J Clin Invest.* 2005;115:2971–2974.
34. Maudsley S. G protein-coupled receptor biased agonism: development towards future selective therapeutics. *Mini Rev Med Chem.* 2012;12:830.
35. Wilden U, Hall SW, Kühn H. Phosphodiesterase activation by photoexcited rhodopsin is quenched when rhodopsin is phosphorylated and binds the intrinsic 48-kDa protein of rod outer segments. *Proc Natl Acad Sci USA.* 1986;83:1174–1178.
36. Kendall RT, Luttrell LM. Diversity in arrestin function. *Cell Mol Life Sci.* 2009;66:2953–2973.
37. Shilton BH, McDowell JH, Smith WC, Hargrave PA. The solution structure and activation of visual arrestin studied by small angle X-ray scattering. *Eur J Biochem.* 2002;269:3801–3809.
38. Violin JD, Dewire SM, Barnes WG, Lefkowitz RJ. G protein-coupled receptor kinase and beta arrestin-mediated desensitization of the angiotensin II type 1A receptor elucidated by diacylglycerol dynamics. *J Biol Chem.* 2006;281: 36411–36419.
39. McDonald PH, Chow CW, Miller WE, et al. Beta-arrestin 2: a receptor-regulated MAPK scaffold for the activation of JNK3. *Science.* 2000;290:1574–1577.
40. Shenoy SK, McDonald PH, Kohout TA, Lefkowitz RJ. Regulation of receptor fate by ubiquitination of activated beta 2-adrenergic receptor and beta-arrestin. *Science.* 2001;294:1307–1313.
41. Kendall RT, Strungs EG, Rachidi SM, et al. The beta-arrestin pathway-selective type 1A angiotensin receptor (AT1A) agonist [Sar1, Ile4, Ile8] angiotensin II regulates a robust G protein-independent signaling network. *J Biol Chem.* 2011;286: 19880–19891.
42. Beaulieu JM, Gainetdinov RR, Caron MG. Akt/GSK3 signaling in the action of psychotropic drugs. *Annu Rev Pharmacol Toxicol.* 2009;49:327–347.
43. Beaulieu JM. Morphine-induced mu-opioid receptor internalization: a paradox solved in neurons. *J Neurosci.* 2005;25:10061–10063.

44. Luttrell LM, Roudabush FL, Choy EW, et al. Activation and targeting of extracellular signal-regulated kinases by β-arrestin scaffolds. *Proc Natl Acad Sci USA*. 2001;98: 2449–2454.
45. Tohgo A, Pierce KL, Choy EW, Lefkowitz RJ, Luttrell LM. β-Arrestin scaffolding of the ERK cascade enhances cytosolic ERK activity but inhibits ERK-mediated transcription following angiotensin AT1a receptor stimulation. *J Biol Chem*. 2002;277: 9429–9436.
46. Povsic TJ, Kohout TA, Lefkowitz RJ. β-Arrestin1 mediates insulin-like growth factor 1 (IGF-1) activation of phosphatidylinositol 3-kinase (PI3K) and anti-apoptosis. *J Biol Chem*. 2003;278:51334–51339.
47. Noma T, Lemaire A, Naga Prasad SV, et al. β-Arrestin-mediated β1-adrenergic receptor transactivation of the EGFR confers cardioprotection. *J Clin Invest*. 2007;117: 2445–2458.
48. Kim J, Ahn S, Rajagopal K, Lefkowitz RJ. Independent β-arrestin2 and Gq/protein kinase C ζ pathways for ERK stimulated by angiotensin type 1A receptors in vascular smooth muscle cells converge on transactivation of the epidermal growth factor receptor. *J Biol Chem*. 2009;284:11953–11962.
49. Imamura T, Huang J, Dalle S, et al. β-Arrestin-mediated recruitment of the Src family kinase Yes mediates endothelin-1-stimulated glucose transport. *J Biol Chem*. 2001; 276:43663–43667.
50. Barlic J, Andrews JD, Kelvin AA, et al. Regulation of tyrosine kinase activation and granule release through β-arrestin by CXCRI. *Nat Immunol*. 2000;1:227–233.
51. Hunton DL, Barnes WG, Kim J, et al. β-Arrestin 2-dependent angiotensin II type 1A receptor-mediated pathway of chemotaxis. *Mol Pharmacol*. 2005;67:1229–1236.
52. Ge L, Ly Y, Hollenberg M, DeFea K. A β-arrestin dependent scaffold is associated with prolonged MAPK activation in pseudopodia during protease-activated receptor-2-induced chemotaxis. *J Biol Chem*. 2003;278:34418–34426.
53. DeFea KA, Vaughn ZD, O'Bryan EM, Nishijima D, Déry O, Bunnett NW. The proliferative and antiapoptotic effects of substance P are facilitated by formation of a β-arrestin-dependent scaffolding complex. *Proc Natl Acad Sci USA*. 2000;97: 11086–11091.
54. Revankar CM, Vines CM, Cimino DF, Prossnitz ER. Arrestins block G protein-coupled receptor-mediated apoptosis. *J Biol Chem*. 2004;279:24578–24584.
55. Ahn S, Kim J, Hara MR, Ren XR, Lefkowitz RJ. β-Arrestin-2 mediates anti-apoptotic signaling through regulation of BAD phosphorylation. *J Biol Chem*. 2009;284: 8855–8865.
56. Gao H, Sun Y, Wu Y, et al. Identification of β-arrestin2 as a G protein-coupled receptor-stimulated regulator of NF-κB pathways. *Mol Cell*. 2004;14:303–317.
57. Kang J, Shi Y, Xiang B, et al. A nuclear function of β-arrestin1 in GPCR signaling: regulation of histone acetylation and gene transcription. *Cell*. 2005;123:833–847.
58. Bryja V, Gradl D, Schambony A, Arenas E, Schulte G. β-Arrestin is a necessary component of Wnt/β-catenin signaling in vitro and in vivo. *Proc Natl Acad Sci USA*. 2007;104:6690–6695.
59. DeWire SM, Kim J, Whalen EJ, Ahn S, Chen M, Lefkowitz RJ. β-Arrestin-mediated signaling regulates protein synthesis. *J Biol Chem*. 2008;283:10611–10620.
60. Martin B, Brenneman R, Becker KG, Gucek M, Cole RN, Maudsley S. iTRAQ analysis of complex proteome alterations in 3xTgAD Alzheimer's mice: understanding the interface between physiology and disease. *PLoS One*. 2008;3:e2750.
61. Chadwick W, Brenneman R, Martin B, Maudsley S. Complex and multidimensional lipid raft alterations in a murine model of Alzheimer's disease. *Int J Alzheimers Dis*. 2010;2010:604792.

62. Martin B, Brenneman R, Golden E, et al. Growth factor signals in neural cells: coherent patterns of interaction control multiple levels of molecular and phenotypic responses. *J Biol Chem*. 2009;284:2493–2511.
63. Hunter T. Signaling–2000 and beyond. *Cell*. 2000;100:113–127.
64. Schlessinger J. Cell signaling by receptor tyrosine kinases. *Cell*. 2000;103:211–225.
65. Diviani D, Baisamy L, Appert-Collin A. AKAP-Lbc: a molecular scaffold for the integration of cyclic AMP and Rho transduction pathways. *Eur J Cell Biol*. 2006;85:603–610.
66. Tapon N, Nagata K, Lamarche N, Hall A. A new rac target POSH is an SH3-containing scaffold protein involved in the JNK and NF-kappaB signaling pathways. *EMBO J*. 1998;17:1395–1404.
67. Dickens M, Rogers JS, Cavanagh J, et al. A cytoplasmic inhibitor of the JNK signal transduction pathway. *Science*. 1997;277:693–696.
68. Shinoda S, Skradski SL, Araki T, et al. Formation of a tumour necrosis factor receptor 1 molecular scaffolding complex and activation of apoptosis signal-regulating kinase 1 during seizure-induced neuronal death. *Eur J Neurosci*. 2003;17:2065–2076.
69. Willoughby EA, Collins MK. Dynamic interaction between the dual specificity phosphatase MKP7 and the JNK3 scaffold protein beta-arrestin 2. *J Biol Chem*. 2005;280:25651–25658.
70. Maudsley S, Davidson L, Pawson AJ, et al. Gonadotropin-releasing hormone functionally antagonizes testosterone activation of the human androgen receptor in prostate cells through focal adhesion complexes involving Hic-5. *Neuroendocrinology*. 2006; 84:285–300.
71. Maudsley S, Naor Z, Bonfil D, et al. Proline-rich tyrosine kinase 2 mediates gonadotropin-releasing hormone signaling to a specific extracellularly regulated kinase-sensitive transcriptional locus in the luteinizing hormone beta-subunit gene. *Mol Endocrinol*. 2007;21:1216–1233.
72. Maudsley S, Martin B, Luttrell LM. G protein-coupled receptor signaling complexity in neuronal tissue: implications for novel therapeutics. *Curr Alzheimer Res*. 2007;4:3–19.
73. Song X, Coffa S, Fu H, Gurevich VV. How does arrestin assemble MAPKs into a signaling complex? *J Biol Chem*. 2009;284:685–695.
74. Daaka Y, Luttrell LM, Ahn S, et al. Essential role for G protein-coupled receptor endocytosis in the activation of mitogen-activated protein kinase. *J Biol Chem*. 1998;273:685–688.
75. Li X, MacLeod R, Dunlop AJ, et al. A scanning peptide array approach uncovers association sites within the JNK/β-arrestin signalling complex. *FEBS Lett*. 2009; 583:3310–3316.
76. Miller WE, McDonald PH, Cai SF, Field ME, Davis RJ, Lefkowitz RJ. Identification of a motif in the carboxyl terminus of β-arrestin2 responsible for activation of JNK3. *J Biol Chem*. 2001;276:27770–27777.
77. Lynch MJ, Baillie GS, Mohamed A, et al. RNA silencing identifies PDE4D5 as the functionally relevant cAMP phosphodiesterase interacting with β-arrestin to control the protein kinase A/AKAP79-mediated switching of the β2-adrenergic receptor to activation of ERK in HEK293B2 cells. *J Biol Chem*. 2005;280:33178–33189.
78. Shenoy SK, McDonald PH, Kohout TA, Lefkowitz RJ. Regulation of receptor fate by ubiquitination of activated β2-adrenergic receptor and β-arrestin. *Science*. 2001; 294:1307–1313.
79. Mangmool S, Shukla AK, Rockman HA. β-Arrestin-dependent activation of Ca2+/calmodulin kinase II after β-adrenergic receptor stimulation. *J Cell Biol*. 2010;189: 573–587.

80. Wang P, DeFea KA. Protease-activated receptor-2 simultaneously directs β-arrestin-1-dependent inhibition and Gαq-dependent activation of phosphatidylinositol 3-kinase. *Biochemistry*. 2006;45:9374–9385.
81. Scott JD, Pawson T. Cell signaling in space and time: where proteins come together and when they're apart. *Science*. 2009;326:1220–1224.
82. Luttrell LM. Location, location, location: activation and targeting of MAP kinases by G protein-coupled receptors. *J Mol Endocrinol*. 2003;30:117–126.
83. Gesty-Palmer D, Flannery P, Yuan L, et al. A beta-arrestin-biased agonist of the parathyroid hormone receptor (PTH1R) promotes bone formation independent of G protein activation. *Sci Transl Med*. 2009;1:1ra1.
84. Luttrell LM, Gesty-Palmer D. Beyond desensitization: physiological relevance of arrestin-dependent signaling. *Pharmacol Rev*. 2010;62:305–330.
85. Schadt EE, Lamb J, Yang X, et al. An integrative genomics approach to infer causal associations between gene expression and disease. *Nat Genet*. 2005;37:710–717.
86. Chen Y, Zhu J, Lum PY, et al. Variations in DNA elucidate molecular networks that cause disease. *Nature*. 2008;45:429–435.
87. Maudsley S, Pierce KL, Zamah AM, et al. The beta(2)-adrenergic receptor mediates extracellular signal-regulated kinase activation via assembly of a multi-receptor complex with the epidermal growth factor receptor. *J Biol Chem*. 2000;275:9572–9580.
88. Martin B, Pearson M, Brenneman R, et al. Conserved and differential effects of dietary energy intake on the hippocampal transcriptomes of females and males. *PLoS One*. 2008;3:e2398.
89. Martin B, Pearson M, Brenneman R, et al. Gonadal transcriptome alterations in response to dietary energy intake: sensing the reproductive environment. *PLoS One*. 2009;4:e4146.
90. Hilsenbeck SG, Friedrichs WE, Schiff R, et al. Statistical analysis of array expression data as applied to the problem of tamoxifen resistance. *J Natl Cancer Inst*. 1999; 91:453–459.
91. Schwartz CM, Tavakoli T, Jamias C, et al. Stromal factors SDF1α, sFRP1, and VEGFD induce dopaminergic neuron differentiation of human pluripotent stem cells. *J Neurosci Res*. 2012;90:1367–1381.
92. Stranahan AM, Martin B, Maudsley S. Anti-inflammatory effects of physical activity in relationship to improved cognitive status in humans and mouse models of Alzheimer's disease. *Curr Alzheimer Res*. 2012;9:86–92.
93. Quackenbush J. Microarray data normalization and transformation. *Nat Genet*. 2002;32:496–501.
94. Zhao Y, Li MC, Simon R. An adaptive method for cDNA microarray normalization. *BMC Bioinformatics*. 2005;6:28.
95. Kepler TB, Crosby L, Morgan KT. Normalization and analysis of DNA microarray data by self-consistency and local regression. *Genome Biol*. 2002;3. RESEARCH0037.
96. Sasik R, Calvo E, Corbeil J. Statistical analysis of high-density oligonucleotide arrays: a multiplicative noise model. *Bioinformatics*. 2002;18:1633–1640.
97. Lee ML, Whitmore GA, Björkbacka H, Freeman MW. Nonparametric methods for microarray data based on exchangeability and borrowed power. *J Biopharm Stat*. 2005;15:783–797.
98. Troyanskaya OG, Garber ME, Brown PO, Botstein D, Altman RB. Nonparametric methods for identifying differentially expressed genes in microarray data. *Bioinformatics*. 2002;18:1454–1461.
99. Li H, Wood CL, Getchell TV, Getchell ML, Stromberg AJ. Analysis of oligonucleotide array experiments with repeated measures using mixed models. *BMC Bioinformatics*. 2004;5:209.

100. Meuwissen TH, Goddard ME. Bootstrapping of gene-expression data improves and controls the false discovery rate of differentially expressed genes. *Genet Sel Evol.* 2004;36:191–205.
101. Reiner A, Yekutieli D, Benjamini Y. Identifying differentially expressed genes using false discovery rate controlling procedures. *Bioinformatics.* 2003;19:368–375.
102. Chadwick W, Martin B, Chapter MC, et al. GIT2 acts as a potential keystone protein in functional hypothalamic networks associated with age-related phenotypic changes in rats. *PLoS One.* 2012;7:e36975.
103. Abdelmohsen K, Srikantan S, Tominaga K, et al. Growth inhibition by miR-519 via multiple p21-inducing pathways. *Mol Cell Biol.* 2012;32:2530–2548.
104. Park SS, Wu WW, Zhou Y, Shen RF, Martin B, Maudsley S. Effective correction of experimental errors in quantitative proteomics using stable isotope labeling by amino acids in cell culture (SILAC). *J Proteomics.* 2012;75:3720–3732.
105. Wu WW, Shen RF, Park SS, Martin B, Maudsley S. Precursor ion exclusion for enhanced identification of plasma biomarkers. *Proteomics Clin Appl.* 2012;6:304–308.
106. Elias JE, Gygi SP. Target-decoy search strategy for increased confidence in large-scale protein identifications by mass spectrometry. *Nat Methods.* 2007;4:207–214.
107. Eng JK, McCormack AL, Yates JR. An approach to correlate tandem mass spectral data of peptides with amino-acid sequences in a protein database. *J Am Soc Mass Spectrom.* 1994;5:976–989.
108. Perkins DN, Pappin DJC, Creasy DM, Cottrell JS. Probability-based protein identification by searching sequence databases using mass spectrometry data. *Electrophoresis.* 1999;20:3551–3567.
109. Clauser KR, Baker P, Burlingame AL. Role of accurate mass measurement (+/− 10 ppm) in protein identification strategies employing MS or MS/MS and database searching. *Anal Chem.* 1999;71:2871–2882.
110. Zhang N, Aebersold R, Schwilkowski B. ProbID: a probabilistic algorithm to identify peptides through sequence database searching using tandem mass spectral data. *Proteomics.* 2002;2:1406–1412.
111. Griffin TJ, Xie H, Bandhakavi S, et al. iTRAQ reagent-based quantitative proteomic analysis on a linear ion trap mass spectrometer. *J Proteome Res.* 2007;6:4200–4209.
112. Dayon L, Pasquarello C, Hoogland C, Sanchez JC, Scherl A. Combining low- and high-energy tandem mass spectra for optimized peptide quantification with isobaric tags. *J Proteomics.* 2012;73:769–777.
113. Lange V, Picotti P, Domon B, Aebersold R. Selected reaction monitoring for quantitative proteomics. *Mol Syst Biol.* 2008;4:222.
114. Kim YJ, Zaidi-Ainouch Z, Gallien S, Domon B. Mass spectrometry-based detection and quantification of plasma glycoproteins using selective reaction monitoring. *Nat Protoc.* 2012;7:859–871.
115. Witze ES, Old WM, Resing KA, Ahn NG. Mapping protein post-translational modifications with mass spectrometry. *Nat Methods.* 2007;4:798–806.
116. Walsh CT, Garneau-Tsodikova S, Gatto Jr GJ. Protein posttranslational modifications: the chemistry of proteome diversifications. *Angew Chem Int Ed Engl.* 2005;44:7342–7372.
117. Kamath KS, Vasavada MS, Srivastava S. Proteomic databases and tools to decipher post-translational modifications. *J Proteomics.* 2011;75:127–144.
118. Macek B, Mann M, Olsen JV. Global and site-specific quantitative phosphoproteomics: principles and applications. *Annu Rev Pharmacol Toxicol.* 2009;49:199–221.
119. Park SS, Maudsley S. Discontinuous pH gradient-mediated separation of TiO_2-enriched phosphopeptides. *Anal Biochem.* 2011;409:81–88.
120. Hunter T. The phosphorylation of proteins on tyrosine: its role in cell growth and disease. *Philos Trans R Soc Lond B.* 1998;353:583–605.

121. Pinkse MW, Uitto PM, Hilhorst MJ, Ooms B, Heck AJ. Selective isolation at the femtomole level of phosphopeptides from proteolytic digests using 2D–nanoLC–ESI–MS/MS and titanium oxide precolumns. *Anal Chem.* 2004;76: 3935–3943.
122. Jensen SS, Larsen MR. Evaluation of the impact of some experimental procedures on different phosphopeptide enrichment techniques. *Rapid Commun Mass Spectrom.* 2007;21:3635–3645.
123. Olsen JV, Vermeulen M, Santamaria A, et al. Quantitative phosphoproteomics reveals widespread full phosphorylation site occupancy during mitosis. *Sci Signal.* 2010;3:ra3.
124. Olsen JV, Blagoev B, Gnad F, et al. Global, in vivo, and site-specific phosphorylation dynamics in signaling networks. *Cell.* 2006;127:635–648.
125. Davidson L, Pawson AJ, López de Maturana R, et al. Gonadotropin-releasing hormone-induced activation of diacylglycerol kinase-zeta and its association with active c-src. *J Biol Chem.* 2004;279:11906–11916.
126. Heck AJ. Native mass spectrometry: a bridge between interactomics and structural biology. *Nat Methods.* 2008;5:927–933.
127. Zhong Y, Hyung SJ, Ruotolo BT. Ion mobility-mass spectrometry for structural proteomics. *Expert Rev Proteomics.* 2012;9:47–58.
128. Zhou Y, Yi T, Park SS, et al. Rapid and enhanced proteolytic digestion using electric field-oriented enzyme reactor. *J Proteomics.* 2011;74:1030–1035.
129. Zhong J, Chaerkady R, Kandasamy K, Gucek M, Cole RN, Pandey A. The interactome of a PTB domain-containing adapter protein, Odin, revealed by SILAC. *J Proteomics.* 2011;74:294–303.
130. Martinez-Ferrando I, Chaerkady R, Zhong J, et al. Identification of targets of c-Src tyrosine kinase by chemical complementation and phosphoproteomics. *Mol Cell Proteomics.* 2012;11:355–369.
131. Gong S, Yoon G, Jang I, et al. PSIbase: a database of protein sructural interactome map (PSIMAP). *Bioinformatics.* 2005;21:2541–2543.
132. Venkatesan K, Rual JF, Vazquez A, et al. An empirical framework for binary interactome mapping. *Nat Methods.* 2009;6:83–90.
133. Lievens S, Peelman F, De Bosscher K, Lemmens I, Tavernier J. MAPPIT: a protein interaction toolbox built on insights in cytokine receptor signaling. *Cytokine Growth Factor Rev.* 2011;22:321–329.
134. Yang CY, Chang CH, Yu YL, et al. PhosphoPOINT: a comprehensive human kinase interactome and phospho-protein database. *Bioinformatics.* 2008;24:14–20.
135. Hvidsten TR, Komorowski J, Sandvik AK, Laegreid A. Predicting gene function from gene expressions and ontologies. *Pac Symp Biocomput.* 2010;6:299–310.
136. Adams MD, Celniker SE, Holt RA, et al. The genome sequence of Drosophila melanogaster. *Science.* 2000;287:2185–2195.
137. Liu M, Liberzon A, Kong SW, et al. Network-based analysis of affected biological processes in type 2 diabetes models. *PLoS Genet.* 2007;3:e96.
138. Hirschman L, Yeh A, Blaschke C, Valencia A. Overview of BioCreAtIvE: critical assessment of information extraction for biology. *BMC Bioinformatics.* 2005;6:S1.
139. Dressman HK, Muramoto GG, Chao NJ, et al. Gene expression signatures that predict radiation exposure in mice and humans. *PLoS Med.* 2007;4:e106.
140. Chadwick W, Boyle JP, Zhou Y, et al. Multiple oxygen tension environments reveal diverse patterns of transcriptional regulation in primary astrocytes. *PLoS One.* 2011;6: e21638.
141. Chadwick W, Zhou Y, Park SS, et al. Minimal peroxide exposure of neuronal cells induces multifaceted adaptive responses. *PLoS One.* 2010;5:e14352.
142. Kim S-Y, Volsky DJ. PAGE: parametric analysis of geneset enrichment. *BMC Bioinformatics.* 2005;6:144–156.

143. Mootha VK, Lindgren CM, Eriksson KF, et al. PGC-1alpha-responsive genes involved in oxidative phosphorylation are coordinately downregulated in human diabetes. *Nat Genet*. 2003;34:267–273.
144. Xu L, Furlotte N, Lin Y, et al. Functional cohesion of gene sets determined by latent semantic indexing of PubMed abstracts. *PLoS One*. 2011;6:e18851.
145. Chadwick W, Keselman A, Park SS, et al. Repetitive peroxide exposure reveals pleiotropic mitogen-activated protein kinase signaling mechanisms. *J Signal Transduct*. 2011;2011:636951.
146. Lima-Fernandes E, Enslen H, Camand E, et al. Distinct functional outputs of PTEN signalling are controlled by dynamic association with β-arrestins. *EMBO J*. 2011;30:2557–2568.
147. Klewe IV, Nielsen SM, Tarpø L, et al. Recruitment of beta-arrestin2 to the dopamine D2 receptor: insights into anti-psychotic and anti-parkinsonian drug receptor signaling. *Neuropharmacology*. 2008;54:1215–1222.
148. Xiao K, Sun J, Kim J, et al. Global phosphorylation analysis of beta-arrestin-mediated signaling downstream of a seven transmembrane receptor (7TMR). *Proc Natl Acad Sci USA*. 2010;107:15299–15304.
149. Xiao K, McClatchy DB, Shukla AK, et al. Functional specialization of beta-arrestin interactions revealed by proteomic analysis. *Proc Natl Acad Sci USA*. 2007;104: 12011–12016.
150. Rose RH, Briddon SJ, Holliday ND. Bimolecular fluorescence complementation: lighting up seven transmembrane domain receptor signalling networks. *Br J Pharmacol*. 2010;159:738–750.
151. Zimmerman B, Beautrait A, Aguila B, et al. Differential β-arrestin-dependent conformational signaling and cellular responses revealed by angiotensin analogs. *Sci Signal*. 2012;5:ra33.
152. Emery AC, DiRaddo JO, Miller E, et al. Ligand bias at metabotropic glutamate 1a receptor: molecular determinants that distinguish β-arrestin from G protein-mediated signaling. *Mol Pharmacol*. 2012;82:291–301.
153. Cottingham C, Chen Y, Jiao K, Wang Q. The antidepressant desipramine is an arrestin-biased ligand at the α(2A)-adrenergic receptor driving receptor downregulation in vitro and in vivo. *J Biol Chem*. 2011;286:36063–36075.
154. Allen JA, Yost JM, Setola V, et al. Discovery of β-arrestin-biased dopamine D2 ligands for probing signal transduction pathways essential for antipsychotic efficacy. *Proc Natl Acad Sci USA*. 2011;108:18488–18493.
155. Blättermann S, Peters L, Ottersbach PA, et al. A biased ligand for OXE-R uncouples Gα and Gβγ signaling within a heterotrimer. *Nat Chem Biol*. 2012;8:631–638.
156. Watts DJ, Strogatz SH. Collective dynamics of 'small world' networks. *Nature*. 1998;393:440–442.
157. Schadt EE, Zhang B, Zhu J. Advances in systems biology are enhancing our understanding of disease and moving us closer to novel disease treatments. *Genetica*. 2009;136:259–269.
158. Schadt EE. Molecular networks as sensors and drivers of common human diseases. *Nature*. 2009;461:218–223.

CHAPTER EIGHTEEN

Arrestin Pathways as Drug Targets

Louis M. Luttrell[*,†,‡]
[*]Department of Medicine, Medical University of South Carolina, Charleston, South Carolina USA
[†]Department of Biochemistry & Molecular Biology, Medical University of South Carolina, Charleston, South Carolina USA
[‡]The Research Service of the Ralph H. Johnson Veterans Affairs Medical Center, Charleston, South Carolina USA

Contents

1. Introduction 470
2. Biased GPCR Agonism 471
   2.1 Single and multiple active state models of GPCR signaling 471
   2.2 Quantifying GPCR ligand "bias" 476
3. The Many Faces of Arrestin-Dependent Biased Agonism 477
   3.1 *In vitro* characterization of arrestin-selective ligand bias 477
   3.2 Arrestin-selective ligand bias *in vivo* 480
   3.3 The Screener's dilemma 483
4. The Conserved Arrestin-Dependent Signaling Repertoire 485
   4.1 GPCR desensitization 485
   4.2 Cell proliferation 486
   4.3 Non-proliferative cell growth 487
   4.4 Cell survival and apoptosis 488
   4.5 Cell migration, chemotaxis, and secretory function 489
5. Conclusions 491
Acknowledgments 492
References 492

Abstract

Our growing appreciation of the pluridimensionality of G protein-coupled receptor (GPCR) efficacy, coupled with the phenomenon of orthosteric ligand "bias," offers the prospect of drugs that selectively modulate different aspects of GPCR function for therapeutic benefit. As the best-studied non-G protein effectors, arrestins have been shown to mediate a wide range of GPCR signals, and arrestin pathway-selective ligands have been identified for several receptors. When viewed from the perspective of short term *in vitro* assays, such "biased" agonists appear to activate a subset of the response profile produced by a conventional agonist. Yet, when examined *in vivo*, the limited data available suggest that biased ligand effects can diverge from their conventional counterparts in ways that cannot be predicted from their *in vitro* efficacy profile. While some widely conserved arrestin-regulated biological processes are becoming apparent, what

Progress in Molecular Biology and Translational Science, Volume 118
ISSN 1877-1173
http://dx.doi.org/10.1016/B978-0-12-394440-5.00018-8

is lacking at present is a rational framework for relating the *in vitro* efficacy of a "biased" agonist to its *in vivo* actions that will aid drug discovery programs in identifying "biased" ligands with the desired biological effects.

1. INTRODUCTION

The last decade of basic research in G protein-coupled receptor (GPCR) biology has seen the emergence of two concepts that together have the potential to foster the development of a new generation of GPCR-based therapeutics: "pluridimensional" efficacy and "functional selectivity" or ligand "bias". Pluridimensional efficacy, a phrase coined by the Michel Bouvier laboratory in 2006,[1] refers to the capacity of GPCRs to signal via multiple G protein and non-G protein effectors.[2] The best-studied non-G protein effectors are the arrestins, which promote homologous desensitization of G protein-mediated signaling while simultaneously acting as ligand-regulated scaffolds that recruit additional catalytically active effectors.[3–5] The simple model is that arrestin binding defines two discrete GPCR signaling modes, with the unbound receptor functioning primarily as a guanine nucleotide exchange factor for heterotrimeric G proteins, and the arrestin-bound "desensitized" receptor serving as the nucleus of a receptor–arrestin "signalsome" complex.

Functional selectivity, a concept developed by leading GPCR theoreticians including Christopolous and Kenakin, beginning in the 1990s,[6,7] embodies the concept that GPCRs can adopt multiple "active" conformations that couple the receptor to downstream effectors with different efficiency, and that chemically distinct ligands can affect the conformational equilibrium of the receptor in ways that produce these distinct active states in different proportions than the native ligand.[8–10] Functional selectivity implies that the ligand can control signaling by coupling the receptor to only a subset of its full complement of effectors, which means that GPCR signaling can be manipulated in ways more complex than simple agonism and antagonism. *In vivo*, "biased" ligands would produce mixed effects, directly activating some downstream pathways while antagonizing the effects of circulating endogenous ligands on others. The promise of biased agonism lies in this potential to qualitatively change GPCR signaling, permitting adaptive or therapeutically beneficial signals to be transmitted while antagonizing maladaptive ones.[9–11]

2. BIASED GPCR AGONISM

Classical receptor theory is grounded in the premise that GPCRs exist in equilibrium between inactive and active conformations and that ligands affect signaling by binding with variable degrees of selectivity to the inactive (R) and active (R^*) receptor states. For "biased" agonsim to be possible, there must be at least two distinct active receptor conformations that can be distinguished functionally, that is, that couple differently to downstream effectors. In this case, ligands could conceivably display selectivity for the different active states, just as they can between the inactive and active states, allowing the ligand to "bias" signaling.

2.1. Single and multiple active state models of GPCR signaling

When only a single readout of receptor activity is considered, the actions of a drug or hormone can be adequately described by two terms: affinity, the equilibrium dissociation constant of the ligand–receptor complex, and maximal system response, which is a function of efficacy.[12,13] In this context, it is necessary to account for only two receptor conformations: an "off" state (R) that is silent in the assay and an "on" state (R^*) that produces the measured response. Early allosteric models of GPCR activation envisioned that receptors exist in spontaneous equilibrium between unique R and R^* conformations, and that ligands possessing intrinsic efficacy act by perturbing this equilibrium so as to increase or decrease the fraction of receptors in the R^* conformation.[14–16] Since, in most cases, spontaneous adoption of the R^* state is a rare event, orthosteric ligands that preferentially stabilized R^* were described as full or partial agonists depending on whether the response at saturating ligand concentration reached the maximal system response, while those that had no apparent efficacy but competitively inhibited agonist effects were classified as antagonists.

The spectrum of ligand activity was broadened by the discovery of constitutively activating GPCR mutations that generate measurable levels of receptor activity in the absence of ligand, since they allowed antagonists to be further divided into those that suppressed basal activity and those that had no effect even in systems with a high basal tone.[17–19] Thus, in two-state models, *agonists* are ligands that stabilize R^*, pulling the equilibrium toward the "on" state, "neutral" *antagonists* bind indiscriminately to R and R^*, producing no measurable response but blocking the action of agonists, and *inverse agonists* are ligands that appear as antagonists when basal receptor activity is low, but that reduce constitutive receptor activity by binding preferentially to R and pulling the equilibrium toward the "off" state. Implicit in two-state models, however, is the assumption that ligand binding affects only the proportion of receptors in the "active" state, not the nature of that state. If the receptor is the sole determinant of activity, then ligand classification must be independent of the assay used to detect R^*, that is, the relative order of potency for a series of ligands cannot vary when two or more assays of receptor activation are employed.

The situation becomes much more complex if one allows for the existence of more than one active receptor conformation.[6–8] Since there is no *a priori* reason to assume that a conformation that enables coupling to one downstream effector will necessarily couple it to all possible downstream effectors, one must consider the possibility of multiple R^* conformations that differ in their signaling capacity. Further, if receptors can adopt more than one R^* conformation, there is no *a priori* reason to assume that chemically distinct ligands will generate an identical distribution of active states or even that the ligand-induced active state will mimic the spontaneously formed active state. As increasingly diverse assays for measuring receptor activity were introduced, it became clear that the relative activity of agonists does not always adhere to the predictions of simple receptor theory, that is, structurally distinct ligands can stabilize different conformational populations and elicit ligand-specific efficacy signatures.[2,9] The first formal model to account for these digressions postulated that it is the ligand–receptor complex, not the receptor alone, that specifies the active state.[6] Whether through conformational selection, that is, choosing from a menu of spontaneously formed R^* states, or conformational induction, that is, forcing novel R^* conformations, each ligand–receptor combination has the potential to "bias" the coupling of the receptor to different signaling pathways.

Figure 18.1A illustrates the differences between conventional full and partial agonism and biased agonism using a hypothetical GPCR with five conformationally distinct "active" states (R^*1–R^*5), each of which couples the receptor to downstream G protein and non-G protein effectors with different efficiency. Note that the 1:1 coupling between "active" state and effector depicted is an oversimplification. In such a system, a full agonist (A) will bind and stabilize R^*1–R^*5 equivalently and shift 100% of the receptor population into an "active" state at full receptor occupancy (relative intrinsic efficacy = 1). Although there are exceptions,[21] native hormones and neurotransmitters are most commonly full agonists, as they

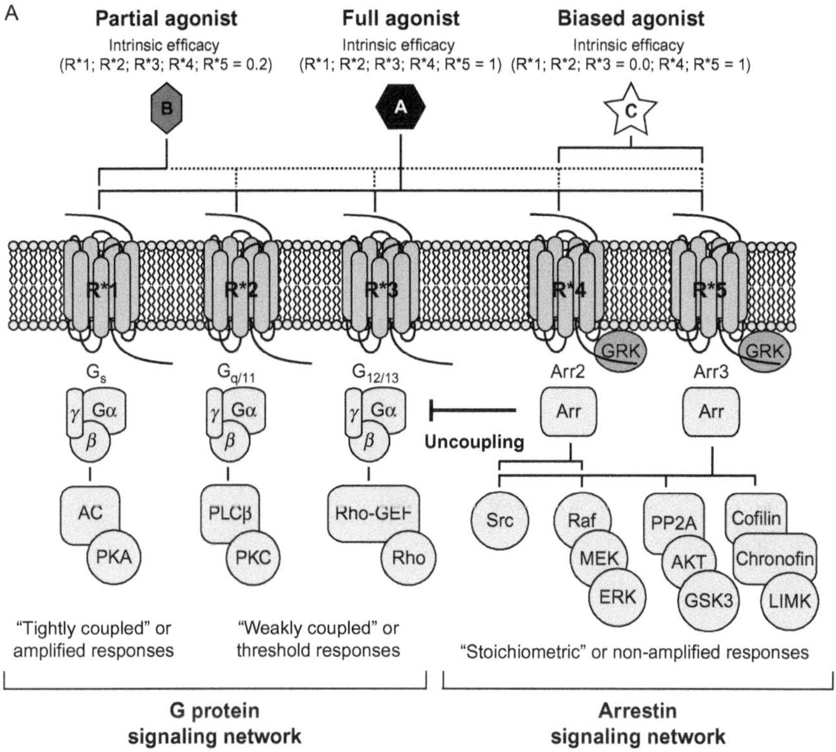

Figure 18.1 Agonism, partial agonism, and "bias" in GPCR signaling. (A) Schematic depicting a hypothetical GPCR with five conformationally distinct "active" states (R*1–R*5), each of which couples the receptor to downstream G protein (G_s; $G_{q/11}$; $G_{12/13}$) and non-G protein [arrestin2 (Arr2); arrestin3 (Arr3) effectors] with different efficiency. In such a system, a conventional full agonist (A) will produce a full system response in all downstream effectors. Due to a lower intrinsic efficacy, a partial agonist (B) may produce a full system response in tightly coupled or highly amplified systems, yet exhibit reduced or absent efficacy toward effectors that are weakly coupled, have high activation thresholds, or are non-amplified. Nevertheless, differences between full and partial agonist signaling reflect the efficiency with which they stabilize receptors in the "active" state (intrinsic efficacy), not their ability to distinguish between different active states. In contrast, "biased" agonists (C) engage different active receptor conformations with variable intrinsic efficacy, a property that permits them to activate some downstream pathways, for example, arrestin-dependent signals, while antagonizing others. The therapeutic potential of "biased" agonism lies in this ability to qualitatively change GPCR signaling.

(Continued)

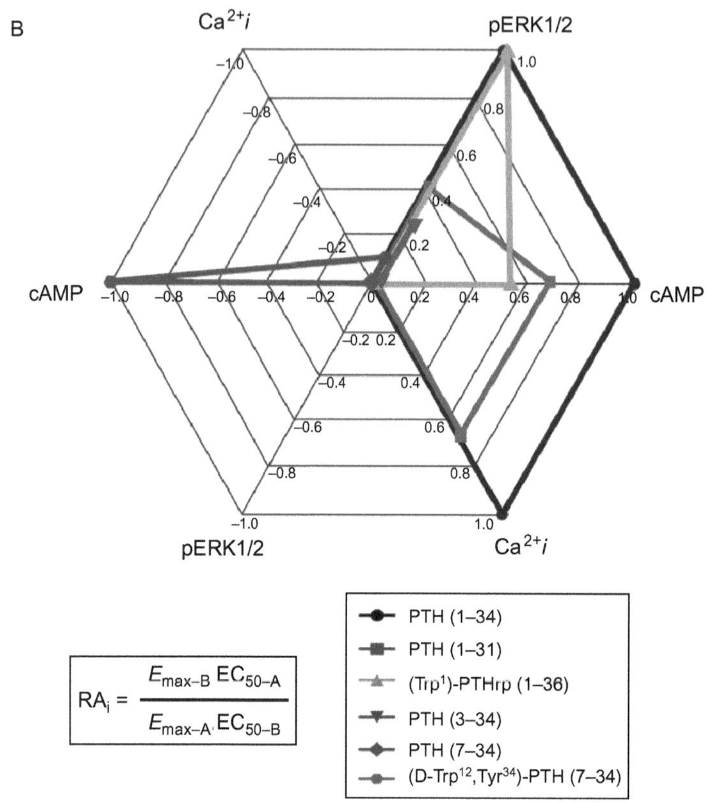

Figure 18.1—cont'd (B) Multiaxial representation of PTH_1R ligand activity in three different assays of receptor activation: cAMP production, intracellular calcium influx, and ERK1/2 phosphorylation. Estimated RA_i values for each ligand are plotted on each axis to represent the magnitude and direction of effects in each signaling response. Ligands with "balanced" efficacy such as hPTH1-34 and hPTH(1-31) show effects of similar amplitude and direction on all three axes, while those demonstrating "bias" show disproportionate activity in one or more pathways or reversal of efficacy. *Reproduced from Ref. 20.* (See Color Insert.)

co-evolved with the receptor to elicit the most physiologically adaptive set of downstream signals. Like a full agonist, a conventional partial agonist (B) does not discriminate between "active" states, but because it possesses a lower intrinsic efficacy, only a fraction of the receptor population will be in an active state at full receptor occupancy. In contrast, a "biased" agonist (C) exhibits selective binding to a subset of the "active" receptor conformations, such that at full receptor occupancy, some active states are favored

(R^*4–R^*5) over others (R^*1–R^*3), that is, its intrinsic efficacy differs between "active" states.

Compared to a conventional agonist that can produce a full system response in all effector systems, both partial and "biased" agonists may evoke only a subset of downstream responses. In the case of a partial agonist, responses that are tightly coupled or highly amplified, that is, where fractional receptor activation can still generate a maximal response, will be preserved. Responses that are weakly coupled or that have an activation threshold, that is, where full receptor activation is necessary for a maximal response, will be disproportionally decreased or absent. Similarly, unamplified responses, for example, those transmitted via stoichiometric receptor–arrestin signalsomes, may be undetectable due to the reduced fraction of receptors in an "active" conformation. Under these conditions, a partial agonist may activate only a subset of the full agonist response profile, and "system factors," such as cell-type-specific variability in receptor density or the expression of signaling intermediates, may cause the response profile to vary between experimental systems. Importantly, variability arising from differences in "signal strength" does not require the existence of multiple active states,[8–10] that is, R^*1–R^*5 may be indistinguishable. Thus, while the measurable downstream responses may vary between systems, the rank order of potency of a series of conventional agonist and partial agonist ligands will not vary when multiple responses are compared, since the intrinsic efficacy of each ligand is the same across all "active" states.

On the other hand, "biased" agonists, because they stabilize only a subset of "active" states, possess the ability to activate some pathways but not others. Reversal of potency, that is, where two ligands exhibit an opposite rank order of potency for two downstream responses measured in the same system, or reversal of efficacy, that is, where a ligand has opposite effects in two different downstream responses, are the hallmarks of ligand "bias" and offer definitive evidence of the existence of multiple active states.[8–10] Importantly, the existence of "biased" agonism means that chemically distinct ligands are activating the same receptor in different ways, and it is the ligand–receptor complex, not the receptor alone, that specifies the cellular response. In effect, the receptor bound to a "biased" ligand is a different functional entity, with different signaling characteristics, compared with the same receptor bound to a conventional agonist or partial agonist. The therapeutic potential of "biased" agonism lies in this ability to produce biological response patterns that cannot be obtained using conventional agonist or antagonists and qualitative change GPCR signaling.

2.2. Quantifying GPCR ligand "bias"

Ligand bias can range from relatively modest deviations in rank order of potency[22,23] to frank reversal of efficacy, wherein the classification of a ligand as agonist, antagonist, or inverse agonist becomes assay dependent.[20,24] In such cases, a more complete characterization of ligand efficacy is required. The most generalizable method for quantifying ligand bias uses the Black/Leff operational model,[12] which relates the equilibrium dissociation constant (K_A) of the ligand, a direct measure of receptor occupancy, to a coupling efficiency factor (τ), which encompasses both the intrinsic efficacy of the agonist and system-dependent factors such as receptor density and coupling efficiency.[10,25] Since the latter factors are constant for any given signaling pathway assayed in the same cell background, the ratio of τ values for any two agonists yields a measure of intrinsic efficacy that is independent of receptor number or coupling efficiency. Once determined for each agonist and pathway of interest, τ/K_A ratios can be used to quantify bias relative to a reference agonist. The advantages of the operational model are its ability to quantify the full range of agonism, from submaximal effects to effects in very sensitive systems with receptor reserve, and that once τ/K_A ratios have been determined in one system they are applicable to all systems without the need to individually quantify functional selectivity in each. Variation in receptor density and coupling efficiency between systems might change the ability of all agonists targeting a given receptor to activate a particular pathway, but it will not change the pathway-selective "bias" of different ligands relative to one another.[9,10]

A simpler, but more limited, approach to visualizing ligand bias is to determine the intrinsic Relative Activity (RA_i) of a series of ligands using multiple readouts of receptor activation. RA_i is a measure of the microscopic affinity constant of an agonist for the active state of the receptor expressed relative to that of a reference agonist, and can be estimated from empirically determined EC_{50} and E_{max} values using the formula shown in Fig. 18.1B.[10,25] This method yields results similar to the operational model, provided that the Hill slope of the ligand concentration–response curves is close to 1.0, that is, that the observed EC_{50} is proportional to the degree of receptor occupancy needed to elicit the response. As an illustration, Fig. 18.1B depicts a multiaxial plot of experimentally determined RA_i estimates using a panel of six different peptide ligands for the human type 1 parathyroid hormone receptor ($hPTH_1R$) in three different assays of receptor activation; cAMP production, intracellular calcium release, and ERK1/2 activation.[20] As shown, the reference agonist hPTH(1–34), which reproduces the response profile of the native 84 amino acid hormone,[26] behaves as a full agonist in all three assays. In this system, the hPTH(1–31) fragment[27,28] appears as a classical partial agonist, that is, its RA_i is less than that of the full agonist, but similar in all assays. The N-terminal truncated fragment hPTH(7–34)[29] lacks any demonstrable efficacy and thus appears as a classical "neutral" antagonist. The other three ligands exhibit signal bias. Trp^{1}-PTHrp(1–36)[24] appears selective for G_s-cAMP signaling; bPTH (3–34) is selective for arrestin-dependent ERK1/2 activation; and (D-Trp^{12},Tyr^{34})-bPTH(7–34)[24,30] exhibits frank reversal of efficacy, appearing as an inverse agonist for cAMP production, neutral for calcium signaling, and a partial agonist for ERK1/2 activation.[24,31]

3. THE MANY FACES OF ARRESTIN-DEPENDENT BIASED AGONISM

The cellular response to an extracellular GPCR ligand originates at a single point of contact between the ligand and its receptor on the plasma membrane. Agonist binding changes the distribution of receptors occupying each of the component microstates that comprise the overall conformational ensemble, shifting the proportion of receptors existing in one or more "active" states.[9,10] The nature of the information that is transmitted across the membrane is specified not only by the receptor, but also by the structure of the ligand and the cellular environment in which the receptor is expressed. Inside the cell, the flow of information proceeds in ever widening circles, like ripples on the surface of a pond that emanate outward from a dropped pebble. Ligand effects can be assayed at many steps along the way, with each measurement providing some, more, or less, useful information about the ultimate biological phenotype.

3.1. *In vitro* characterization of arrestin-selective ligand bias

The most immediate measurable consequence of ligand binding is a change in receptor conformation, and intramolecular fluorescence probes can be used to monitor how orthosteric ligand binding affects receptor conformation at different points within the molecule.[32,33] A recent study by Liu *et al.* using site-specific 1D ^{19}F NMR spectroscopy offers insight into how conventional and biased ligands affect the conformation of a population of receptors.[34] They found that the β_2 adrenergic receptor undergoes major conformational shifts at C265 (helix VI) and C327 (helix VII) upon binding full agonists such as isoproterenol and formoterol, whereas the neutral antagonist albuterol and the inverse agonist carazolol either fail to perturb the conformational equilibrium or shift helix VI independently of helix VII. The partial agonists norepinephrine, tolubuterol, and clenbuterol, that lack or have weaker polar features on their aromatic heads, produce smaller effects on helix VI but shift helix VII as effectively as full agonists. Comparing this with the arrestin-selective β_2 adrenergic receptor agonist carvedilol and the arrestin-enhanced agonist isoetharine, they find that G protein activation correlates with movement of helix VI, whereas β-arrestin-biased ligands predominantly impact the conformational state of helix VII. Importantly, helix VI and helix VII were found to move independently, providing a physical basis for biased agonism and a conformational "signature" predictive of arrestin selectivity.

The initial transfer of information within the cell occurs when the ligand-bound GPCR physically touches its proximal effectors, for example, G proteins and arrestins. The development of bioluminescence resonance energy transfer (BRET)-based technologies to measure protein–protein interactions has enabled these events to be monitored in real time in live cells.[35,36] The most comprehensive study of biased agonism reported to date using these approaches is by Sauliere *et al.*, who compared the native angiotensin AT_{1A} receptor agonist, angiotensin II (AngII) to the arrestin-biased agonist, [Sar^1,Ile^4,Ile^8]-AngII (SII), using a panel of BRET-based sensors encompassing all heterotrimeric G protein families and arrestins.[37] With respect to heterotrimeric G protein activation, they found that SII behaves as a weak partial agonist, activating all G protein species affected by AngII, albeit with much lower efficacy than the native ligand. SII did not qualitatively change the G protein coupling of the receptor; the SII response was proportional to AngII across the board. Reflective of its arrestin signaling bias, SII recruited arrestin3 to the AT_{1A} receptor with nearly comparable efficacy as the native ligand. These data support the model that biased ligands select receptor conformations that couple the receptor to downstream effectors with different efficiency that the native ligand, but do not drive the receptor to engage otherwise unrecognized effectors.

The next level of signal propagation occurs as G proteins, arrestins, and potentially other non-G protein effectors, activate second messenger pathways, and protein phosphorylation cascades. This is the level at which much high-throughput pharmaceutical screening occurs. To date, most studies of short term downstream signaling (e.g., Refs. 24,38,39) have failed to find unique signaling events produced by biased agonists. Rather, biased ligands appear to activate a subset of the downstream second messenger and protein kinase cascades regulated by their native counterparts. In their comprehensive survey, Sauliere *et al.*[37] were able to discern relatively subtle and cell-type specific differences in downstream signaling between AngII and SII. For example, they found that SII produced $G_{i/o}$-dependent inhibition of forskolin-stimulated cAMP production in HEK293T cells, while AngII caused $G_{q/11}$-dependent potentiation of cAMP forskolin-stimulated cAMP production that was further enhanced by pertussis toxin treatment. Such differences probably arise from signal strength effects, in that some responses, for example, $G_{i/o}$-mediated inhibition of adenylyl cyclase, are graded, while others, for example, $G_{q/11}$-mediated calcium flux, have a discrete activation threshold. In such cases, a weak partial agonist may measurably affect a tightly coupled or graded response while failing activate a less well-coupled response or one with a high activation threshold (Fig. 18.1A). Nonetheless, these results at least admit the possibility that biased agonists can qualitatively change information flow as signals propagate by coupling the receptor to downstream effectors with different efficiencies.

Of course, the failure to find novel biased agonist responses may be explained by the fact that most work to date has focused on determining which effects of native hormones are arrestin dependent, an approach that guarantees that biased agonist responses will be defined as a subset of the whole. Still, the limited number of discovery driven comparisons of native and arrestin-selective agonism to date support the idea that biased agonist effects comprise a subset of native agonist responses. Using a gel-based whole cell phosphoproteomic approach, Kendall *et al.* uncovered novel arrestin-dependent $AT_{1A}R$ pathways leading to regulation of protein phosphatase (PP)2A–Akt–glycogen synthase kinase (GSK)3β signaling and stimulation of prostaglandin E2 synthesis by prostaglandin E synthase type 3.[40] But validation experiments demonstrated that AngII, as well as SII, activated these novel pathways. The most comprehensive side-by-side global phosphoproteomic comparison of AngII and SII, performed by Christensen *et al.*, detected over 1183 regulated protein phosphorylation sites out of 10,000 sites surveyed using LTQ-Orbitrap high resolution mass spectrometry.[41] Of these, 756 (64%) were unique to AngII, 369 (34%) were regulated by both AngII and SII, and only 58 (5%) were unique to SII. While their study clearly demonstrated that G proteins and arrestins regulate robust, and largely distinct, kinase cascades, they confirm that, at least in the short term, biased agonists work by selecting from within the menu of native ligand responses.

The same story holds for the effects of conventional and arrestin pathway-selective biased agonists on integrated biological responses such as cell proliferation or migration. For example, both AngII and SII have been reported to produce positive inotropic and lusitropic effects in isolated murine cardiomyocytes that are arrestin dependent.[42] Likewise, the reported ability of SII to stimulate proliferation of neonatal rat cardiomyocytes is a property shared with AngII; its lack of a hypertrophic response reflects its inability to activate $G_{q/11}$-dependent pathways, not its ability to engender novel responses.[43] Similarly, the capacity to promote arrestin-dependent cell survival and stimulate migration in primary murine osteoblasts is a property shared by (D-Trp^{12},Tyr^{34})-PTH(7–34) and hPTH(1–34).[44] Both the native and biased ligands couple the receptor to arrestins and activate similar arrestin-dependent biological responses.

3.2. Arrestin-selective ligand bias *in vivo*

If, as available *in vitro* data suggest, arrestin pathway-selective biased ligands activate a subset of the signaling repertoire activated by a full agonist, then the simple expectation would be that *in vivo*, these ligands would reproduce part of the biological activity of the corresponding native hormone. Surprisingly, the limited data available suggest this is not necessarily the case, and that emergent biological properties may arise when arrestin-selective biased agonists are used *in vivo*.

The actions of the arrestin pathway-selective PTH_1R agonist, (D-Trp^{12}, Tyr^{34})-bPTH(7–34), may be the most informative in this regard, in that its *in vivo* effects at the genomic level have been characterized and correlated with its phenotypic response (Refs. 31,44; see Chapter 13). Parametric geneset enrichment analysis performed on cDNA microarray data from calvarial bone of wild-type C57BL/6 mice treated with vehicle, the conventional PTH_1R agonist, hPTH(1–34), or (D-Trp^{12},Tyr^{34})-bPTH(7–34), found a striking degree of non-overlap in the genome level responses to the two ligands.[44] As expected of a conventional PTH_1R agonist based on study of the actions of native PTH in bone,[45–47] hPTH(1–34) affected gene clusters classically associated with embryologic skeletal patterning and PTH actions in bone, including Wnt/β-catenin signaling, bone morphogenic protein signaling, TGF-β signaling, phosphatidylinositol 3-kinase (PI3K)/AKT signaling, and ERK/MAPK signaling,[48,49] translating into regulation of processes linked to the control of skeletal morphogenesis, osteoblast differentiation, matrix biosynthesis, and mineralization, and bone turnover. In contrast,

(D-Trp^{12},Tyr^{34})-bPTH(7–34) primarily affected pathways regulating apoptosis, for example, ataxia telangiectasia mutated signaling and p53 signaling; cell survival, for example, PI3K/AKT signaling; and cell cycle checkpoint regulation. The effects of (D-Trp^{12},Tyr^{34})-bPTH(7–34) appeared to reflect arrestin-dependent signaling, as opposed to antagonism of G protein signaling by endogenous PTH, in that 78% of the significant changes in gene expression observed in wild-type animals exposed to the biased ligand were not observed in identically treated arrestin3 null mice, and there was a high degree of correlation between the pathways analysis performed using the complete list of (D-Trp^{12},Tyr^{34})-bPTH(7–34) regulated genes and the list of genes whose regulation was unique to the wild type background.[44] In contrast, 54% of genes regulated by hPTH(1–34) in wild-type mice were also regulated in arrestin3 null mice, and there was much less concordance when the hPTH (1–34) pathways analysis was repeated using the list of wild-type unique genes. As expected, most of the effects of intermittent hPTH(1–34) in bone, including its ability to activate osteoclastic bone resorption, derived from G protein-dependent signals that do not require arrestin3, so many of its biological actions were preserved in the arrestin3 null background.

The ultimate expression of ligand action is a change in biological phenotype at the tissue or organ level. In bone, PTH directly stimulates bone forming osteoblasts, promoting deposition of new bone matrix and accelerating the rate of mineralization by increasing both osteoblast number and activity. At the same time, it accelerates bone turnover by stimulating osteoblasts to secrete soluble osteoclast-activating factors, for example, receptor activator of NFκB ligand, that increase the number and activity of bone resorbing osteoclasts. The effects of PTH on osteoblasts and osteoclasts are coupled, such that the net effect of PTH on bone mass is a function of the magnitude and timing of PTH exposure.[50] Continuous exposure to excess PTH leads to hypercalcemia and net bone resorption, whereas intermittent exposure stimulates bone formation in excess of resorption and a net increase in bone mass. Thus, wild-type mice given daily injections of hPTH(1–34) for 8 weeks experience a net increase in bone mass.[31,46,47] Trabecular bone volume and cortical thickness increase, as do osteoblast number, osteoid surface, serum osteocalcin level, and mineral apposition rate. Reflective of osteoblast–osteoclast coupling, osteoclast numbers also increase, along with urinary deoxypyrodiniline, a marker of bone resorption. Twenty-four hour urine calcium excretion rises, reflecting the net effect of PTH on intestinal calcium absorption and renal tubular calcium resorption. Interestingly, arrestin3 null mice exhibit an impaired anabolic response to hPTH(1–34), with blunted increases in trabecular bone volume and no change in cortical thickness.[31,51]

Traditionally, the actions of PTH in bone have been attributed to cAMP-G_s signaling,[52] so the expectation would be that systemic administration of a ligand like (D-Trp^{12},Tyr^{34})-bPTH(7–34), which antagonizes PTH_1R-G_s coupling, would at best have no effect on bone mass. Indeed, treatment of rats with the neutral PTH_1R antagonist, hPTH(7–34) (Fig. 18.1B), reportedly does not affect bone mass in rats.[53] Paradoxically, wild-type mice treated with (D-Trp^{12},Tyr^{34})-bPTH(7–34) exhibited increased bone formation with greater trabecular bone volume and increased osteoblast number, osteoid surface, serum osteocalcin, and mineral apposition. Although similar to the response to hPTH(1–34), some differences were apparent. In trabecular bone, (D-Trp^{12},Tyr^{34})-bPTH(7–34) primarily increased trabecular number, while hPTH(1–34) had larger effects on trabecular thickness. More dramatically, (D-Trp^{12},Tyr^{34})-bPTH(7–34) had no effect on osteoclast number or bone turnover markers. All responses to (D-Trp^{12},Tyr^{34})-bPTH(7–34) were either absent or reversed in arrestin3 null mice, suggesting that *in vivo*, arrestin signaling is sufficient to promote osteoblastic bone formation but not to stimulate bone resorption. The key to this unexpected phenotype apparently lies in the ability of (D-Trp^{12},Tyr^{34})-bPTH(7–34) to expand the osteoblast pool through arrestin-dependent cell cycle regulation and anti-apoptotic signaling, while "uncoupling" the PTH_1R from G_s-cAMP dependent activation of osteoclasts.[31,44]

Other evidence that unexpected *in vivo* phenotypes may emerge from systemic administration of arrestin-selective ligands comes from the study of biased AT_{1A} receptor ligands in heart failure. *In vitro*, SII produces positive inotropic and lusitropic effects on isolated murine cardiomyocytes.[42] Another arrestin pathway-selective AT_{1A} receptor agonist, TRV120027 [Sar-Arg-Val-Tyr-Ile-His-Pro-(D)-Ala-OH], which stimulates arrestin-dependent activation of Src, ERK1/2, and endothelial nitric-oxide synthase *in vitro*, similarly improves cardiomyocyte contractility.[54] Administration of TRV120027 to rats reduces mean arterial blood pressure, as do the non-peptide AT_{1A} receptor antagonists losartan and telmisartan, but unlike the neutral antagonists, which decrease cardiac performance, TRV120027 increases cardiac performance and preserves cardiac stroke volume. Unexpectedly, however, the principal effect of SII in a murine model of heart failure appears to be to promote aldosterone production via arrestin2- and ERK1/2-dependent upregulation of steroidogenic acute regulatory protein (StAR), a steroid transport protein that is the rate-limiting step in aldosterone biosynthesis. In this model, the resultant elevation in circulating aldosterone levels contributes salt and volume retention, adverse cardiac remodeling, and heart failure progression.[55]

3.3. The Screener's dilemma

Results such as these help validate the concept that biased ligands can be used to produce biological effects not attainable through conventional agonism or antagonism, but they also highlight a current barrier to the rational design of arrestin pathway-selective drugs. It would not have been possible to predict that *in vivo*, (D-Trp^{12},Tyr^{34})-bPTH(7–34) would increase bone mass, or SII would promote aldosterone biosynthesis and salt retention, based on our prior knowledge of their *in vitro* efficacy profile or the *in vivo* mechanism of action of the parent hormones.

Figure 18.2 illustrates the point. At each *in vitro* level for which at least some data are available, arrestin-selective biased agonists activate a subset of the pathways affected by the native agonist. Biophysical measurements of receptor conformation are most consistent with the interpretation that ligands stabilize different active receptor conformations in different proportions.[34] Biased ligands may favor receptor conformations that normally represent a minor proportion of the conformational population produced by the conventional agonist, but they are inevitably subject to the steric constraints imposed by receptor structure and are unlikely to couple the receptor to novel non-native signaling processes. Direct measurements of biased ligand effects on receptor–effector coupling, second messenger generation, and cell behavior are consistent with the hypothesis that biased ligand effects comprise a subset of those produced by the native ligand. Even global proteomic approaches have thus far failed to find unique biology in the short term signaling response to biased ligands.[41] *In vivo*, however, the available data suggest that biased ligands can affect gene transcription and tissue phenotype in unpredictable ways, analogous to the "butterfly effect" of chaos theory, where a small change at one place in a deterministic nonlinear system can produce outputs so complex that the results mimic random behavior.[57]

While it is clearly possible to characterize even subtle degrees of ligand bias *in vitro*,[23,25] the "screener's dilemma" arises from not knowing what information is necessary to adequately predict the *in vivo* efficacy of a potential biased therapeutic. Because efficacy is pluridimensional and ligands are biased, drug discovery programs must at least consider the possibility of unpredicted "on-target" effects, not "off-target" effects arising from

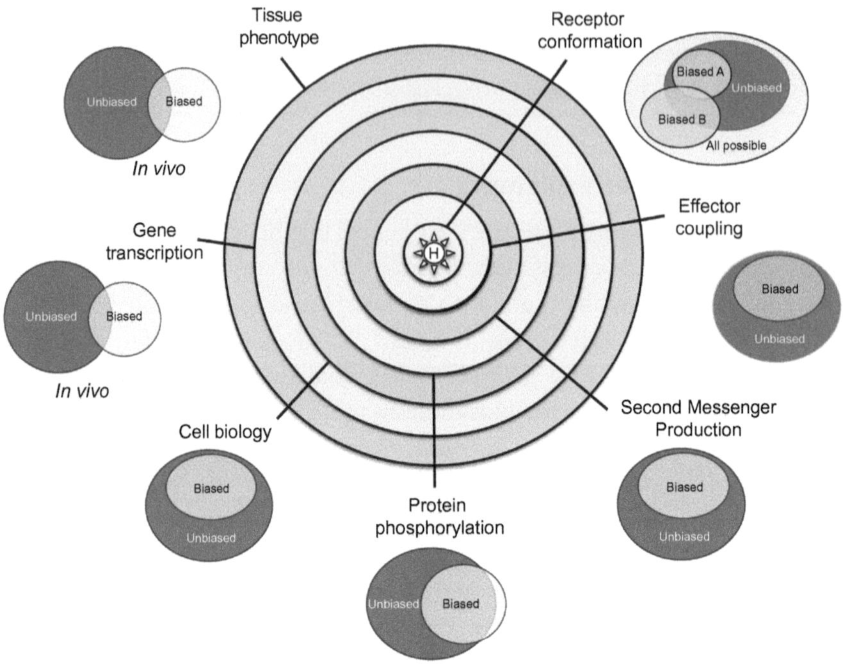

Figure 18.2 Conceptual representation of biased agonism viewed from multiple levels within a biological system. At the initial point of contact, ligand (H) binding changes the distribution of receptor conformations leading to stabilization of one or more active states. As signals radiate outward from the ligand–receptor complex, receptors engage proximal effectors such as G proteins and β-arrestins (seconds); second messenger and protein kinase cascades are activated (seconds–minutes); and changes in cell behavior, for example, proliferation, migration, or apoptosis, result (hours–days). When assayed *in vitro*, biased ligands appear to affect a subset of the pathways activated by the native ligand (*represented as Venn diagrams*). But when the *in vivo* effects of native and biased agonists are compared, the limited data currently available suggest that biased agonists have effects that cannot be predicted from the knowledge of their *in vitro* efficacy or the physiological actions of the native ligand. This "disconnect" between the *in vitro* and *in vivo* efficacy of biased ligands is a potential barrier to the rational development of biased GPCR therapeutics. *Reproduced from Ref. 56.*

unexpected drug–protein interactions, but effects arising from a drug affecting its intended target in unintended ways. When dealing with conventional agonism or antagonism, a reasonable prediction of the on-target effects of a drug can be derived from study of the native ligand. For example, the ability of β-adrenergic receptor agonists to cause bronchodilation and tachycardia, and of β-blockers to promote bronchoconstriction and bradycardia, can be derived from physiological studies of the actions of epinephrine and norepinephrine. Biased agonists, on the other hand, activate GPCRs in "unnatural" ways, and their biological effects may be less predictable. The same properties that make them valuable as potential therapeutics, that is, the ability to qualitatively change GPCR signaling, may also make them more prone to produce unanticipated effects.

4. THE CONSERVED ARRESTIN-DEPENDENT SIGNALING REPERTOIRE

In the native context, GPCRs and their cognate ligands co-evolved to control signaling in the most physiologically adaptive manner. Signaling, reflected in the distribution of active states stabilized by the native ligand and the efficiency with which these states couple the receptor to each of its downstream effectors, is therefore "balanced" to meet the needs of the organism. Biased agonists, on the other hand, alter the distribution of active states. The resulting signals are not qualitatively different, but are "unbalanced" compared to the native ligand.[11] As these unbalanced signals propagate, differences in the tissue response may arise from relatively subtle differences in signal strength or the kinetics of pathway activation and deactivation. Certainly, the presence of disease can cause adaptive responses to become maladaptive, as in the case of endogenous catecholamine signaling in the setting of congestive heart failure,[58] and this is one arena in which biased agonists hold promise as therapeutics.[42,54] The ability to recalibrate GPCR signaling to adapt to the needs of a perturbed system is something that cannot be achieved using conventional agonist or antagonists.

Part of the solution to the screener's dilemma may lie in the study of conserved arrestin signaling functions. If arrestin signaling affects a core set of biological processes in consistent ways, it may become easier to identify settings in which accentuating or diminishing arrestin signaling would help to "recalibrate" GPCR signaling to restore adaptive function.[11] To date, the hypothesis that aspects of the arrestin signaling repertoire are conserved across different GPCRs and tissues has not been systematically investigated. However, a review of the current literature (as presented in the preceding chapters of this volume) does point to certain commonalities that help to define a framework for understanding arrestin signaling at the biological process level.

4.1. GPCR desensitization

The most obvious conserved arrestin function is the negative regulation of G protein signaling, through direct steric hindrance of the GPCR–G protein interaction, mediation of receptor sequestration, and recruitment of second messenger-degrading enzymes (see Chapter 3). Hence, G protein-selective "biased" agonists have potential as non-desensitizing ligands. Arrestin3 null mice demonstrate prolonged opiate analgesia,[59,60] and a recently reported G protein-biased mu opioid receptor agonist, TRV130, appears to mimic the effects of arrestin3 knockout, producing potent analgesia with reduced gastrointestinal dysfunction and respiratory depression than morphine in rats and mice.[61] Another example is, the niacin receptor, GPR109A, which has been shown to decrease serum free fatty acids by activating $G_{i/o}$ while at the same time producing cutaneous flushing through arrestin-dependent activation of phospholipase A_2.[62] Predictably, a G protein pathway-selective GPR109A agonist, MK-0354, has been shown to decrease serum free fatty acids *in vivo* without causing flushing, making it a potentially useful agent for treating hyperlipidemia.[63]

4.2. Cell proliferation

Arrestins interface with several pathways that regulate progression of the cell cycle (see Chapters 5 and 15). In some settings, arrestin–Src complexes mediate the ability of GPCRs to transactivate the epidermal growth factor (EGF) receptor tyrosine kinase pathway, stimulating EGF receptor-dependent ERK1/2 activity and cell proliferation.[64] Arrestin2–Src also increases cell proliferation by activating an Rb–Raf-1 pathway that promotes Rb dissociation from E2F-responsive proliferative promoters, leading to increased E2F1 binding, transcription of S-phase genes, and cell cycle progression.[65] Conversely, direct scaffolding of the ERK1/2 cascade by arrestins promotes cytosolic retention of active ERK1/2 and inhibits GPCR-induced transcription and cell proliferation.[66–68]

Whether arrestin-dependent signaling promotes or inhibits cell cycle progression appears to vary between systems, perhaps depending on which signaling pathway dominates. Primary calvarial preosteoblasts from arrestin3 null mice proliferate much faster that wild preosteoblasts, and the arrestin-selective biased agonist (D-Trp^{12},Tyr^{34})-bPTH(7–34) slows preosteoblast proliferation in an arrestin3-dependent manner, suggesting that arrestins restrain cell proliferation.[44] Conversely, neointimal hyperplasia following carotid endothelial injury is diminished in arrestin3 null mice. Loss of arrestin3 is associated with decreased GPCR-stimulated vascular smooth muscle cell proliferation, ERK1/2 activation, and migration, consistent with a role for arrestin3-signaling in mediating the proliferative response.[64] In this system, knockout of arrestin2 has the opposite effect, suggesting that arrestin2 and 3 play opposing roles in regulation of vascular smooth muscle proliferation. In primary vascular smooth muscle, SII induces ERK1/2 activation and proliferation by promoting AT_{1A} receptor-dependent EGF receptor transactivation.[69] Ang II and SII both stimulate Src-dependent EGF receptor phosphorylation on Tyr^{845}, an effect that is lost when arrestin3 is downregulated by RNA interference.[70]

Arrestin signaling also promotes cell proliferation in some forms of cancer. Human bladder cancer cells express high levels of both the thromboxane (TP)-β receptor isoform and arrestin3, and the degree of TP-β upregulation correlates with poorer prognosis.[71] TP-α and TP-β are splice variants that differ only in the C-terminus, with TP-β carrying a longer tail that allows it to engage arrestin3 and undergo agonist-dependent internalization.[72] Expressing TP-β in nontransformed SV-HUV urothelial cells, confers agonist-dependent ERK1/2 and FAK phosphorylation, and enhances cell proliferation, migration, and invasion *in vitro*, responses that are lost when arrestin3, but not arrestin2, is downregulated by RNA inference. Similarly, arrestin2-dependent activation of c-Src and EGF receptor appears to contribute to the tumor promoting effects of Prostaglandin E (EP)2 receptors in papilloma formation.[73] In non-small cell lung cancer, arrestin2 mediates nicotine-induced epithelial cell proliferation through activation of the MAPK and Rb–Raf-1 pathways.[65]

4.3. Non-proliferative cell growth

Some evidence suggests that control of non-proliferative cell growth, that is, hypertrophy, may be a conserved arrestin-dependent signaling function. Arrestin scaffolding of the ERK1/2 cascade allows it to preferentially target membrane and cytosolic substrates including the ribosomal S6 kinase, p90RSK,[43] and MAP kinase-interacting kinase 1 (Mnk1), a regulator of the ribosomal protein translation initiation complex. Arrestin3-dependent ERK1/2 activation by the AT_{1A} receptor mediates phosphorylation of Mnk1 and eukaryotic translation initiation factor 4E (eIF4E), increasing rates of mRNA translation.[74] In addition, arrestin scaffolding of the PP2A–AKT–GSK3 complex may permit GPCR crosstalk with PI3K–AKT regulation of mammalian target of rapamycin (mTOR), a critical coordinator of the nutrient-sensing and cellular growth responses.[40]

In vitro, a mutant angiotensin AT_{1A} receptor with a deletion in its second intracellular loop that inhibits G protein coupling (AT_1–i2m) activates an Src–Ras–ERK1/2 pathway leading to cytosolic ERK1/2 and p90RSK activation.[75] *In vivo,* cardiomyocyte-specific overexpression of AT_1–i2m leads to greater cardiomyocyte hypertrophy, bradycardia, and fetal cardiac gene expression than comparable overexpression of the wild-type receptor. Conversely, overexpressed wild-type AT_{1A} receptors produce greater cardiomyocyte apoptosis and interstitial fibrosis than the G protein-uncoupled mutant, suggesting that G protein-dependent and arrestin-dependent AT_{1A} receptor signals mediate different aspects of the hypertrophic response.[76]

4.4. Cell survival and apoptosis

A widely conserved and apparently critical set of arrestin-mediated signals involves the regulation of pro-survival and anti-apoptotic signaling. In addition to chronically dampening β-catenin signaling (see Chapter 11), the capacity of arrestin3 to scaffold a PP2A–AKT–GSK3 complex probably allows it to serve as a positive regulator of PI3K/AKT-dependent survival signaling. Activation of AT_{1A} receptors by SII promotes G protein-independent phosphorylation of the PP2A inhibitor, I2PP2A, leading to inhibition of arrestin-bound PP2A and increased phosphorylation of Thr308, the activating phosphorylation site on AKT.[40] Arrestin-dependent activation of the ERK1/2 substrate p90RSK also reportedly acts in concert with PI3K–AKT to downregulate phospho-BAD, inducing anti-apoptotic cytoprotective effects in rat vascular smooth muscle cells.[77] Finally, arrestin scaffolding of the ubiquitin ligase, Mdm2, modulates p53-dependent apoptosis. Besides controlling GPCR endocytosis through arrestin ubiquitination (see Chapter 7), Mdm2 is a major negative regulator of the p53 tumor suppressor, since ubiquitination of p53 promotes its degradation by the proteosome. Arrestin binding to Mdm2 prevents self-ubiquitination and p53 ubiquitination. Decreasing p53 ubiquitination increases its abundance, leading to enhanced apoptotic p53 signaling. As a result, overexpressing arrestin3 enhances, and downregulating arrestin3 expression attenuates, p53-mediated apoptosis.[78]

Functional genomic analysis of the actions of (D-Trp^{12},Tyr^{34})-bPTH (7–34) on bone *in vivo* suggest that arrestin-dependent anti-apoptotic signaling underlies the observed expansion of the osteoblast pool and increase in trabecular bone mass.[31,44] *In vitro*, both (D-Trp^{12},Tyr^{34})-bPTH(7–34) and PTH(1–34) protect wild type, but not arrestin3-null, osteoblasts from a pro-apoptotic etoposide challenge.[44] In addition to its proliferative effects, the arrestin-selective AT_{1A} receptor agonist, SII, is likewise anti-apoptotic in primary vascular smooth muscle, a feature that probably contributes to arrestin-dependent neointimal hyperplasia.[77] Arrestin signaling may also protect cardiomyocytes from undergoing apoptosis. *In vitro*, a mutant β_1 adrenergic receptor that lacks the GRK phosphorylation sites in its C-terminal tail (-GRKβ1) and cannot undergo arrestin-dependent desensitization, fails to transactivate the EGF receptors despite exaggerated G protein activation.[79] In response to chronic isoproterenol stimulation, transgenic mice expressing the -GRKβ1 receptor develop more severe dilated cardiomyopathy with significantly increased LV end-diastolic dimension, decreased fractional shortening, and increased myocardial apoptosis, than wild-type β1 receptor transgenic mice, suggesting that arrestin coupling promotes cardiomyocyte survival.

4.5. Cell migration, chemotaxis, and secretory function

Extensive data support the role of arrestin scaffolds in cytoskeletal rearrangement and biological processes that depend on it, including vesicle endocytosis, exocytosis, chemotaxis, migration, and metastasis (see Chapters 4, 8, 14, and 15). Arrestins regulate several small G proteins involved in vesicle trafficking and cytoskeletal reorganization; including RalA/B; the Rho family GTPases Rho, Rac, and Cdc42; several Rab isoforms; and ARF6 (see Chapter 6). Arrestin scaffolding of Src family nonreceptor tyrosine kinases regulates GRK2, dynamin, and adaptor protein (AP)2 function during GPCR endocytosis,[80–84] and plays a role in secretory granule release and exocytic vesicle dependent translocation of the glucose transporter, Glut4.[85,86] Arrestin-dependent regulation of ERK1/2 and cortical actin assembly at the leading edge is required for GPCR-mediated chemotaxis. During protease-activated receptor (PAR2) stimulated chemotaxis, assembly of a PAR2–arrestin–ERK1/2 complex at the leading edge activates actin cytoskeleton reorganization.[87,88] In addition, arrestins scaffold a GPCR-based complex containing the actin filament-severing protein, cofilin, LIM kinase, and the cofilin-specific phosphatase, chronophin, which is required for the dephosphorylation and activation of cophillin.[89] Arrestin-bound cophillin generates the free barbed ends on actin filaments that permit filament extension.

In addition to its roles in promoting cancer cell proliferation and survival, arrestin-dependent cell migration may contribute to invasive cancer phenotypes. As mentioned previously, arrestin signaling by the TP-β receptor promotes cell migration and invasion *in vitro*, responses that are lost when arrestin3 is downregulated.[71] Similarly, the LPA1 and LPA2 lysophosphatidic acid receptors, arrestins 2 and 3, Ral and Ral-GDS are all upregulated in more advanced stages of human breast cancer. *In vitro*, LPA stimulates the migration and invasion of MDA-MB-231 breast cancer cells, and downregulating arrestin or Ral expression inhibits the response.[90] Endothelin ET-A receptors are expressed in a majority of human ovarian cancers and its co-expression with arrestin2 increases with advancing tumor grade. In a tumor xenograft model, ovarian cancer cells expressing the arrestin2 S412D mutant, which impairs arrestin-dependent ET-A receptor signaling, develop fewer metastases, suggesting that arrestin signaling contributes to ovarian tumor progression and metastatic potential.[91]

Seemingly opposing arrestin-dependent regulation of chemotaxis and production of inflammatory cytokines leads to complex effects in the immune system. During the immune response, arrestin scaffolding appears to be required for the establishment of cell polarity and to facilitate degranulation at the site of injury. Splenocytes derived from arrestin3-null mice exhibit strikingly impaired chemotactic responses to stromal cell-derived factor-1, CXCL12.[92] Similar observations apply to other GPCRs expressed in leukocytes, including CXCR1 (IL-8) receptors in RBL cells,[93] CXCR2 (IL-8) receptors in neutrophils,[94] and CXCR3 (CCL-9 and CCL-10), CXCR4 (stromal cell-derived factor 1α), and CCR5 (RANTES) receptors expressed in HEK293 cells.[95–97] On the other hand, arrestins function as negative regulators of NFκB signaling,[98–100] and can limit the intensity of the chemotactic stimulus itself by inhibiting the local production of pro-inflammatory cytokines. Thus, isolated peritoneal neutrophils from arrestin3 null animals exhibit increased basal and LPS-stimulated TNFα and IL-6 production.[101]

Which of these potentially opposing functions of arrestins dominates *in vivo* appears to vary. In allergen-sensitized arrestin3-null mice, T helper (Th)2 cell chemotaxis into the lungs is blunted and levels of the Th-derived cytokines IL-4, IL-5, and IL-13 in bronchoalveolar lavage fluid are markedly reduced, suggesting that impaired chemotaxis leads to a diminished immune response.[102] In contrast, an *in vivo* study of CXCR2-mediated chemotaxis found that neutrophil recruitment to sites of inflammation in both a dorsal air pouch and an excisional wound model was increased in arrestin3 null mice despite impaired receptor desensitization and endocytosis,[103] prompting the authors to conclude that arrestin3 is an overall negative regulator of CXCR2 signaling *in vivo*. Similar results were reported in a cecal ligation and puncture model of neutrophil infiltration, where arrestin3 null mice showed significantly higher peritoneal and pulmonary neutrophil accumulation than wild-type controls.[101] Thus, while it is clear that arrestins are critical mediators of chemotaxis in cells of the immune system, their overall functions in modulating immune responses are quite complex.

5. CONCLUSIONS

Conventional receptor theory predicts that GPCRs exist in spontaneous equilibrium between "inactive" and "active" states that are stabilized by ligand binding, leading to increases or decreases in receptor coupling to its intracellular effectors. Agonists increase the proportion of receptors in the active state and antagonists decrease it, but the receptor alone specifies which pathways are activated. What the concepts of pluridimensional efficacy and functional selectivity have taught us is that receptors have multiple active states with different signaling capacity, and structurally distinct ligands have the ability to stabilize these active states in different proportions compared with the native ligand. In other words, the ligand–receptor complex, not the receptor alone, determines signal output. And if the ligand–receptor complex defines the signaling, then biased ligands can create, in effect, new receptors whose functionality is both unknown and possibly unique to each ligand–receptor pair.

At present, we face a gap between the *in vitro* characterization of ligand bias and the unpredictability of its *in vivo* effects. The physiological effects of conventional agonist or antagonist ligands can largely be predicted based on the physiological properties of the native ligand. But if every biased ligand-GPCR pair is a functionally distinct entity with the potential to influence biological systems in unique ways, then the discovery of therapeutically useful biased agonists becomes a matter of trial and error. The screener cannot know which *in vitro* properties of a biased agonist will predict the desired *in vivo* biological response. With respect to arrestin-selective biased ligands, additional work is needed to determine the most physiologically relevant arrestin-mediated pathways *in vivo,* and the extent to which these pathways are conserved between different GPCRs and different tissues. The facts that there are only two non-visual arrestins, they are ubiquitously expressed outside the retina, and they interact with the vast majority of GPCRs, at least offer hope that it will become possible to relate the effects of arrestin-selective agonists *in vitro* and *in vivo*. Within such a framework, it may be easier to capitalize on the promise of biased agonism.

ACKNOWLEDGMENTS

Supported in part by National Institutes of Health Grant DK55524 (L. M. L.) and the Research Service of the Charleston, SC Veterans Affairs Medical Center. The contents of this chapter do not represent the views of the Department of Veterans Affairs or the United States Government.

REFERENCES

1. Galandrin S, Bouvier M. Distinct signaling profiles of beta1 and beta2 adrenergic receptor ligands toward adenylyl cyclase and mitogen-activated protein kinase reveals the pluridimensionality of efficacy. *Mol Pharmacol.* 2006;70:1575–1584.
2. Maudsley S, Martin B, Luttrell LM. The origins of diversity and specificity in G protein-coupled receptor signaling. *J Pharmacol Exp Ther.* 2005;314:485–494.
3. Luttrell LM, Gesty-Palmer D. Beyond desensitization: physiological relevance of arrestin-dependent signaling. *Pharmacol Rev.* 2010;62:305–330.
4. Shenoy SK, Lefkowitz RJ. β-Arrestin-mediated receptor trafficking and signal transduction. *Trends Pharmacol Sci.* 2011;32:521–533.
5. Whalen EJ, Rajagopal S, Lefkowitz RJ. Therapeutic potential of β-arrestin- and G protein-biased agonists. *Trends Mol Med.* 2011;7:126–139.
6. Kenakin T. Agonist-receptor efficacy. II. Agonist-trafficking of receptor signals. *Trends Pharmacol Sci.* 1995;16:232–238.
7. Christopoulos A, Kenakin T. G protein-coupled receptor allosterism and complexing. *Pharmacol Rev.* 2002;54:323–374.
8. Kenakin TP. Functional selectivity through protean and biased agonism: who steers the ship? *Mol Pharmacol.* 2007;72:1393–1401.
9. Kenakin T, Miller LE. Seven transmembrane receptors as shapeshifting proteins: the impact of allosteric modulation and functional selectivity on new drug discovery. *Pharmacol Rev.* 2010;62:265–304.
10. Luttrell LM, Kenakin TP. Refining efficacy: allosterism and bias in G protein-coupled receptor signaling. *Methods Mol Biol.* 2011;756:3–35.
11. Maudsley S, Patel SA, Park S-S, Luttrell LM, Martin B. Functional signaling biases in G protein-coupled receptors: game theory and receptor dynamics. *Mini Rev Med Chem.* 2012;12:831–840.
12. Black JW, Leff P. Operational models of pharmacological agonist. *Proc R Soc Lond (Biol).* 1983;220:141–162.
13. Black JW, Leff P, Shankley NP. An operational model of pharmacological agonism: the effect of E/[A] curve shape on agonist dissociation constant estimation. *Br J Pharmacol.* 1985;84:561–571.
14. Karlin A. On the application of "a plausible model" of allosteric proteins to the receptor for acetylcholine. *J Theor Biol.* 1967;16:306–320.
15. Thron CD. On the analysis of pharmacological experiments in terms of an allosteric receptor model. *Mol Pharmacol.* 1973;9:1–9.
16. Kenakin TP. Receptor conformational induction versus selection: all part of the same energy landscape. *Trends Pharmacol Sci.* 1996;17:190–191.
17. Costa T, Herz A. Antagonists with negative intrinsic activity at delta opioid receptors coupled to GTP-binding proteins. *Proc Natl Acad Sci USA.* 1989;86:7321–7325.
18. Samama P, Cotecchia S, Costa T, Lefkowitz RJ. A mutation-induced activated state of the beta 2-adrenergic receptor. Extending the ternary complex model. *J Biol Chem.* 1993;268:4625–4636.
19. Weiss JM, Morgan PH, Lutz MW, Kenakin TP. The cubic ternary complex receptor-occupancy model. III. Resurrecting efficacy. *J Theor Biol.* 1996;181:381–397.

20. Appleton KM, Lee MH, Alele C, et al. Biasing the parathyroid hormone receptor: relating in vitro ligand efficacy to in vivo biological activity. *Methods Enzymol.* 2013;522:229–262.
21. Kohout TA, Nicholas SL, Perry SJ, Reinhart G, Junger S, Struthers RS. Differential desensitization, receptor phosphorylation, beta-arrestin recruitment, and ERK1/2 activation by the two endogenous ligands for the CC chemokine receptor 7. *J Biol Chem.* 2004;279:23214–23222.
22. Drake MT, Violin JD, Whalen EJ, Wisler JW, Shenoy SK, Lefkowitz RJ. beta-Arrestin-biased agonism at the beta2-adrenergic receptor. *J Biol Chem.* 2008;283: 5669–5676.
23. Rajagopal S, Ahn S, Rominger DH, et al. Quantifying ligand bias at seven-transmembrane receptors. *Mol Pharmacol.* 2011;80:367–377.
24. Gesty-Palmer D, Chen M, Reiter E, et al. Distinct conformations of the parathyroid hormone receptor mediate G protein and beta-arrestin dependent activation of ERK1/2. *J Biol Chem.* 2006;281:10856–10864.
25. Kenakin TP. 7TM receptor allostery: putting numbers to shapeshifting proteins. *Trends Pharmacol Sci.* 2009;30:460–469.
26. Bringhurst FR, Juppner H, Guo J, et al. Cloned, stably expressed parathyroid hormone (PTH)/PTH-related peptide receptors activate multiple messenger signals and biological responses in LLC-PK1 kidney cells. *Endocrinology.* 1993;132:2090–2098.
27. Jouishomme H, Whitfield JF, Gagnon L, et al. Further definition of the protein kinase C activation domain of the parathyroid hormone. *J Bone Miner Res.* 1994;9:943–949.
28. Friedman PA, Gesek FA, Morley P, Whitfield JF, Willick GE. Cell specific signaling and structure-activity relations of parathyroid hormone analogs in mouse kidney cells. *Endocrinology.* 1999;140:301–309.
29. Sneddon WB, Magyar CE, Willick GE, et al. Ligand-selective dissociation of activation and internalization of the parathyroid hormone receptor. Conditional efficacy of PTH peptide fragments. *Endocrinology.* 2004;145:2815–2823.
30. Gardella TJ, Luck MD, Jensen GS, Schipani E, Potts Jr JT, Juppner H. Inverse agonism of amino-terminally truncated parathyroid hormone (PTH) and PTH-related peptide (PTHrP) analogs revealed with constitutively active mutant PTH/PTHrP receptors. *Endocrinology.* 1996;137:3936–3941.
31. Gesty-Palmer D, Flannery P, Yuan L, Spurney R, Lefkowitz RJ, Luttrell LM. A beta-arrestin biased agonist of the parathyroid hormone receptor (PTH1R) promotes bone formation independent of G protein activation. *Sci Transl Med.* 2009;1:1ra1.
32. Yao X, Parnot C, Deupi X, et al. Coupling ligand structure to specific conformational switches in the beta2-adrenoceptor. *Nat Chem Biol.* 2006;2:417–422.
33. Yao XJ, Vélez Ruiz G, Whorton MR, et al. The effect of ligand efficacy on the formation and stability of a GPCR-G protein complex. *Proc Natl Acad Sci USA.* 2009;106:9501–9506.
34. Liu JJ, Horst R, Katritch V, Stevens RC, Wüthrich K. Biased signaling pathways in β_2-adrenergic receptor characterized by ^{19}F-NMR. *Science.* 2012;335:1106–1110.
35. Coulon V, Audet M, Homburger V, et al. Subcellular imaging of dynamic protein interactions by bioluminescence resonance energy transfer. *Biophys J.* 2008;94: 1001–1009.
36. Lohse MJ, Nuber S, Hoffmann C. Fluorescence/bioluminescence resonance energy transfer techniques to study G-protein-coupled receptor activation and signaling. *Pharmacol Rev.* 2012;64:299–336.
37. Saulière A, Bellot M, Paris H, et al. Deciphering biased-agonism complexity reveals a new active AT1 receptor entity. *Nat Chem Biol.* 2012;8:622–630.
38. Wei H, Ahn S, Shenoy SK, et al. Independent G protein and beta-arrestin2 mediated activation of ERK by angiotensin. *Proc Natl Acad Sci USA.* 2003;100:10782–10787.

39. Azzi M, Charest PG, Angers S, et al. Beta-arrestin-mediated activation of MAPK by inverse agonists reveals distinct active conformations for G protein-coupled receptors. *Proc Natl Acad Sci USA*. 2003;100:11406–11411.
40. Kendall RT, Strungs EG, Rachidi SM, et al. The β-arrestin pathway-selective angiotensin AT_{1A} receptor agonist, $Sar^1Ile^4Ile^8$-AngII, regulates a robust G protein-independent signaling network. *J Biol Chem*. 2011;286:19880–19891.
41. Christensen GL, Kelstrup CD, Lyngsø C, et al. Quantitative phosphoproteomics dissection of seven-transmembrane receptor signaling using full and biased agonists. *Mol Cell Proteomics*. 2010;9:1540–1553.
42. Rajagopal K, Whalen EJ, Violin JD, et al. Beta-arrestin2-mediated inotropic effects of the angiotensin II type 1A receptor in isolated cardiac myocytes. *Proc Natl Acad Sci USA*. 2006;103:16284–16289.
43. Aplin M, Christensen GL, Schneider M, et al. Differential extracellular signal-regulated kinases 1 and 2 activation by the angiotensin type 1 receptor supports distinct phenotypes of cardiac myocytes. *Basic Clin Pharmacol Toxicol*. 2007;100:296–301.
44. Gesty-Palmer D, Yuan L, Martin B, et al. β-arrestin-selective G protein-coupled receptor agonists engender unique biological efficacy in vivo. *Mol Endocrinol*. 2013;27:296–314.
45. Tam CS, Heersche JN, Murray TM, Parsons JA. Parathyroid hormone stimulates the bone apposition rate independently of its resorptive action: differential effects of intermittent and continuous administration. *Endocrinology*. 1982;110:506–512.
46. Hock JM, Gera I. Effects of continuous and intermittent administration and inhibition of resorption on the anabolic response of bone to parathyroid hormone. *J Bone Miner Res*. 1992;7:65–72.
47. Dobnig H, Turner RT. Evidence that intermittent treatment with parathyroid hormone increases bone formation in adult rats by activation of bone lining cells. *Endocrinology*. 1995;136:3632–3638.
48. Deng ZL, Sharff KA, Tang N, et al. Regulation of osteogenic differentiation during skeletal development. *Front Biosci*. 2008;13:2001–2021.
49. Lian JB, Stein GS, Javed A, et al. Networks and hubs for the transcriptional control of osteoblastogenesis. *Rev Endocr Metab Disord*. 2006;7:1–16.
50. Qin L, Raggatt LJ, Partridge NC. Parathyroid hormone: a double-edged sword for bone metabolism. *Trends Endocrinol Metab*. 2004;15:60–65.
51. Ferrari SL, Pierroz DD, Glatt V, et al. Bone response to intermittent parathyroid hormone is altered in mice null for beta-arrestin2. *Endocrinology*. 2005;146:1854–1862.
52. Mohan S, Kutilek S, Zhang C, et al. Comparison of bone formation responses to parathyroid hormone(1–34), (1–31), and (2–34) in mice. *Bone*. 2000;27:471–478.
53. Sebastian EM, Suva LJ, Friedman PA. Differential effects of intermittent PTH(1–34) and PTH(7–34) on bone microarchitecture and aortic calcification in experimental renal failure. *Bone*. 2008;43:1022–1030.
54. Violin JD, DeWire SM, Yamashita D, et al. Selectively engaging β-arrestins at the angiotensin II type 1 receptor reduces blood pressure and increases cardiac performance. *J Pharmacol Exp Ther*. 2010;335:572–579.
55. Lymperopoulos A, Rengo G, Zincarelli C, Kim J, Soltys S, Koch WJ. An adrenal beta-arrestin 1-mediated signaling pathway underlies angiotensin II-induced aldosterone production in vitro and in vivo. *Proc Natl Acad Sci USA*. 2009;106:5825–5830.
56. Appleton KM, Luttrell LM. Emergent biological properties of arrestin pathway-selective biased agonism. *J Rec Signal Trans*. 2013.
57. Bassingthwaite JB, Liebovitch LS, West BJ. *Fractal Physiology*. New York: Oxford University Press Inc.; 1994.

58. Grassi G, Seravalle G, Quarti-Trevano F, Dell'oro R. Sympathetic activation in congestive heart failure: evidence, consequences and therapeutic implications. *Curr Vasc Pharmacol.* 2009;7:137–145.
59. Bohn LM, Lefkowitz RJ, Gainetdinov RR, Peppel K, Caron MG, Lin F-T. Enhanced morphine analgesia in mice lacking beta-arrestin 2. *Science.* 1999;286:2495–2498.
60. Bohn LM, Lefkowitz RJ, Caron MG. Differential mechanisms of morphine antinociceptive tolerance revealed in beta-arrestin-2 knock-out mice. *J Neurosci.* 2002;22:10494–10500.
61. Dewire SM, Yamashita DS, Rominger DH, et al. A G protein-biased ligand at the μ-opioid receptor is potently analgesic with reduced gastrointestinal and respiratory dysfunction compared with morphine. *J Pharmacol Exp Ther.* 2013;344:708–717.
62. Walters RW, Shukla AK, Kovacs JJ, et al. beta-Arrestin1 mediates nicotinic acid-induced flushing, but not its antilipolytic effect, in mice. *J Clin Invest.* 2009;119:1312–1321.
63. Semple G, Skinner PJ, Gharbaoui T, et al. 3-(1H-tetrazol-5-yl)-1,4,5,6-tetrahydro-cyclopentapyrazole (MK-0354): a partial agonist of the nicotinic acid receptor, G-protein coupled receptor 109a, with antilipolytic but no vasodilatory activity in mice. *J Med Chem.* 2008;51:5101–5108.
64. Kim J, Zhang L, Peppel K, et al. Beta-arrestins regulate atherosclerosis and neointimal hyperplasia by controlling smooth muscle cell proliferation and migration. *Circ Res.* 2008;103:70–79.
65. Dasgupta P, Rastogi S, Pillai S, et al. Nicotine induces cell proliferation by beta-arrestin-mediated activation of Src and Rb-Raf-1 pathways. *J Clin Invest.* 2006;116:2208–2217.
66. DeFea KA, Zalevsky J, Thoma MS, Dery O, Mullins RD, Bunnett NW. beta-Arrestin-dependent endocytosis of proteinase-activated receptor 2 is required for intracellular targeting of activated ERK1/2. *J Cell Biol.* 2000;148:1267–1281.
67. Tohgo A, Pierce KL, Choy EW, Lefkowitz RJ, Luttrell LM. barrestin scaffolding of the ERK cascade enhances cytosolic ERK activity but inhibits ERK mediated transcription following angiotensin AT1a receptor stimulation. *J Biol Chem.* 2002; 277:9429–9436.
68. Ahn S, Shenoy SK, Wei H, Lefkowitz RJ. Differential kinetic and spatial patterns of beta-arrestin and G protein-mediated ERK activation by the angiotensin II receptor. *J Biol Chem.* 2004;279:35518–35525.
69. Miura S, Zhang J, Matsuo Y, Saku K, Karnik SS. Activation of extracellular signal-activated kinase by angiotensin II-induced Gq-independent epidermal growth factor receptor transactivation. *Hypertens Res.* 2004;27:65–77.
70. Kim J, Ahn S, Rajagopal K, Lefkowitz RJ. Independent beta-arrestin2 and Gq/protein kinase Czeta pathways for ERK stimulated by angiotensin type 1A receptors in vascular smooth muscle cells converge on transactivation of the epidermal growth factor receptor. *J Biol Chem.* 2009;284:11953–11962.
71. Moussa O, Ashton AW, Fraig M, et al. Novel role of thromboxane receptors beta isoform in bladder cancer pathogenesis. *Cancer Res.* 2008;68:4097–4104.
72. Parent JL, Labrecque P, Orsini MJ, Benovic JL. Internalization of the TXA2 receptor and β isoforms. Role of the differentially spliced cooh terminus in agonist-promoted receptor internalization. *J Biol Chem.* 1999;274:8941–8948.
73. Chun KS, Lao HC, Trempus CS, Okada M, Langenbach R. The prostaglandin receptor EP2 activates multiple signaling pathways and beta-arrestin1 complex formation during mouse skin papilloma development. *Carcinogenesis.* 2009;30:1620–1627.
74. DeWire SM, Kim J, Whalen EJ, Ahn S, Chen M, Lefkowitz RJ. Beta-arrestin-mediated signaling regulates protein synthesis. *J Biol Chem.* 2008;283:10611–10620.

75. Seta K, Nanamori M, Modrall JG, Neubig RR, Sadoshima J. AT1 receptor mutant lacking heterotrimeric G protein coupling activates the Src-Ras-ERK pathway without nuclear translocation of ERKs. *J Biol Chem.* 2002;277:9268–9277.
76. Zhai P, Yamamoto M, Galeotti J, et al. Cardiac-specific overexpression of AT1 receptor mutant lacking G alpha q/G alpha i coupling causes hypertrophy and bradycardia in transgenic mice. *J Clin Invest.* 2005;115:3045–3056.
77. Ahn S, Kim J, Hara MR, Ren XR, Lefkowitz RJ. beta-Arrestin-2 mediates anti-apoptotic signaling through regulation of BAD phosphorylation. *J Biol Chem.* 2009;284:8855–8865.
78. Wang P, Gao H, Ni Y, et al. Beta-arrestin 2 functions as a G-protein-coupled receptor-activated regulator of oncoprotein Mdm2. *J Biol Chem.* 2003;278:6363–6370.
79. Noma T, Lemaire A, Naga Prasad SV, et al. Beta-arrestin-mediated beta1-adrenergic receptor transactivation of the EGFR confers cardioprotection. *J Clin Invest.* 2007;117:2445–2458.
80. Penela P, Elorza A, Sarnago S, Mayor Jr F. Beta-arrestin- and c-Src-dependent degradation of G-protein-coupled receptor kinase 2. *EMBO J.* 2001;20:5129–5138.
81. Ahn S, Maudsley S, Luttrell LM, Lefkowitz RJ, Daaka Y. Src-mediated tyrosine phosphorylation of dynamin is required for beta2-adrenergic receptor internalization and mitogen-activated protein kinase signaling. *J Biol Chem.* 1999;274:1185–1188.
82. Ahn S, Kim J, Lucaveche CL, et al. Src-dependent tyrosine phosphorylation regulates dynamin self-assembly and ligand-induced endocytosis of the epidermal growth factor receptor. *J Biol Chem.* 2002;277:26642–26651.
83. Fessart D, Simaan M, Zimmerman B, et al. Src-dependent phosphorylation of beta2-adaptin dissociates the beta-arrestin-AP-2 complex. *J Cell Sci.* 2007;120:1723–1732.
84. Zimmerman B, Simaan M, Lee M-H, Luttrell LM, Laporte SA. c-Src-mediated phosphorylation of AP-2 reveals a general mechanism for receptors internalizing through the clathrin pathway. *Cell Signal.* 2009;21:103–110.
85. Barlic J, Andrews JD, Kelvin AA, et al. Regulation of tyrosine kinase activation and granule release through beta-arrestin by CXCRI. *Nat Immunol.* 2000;1:227–233.
86. Imamura T, Huang J, Dalle S, et al. beta-Arrestin-mediated recruitment of the Src family kinase Yes mediates endothelin-1-stimulated glucose transport. *J Biol Chem.* 2001;276:43663–43667.
87. Ge L, Ly Y, Hollenberg M, DeFea K. A beta-arrestin-dependent scaffold is associated with prolonged MAPK activation in pseudopodia during protease-activated receptor-2-induced chemotaxis. *J Biol Chem.* 2003;278:34418–34426.
88. Ge L, Shenoy SK, Lefkowitz RJ, DeFea K. Constitutive protease-activated receptor-2-mediated migration of MDA MB-231 breast cancer cells requires both beta-arrestin-1 and −2. *J Biol Chem.* 2004;279:55419–55424.
89. Zoudilova M, Kumar P, Ge L, Wang P, Bokoch GM, DeFea KA. Beta-arrestin-dependent regulation of the cofilin pathway downstream of protease-activated receptor-2. *J Biol Chem.* 2007;282:20634–20646.
90. Li TT, Alemayehu M, Aziziyeh AI, et al. Beta-arrestin/Ral signaling regulates lysophosphatidic acid-mediated migration and invasion of human breast tumor cells. *Mol Cancer Res.* 2009;7:1064–1077.
91. Rosanò L, Cianfrocca R, Masi S, et al. Beta-arrestin links endothelin A receptor to beta-catenin signaling to induce ovarian cancer cell invasion and metastasis. *Proc Natl Acad Sci USA.* 2009;106:2806–2811.
92. Fong AM, Premont RT, Richardson RM, Yu YR, Lefkowitz RJ, Patel DD. Defective lymphocyte chemotaxis in beta-arrestin2- and GRK6-deficient mice. *Proc Natl Acad Sci USA.* 2002;99:7478–7483.
93. Barlic J, Khandaker MH, Mahon E, et al. beta-Arrestins regulate interleukin-8-induced CXCR1 internalization. *J Biol Chem.* 1999;274:16287–16294.

94. Richardson RM, Marjoram RJ, Barak LS, Snyderman R. Role of the cytoplasmic tails of CXCR1 and CXCR2 in mediating leukocyte migration, activation and regulation. *J Immunol.* 2003;170:2904–2911.
95. Colvin RA, Campanella GS, Sun J, Luster AD. Intracellular domains of CXCR3 that mediate CXCL9, CXCL10, and CXCL11 function. *J Biol Chem.* 2004;279: 30219–30227.
96. Sun Y, Cheng Z, Ma L, Pei G. Beta-arrestin2 is critically involved in CXCR4-mediated chemotaxis, and this is mediated by its enhancement of p38 MAPK activation. *J Biol Chem.* 2002;277:49212–49219.
97. Lagane B, Ballet S, Planchenault T, et al. Mutation of the DRY motif reveals different structural requirements for the CC chemokine receptor 5-mediated signaling and receptor endocytosis. *Mol Pharmacol.* 2005;67:1966–1976.
98. Witherow DS, Garrison TR, Miller WE, Lefkowitz RJ. beta-Arrestin inhibits NF-kappaB activity by means of its interaction with the NF-kappaB inhibitor IkappaBalpha. *Proc Natl Acad Sci USA.* 2004;101:8603–8607.
99. Luan B, Zhang Z, Wu Y, Kang J, Pei G. Beta-arrestin2 functions as a phosphorylation-regulated suppressor of UV-induced NF-kappaB activation. *EMBO J.* 2005;24: 4237–4246.
100. Gao H, Sun Y, Wu Y, et al. Identification of beta-arrestin2 as a G protein-coupled receptor-stimulated regulator of NF-kappaB pathways. *Mol Cell.* 2004;14:303–317.
101. Basher F, Fan H, Zingarelli B, et al. beta-Arrestin 2: a negative regulator of inflammatory responses in polymorphonuclear leukocytes. *Int J Clin Exp Med.* 2008;1:32–41.
102. Walker JK, Fong AM, Lawson BL, et al. Beta-arrestin-2 regulates the development of allergic asthma. *J Clin Invest.* 2003;112:566–574.
103. Su Y, Raghuwanshi SK, Yu Y, Nanney LB, Richardson RM, Richmond A. Altered CXCR2 signaling in β-arrestin-2-deficient mouse models. *J Immunol.* 2005; 175:5396–5402.

INDEX

Note: Page numbers followed by "*f*" indicate figures, and "*t*" indicate tables.

A

α_1-Adrenergic receptors (α_1ARs), 297–298
α_2-Adrenergic receptors (α_2ARs)
  desensitizers, 298–299, 300*t*
  signal transducers, 299–303
α_1ARs. *See* α_1-Adrenergic receptors (α_1ARs)
α_2ARs. *See* α_2-Adrenergic receptors (α_2ARs)
ACTH receptor. *See* Adrenocorticotropic hormone (ACTH) receptor
Actin assembly
  cofilin pathway, 210–212
  cortical actin filaments, 209–210
  creation, seeds, 209–210
  filamin and actin-binding proteins, 212
  GPCRs, 209–210
  RhoA GTPases, 212–213
Adaptive immune system
  β-arrestins and lymphocytes, 362
  lymphocyte survival and apoptosis, 363–364
  migration and chemotaxis, 363
  T-cell activation, 363
  TH1, TH2 and TH17 cells, 364–365
Adenosine diphosphate (ADP) ribosylation factors (ARFs)
  Arf family GTPases
    ARF6 and receptor endocytosis, 165–166
    β-arrestin and ARF6, 166
    isoforms, 164–165
    small ras-related G proteins, 164–165
  β-arrestin and small GTPase family, 151–152, 153*t*
Adrenal gland
  AngII-dependent aldosterone production/secretion, 324–325
  neurohormonal control, circulatory system, 296
β_2-Adrenergic receptor
  ARF6 activation, 165–166
  β-arrestin, 161
  HEK293 cells, 165–166
  Rab5a-positive endosomes, 163
Adrenergic receptors (ARs)
  β-AR, 279, 280
  β_2AR expression, 280
  α_2AR subtypes, 278–279
  cardiovascular and βarrs
    α_1ARs, 297–298
    α_2ARs, 298–303
    βARs, 303–307
  classes, 278
  ERK1/2 signaling, 280
  GRK, 279
  internalization, 279–280
  sedation, 281
Adrenocorticotropic hormone (ACTH) receptor, 283–284
Amoebal arrestin domain-containing (ADC) protein
  Dictyostelium arrestin domain-containing protein AdcA, 35, 36*f*
  N-terminal side triple-polyhistidine cluster, 35
  protein–protein/protein–lipid interaction, 35
Anaphase-promoting complex (APC), 180
Angiotensin-II-type 1A-receptor ($AT_{1A}R$), 367
Angiotensin-II-type 1 receptors (AT_1Rs)
  adrenal and βarrs, 311–312
  desensitizers, 307–308
  signal transducers, 308–310
  vascular and βarrs, 310–311
Angiotensin receptor blockers (ARBs), 11
Angiotensin type 2 receptors (AT_2Rs), 312–313
ARBs. *See* Angiotensin receptor blockers (ARBs)

ARRBs. *See* β-Arrestins (ARRBs)
ARRDCs. *See* Arrestin domain-containing (ARRDCs) proteins
Arrestin domain-containing (ARRDCs) proteins
  β_2AR, 197–198
  budding process, viruses, 45
  definition, 197–198
  endocytic vesicles, 197–198
  functional characteristics, 23
  mammals (*see* Mammalian ARRDCs)
  overexpression, 197–198
Arrestin-fold proteins
  adaptin and clathrin-binding sites, 42
  β-adrenergic receptor, 23
  amoebal ADCs, 35
  ARRDCs, 22–23
  ARTs, yeast and fungi, 33–34
  Aspergillus, 45
  binding sites, inositol phosphates, 28–30
  cell polarity and migration, 24
  conservation analysis, visual and β-arrestins, 30
  DSCR3, 39–40
  endocytic machinery, binding sites, 26–27
  ESCRT, 24
  functional conservation, 24–25, 25*f*
  functional context, 47, 48*f*
  G-protein signaling, 42
  β-integrins, 44
  mammalian ARRDCs, 30–33
  MAPK, 24
  membrane-embedded proteins, 46–47
  multivesicular body, 45
  posttranslational modification sites, 27–28
  pseudosymmetrical N-and C-domains, 22
  receptor-dependent signaling, 23–24
  receptor ubiquitination, 44
  recognition, activated receptors, 25–26
  recycling/lysosomal degradation pathway, 23
  retromer–cargo interactions, 45–46
  RGP1, 40–42
  sequence conservation, 42
  similarities and differences, arrestin clan, 42, 43*t*
  steric interference, G-proteins, 44
  tertiary topology, 49
  ubiquitin-dependent degradation, lysosome, 45
  VPS26, 35–39
  yeast transporters, 44
Arrestin-related trafficking adaptors (ARTs)
  arrestin fold protein, 42
  scaffold E3 ubiquitin ligases, 44
  yeast and fungi (*see* Yeast and fungi, ARTs)
Arrestins
  adaptors, 176–178
  and arrestin-fold proteins (*see* Arrestin-fold proteins)
  arrestin1, retina
    calmodulin, 255
    description, 252–253
    enolase1, 256–257
    kinases, 254
    microtubules, 253–254
    NSF, 255–256
    ubiquitin ligases, 254–255
  Aspergillus nidulans, 194–195
  bar code hypothesis, 12–13
  biased signaling, 10–11
  biology, 14, 183–186
  bone (*see* Bone)
  cardiovascular system (*see* Cardiovascular system)
  cellular and *in vivo* experiments, 14
  description, 4
  desensitization, 8
  diseases, 14
  and endocytosis, 8–9
  functions and activities, 4, 186–187
  GPCR (*see* G-protein-coupled receptors (GPCRs))
  GRK (*see* G protein-coupled receptor kinases (GRKs))
  heterotrimeric G proteins, 116
  kinases, 125–126
  mammalian arrestin domain-containing proteins, 196–198
  mechanisms, 182–183
  pathways
    arrestin-dependent biased agonism, 475–483

biased GPCR agonism, 469–475
cell migration, chemotaxis and secretory function, 487–489
cell proliferation, 484–485
GPCR desensitization, 483–484
non-proliferative cell growth, 485–486
survival and apoptosis, cell, 486–487
receptors and functions, 13
regulation, kinase and phosphatase pathways (*see* Phosphatase pathways)
Saccharomyces cerevisiae, 196
S-antigen, 4–5
second messenger-dependent protein kinases, 124–125
signaling, 9–10, 118
Src family nonreceptor tyrosine kinases, 116–118
structural determinants
β2AR and rhodopsin, 58–59
crystal structure, rhodopsin, 59
description, 83
interactions, signaling proteins (*see* Signaling proteins, arrestins)
receptors (*see* Receptors)
S-antigen, 58
signaling-biased arrestin mutants, designing (*see* Signaling-biased arrestins)
size, 58
subtypes, 59, 60*f*
X-ray crystal structure, 59
7TMRs, 5, 193–194
β-Arrestins (ARRBs)
actin assembly (*see* Actin assembly)
β-arrestin 1 and 2, 167, 226–228
bladder cancer, 401–402
bone formation and turnover, 343–344
breast cancer, 398–400
and cancer-associated cell phenotypes
apoptosis and survival, 397–398
cell migration and invasion, 396–397
cell proliferation, 395–396
ERK activation, 395
thromboxane receptor β isoform (TPβ), 395
angiotensin II receptors type 1A, 367
carbohydrate regulation and lipid homeostasis, 415–421
and cardiovascular (*see* Cardiovascular system)
cell migration, 206
chemokine receptors (*see* Chemokine receptors)
and chemotaxis (*see* Chemotaxis)
colorectal cancer, 400
description, 405
Drosophila homolog, 229–230
ERK signaling, 394
E3 ubiquitin ligase adaptors, 198–199
focal contacts, 206
glioblastoma multiforme, 404
gene transcription, 420–421
glucocorticoid receptors, lung epithelium, 226–228
GPCR (*see* G-protein-coupled receptors (GPCRs))
GTPases families (*see* GTPase-activating proteins (GAPs))
hematological malignancies, 402–403
histone modification, 420–421
identification, 168–169
invertebrate species, 226–228, 227*f*
kinase activities (*see* Kinase activities, β-arrestins)
knockout mice, 226
lung cancer, 400–401
mammalian multifunctional adaptor protein, 198
mechanism, GTPase activation, 168, 168*f*
model organisms, 228–229
molecular
mechanisms, 394–395
modification, 198
pancreatic cancer, 403–404
prostate cancer, 403
proteomic analysis, 169
receptor structure and translocation, 168
receptor turnover and actin assembly, 206, 207*f*
7TMRs, 150, 167
vertebrate development (*see* Vertebrate development, βarrs)
whole-body energy balance and body weight control, 413–415
βArrs. *See* β-Arrestins (ARRBs)
ARs. *See* Adrenergic receptors (ARs)

ARTs. *See* Arrestin-related trafficking adaptors (ARTs)
Aspergillus nidulans, 194–195
Asthma, 379–380
$AT_{1A}R$. *See* Angiotensin-II-type 1A-receptor ($AT_{1A}R$)
Atherosclerosis
development, 310
ET receptors, 313
factors, 315–316
vascular remodeling and occlusion, 313–314
vasculature, 324–325
AT_1Rs. *See* Angiotensin II type 1 receptors (AT_1Rs)
AT_2Rs. *See* Angiotensin type 2 receptors (AT_2Rs)
Autoimmune diseases, 258–259

B

β-Adrenergic receptor kinase (βARK)
β2AR, 7
cDNA, 7
description, 5–6
purification, 6
β2-Adrenergic receptors (β2ARs)
activation, monocytes, 369–370
desensitization, 367
β-Adrenergic receptors (βARs)
catecholamines, 303
desensitization, 304
heart failure, 303–304
signal transducers, 305–306
vascular smooth muscle, 306–307
Bar code hypothesis, 12–13
βARK. *See* β-Adrenergic receptor kinase (βARK)
βARs. *See* β-Adrenergic receptors (βARs)
β2ARs. *See* β2-Adrenergic receptors (β2ARs)
Biased agonism
arrestin-dependent
arrestin-selective ligand bias *in vivo*, 478–481
in vitro characterization, arrestin-selective ligand bias, 475–478
screener's dilemma, 481–483, 482*f*
cardiovascular GPCR ligands, 320–324, 321*f*
GPCR (*see* G-protein-coupled receptors (GPCRs))
PTH receptor, 342–343, 342*t*
Bioinformatics
combinatorial informatics approach, 452–453
dataset enrichment, 449–451
data texturization applications, 452–453
functional signaling responses, 449–451
gene/protein networks, 449–451
geneset enrichment analysis, 451–452
mass analytical cell signaling data, 449
signal transduction research, 452–453
systems-level cellular signaling analysis, 452–453, 454*f*
Bioluminescence resonance energy transfer (BRET), 183–185
Bladder cancer, 401–402
BMD. *See* Bone mineral density (BMD)
Body energy balance
hypothalamic signals, peripheral tissues, 413–414
peripheral signals, 414–415
Bone
β-arrestin 2, bone formation and turnover, 343–344
drug design, 350
effects, PTH arrestin signaling, 344–345
functional selectivity, 339–343
GPCR-signaling
active and inactive states, 336
arrestins, 338–339
functional selectivity, 337–338
heptahelical GPCRs, 335–336
protein-protein interactions, 337
"two-state" model, 336–337
osteoporosis/BMD, 334–335
PTH, 335
skeletal effects, 345–347, 346*f*, 348*f*
Bone mineral density (BMD), 334–335, 341
Breast cancer, ARRBs
LPA receptor, 399
mRNA expression, 399
PAR-2 expression, 398–399
tumor cells and stromal expression, 399–400
Bric-a-brac (BTB) families, 180

C

Calmodulin, 255
Cannabinoid receptors, 285–286
Carbohydrate regulation and lipid homeostasis
glucagon and GPCR signals, 418–419
insulin signaling, 415–417
PPARs, 419–421
signals mediating insulin secretion, 417–418
Cardiovascular system
angiotensin II (Ang II) receptors and βarrs, 307–313
βarr isoform, 324–325
ARs and βarrs, 297–307
biased agonism/antagonism, GPCR ligands, 320–324, 321*f*
cardiac function, 296
GPCR and βarrs, 313–318
non-GPCRs and βarrs, 318–320
7TMRs, 296–297
Casein kinase 2 (CK2), 127*t*, 136
CDK inhibitor. *See* Cyclin-dependent kinase (CDK) inhibitor
Cell migration
cellular activities, 219–220
chemotaxis and secretory function, 487–489
cofilin, 210–211
CXCR4, 208–209
ERK1/2 and β-arrestins, 214–215
filamin, 212
inhibition, 208–209
Cell proliferation, 484–485
Central nervous system (CNS)
βarrs
cannabinoid receptors, 285–286
CRF receptors, 283–285
drug development, 286–287
ethanol preference, 286
GRK and arrestins, 268–275
MOR (*see* Mu-opioid receptors (MOR))
noradrenergic neurotransmission (*see* Noradrenergic receptors)
psychiatric drugs, 286–287
serotonergic neurotransmission (*see* Serotonin (5HT))
α_2ARs function, 299
Chemokine receptors
β-arrestins, 208–209
cancer and immune cells, 206–207
CC, 366
CXC, 365–366
CXCL12 monomers, 208–209
CXCR4, 208–209
defined, 365
oligomerize and oligomers, 206–207
receptor turnover, 208
signaling, 366
Chemotaxis
health and disease, 218–219
in vivo studies, 217
CNS. *See* Central nervous system (CNS)
Cofilin-chronophin-LIM kinase, 127*t*, 138–139
Colorectal cancer, 400
Cone photoreceptors
arrestin1 *vs.* arrestin 4, 247
pinealocytes, arrestin 4, 246–247
S-cones, 246–247
Corticotropin-releasing factor (CRF) receptors
ACTH receptor, 283–284
CRF_1 and CRF_2, 284
GRKs, 284–285
HEK cells and mouse cortical neurons, 285
Porsolt swim test and tail suspension test, 284
stress and anxiety, 284
UCN2, 285
CRF receptors. *See* Corticotropin-releasing factor (CRF) receptors
C-terminus of Hsc70-interacting protein (CHIP), 193
Cyclin-dependent kinase (CDK) inhibitor, 237–238
Cystic fibrosis, 380

D

DA receptors. *See* Dopamine (DA) receptors
Deubiquitinases (DUBs)
β-arrestin-dependent cellular effects, 189

Deubiquitinases (DUBs) (*Continued*)
description, 181
in vitro assays, 189
posttranslational modifications, 7TMR signals, 188*f*, 189
seven-transmembrane receptor, 193–194
USP33, 187
Diacylglycerol kinase (DGK), 124–125, 139
Dopamine (DA) receptors
AKT phosphorylation, GSK3, 272–274, 273*f*
β-arrestin 2-dependent D2R signaling pathway, 274
desensitization and internalization, 272
GRK6, 271
mood stabilizers, lithium, 274–275
neurological and psychiatric disorders, 270
pharmacologic treatments, 271
δ-Opioid receptors (DOR), 273*f*, 283
DOR. *See* δ-Opioid receptors (DOR)
Down syndrome critical region gene 3 (DSCR3)
Dictyostelium, 39–40
signaling platform, 40
trisomy, chromosome 21, 39
Drosophila
βarr_2, 236–237
frizzled proteins, 231–232
nonvisual βarr, 229
single βarr gene, 228
transcription factors Ci, 234
Drug design, 350
DSCR3. *See* Down syndrome critical region gene 3 (DSCR3)

E

EAE. *See* Experimental autoimmune encephalomyelitis (EAE)
EGFRs. *See* Epidermal growth factor receptors (EGFRs)
Endocytosis, 8–9
Endosomal sorting complex required for transport (ESCRT), 24
Endothelin (ET) receptors, 313–314
Epidermal growth factor receptors (EGFRs), 129–130, 132
ERK1/2 activation, β-arrestin
actin nucleation, 215–216
chemotaxis, 215
CXCR4, 215
phosphorylation, actin assembly proteins, 216
ESCRT. *See* Endosomal sorting complex required for transport (ESCRT)
ET receptors. *See* Endothelin (ET) receptors
Experimental autoimmune encephalomyelitis (EAE), 379
Extracellular signal-regulated kinases 1 and 2 (ERK1/2). *See* ERK1/2 activation, β-arrestin

F

Fibrosis
β-arrestins, TGF-β signaling, 375, 376*f*
cirrhosis, 377
destruction, alveolar basement membrane, 376–377
Fluorescent transcriptomic array analysis, 438
fMLP receptor. *See* Formyl-Met-Leu-Phe (fMLP) receptor
Formyl-Met-Leu-Phe (fMLP) receptor
neutrophils and LPA, cancer cells, 213
Ral-GDS mutant, 213, 214*f*

G

GAPs. *See* GTPase-activating proteins (GAPs)
GARP. *See* Golgi-associated retrograde protein (GARP)
GBM. *See* Glioblastoma (GBM)
GDP. *See* Guanosine diphosphate (GDP)
GEFs. *See* Guanine exchange factors (GEFs)
Gene set enrichment analysis (GSEA), 451–452
Ghrelin, 415
GIP. *See* Glucose-dependent insulinotropic polypeptide (GIP)
Glioblastoma (GBM), 404
Global proteomic analysis, receptor signaling systems
cellular signaling proteomics, 442–443
computational approaches and software tools, 442
disease processes, 441–442

fragmentation ion spectrum, 442
gene array, 442
genomic transcription, 441–442
integrated analytical approach, 441–442
liquid chromatography, 442–443
peptide identification, 442
protein quantitation techniques, 442–443
quantitative protein data, 443–444
translation, protein, 441–442
Glucagon
α cells, islets, 418–419
insulin and endothelin type A (ETA), 419
pineal hormone melatonin, 419
Glucagon-like peptide (GLP)
anti-diabetic therapy, 417
cAMP production and insulin secretion, cells, 417
polypeptide hormone, 417
Glucose-dependent insulinotropic polypeptide (GIP), 418
Glycogen synthase kinase 3β (GSK3β)
GPCR-signaling networks, 117*f*
phosphorylates β-catenin, 125–126
and PP2A-AKT signaling, 125–126, 127*t*, 136–138
Golgi-associated retrograde protein (GARP), 39
GPCRs. *See* G-protein-coupled receptors (GPCRs)
G protein-coupled receptor kinases (GRKs)
β-arrestin and X-arrestin, 6–7
arrestin proteins and isoforms, 268
βarr2 knockdown, 235–236
bar code hypothesis, 12–13
βARK purification, 6
casein kinase CK1α, 236
cDNA, 7
cloning, β2AR, 7
CRF receptors, 284–285
DA receptors (*see* Dopamine (DA) receptors)
description, 4
Drosophila homolog, 229–230
G protein-mediated signaling, 268–269, 269*f*
GRK1-7, 7–8
GRK3 expression, 271
Hedgehog (Hh) signaling, 229–230
5HT receptors, 276
48K protein, 6
neurotransmitters, 269–270
noradrenergic receptors, 279
rhodopsin phosphorylation and β2AR, 5–6
smoothened (Smo) internalization, 235–236
structures, 8
"turnoff" mechanism, 7
G-protein-coupled receptors (GPCRs)
adaptors/scaffolds, 118–121
angiotensin II-stimulated ERK1/2 activation, 118
arrestin-regulated kinome, 123–124
arrestin-signaling scaffolds, 121–122
$AT_{1A}R$ and PAR2, 209–211
biased agonism/antagonism, 320–324, 321*f*
binding, β-arrestins, 412
cardiovascular and βarrs, 313–318
cofilin dephosphorylation, 210–211, 211*f*
CRF receptors, 284–285
desensitization, 483–484
energy homeostasis, 413
functional selectivity, 468
and glucagon (*see* Glucagon)
inflammation and β-arrestins, 367–369
internalization, 212, 358
kinase activity, spatial and temporal control, 122–123
neurotransmitters, 269–270
noradrenergic receptors (*see* Noradrenergic receptors)
quantification, ligand "bias", 474–475
regulation, insulin secretion, 417
signaling
agonism, partial agonism and "bias", 470–473, 471*f*
β-arrestin-dependent regulation, 268–269, 269*f*
bone, 335–339
development, drug compounds, 430–431
diversity, 431–436
downstream effectors, 470
drug/hormone, 469
functional diversity, 430–431
physiological processes, 430–431

G-protein-coupled receptors (GPCRs) (*Continued*)
pluridimensional networks, 116–118, 117*f*
protein–protein interactions, 430–431
responses, 473
spectrum, ligand activity, 469–470
trafficking
agonist-induced sequestration, 99–100
cell signaling, 94–95
cellular processes, 94–95
cytosolic proteins, 107
desensitization, 95–99
intracellular accessory proteins, 94–95
molecular protein interactions, 94
physical and chemical extracellular stimuli, 93–94
postendocytic vesicular trafficking (*see* Postendocytic vesicular trafficking, GPCRs)
receptor sequestration, 99–100
sequestration (*see* Sequestration, GPCRs)
tyrosine kinase, 216
Green fluorescent protein (GFP), 183–185, 186, 187
GRKs. *See* G protein-coupled receptor kinases (GRKs)
GSEA. *See* Gene set enrichment analysis (GSEA)
GTPase-activating proteins (GAPs)
and ARF, 165–166
Arf family (*see* ADP-ribosylation factors (ARFs))
ARFGAP21, 159–160
βarrestin, 169
and GEFs, 152
GTP hydrolysis, 164–165
Rab family (*see* Rab family GTPases)
Ran family, 166–167
Ras family (*see* Ras family GTPases)
Rho family (*see* Rho family GTPases)
Guanine exchange factors (GEFs)
βarrestin/ARF6 signaling axis, 166
and GAPs, 152
GTPases activation, 150–151, 151*f*
RalGEFs, 155–157
RhoGEFs, 158
Guanosine diphosphate (GDP)
ARF6, 165–166
and GTP, 150–151, 151*f*, 168, 168*f*

H

Heart failure (HF)
acute, 304
aldosterone, 311
βarrs, 299
α_1ARs, 297–298
βARs, 303–304
congestive, 320
post-MI, 307–308, 309
treatment, 312
Hedgehog (Hh) signaling, βArrs
βarr1, 237–238
βarr2, 235–237
embryonic development, 234
Smo trafficking, 234
target gene expression, 234, 235*f*
Heterotrimeric G proteins, 468, 476
HF. *See* Heart failure (HF)
HIF. *See* Hypoxia inducible factor (HIF)
Homologous to E6AP *C*-terminus (HECT), 180, 190–191, 196
hPTH. *See* Human parathyroid hormone (hPTH)
Human parathyroid hormone (hPTH)
and bPTH, 480
genes regulation, 478–479
Hypothalamic signals, peripheral tissues
agouti and agouti-related peptide (AgRP), 413–414
energy homeostasis, 413
food intake and energy expenditure, 413
hypocretins, 414
melanocortin receptors (MC3R and MC4R), 413–414
organismal metabolic reactions, 413
Hypoxia inducible factor (HIF), 420–421

I

IFT. *See* Intraflagellar transport (IFT)
IMAC. *See* Immobilized metal affinity chromatography (IMAC)
Immobilized metal affinity chromatography (IMAC), 444–446
Immune system, β-arrestins

activation, effector pathways, 358
acute rejection, organ transplant, 379
and adaptive immunity, 362–365
asthma, 379–380
cardioprotection, 380–381
and chemokine receptors, 365–366
cystic fibrosis, 380
disease models, 372, 373*t*
EAE (*see* Experimental autoimmune encephalomyelitis (EAE))
fibrosis, 375–377, 376*f*
GPCR signaling, 358
and GPCRs, inflammation, 367–369
innate immunity, 359–362
mammalian hematopoiesis, 359
multiple sclerosis, 378
non-GPCR signaling pathways, 371–372
pathogen recognition and clearance, 372–375
rheumatoid arthritis, 377–378
and structural cells, 365
therapeutic development, inflammatory diseases, 381–382
and TLR signaling, 369–370
Innate immune system and β-arrestins
and complement system, 361–362
and macrophages, 359–360
and mast cells, 361
and NK cells (*see* Natural killer (NK) cells)
and polymorphonuclear leukocytes, 360–361
Insulin-like growth factor 1 receptor (IGF-1R), 191–192
Insulin signaling
chronic insulin treatment, 416–417
glucose transport, 415–416
IRS proteins, 415–416
type 2 diabetes phenotypes *in vivo*, 416
Intraflagellar transport (IFT), 234, 236
Inverse agonists, 336

J

c-Jun-N-terminal kinase 3 (JNK3)
AT_{1A} receptors, 133–134
arrestin3-dependent activation, 133–134
β-arrestins, 193
and ASK1-MKK4 cascades, 121, 122, 127*t*
components, 133
COS-7 cells, 119
ERK1/2 activation, 133–134
neuronal MAPK, 119
stress kinases activation, 133, 134

K

Kinase activities, β-arrestins
chemotaxis and Src complexes, 216
ERK1/2 activity, 215–216
Src and non-receptor-tyrosine kinases, 214–215

L

LC. *See* Liquid chromatography (LC)
LH receptor. *See* Luteinizing hormone(LH) receptor
LIMK. *See* LIM kinase (LIMK)
LIM kinase (LIMK), 210–211, 211*f*
Lipid kinases
diacylglycerol kinase, 139
GPCR-signaling pathways, 126, 127*t*
phosphatidylinositol 3-kinase, 139–140
phosphatidylinositol 4-phosphate 5-kinase, 140
Lipopolysaccharide (LPS)
β-arrestin-2-deficient mice, 364
LPS–TLR receptor, 369
Liquid chromatography (LC), 442–443
LPA receptor. *See* Lysophosphatidic acid (LPA) receptor
LPS. *See* Lipopolysaccharide (LPS)
Lung cancer
human NSCLC tumors, 401
nicotine, 400–401
prostaglandins, 401
smoking, 400–401
Luteinizing hormone (LH) receptor, 129–130
Lysophosphatidic acid (LPA) receptor
arrestin2/3 null MEFs, ERK1/2 activation, 130
β2-adrenergic, 132
ERK1/2 activation, 132

M

Mammalian ARRDCs
C-terminal cysteine, 31

Mammalian ARRDCs (*Continued*)
endocytosis, glucose transporter, 31
eukaryotes outside mammals, 30
human isoforms, 31
module β-strand sandwich, 30–31, 32*f*
X-ray analysis, 30–31
MAPK. *See* Mitogen-activated protein kinases (MAPK)
Matrix metalloprotease (MMP), 129–130
Mdm2. *See* Murine double minute 2 (Mdm2)
MEFs. *See* Murine embryo fibroblasts (MEFs)
Messenger RNA (mRNA)
osteoid surface and osteocalcin levels, 343–344
RANKL expression, 344, 347, 350
Metabolic regulation
β-arrestins (*see* β-arrestins (ARRBs))
cell-surface receptors, 412
metabolic syndromes, 412
modulation, enzyme activities, 411–412
nuclear receptors, 412
MI. *See* Myocardial infarction (MI)
Mitogen-activated protein kinases (MAPK)
arrestin-3–KNC mutant, 82
ASK1, MKK4 and JNK3, 82
binding sites, 81–82
ERK1/2 and MEK1, 81–82
extracellular signal-regulated kinases, 130–133
JNK3 activation, 81
c-Jun N-terminal kinase 3, 133–134
mutations, 81
and NFκB-dependent signaling pathways, 341–342
p38 MAP kinase, 134–135
PTH-stimulated pathway activation, 340
Ras proteins
activation, 154, 155
βarrestins, 155
BRET-based Ras probes, 155
HEK293 cells, 155
human bronchial epithelium, 154–155
7TM receptors activate mitogenic pathways, 154–155
Saccharomyces cerevisiae, 119
Ser/Thr protein kinases, 127*t*
shape, arrestin, 73
upstream MAPKKKs and MAPKKs, 80–81
WT arrestin-3, 82
MOR. *See* Mu-opioid receptors (MOR)
mRNA. *See* Messenger RNA (mRNA)
Multivesicular body (MVB), 45
Mu-opioid receptors (MOR)
β-arrestin 2 deletion, 281–282
DOR activation, β-arrestin 1, 273*f*, 283
morphine treatment, 282
physiological effects, 281
reinforcing and psychomotor effects, morphine, 283
stimulation, agonists, 282–283
Murine double minute 2 (Mdm2)
catalytic domain, 182
E3 ubiquitin ligases, 182–183
GRK2, 192–193
Murine embryo fibroblasts (MEFs), 130, 132, 134
MVB. *See* Multivesicular body (MVB)
Myocardial infarction (MI)
experimental, 312
post-MI HF progression, 312

N

Natural killer (NK) cells, 361
Neisseria meningitidis, 372–373
Nephrogenic syndrome of inappropriate antidiuresis (NSIAD), 315
N-ethylmaleimide-sensitive factor (NSF), 255–256
Niacin receptor (GPR109A), 315–316
Noradrenergic receptors
AR (*see* Adrenergic receptors (ARs))
β_2AR expression, 280
ERK1/2 signaling, 280
and GRK, 279
internalization, 279–280
sedation, 281
NSF. *See* *N*-ethylmaleimide-sensitive factor (NSF)
NSIAD. *See* Nephrogenic syndrome of inappropriate antidiuresis (NSIAD)

O

Oguchi disease, 257–258

P

PACAP. *See* Pituitary adenylate cyclase-activating polypeptide (PACAP)
Pancreatic cancer, 403–404
Parathyroid hormone (PTH)
  biased agonism, receptor, 342–343, 342*t*
  pathway-selective, 351
  receptor-mediated G protein signaling, bone, 340
  regulation, bone metabolism, 339–340
PARs. *See* Protease-activated receptors (PARs)
Pathogen recognition and clearance, β-arrestins
  cryptococcal meningitis, 374
  N. meningitidis, 372–373
  polymicrobial sepsis, 374
  S. pneumoniae, 374
  viral responses, 375
Peroxisome proliferator-activated receptors (PPARs)
  adipogenic genes, 421
  gene expression, fatty acids, 419–420
  HIF, 420–421
  muscle's oxidative capacity, 420
  rosiglitazone-stimulation, 421
  RXR, 419–420
  Thiazolidinediones, 420
Phosphatase pathways
  GPCR-G protein-effector pathways, 126
  lipid kinases (*see* Lipid kinases)
  mitogen-activated protein kinases (*see* Mitogen-activated protein kinases (MAPK))
  protein (*see* Protein phosphatases)
  Ser/Thr kinases (*see* Ser/Thr kinases)
  tyrosine protein kinases (*see* Tyrosine protein kinases)
Phosphate-binding residues and sensor
  Arg175Asn mutation, 66–67
  crystal structure, Arg175, 60*f*, 66–67
  elimination, positive charges, 65–66
  light-activated phosphorhodopsin (P-Rh*), 65–66
  mammals and *Drosophila*, 67
  mutagenesis, 67
  rhodopsin C-terminus, 66–67
  β-strands I and II, 67
Phosphatidylinositol 4-phosphate 5-kinase (PIP5K), 140
Phosphodiesterases (PDE), 193
Phospholipase C (PLC), 251–252
Physical interactomics
  analytical fluidic phase, 448
  antibody-based interactome purification, 448
  functional protein complex map, 448–449
  MS-based functional interactomic analysis, 448
  SILAC, 448–449
  treatment-induced alterations, 448–449
Pituitary adenylate cyclase-activating polypeptide (PACAP), 417–418
PKC. *See* Protein kinase C (PKC)
PLC. *See* Phospholipase C (PLC)
p38 MAP kinase, Ser/Thr protein kinases, 127*t*, 134–135
PMN. *See* Polymorphonuclear leukocytes (PMN)
Polymorphonuclear leukocytes (PMN), 360–361
Postendocytic vesicular trafficking, GPCRs
  β-arrestin-mediated postendocytic sorting, 106–107
  clathrin-mediated endocytosis machinery, 106
  endocytic adaptor proteins, 104–106
  receptor/arrestin complexes, 104–106, 105*f*
  recycling/adaptor proteins, 104–106
  somatostatin (sst) receptors, 106
Posttranslational modification (PTM)
  cellular signaling pathways, 444–446
  functional spectrum, 444
  global profiling, phosphoproteomes, 446–447
  IMAC, 444–446
  mass spectrometers, 444–446
  metal affinity chromatography, 444–446
  pH-gradient elution technique, 446–447
  protein phosphorylation, 444–446
  types, 444–446

PPARs. *See* Peroxisome proliferator-activated receptors (PPARs)
Prostaglandin E2 (PGE2) receptor, 216
Prostate cancer, 403
Protease-activated receptor-2 (PAR2), 209–211
Protease-activated receptors (PARs)
  βarr1KO mouse platelets, 318
  types, 318
Protein kinase C (PKC)
  activation, arrestin1 mobility, 252
  arrestin1 translocation, eyes, 251–252
Protein phosphatases
  cofilin-chronophin-LIM kinase, 138–139
  GPCR-signaling pathways, 126, 127*t*
  PP2A-AKT-GSK3β, 136–138
  tyrosine phosphatases, 139
PTH. *See* Parathyroid hormone (PTH)
PTM. *See* Posttranslational modification (PTM)
PXXP motifs, 126–128
P2Y receptors, 316–317

R

Rab family GTPases
  function, 164
  membrane trafficking, 162–163
  rab gene, 162–163
  receptors and β-arrestin, 163
RA_i. *See* Relative activity, intrinsic (RA_i)
Ran family GTPases, 166–167
RANKL. *See* Receptor activator of NFκB ligand (RANKL)
RANTES. *See* Regulated upon activation T cell expressed and secreted (RANTES)
Ras family GTPases
  MAPK pathway, 154–155
  RalA multifunction, 155–157
Really Interesting New Gene (RING), 180
Receptor activator of NFκB ligand (RANKL)
  mRNA expression, 344, 347, 350
  osteoblasts' surface-bound, 339–340
Receptors
  and arrestins
    amino acid substitutions, 71
    arrestin–rhodopsin interaction, 63
    clam-like movement, 63
    cone arrestin-4, 69–70
    conformation, receptor-bound arrestin
      biophysical methods, 69
      EPR technique, 68–69
      "finger loop", movement, 67–68
      fluorescent labels and NMR, 68–69
      nonreceptor partners, 68
      plastic loops, 68–69
    cytoplasmic tips, GPCRs, 63–64
    effect, phosphates, 60–61
    "footprint", 60–61, 62*f*
    GPCR (*see* G-protein-coupled receptors (GPCRs))
    "139 loop" and "finger loop", 64–65
    M2 muscarinic receptor, 70
    phosphate-binding residues and phosphate sensor, 65–67
    photoreceptors, 63–64
    receptor-binding residues, 61–63, 62*f*
    replacement, alanines, 70–71
    and rhodopsin expression, 63–64
    structure comparison, β2AR, 64–65
    WT arrestin-3, 71
  desensitization
    agonist-dependent isomerization, 95–97
    β-arrestin selectivity, 97–98
    cardiac myocytes, 99
    classes, regulatory molecules, 95–97
    classical model, 95–97, 96*f*
    Drosophila and mice, 99
    μ-opioid receptors, 98–99
    rhodopsin, 97–98
  signaling systems
    bioinformatic interpretation, 449–454
    Brownian modes of motion, 436–438
    global proteomic analysis, 441–444
    pharmacological scenarios, 438
    physical interactomics, 447–449
    posttranslational proteomic analysis, 444–447
    signaling network, 436–438
    substrate-specific isoforms, 436–438
    transcriptomic analysis, 438–441
Regulated upon activation T cell expressed and secreted (RANTES), 208
Relative activity, intrinsic (RA_i), 474–475

Retinitis pigmentosa, 258
Retinoid X Receptor (RXR), 419–420
Retrograde Golgi transport protein (RGP1)
BackPhyre, 40
Clustal Omega, 40, 41*f*
golgi membrane proteins, endosomes, 40
phylogenetic tree, 40
retrograde transport pathway, 40–42
RGP1. *See* Retrograde Golgi transport protein (RGP1)
Rheumatoid arthritis, 377–378
Rho-activated kinase (ROCK), 216
Rho family GTPases
βarrestin modulation, stress fiber formation, 158–160
Cdc42 and filopodia formation, 161–162
function, 157–158
membrane ruffling and protrusions, 160–161
Rac regulates formation, 157–158
RhoA
βarrestin-1-dependent p38MAPK activation, 212–213
cancer cells, 213
fMLP/LPA receptor, 213
PI3K activity, 213, 214*f*
Ral-GDS mutants, 213, 214*f*
RhoA, Rac-1 and Cdc42, 212–213
Rho GTPase activity, 213
TβRIII, 212–213
ROCK. *See* Rho-activated kinase (ROCK)
RXR. *See* Retinoid X Receptor (RXR)

S

Saccharomyces cerevisiae, 119, 196
Selective reaction monitoring (SRM), 442–443
Sequestration, GPCRs
agonist responsiveness, 100
angiotensin II type 1 receptor, 101–103
cellular distribution, β-arrestin, 101–103
endocytic machinery, 100–101
GRK2-mediated phosphorylation, 100
phosphatidylinositol 4-phosphate 5-kinase, 104
receptor domains, 103
Serotonin (5HT)
description, 275
elevation, rodent brains, 276
ERK1/2 signaling pathway, 278
GRK, 276
head twitch response, 5HTP, 277
HEK293 cells, 5HT4R, 276–277
5HT2A receptors, 275–276, 277
interaction, β-arrestin and 5HT2A receptors, 277
Ser/Thr kinases
casein kinase 2, 136
GPCR-signaling pathways, 126, 127*t*
nuclear factor-κB, 135–136
Seven transmembrane receptors (7TMRs)
activate mitogenic pathways, 154–155
agonists acting, 158
agonist stimulation, 184*f*
Apelin receptor, 163
and ARF6, 165–166
β2AR mutants, 190
βarrestin/ARNO/ARF6/ELMO signaling axis, 166
"β-arrestin-mediated" signaling, 10
β-arrestins, 191
carboxyl tail, 189
Cdc42 activation, 162
CXCR4, 190–191
E3 ligase adaptors, 189–190
endocytosis, clathrin, 9–10
family, 5
G protein and β-arrestin-dependent signaling, 176, 177*f*
GTPase activation, 168, 168*f*
non-7TM receptors, 162
non-7TMRs, 189–190
μ-opioid receptor, 191
receptor ubiquitination, 190
secondary adaptors, 197–198
VE-cadherins, 161
WW domains, 190
Signaling-biased arrestins
bovine, mouse and human arrestin-1, 79–80
enhanced phosphorylation-independent mutants, 78–79
IP6 binding sites, 80
MAPK signaling, 80–82
oligomerization, arrestin-3, 80
self-association, arrestin-1, 79

Signaling proteins, arrestins
  Arg196–Ser197 and Lys195–Pro196, 76–77
  clathrin, 72
  clathrin and AP2 interaction sites, 72–73
  conformational preferences
    "active" receptor-bound arrestin, 73–74
    arrestin-3, 74
    ERK1/2 and c-Raf1, 74
    expression, arrestin domains, 75
    MEK1, 75–76
    ubiquitin ligase Mdm2, 74–75
  JNK3 phosphorylation, 76
  MAP kinases, 73
  MEK1 and c-Raf1 binding, 73
  microtubules and Ca^{2+}-liganded calmodulin, 72
  Val343Thr, mutation, 76–77
SILAC. *See* Stable isotope labeling by amino acids in cell culture (SILAC)
siRNA. *See* Small interfering RNA (siRNA)
Skp1-Cullin-F-box (SCF), 180
Small GTPase family
  activation, 150–151, 151*f*
  ARF isoforms, 151–152
  βarrestins, 151–152, 153*t*, 168–169
  function, 152
  GDP, 150–151
  GEFs and GAPs, 152
  guanine nucleotide binding, 150–151
  Ras proteins, 151–152
  regulatory proteins, 152–154
  signaling pathways, 152–154
Small interfering RNA (siRNA), 154–155, 162, 166
SRM. *See* Selective reaction monitoring (SRM)
Stable isotope labeling by amino acids in cell culture (SILAC), 448–449
Steroidogenic acute regulatory (StAR) protein, 311
Streptococcus pneumoniae, 374
SUMOylation, 182–183
Suppressor of cytokine signaling (SOCS), 180
Systems analysis, arrestin pathway functions
  β-arrestin signaling, 455
  bone density regulation, 455
  cell signaling research, 457
  coherent connectivity, 456–457
  GPCR (*see* G-protein-coupled receptors (GPCRs))
  hypothesis-driven linear science, 455–456
  network theories, 456–457
  physiological systems, 456–457
  receptor signaling systems, 436–454
  systems-level signaling approaches, 455–456

T

TβRIII. *See* Type III TGF-β receptor (TβRIII)
TGF-β. *See* Transforming growth factor beta (TGF-β)
TGN. *See* Trans-Golgi network (TGN)
TLRs. *See* Toll-like receptors (TLRs)
Toll-like receptors (TLRs)
  β-arrestins, 369–370, 370*f*
  defined, 369
  LPS–TLR, 369
Transcriptomic analysis
  false discovery rate, 440–441
  fluorescent labeling, 438–440
  linear/logarithmic regression techniques, 438–440
  normalization processes, 438–440
  parametric analytical methods, 440–441
  serum albumin, 438–440
  statistical informatic annotation, 440–441
Transforming growth factor beta (TGF-β)
  β-arrestins, signaling, 375, 376*f*
  binding, 371–372
  tissue injury, 375
Trans-Golgi network (TGN), 35–37
Transient receptor potential (TRP), 192
Translocation, visual arrestins
  biochemical components, phototransduction, 247–248
  C-terminus, arrestin1, 250
  cytoskeletal components, 250–251
  depolymerization, microtubules, 250–251
  diffusion model, 249
  dynamic sensitivity, photoreceptors, 248

G-protein activation, 252
light adaptation, 247–248
PLC and PKC, 251–252
preservation, rod photoreceptors, 248
selective partitioning, arrestin1, 249–250
signaling cascade, 251
signaling inactivation, 248
transition zone, cilium, 249
Xenopus, 249
Type III TGF-β receptor (TβRIII), 212–213
Tyrosine protein kinases
GPCR-signaling pathways, 126, 127*t*
receptor, 129–130
Src family nonreceptor, 126–129

U

Ubiquitination
arrestin
biology, 183–186
functional effects, 186–187
mechanisms, 182–183
CHIP activation, 193
chronic β_2AR activation, 193
definition, 178
enzymes, 192–193
IGF-1R, 191–192
ion channels, 192
MAPK activation, 193
pleiotropic posttranslational modification, 178–181
7TMRs, 190–191
Ubiquitin ligases, 254–255
Ubiquitin-specific protease 33 (USP33)
"active" β-arrestin2, 189
β_2AR, 193–194
β-arrestin ubiquitination and deubiquitination, 187
and USP20 counteracts, 193–194, 195*f*
UCN2. *See* Urocortin neuropeptide 2 (UCN2)
Urocortin neuropeptide 2 (UCN2), 285

V

Vacuolar protein sorting 26 (VPS26)
cargo-selective retromer complex, 38
differential cargo recognition, 38
DSCR3, 39–40
Entamoeba histolytica, 38–39
formation, endosomal vesicles, 39
GARP, 39
lipid specificity, 37–38
protein partner recognition functions, 37
retrograde transport pathway, 39
trans Golgi network, 35–37
Vascular smooth muscle (VSM) cells
AT_1R, 300*t*, 310, 319–320
βarrs, 306–307, 313–314
cardiovascular α_1ARs, 297
cell-residing P2Y receptors, 317
GPCR-stimulated proliferation, 310
hyperplasia and hypertrophy, 313
vasopressin receptor, 314
Vasopressin receptors, 314–315
Vertebrate development, βarrs
βArr1, hematopoiesis, 230–231
βArr2, Wnt signaling, 231–234
Hh signaling, 234–238
Visual arrestins
arrestin1 (*see* Arrestins)
autoimmune diseases, 258–259
cone photoreceptors, 246–247
description, 259
nomenclature, 244
Oguchi disease, 257–258
quenching phototransduction
all-*trans* retinal chromophore, 245
G-protein-coupled receptor, 244–245
rhodopsin kinase phosphorylates, 245
retinitis pigmentosa, 258
splice variants, arrestin1, 245–246
translocation (*see* Translocation, visual arrestins)
VPS26. *See* Vacuolar protein sorting 26 (VPS26)
VSM cells. *See* Vascular smooth muscle (VSM) cells
V_2 vasopressin receptor (V_2R)
activation, 188*f*
"active" β-arrestin2, 189
β_2AR, 183–185
β-arrestin ubiquitination, 185–186
ubiquitination, 191

W

WASps. *See* Wiscott–Aldrich syndrome proteins (WASps)
Weibel–Palade bodies (WPB), 157
Wiscott–Aldrich syndrome proteins (WASps)
  and Arp2/3 complex components, 215–216
  βarrestin-dependent phosphorylation, 215–216
  ERK1/2 phosphorylation, 215–216
Wnt/JNK pathway, 234
Wnt signaling, βArr2
  canonical, 231–232
  Frizzled proteins, 231–232
  noncanonical, 231–234
WPB. *See* Weibel–Palade bodies (WPB)

Y

Yeast and fungi, ARTs
  Aspergillus nidulans, 33–34
  cortical punctate structures, 33–34
  E3 ubiquitin ligases, 33
  Saccharomyces genome, 33
  ubiquitination, 34
  and VPS26 families, 33, 34*f*

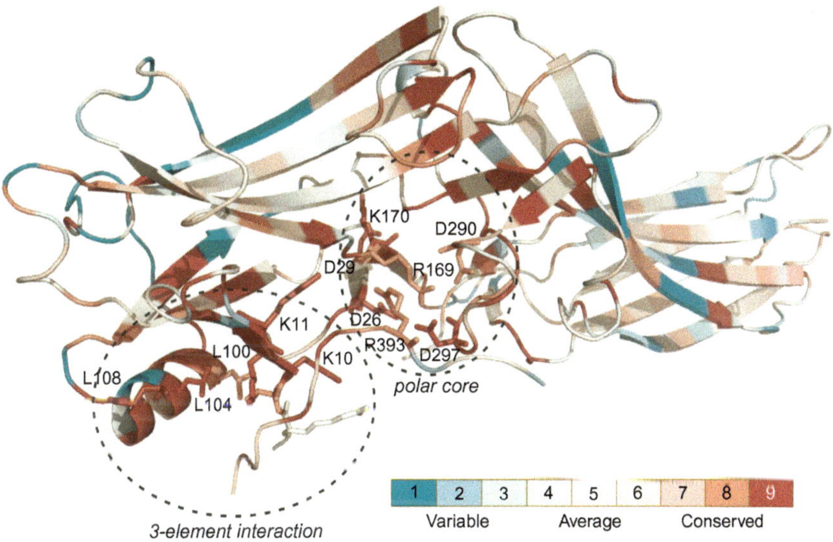

Aubry and Klein, Figure 2.1

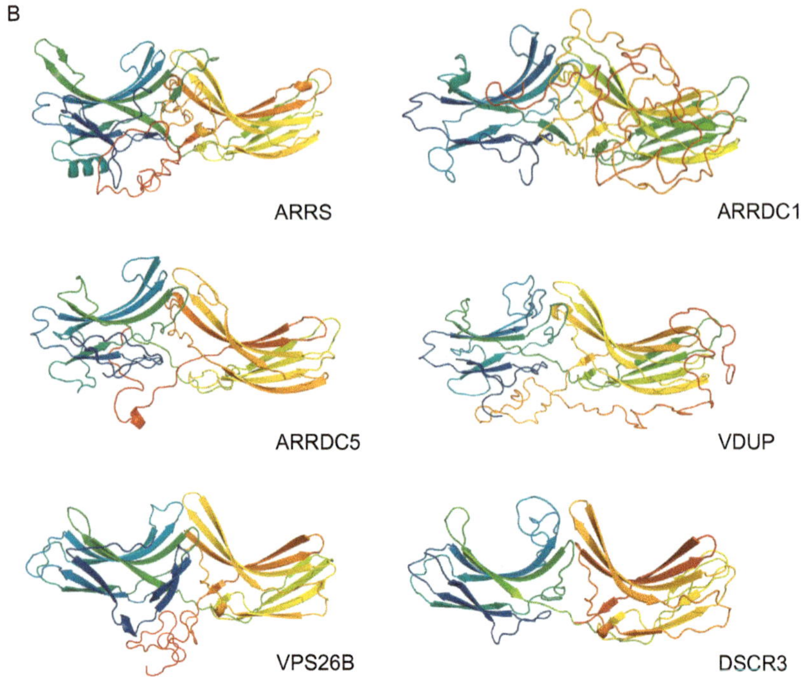

AUBRY AND KLEIN, FIGURE 2.3

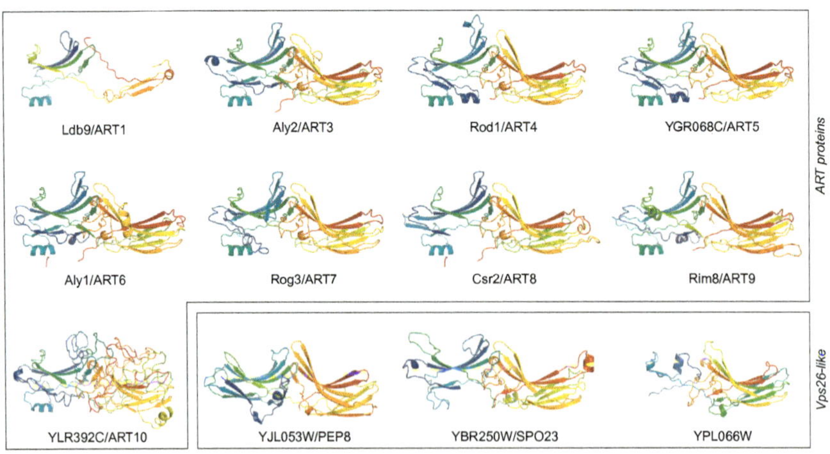

AUBRY AND KLEIN, FIGURE 2.4

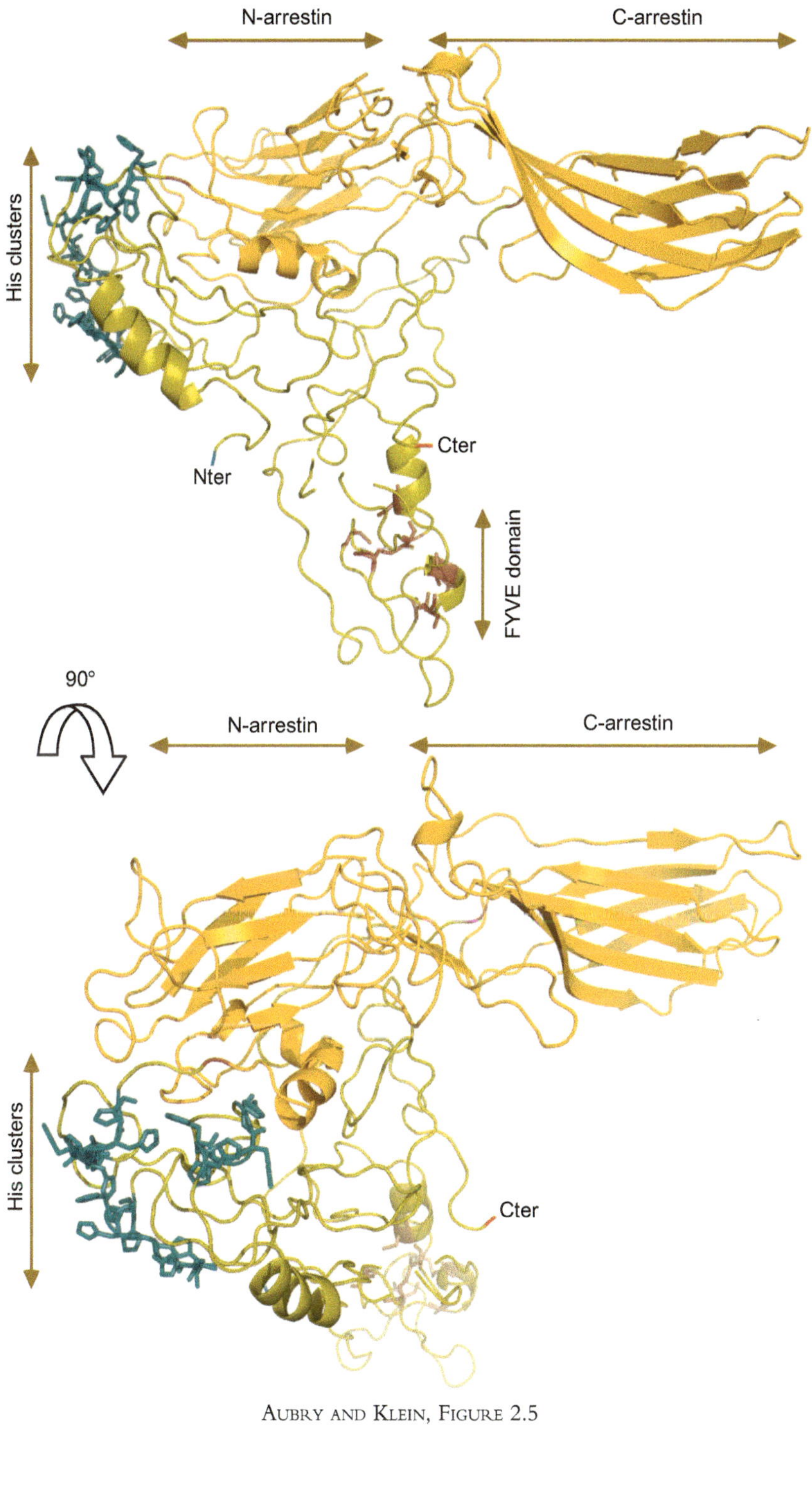

Aubry and Klein, Figure 2.5

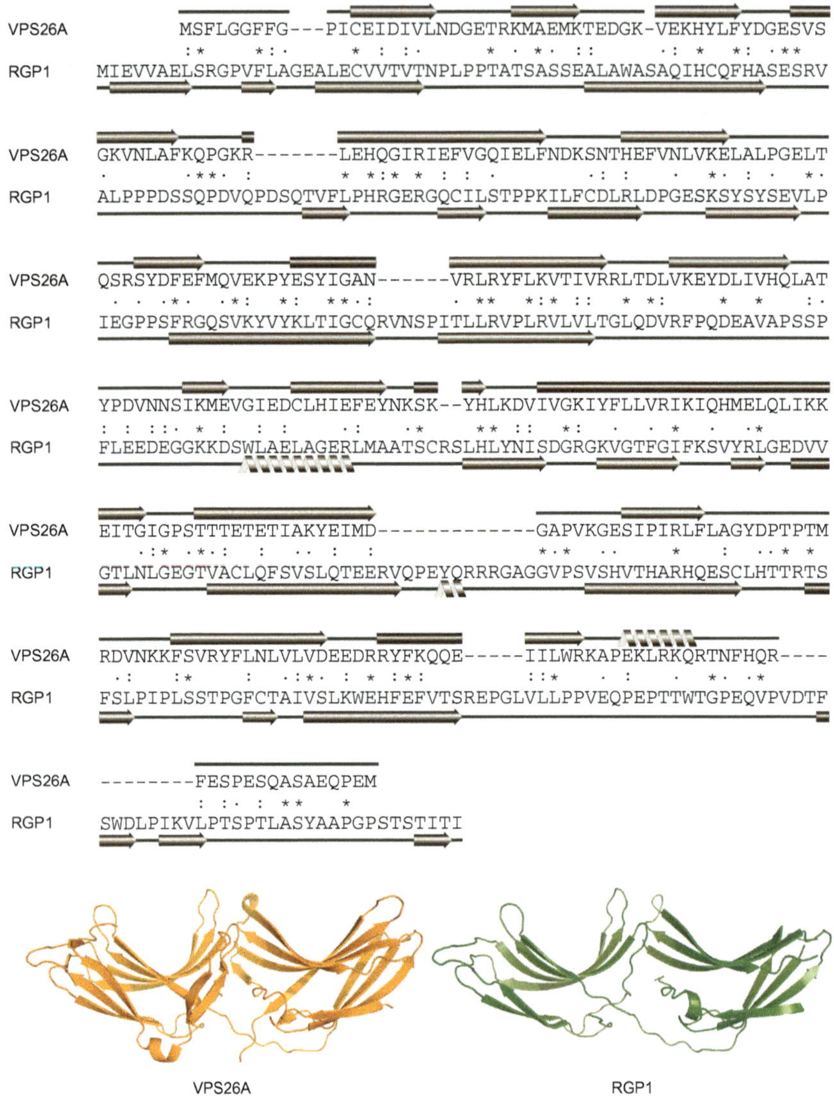

Aubry and Klein, Figure 2.6

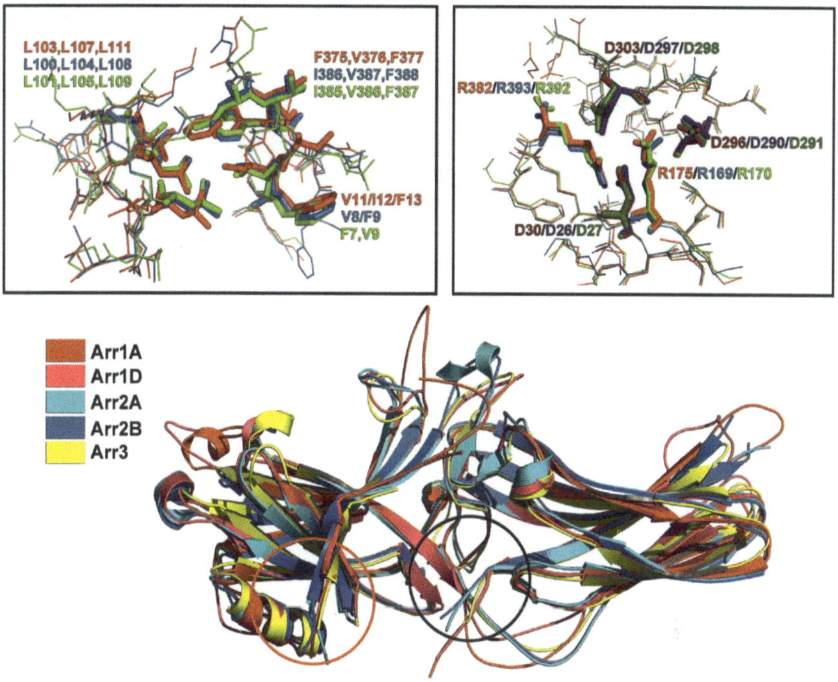

Gurevich and Gurevich, Figure 3.1

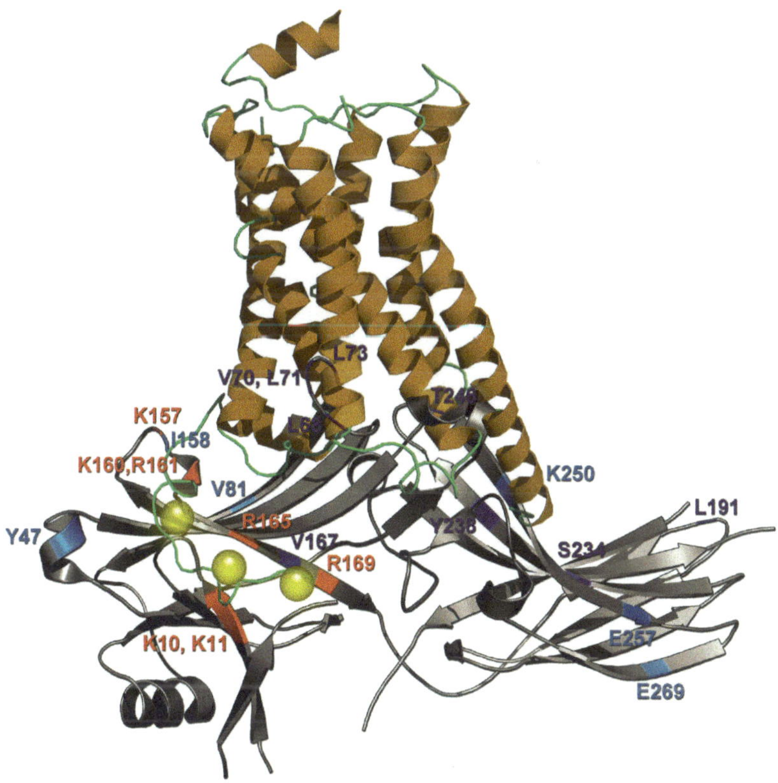

Gurevich and Gurevich, Figure 3.2

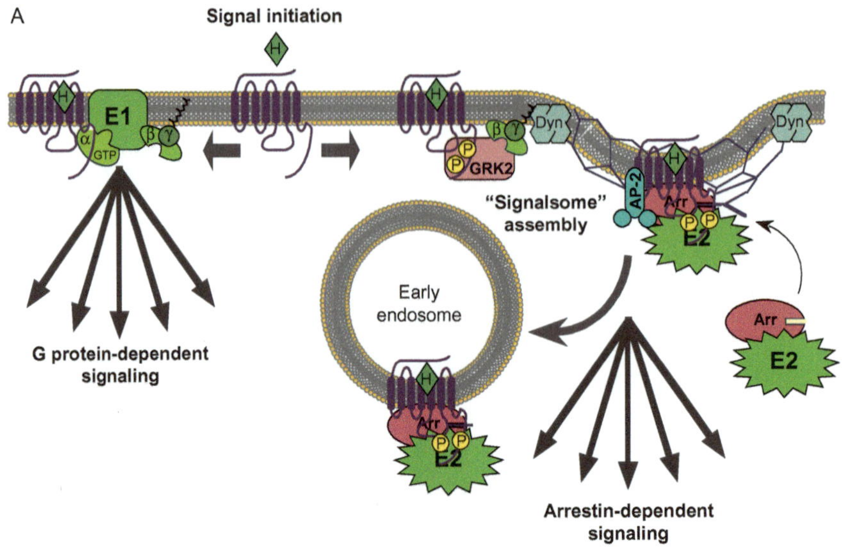

Luttrell and Miller, Figure 5.2

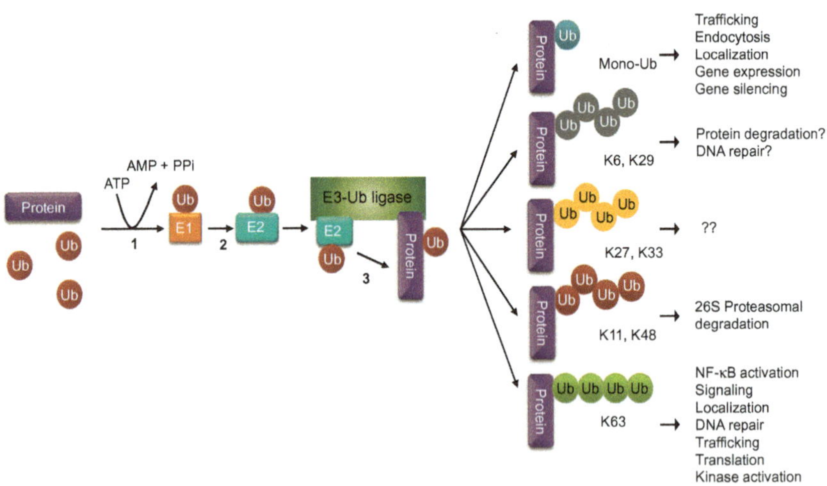

Kommaddi and Shenoy, Figure 7.2

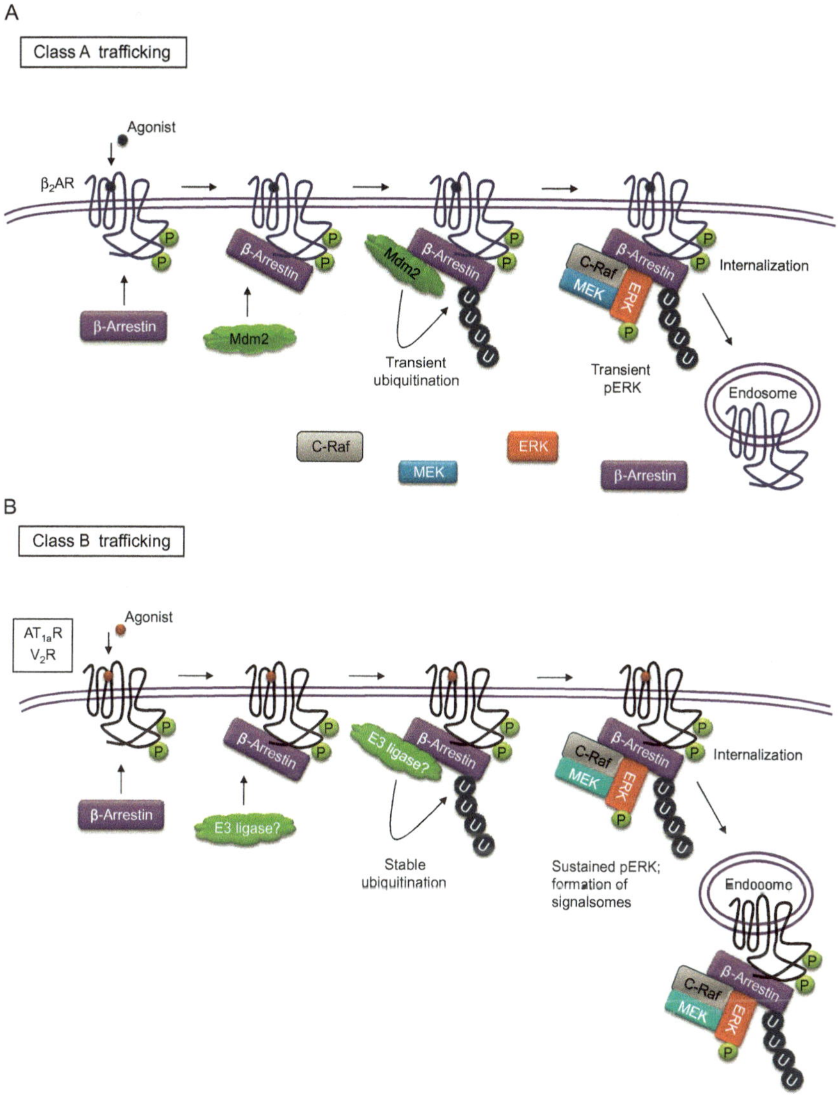

Kommaddi and Shenoy, Figure 7.3

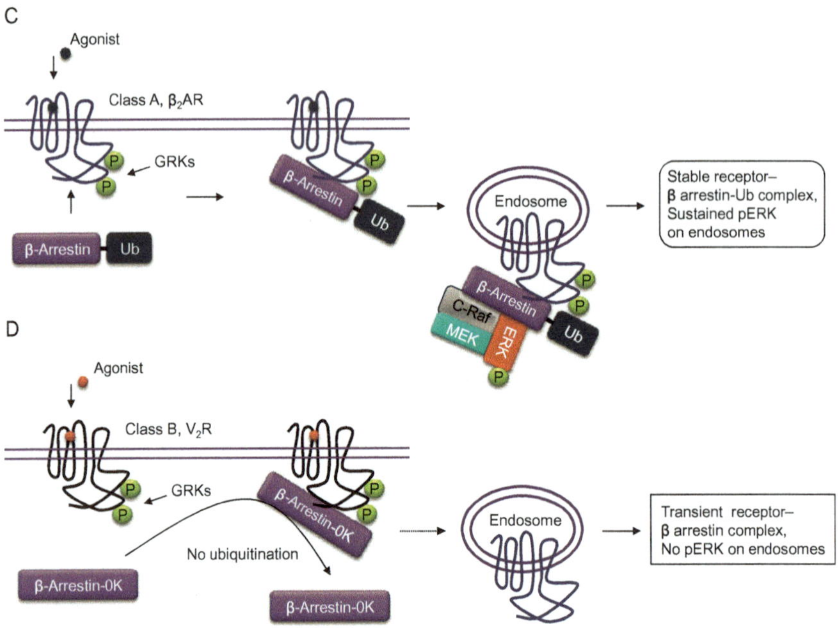

KOMMADDI AND SHENOY, FIGURE 7.3

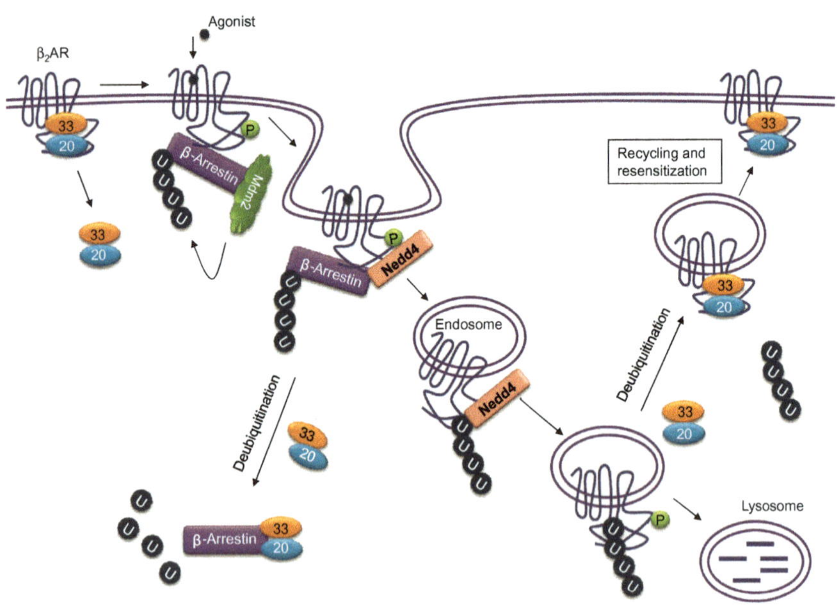

KOMMADDI AND SHENOY, FIGURE 7.5

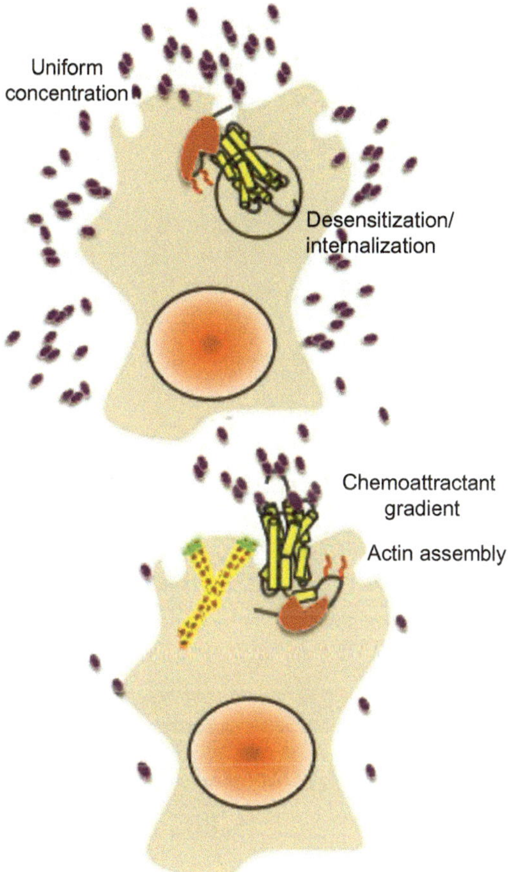

DeFea, Figure 8.1

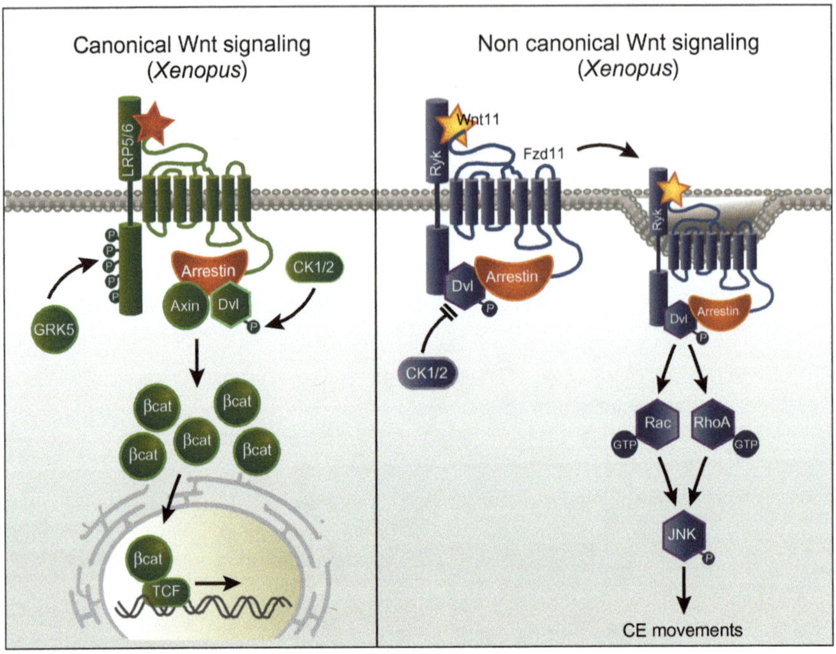

PHILIPP *ET AL.*, FIGURE 9.2

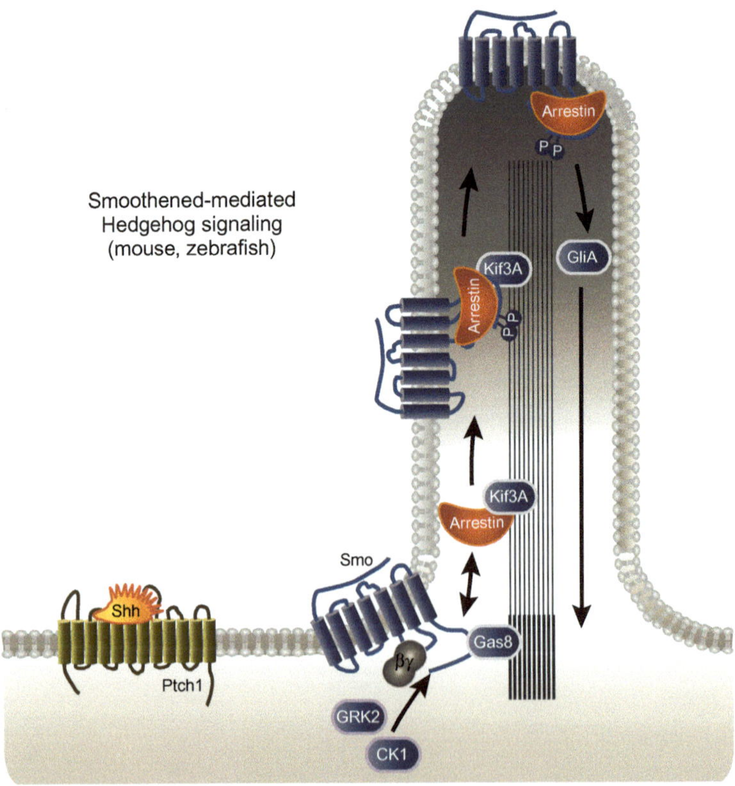

PHILIPP *ET AL.*, FIGURE 9.3

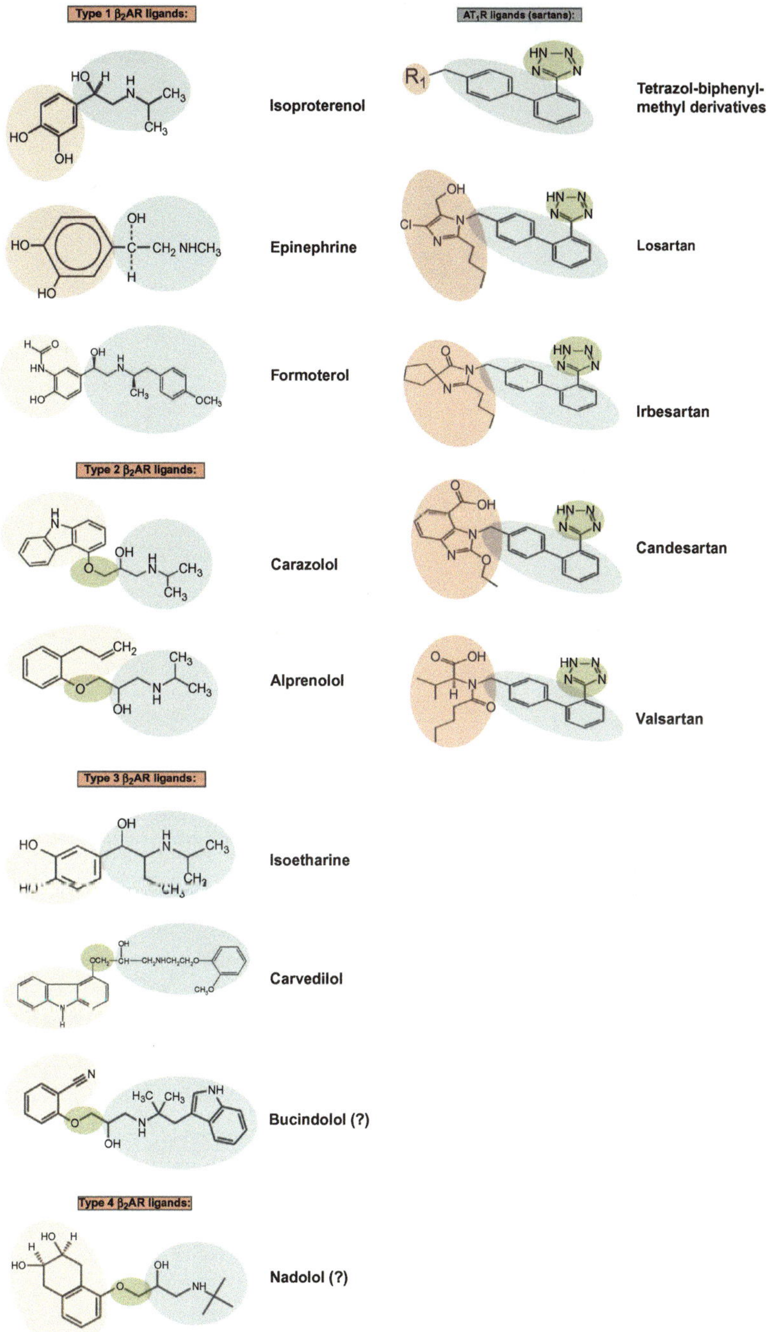

Lymperopoulos and Bathgate, Figure 12.1

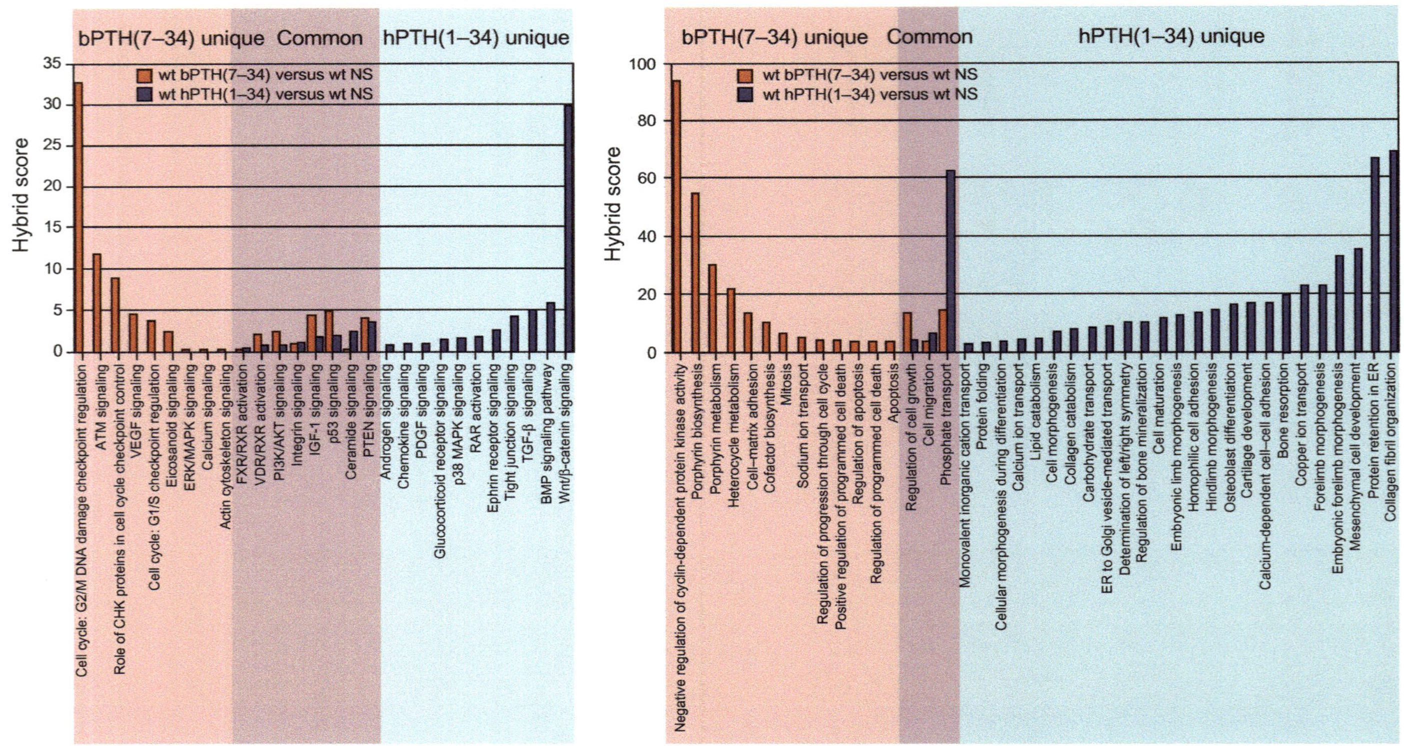

Bohinc and Gesty-Palmer, Figure 13.3

A

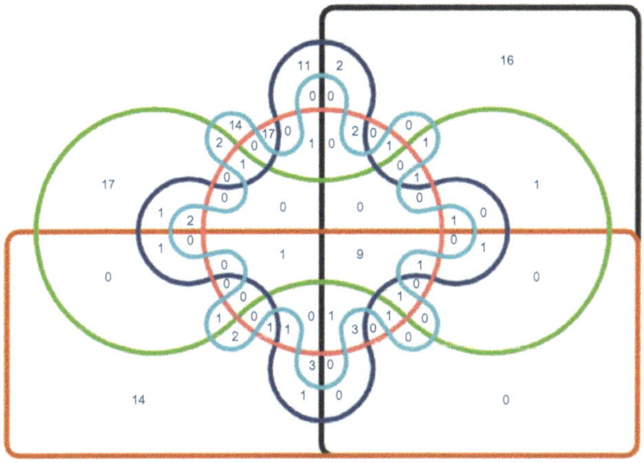

VENNTURE

- Parallel analyses -

B

Omnimorph

- Data texturization -

Maudsley *et al.*, Figure 17.1

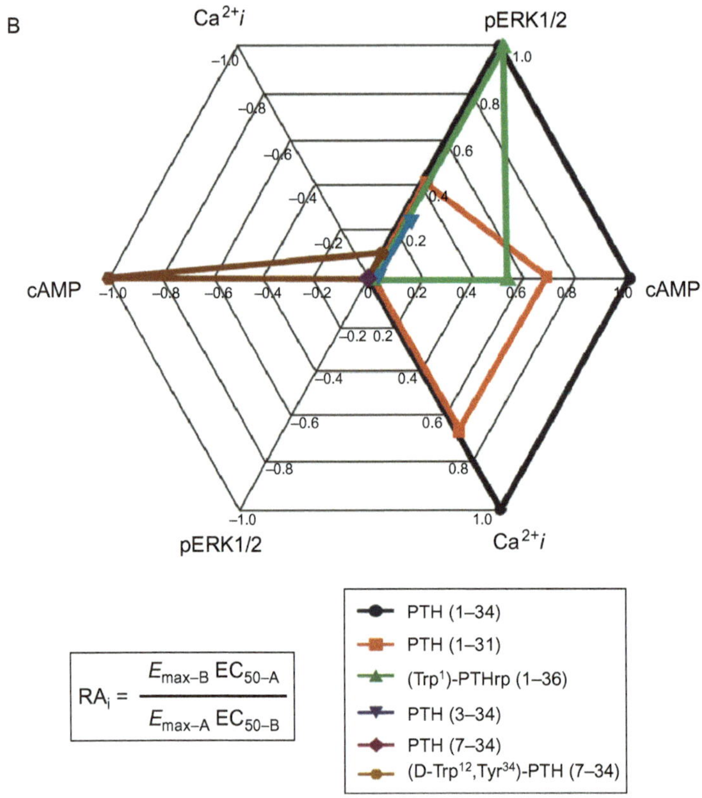

LUTTRELL, FIGURE 18.1

FSC
www.fsc.org
MIX
Papier | Fördert
gute Waldnutzung
FSC® C083411

Kolibri 360 GmbH
Ferdinand-Jühlke-Straße 7
99095 Erfurt, Deutschland
produktsicherheit@kolibri360.de